Symmetry
and
Structural
Properties
of
Condensed
Matter

Proceedings of

the 7th

International

School on

Theoretical

Physics

Symmetry and
Structural
Properties of
Condensed
Matter

Myczkowce, Poland
11 – 18 September 2002

Editors

T. Lulek

B. Lulek

A. Wal

University of Rzeszów
Rzeszów, Poland

World Scientific
New Jersey • London • Hong Kong • Singapore

Published by

World Scientific Publishing Co. Pte. Ltd.
5 Toh Tuck Link, Singapore 596224
USA office: Suite 202, 1060 Main Street, River Edge, NJ 07661
UK office: 57 Shelton Street, Covent Garden, London WC2H 9HE

British Library Cataloguing-in-Publication Data
A catalogue record for this book is available from the British Library.

ISBN 981-238-272-0

Printed by FuIsland Offset Printing (S) Pte Ltd, Singapore

Preface

This is the seventh's volume of the biannual series of proceedings of the summer schools on theoretical physics under the title "Symmetry and Structural Properties of Condensed Matter" (SSPCM). The schools started in 1990 in Zajączkowo, a village near Poznań, continued there five times, and the last two meetings were held in Myczkowce near Lesko, in the Undercarpatia Region in the south - eastern part of Poland. The last SSPCM school was held since 11 till 18 September 2002, and was organized jointly by two Institutes of Physics, from the newly formed Rzeszów University, and Adam Mickiewicz University of Poznań. We are happy to provide the Reader with the results of this last school in the present Proceedings.

The main goal of the whole series of SSPCM schools is promotion of advanced mathematical methods of physics of condensed matter, which are directed towards searching for new symmetries of structural and physical properties of solids. We were dealing hitherto with Wigner - Racah calculus, crystallography and its extentions to quasicrystals, liquid crystals, incommensurate phases, non-rigid molecules, self-similar structures, Aharonov-Bohm phenomena, quantum Hall effect, magnetic translation groups, statistics of particles, band structure in solids, broken symmetries and phase transitions, Bethe Ansatz, and various other aspects of an adequate description of condensed matter. We were considering not only some specific physical phenomena themselves, but also were interested in appropriate mathematical tools, like group actions on sets, representation theory for symmetric and unitary groups, Hecke algebras, algebraic combinatorics, conformal symmetries, quantum groups etc. The last school was devoted to focus on the following three main subjects:

A. Conformal symmetry, central charge, condensation of flux;

B. Rigged string configurations, Yang - Baxter equations, and their applications in solid state physics;

C. Energy band structure in solids.

The proceedings are divided into three parts accordingly, but particular topics can overlap between the main subjects, and even with the problems pursued in previous SSPCM schools.

Part A starts with the general lecture of Prof. Brian G. Wybourne who makes some historical and philosophical remarks on the scales in physics from the small like $10^{(-35)}$ m till the large up to $10^{(26)}$ m, points out mutual interrelations between particle physics and cosmology, prophets further discoveries at both extremes, and encourages young researches to push forward "the circle of our understanding", even if it implies the increase of "the circumference of our ignorance". We also keep our fingers towards such new developments!

Profs. McCabe and Wydro, in the article encompassing their series of lectures, provide a review of their investigations on Yang - Lee edge singularities for lattice spin models. The article involves conformal field theory, finite size scaling, and reports some new results for two - dimensional systems. Prof. Ohnuki presents path integral representations of the Dirac algebra for a quantum - mechanical system, constrained to a manifold, diffeomorphic to a D-dimensional sphere. He shows that for D larger than one the representation is unique, whereas for $D = 1$ there exists a family of unitarily inequivalent irreps, specified by a continuous parameter, related to Aharonov - Bohm gauge potential. He also expresses a desire to solve the corresponding problem for the central potential for $D = 2$, with an adequate description of radial degrees of freedom. Prof. Vourdas provides us, in his lecture on applications of finite quantum systems to quantum information processes, with an idea of combining the famous eigenspaces of angular momentum theory with the Fourier transform by displacement operators. His scheme can be concisely expressed in terms of factorization of the so called "qudits", i.e. multidimensional generalizations of qubits, the two-dimensional quantum systems. In this way, large qudits can be factorized by smaller ones, with the exploitation of an elegant combinatorics, including Chinese reminder theorem. We believe that it opens an exciting field of investigations in quantum computing. Prof. Penson presents some combinatorial results on Stirling and Bell numbers in the context of coherent states, normal ordering, Dirac combs, and positive measures. His lecture has proved to be of a great interest for the audience dealing with boson polynomials. Also Prof. Wybourne raises a combinatorial problem, related to evaluation of the Laughlin wavefunction for quantum Hall effect as the expansion of Slater determinants. The latter are labeled by special admissible partitions of $n = N(N - 1)$. The article is dedicated to the memory of Prof. Itzykson, who raised the problem ten years ago. Later on, a group of researchers formulated a conjecture which should be valid until any coefficient accidentally vanishes. Prof. Wybourne shows up that it is the case, and points out the first such accident, for $N = 8$ (the problem is combinatorially explosive), and is ready to grant a bottle of New Zeeland wine for anybody who will provide a full resolution of the problem.

The lecture of prof. Zipper deals with carbon nanotubes, their structural, electronic, magnetic and mechanical properties. Such a nanotube can be looked at as a graphene, i. e. a single sheet of graphite, rolled up according to a prescription given by a two - component chiral vector. Despite of the same material, i. e. carbon atoms, and identical general structure of a sheet, properties of nanotubes with different chiral vectors vary substantially from metallic to semiconducting. In other words, electronic properties, including persistent currents and trapping of quanta of magnetic flux, are controlled in a way by the geometry of rolling. Dr Wójs presents results of investigations of the Wrocław group of prof. Jacak on an application of the notion of composite fermions for a proper resolution of incompressible spectra of fractional quantum Hall effect. This lecture pursues some earlier topics of the SSPCM series, in particular fractional statistics. The contribution given by Dr Tsomokos describes the interaction of electrons in a mesoscopic ring with microwaves which are quantized. It yields a nonlinear device which demonstrates the "quantum alternating current - Aharonov-Bohm effect". Effects of quantum coherence were also dealt with in the lecture of Mrs. Paśko, a young collaborator of Prof. Tralle, with the special emphasis on the phase coherence of the spin part of the electronic wavefunction. Due to the weakness of the spin-orbit interaction, the spin coherence length is much larger than the orbital one, which opens the way to observe some new Aharonov-Bohm like interferences, in particular at edges of the sample. Mr. Lijnen, from the team of prof. Ceulemans, presented some geometric methods of visualization of multidimensional structures related to intramolecular rearrangements. The methods, referred to as the polyhedral representation, are associated with equipping of graphs with the structure of polyhedra. Prof. Blackman gives a review of scaling behaviour, present in such diverse fields as nanotechnology, atmospheric sciences, and cosmology. It happens when some basic ingredients are introduced into a system at a constant rate - the invariant of a scaling. He considers a model of growing and coalescing droplets, and discusses limits of validity of some analytic solutions. Prof. Zieliński studies propagation of acoustic waves in crystals with anharmonic defects, in particular surfaces. He thus continues the subject of floppy (non-rigid) molecules and crystals, raised in the first SSPCM meeting. He shows that the assumption of local anharmonicity gives rise to a variety of phenomena like extended and solitary waves, generation of higher harmonics, deterministic chaos, surface structural instabilities and so on.

As we can see, part A is a collection of articles, related mostly to nanoscopic physics, with the scaling symmetry and gauge invariants as the main common entities. We believe that some novel results in this area will appear in a near future. hopefully, they will be reported in the next SSPCM meetings.

Part B starts with the article of Prof. Schilling, a compendium of her series of lectures on algebraic and combinatoric Bethe Ansatz. She provides an algebraic derivation of the system of Bethe equations from principal Yang - Baxter relations for monodromy matrix for the Heisenberg magnetic ring, and describes a combinatorial scheme of classification of solutions in terms of rigged string configurations. The classification bases on the hypothesis of strings, which is believed to hold in the asymptotic case of an infinite number of spin nodes. Essentially distinct solutions of Bethe equations constitute configurations of strings, weighted by appropriate quantum numbers called riggings, which measure pseudomomenta (cocharges of Schutzenberger and Lascoux) of these strings. This whole construction, presented first by Kerov, Kirillov and Reshetikhin, yields a complete system of solutions of Bethe Ansatz equations for Heisenberg magnetic rings, along the general Weyl duality scheme between the symmetric group acting on nodes and the unitary group of the single-node quantum space. In this context, a bijection between paths, representing coupled multiparticle states, and rigged configurations, related to crystal (Kashiwara) bases of simply laced Lie algebras, is described. In this way, a nice and deep combinatorics proves to be essential in description of magnetic states.

Prof. Caspers reports in his article some efforts towards classification of solutions for systems of Bethe equations for finite rings from first principles, without invoking to the hypothesis of strings. In his approach the solutions are labeled by the sequences of winding numbers, and one meets the problem with the redundancy: various sequences yield the same solutions. He makes an essential progress in realizing this program by proposing a classification relying only on arithmetic relations between consecutive winding numbers. Within the framework of resulting types of solutions, he proves to be able to discuss various singularities, limiting and critical points, and other peculiarities resulting from the finite size of the crystal. For example, a critical point happens for such a number of nodes when a bound state of two Bethe pseudoparticles, represented by complex rapidities, converts into a pair of real ones. Finite size effects in a square lattice are also reported in the article of Dr Cojocaru, who associates them with non-string solutions in one dimension. It is a challenging problem to obtain a complete classification of Bethe Ansatz equations for finite crystals.

Prof. Louck presents his new contribution to combinatorial aspects of the Weyl duality, which consists in an extension of the well known bijection between semistandard Young - Weyl taleaux and Gelfand-Tsetlin patterns to skew Young diagrams. The notion of a word of a skew tableau paves the way for enumerating lattice permutations and, at the same time, for evaluation of Littlewood - Richardson numbers. Prof. Mendez proposes in his article a combinatorial description of representation matrices of both groups of the Weyl duality in terms of directed graphs - the most natural analog of a matrix. He demonstrates that eulerian bijections between graphs nicely reproduce the multiplication property of matrices in terms of adjacency of digraphs. Interesting combinatorial ideas were put also in unpublished lecture of Prof. Chen. It is fascinating that the Weyl duality is so deeply related to Bethe Ansatz, together with the Kerov - Kirillov - Reshetikhin construction.

In part B we put also some results of investigations on Bethe Ansatz done by our local group from Rzeszów, including presentation of a two-magnon classical configuration space as a Möbius strip, recognition of cocharge of Schutzenberger and Lascoux as a quasimomentum in the Brillouin zone, the basis of wavelets for solutions of Bethe equations, the role of Jucys - Murphy operators within the scheme of plactic monoid, and magnetic interpretation of Robinson - Schensted - Knuth algorithm.

Part C is devoted to the problem of energy bands in solids. This problem has been discussed since the fourth SSPCM school, and culminated at the notion of an elementary energy band in lectures of profs. Michel and Zak. Here we notify a continuation in the lecture of Dr Sznajder, who demonstrates elementary energy bands in strongly anisotropic structures, together with the topology over the Brillouin zone, compatibility relations in high symmetry points, and numerical values of Davydov gap parameters. Prof. Karwowski and his collaborators from Toruń present some results on the system of two relativistic electrons, confined by an oscillator potential, called harmonium. Dr Zeiner deals with incommensurate crystals, and proposes a new definition of metric in the superspace which admits a unique crystallographic definition of their symmetry groups. The key point in comparison with previously established definitions is that the internal space is not necessarily orthogonal to the physical, allowing thus to treat different incommensurate subsystems on equal footing. Prof. Wojciechowski in his lecture on symmetries of yttrium iron garnets discusses the structure of energy bands and compatibility relations for a nanostructure of two hybridizing iron-oxygen octahedral and tetrahedral clusters. These symmetry consideration allow him to point out a microscopic origin of a strong magnetic anisotropy of the charge transfer. Prof. Barnaś gives a lecture on resistance of domain walls in ferromagnets. He considers the two cases: diffusive and ballistic. The first applies to thick walls and can be treated semiclassically, whereas the second works for nanostructures and predicts a huge resistance in some cases, which is confirmed by recent experiments. Dr Spisak reports on the influence of spin-orbit coupling to transport properties of electrons in disordered media. Prof. Paszkiewicz discusses propagation of acoustic waves through media with low elastic symmetries, and determines the number of independent parameters and distinct acoustic axes.

We like to thank to all participants of the seventh's SSPCM meeting, and to all persons who contributed in any way to organization. Special thanks are due to all lecturers for presentations and preparation of manuscripts. We like to thank to all chairman for their duties, and for patience to the auditory and organizers. Thanks are also due to all members of our International Advisory Committee. Let us thank in a particular way to all students who were listening carefully to the lectures, and were not afraid to give questions. We also thank to all our sponsors, and to persons who helped us in the intermediate stop in Krakow: prof. Antonina Kowalska and prof. Wiesława Sikora. We acknowledge with thanks the main support from both Institutes of Physics: Adam Mickiewicz University of Poznań and the newly established University of Rzeszów, and from the Polish Committee for Scientific Research (KBN), and Polish Ministry of Education and Sport (MENiS). We are also grateful to local authorities: to the Marschall of Undercarpatia Region, Mr. Bogdan Rzońca, and the Prefect of Przemyśl, Mr. Marian Majka. We also appreciate the hospitality of Mr. Grzegorz Szabla, the head of the hotel "Energetyk" in Myczkowce, together with the whole His Team.

Rzeszów, February 2003

Tadeusz Lulek
Andrzej Wal
Barbara Lulek

INTERNATIONAL ADVISORY COMMITTEE

Willem J. Caspers (Enschede, The Netherlands)
Rainer Dirl (Wien, Austria)
Jacek Karwowski (Toruń, Poland)
Adalbert Kerber (Bayreuth, Germany)
Maurice Kibler (Lyon, France)
James D. Louck (Los Alamos, USA)
Janusz Morkowski (Poznań, Poland)
Jan Mozrzymas (Wrocław, Poland)
Yoshio Ohnuki (Nagoya, Japan)
Brian G. Wybourne (Toruń, Poland)

ORGANISING COMMITTEE

Tadeusz LULEK, Chairman (University of Rzeszów)
Andrzej DOBEK, Honorary Chairman (A. Mickiewicz University, Poznań)
Andrea LEHMANN-SZWEYKOWSKA (A. Mickiewicz University, Poznań)
Barbara LULEK (University of Rzeszów)
Przemysław SZLACHETKA (A. Mickiewicz University, Poznań)
Paweł JAKUBCZYK, (University of Rzeszów)
Andrzej WAL, Secretary (University of Rzeszów)

SPONSORS

Polish State Committee for Scientific Research (KBN)
European Physical Society
The Marschall of Undercarpatia Region, Mr. Bogdan Rzońca
The Prefect of Przemyśl, Mr. Marian Majka

Seventh International School on Theoretical Physics

"Symmetry and Structural Properties
of Condensed Matter"

(SSPCM'2002)

11 September - 18 September 2002, Myczkowce, Poland

organized by

University of Rzeszów, Institute of Physics, Rzeszów, Poland
Adam Mickiewicz University, Faculty of Physics, Poznań, Poland

CONTENTS

PART B. RIGGED STRING CONFIGURATIONS, YANG-BAXTER EQUATIONS AND THEIR APPLICATIONS IN SOLID STATE PHYSICS.

PART C. ENERGY BAND STRUCTURE IN SOLIDS.

Part A

CONFORMAL SYMMETRY
CENTRAL CHARGE
CONDENSATION OF FLUX

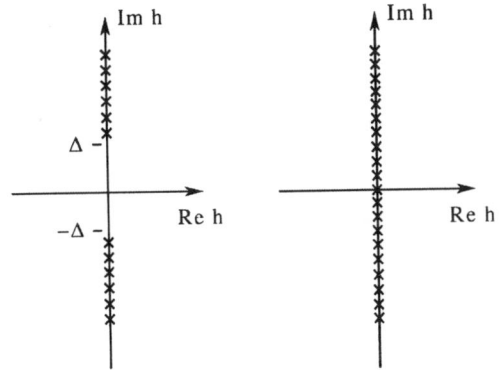

THE SMALL AND THE LARGE IN PHYSICS

BRIAN G WYBOURNE

Instytut Fizyki,
Uniwersytet Mikołaja Kopernika,
87-100 Toruń,
Poland
E-mail: bgw@phys.uni.torun.pl
WEB: http://www.phys.uni.torun.pl/~bgw

> *So, naturalists observe, a flea*
> *Has smaller fleas that on him prey;*
> *And these have smaller still to bite 'em;*
> *And so proceed ad infinitum.*
> *Poetry, a Rhapsody. Jonathan Swift (1726)*

An introduction to the range of physics from the very small to the very large in a historical perspective for a general audience.

1. Introduction

The poem of Jonathan Swift epitomizes the question as to whether there is a limit to smallness, or indeed largeness. By the 1700's the applications of microscopes and telescopes to the ever smaller, and larger, were developing fast. Here we first trace some of the early history, emphasizing the close connection between science and technology in all these developments up to modern times.

2. Lenses, Magnifiers and Microscopes

Seneca ~100 AD noted that *"Letters, however small and indistinct, are seen enlarged and more clearly through a globe of glass filled with water"* while at about the same time **Claudius Ptolemy** observed that a stick appears bent in water and measured, and calculated, the refractive index of water. By **1267 AD Bacon** was noting that *"Great things can be performed by refracted vision. If the letters of a book, or any minute object, be viewed* *through a lesser segment of a sphere of glass or crystal, whose plane is laid upon them, they will appear far better and larger.*

The above early observations gave rise to the idea of magnification by a single lens to aid in seeing small objects, such as Jonathan Swift's fleas. The limits of a single lens as a magnifier were soon reached. The possibility of seeing still smaller objects required a new idea. The use of more than one lens and thus of a compound microscope has been associated with Zacharias Jansen of Middelburg, Holland about 1595. Much development of optics was required to overcome the problems associated with spherical and chromatic aberrations which became more serious as magnifications became greater. It was realized that the resolving power of an optical microscope was proportional to the wavelength of the light being used and hence higher resolution could be obtained by going to shorter wavelengths. By the beginning of the 20th century the limits of microscopes using light were being reached. Further progress in seeing smaller objects would require new physics and new technologies

3. Lenses, Mirrors and Telescopes

In October 1608 Hans Lipperhey, spectacle-maker from Middelburg, Holland applied for a patent for a device for *"seeing faraway things as though nearby"*. News of the telescope spread quickly and at 9pm 26 July 1609 Thomas Harriot observed the moon and made lunar sketches and thus disappeared the *"Man in the Moon"*. More significantly, telescopes became the revolutionary tool of astronomy. October/November 1609 Galileo observed the moon and also discovered four satellites of Jupiter. December 1610 Harriot observed sunspots with his telescope and noted that the sun rotates.

The telescope, like the microscope, suffered from spherical and chromatic aberrations. In 1671 Isaac Newton introduced the reflecting telescope eliminating the chromatic aberrations of lenses at the price of having the aberrations of the sphericity of his mirrors. The latter aberrations were greatly reduced once it had been learnt how to convert a spherical surface into that of the paraboloid. The resolution of a telescope is proportional to the diameter of its lens or mirror and hence telescopes of ever increasing diameter were constructed. Even so it was not possible to even measure the diameter of any stars. Without new physics and new technologies progress must stop.

4. Quantum Physics and Relativity

The new physics and new technologies came from totally unexpected directions - namely quantum physics and relativity - probably the two most significant developments of the past century. They were to form the basis of the new technologies that became so apparent in the closing decade of the last century. I remarked that to increase the resolution of microscopes we need to go to shorter wavelengths. The vital new concept came in 1923 with de Broglie's postulate that particles may exhibit wave-like properties via his celebrated wavelength formula

$$\lambda = \frac{h}{mv}.$$

Indeed by 1927 Davisson, Germer and Thomson had demonstrated diffraction and interference phenomena for electrons. As a result electron microscopes of hitherto undreamt of resolution became possible. Not immediately as a long learning curve was required to make electromagnetic analogues of the glass lenses of the earlier optical microscopes. Problems of lens aberrations had to be overcome in the construction of electromagnetic lenses to control the electron beams. The electrons were accelerated making the application of relativistic kinematics essential in the design of electromagnetic lenses. Likewise developments in high vacuum technology and even superconductivity technology were required to perfect the electron microscope. Vision at the nanometre $10^{-9}m$ scale has become possible with even atoms becoming "visible". Electron beam diameters smaller than that of a hydrogen atom have recently been made allowing, for example, one to "see" single gold atoms on a carbon film and to "watch" pairs of gold atoms interacting as they approach each other.

5. Interferometry, Precision and Measurement

The closing decade of the last century saw the development of telescopes of up to 10metre diameter with the beginning of the new millennium seeing the successful construction of a quartet of 8.2metre Very Large Telescopes (VLT) at the European Southern Observatory (ESO) in Chile. The introduction of adaptive optics, computers and laser techniques has revolutionized ground base astronomy so that now The Hubble Telescope is often used for preliminary work with the ground based telescopes pursuing finer detail and exploring further into the depths of the universe. At the time of introduction of the Hubble Telescope many had felt it spelt the death knell of ground based astronomy. The new physics that was to radically change the use of the telescope in extending our knowledge of the very large started with Michelson's addition of the optical interferometer to the 100" Mt Wilson telescope. The addition of his 20foot ($\sim 6m$) gave him in 1927 measurements of the diameter of the star Betelgeuse that would have hitherto required a 6m telescope. That was the start of Very Long Baseline astronomy that was later to dominate the field of radioastronomy. Michelson's optical interferometric telescopes are really just starting with this years ESO success in interferometrically coupling, at this stage each pair of the 8.2m VLTs and shortly all four VLTs will be coupled. The current pairs of VLTs work with an effective baseline of 102m.

These telescopes are in essence time machines - the further the reach into the depths of the universe the further back in time they are looking. But telescopes cannot look as far back in time as particle accelerators!

6. Particle Accelerators as Telescopes and Time Machines

The development of particle accelerators throughout the past century have allowed physicists to probe the properties of matter at every decreasing distances, almost paradoxically, as the energy of the accelerators have greatly increased and their dimensions to tens of kilometres. In a sense they have become the ultimate high resolution microscopes of our time. At the same time they give information about processes that could only have occurred at the very earliest of times, say in the first ten millionth of a second after the Big Bang. In that sense they let us look backward in time, much further back than any optical telescope. This is seen, for example, in the Relativistic Heavy Ion Collider (RHIC) at Brookhaven where gold atoms are stripped of all of their electrons and beams of the resultant gold nuclei are collided head-on at relativistic speeds ($\sim 99.95\%$ of the speed of light) to produce highly compressed nuclear matter where for a short instant of time the protons and neutrons "melt" to produce a Quark Gluon Plasma (QGP). The temperatures and pressures produced are more extreme than even those existing in the hottest of stars.

7. The Sizes of Things

The range of sizes of things considered in physics covers more than 60 orders of magnitude, from the very smallest string to the radius of the observable universe. Let us recall some basic units. We associate nuclei with the *fermi* with $1fm = 10^{-15}m$, atoms with the *Ångström* with $1\text{Å} = 10^{-10}m$ or in terms of much modern technology the *nanometre* with $1nm = 10^{-9}m$. Moving to the very small we have the *Planck length* with $\ell_p = (\frac{\hbar G}{c^3})^{3/2} \sim 10^{-35}m$ and to the very large the *light year* with $1ly \sim 10^{16}m$.

The smallest object conceived of in physics are *strings* of the dimension of the Planck length. Typically nuclei having A nucleons (protons and neutrons) have a nuclear radius $r \sim 1.2A^{1/3}\,fm$ while for atoms we have the typical atomic radii $H \sim 2.08$Å, $Ne \sim 0.51$Å and $Fr \sim 2.7$Å. Still larger we have the *Earth Radius* $\sim 6.4 \times 10^6 m$, the *Sun Radius* $\sim 7 \times 10^8 m$ and ultimately the *Observable Universe Radius* $\sim 10^{26} m$.

The range and strength of the four fundamental forces also exhibit striking differences in magnitude.

Force	Relative Strength	Range
Strong	1	$\sim 10^{-15} m$
Electromagnetic	$\frac{1}{137}$	infinite
Weak Interaction	$\sim 10^{-5}$	$10^{-17} m$
Gravitational	$\sim 6 \times 10^{-39}$	infinite

Note the difference in strength of almost 40 orders of magnitude and that the gravitational force is the weakest of all known forces and perhaps the most mysterious. Almost nothing is known about gravity at distances shorter than 200microns ($200 \times 10^{-6} m$) which is 16 orders of magnitude worse than the other fundamental forces which have been tested down to $\sim 10^{-19} m$. Gravity is also poorly tested on cosmological scales.

8. From the very small to the very LARGE

In the preceding we have considered lengths ranging from $10^{-35} m$ to $10^{26} m$. In earlier times particle physics and cosmology were seen as completely unrelated subjects of study. Likewise the various forces of nature were studied as distinct and unrelated subjects. The unification of the forces of nature started in the 19th century with Maxwell's development of electromagnetism, uniting magnetism and electricity and continued through the 20th century with the unification of electromagnetic and and weak interaction forces. The task of the complete unification of the four fundamental forces remains as a prime task of the 21st century. There are tantalising glimpses of such a theory coming from string theory. One thing is clear, it is no longer possible to consider particle physics, the physics of the very small, and cosmology, the physics of the very large, as unrelated subjects - both depend on each other. Undoubtedly much remains to be discovered. It is dangerous to predict the future, history is full of the unexpected.

At this stage I am reminded of Augustus De Morgan's extension of Jonathan Swift's poem to

Great fleas have little fleas upon their backs to bite 'em,
And little fleas have lesser fleas, and so ad infinitum.
And the great fleas themselves, in turn, have greater fleas to go on;
While these again have greater still, and greater still, and so on.
De Morgan:A Budget of Paradoxes, p. 377

9. Concluding Remarks

In the preceding I have tried to indicate the way in which the range of physics has changed over the centuries as science and technology have developed in such a way that it has become possible to explore things on both increasing and decreasing scales and to make the point that studies in both directions are essential to further progress in understanding the incredible universe that we occupy. Ultimately the small and the large become so dependent upon each other that it becomes impossible to consider one without the other. Each time we appear to have answered a question we are confronted with new questions, often of a totally unexpected nature. As Erwin Chargaff has noted *The greater the circle of our understanding becomes, the greater the circumference of surrounding ignorance.* I personally believe that many of the problems to be solved will be solved by young people and it is to them we must look and encourage. The advent of the 20th century was approached with great optimism as seen in Henry Rowland's address to the American Physical Society in 1899. Unfortunately his dream of a better 20th century was not realized. Recalling his words could we substitute the twentyfirst century for his twentieth century vision?

...where in the world is the institute of pure research in any department of science with an income of $100,000,000 per year. ... But $100,000,000 per year is but the price of an army or of a navy designed to kill other people. Just think of it, that one per cent of this sum seems to most people too great to save our children and descendants from misery and even death!

But the twentieth century is near - may we not hope for better things before its end? May we not hope to influence the public in this direction?

Henry A. Rowland *The Highest Aim of the Physicist* Presidential Address to the American Physical Society, 28 October 1899.

Acknowledgments

It is a real pleasure to take the opportunity to thank Professors Barbara and Tadeusz Lulek and their splendid team for making such a stimulating meeting possible. My research is supported by a grant from the Polish KBN (Contract No. 5P03B 057 21).

YANG-LEE SINGULARITY OF TWO DIMENSIONAL ISING AND POTTS MODELS

TOMASZ WYDRO

Laboratoire de Physique Moléculaire et des Collisions
Université de Metz, 1 bvd Arago, 57078 Metz FRANCE

JOHN F. MCCABE

412 Morris Ave., Summit, NJ 07901 USA

This presentation discusses numerical studies of Yang-Lee edge singularities in 2D spin models. The studies use the finite-size scaling behavior of quantum spin chains to obtain information on the Yang-Lee edge singularities of discrete spin models. The studies provide values for Fisher's σ exponent and candidate conformal field theories. For the 2D Ising model, known results are reviewed. For the 2D 3-state Potts model, new results are presented.

1. Brief Introduction

Yang-Lee edge singularities are accumulation points of the zeros of a statistical model's partition function in the complex fugacity plane [1]. Yang-Lee edge singularities originally provoked interest, because their movements towards real positive values of the fugacity are signs of a phase transition [1-3]. Outside of this interest, the edge singularities are also points where a statistical model has a behavior characteristic of a critical point [4]. In 2-dimensions, a Yang-Lee edge singularity also corresponds to a non-unitary conformal minimal model [5].

The above connections with phase transitions and critical points stimulated an enormous interest in the properties of Yang Lee edge singularities. These properties are characterized by an exponent σ that describes the density of zeros [2-3]. In 1-dimension (1D), exact results have shown that $\sigma = \frac{1}{2}$ for the Ising and Potts models [3, 6]. In 2D, resummation of high temperature expansions has provided the value for σ in the Ising model, i.e., $\sigma = -0.163 \pm 0.003$ [4, 7]. The value of σ also enabled an identification of a conformal minimal model that describes the singularity [5]. In various dimensions, exact results have provided values for σ's of Yang-Lee edge singularities in continuous spin models such as Heisenberg and spherical models [6, 8]. Direct experimental measurements in high magnetic fields have confirmed the value of σ for the Yang-Lee edge singularity of at least one of these models [9].

The Yang-Lee edge singularities of many other models have not been studied. Notably, those of discrete spin models such as the 3-state and 4-state Potts models have ot been studied in more than 1D. In 2D, this is unfortunate, because the Yang-Lee edge singularities of, at least, some 2D discrete spin models are very probably related to the known non-unitary conformal minimal models.

Finite-size scaling behavior of a quantum spin chain has been used to study the non-unitary conformal model associated with the Yang-Lee edge singularity of the 2D Ising model [10]. We will explain and extend that study. In particular, we will explain how standard finite-size scaling manifests itself at Yang-Lee edge singularities. As an example, we will use scaling of a quantum spin chain to evaluate the exponent σ for the Yang-Lee edge singularity of the 2D Ising model. We will also use the finite-size scaling behavior of a quantum spin chain to evaluate properties of the Yang-Lee edge singularity of the 2D 3-state Potts model. In particular, we will partially identify the conformal minimal model that is associated with this Yang-Lee edge singularity.

Our presentation has two sections. The first section reviews the Yang- Lee theory of edge singularities, finite-size scaling, and relevant aspects of conformal field theory. This section also includes an example, i.e., the Yang-Lee edge singularity the 1D Ising model. The second section discusses quantum spin chains, numerical methods, and finite-size scaling results for the Yang-Lee edge singularities of the 2D Ising and 3-state Potts models.

2. Finite Size Scaling

2.1. *Yang-Lee Theory*

For the hard core gas, Yang and Lee related the distribution of zeros of the grand canonical partition function to the equilibrium phase structure [1]. The relations are most easily illustrated for the lattice model of the hard core gas. In the lattice model, either one or no gas atoms occupy a site of the lattice. Thus, the grand canonical partition function of the lattice gas, $Z_{lattice\ gas}$, is a polynomial of degree N in the fugacity, z, where N is the number of lattice sites. Since the partition function is a polynomial, it can be completely factorized as:

$$Z_{lattice\ gas} = \prod_{j=1,\ldots,N} (z - z_j), \tag{1}$$

where the z_j are the N complex zeros of Z. In the thermodynamic limit, $V \to \infty$, the discrete zeros, z_j, condense on a curve in the complex fugacity plane.

For real and positive values of the fugacity, the partition function is a sum of positive terms, which is defined by

$$Z = \sum_{N=0,1,2,\dots} \sum_{\text{state } j} \exp(-\beta E_j) z^N. \tag{2}$$

Thus, no zero, z_j, is at a positive real value of the fugacity. As $N \to \infty$, zeros can however, condense towards the real axis.

For the hard core gas, it has been shown that the zeros of the partition function of a hard core gas are located on the unit circle in the complex fugacity plane [3]. For a finite lattice, one can rewrite

$$\ln(Z) = \sum_{i=1}^{N} \ln(z - z_i), \tag{3}$$

using a density function $g(x)$ for the zeros [3, 6]. The sum over the zeros, z_i, becomes an integral over a curve B in the thermodynamic limit.

$$\ln(Z) = \int_B dx \ln(z - x) g(x). \tag{4}$$

The form of the density function $g(x)$ determines the value of the order parameters near the system's condensation transition and near the Yang-Lee edge singularity (see below).

Since the free energy is proportional to a logarithm of the grand canonical partition function, a zero of the partition function is a point of non-analyticity for the free energy. The free-energy is a non-analytic function of a physical variable such as density, ρ, or temperature, T, at a phase transition. For example, non-analytic points on pressure-density (P-ρ) isothermals of a gas are associated with 1st order phase coexistence and 2nd order phase transitions, respectively. Though the grand canonical partition function's zeros are at unphysical complex fugacities, one might suspect that an accumulation of zeros at a physical fugacity corresponds to a phase transition.

For the hard core gas, Yang and Lee showed that this suspicion is correct [1]. The positivity of the partition function for real positive fugacity insures that no zeros are on the positive real z-axis. If the system has no phase transition, no zeros enter a region surrounding the axis of positive real fugacities as $V \to \infty$. Then, the free energy and all its z-derivatives are analytic on the entire positive real z-axis. If the system has a phase transition, zeros accumulate in a neighborhood of a positive real value, z_0, of the fugacity. Then, the free energy and all its z-derivatives are non-analytic at $z = z_0$ unless z_0 is an accumulation point for the complex zeros.

A non-analyticity in the free energy or pressure P indicates the presence of a phase transition. For example, a jump in the z-derivative of the density, $\rho(z)$, leads to a $P(\rho)$ isotherm with a 1st-order transition (Fig. 1a). Similarly, an infinite z-derivative, leads to a $P(\rho)$ isotherm of a 2nd-order transition (Figs. 1b). Yang and Lee's results indicate that the density of a gas, $\rho(z)$, will have such non-analytic behavior only where complex zeros in the fugacity accumulate as $V \rightarrow \infty$.

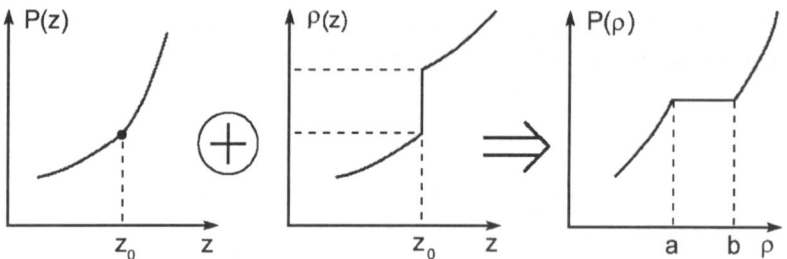

Figure 1a. Illustration of how non-analytic behavior in fugacity produces non-analyticity associated with first order phase transition.

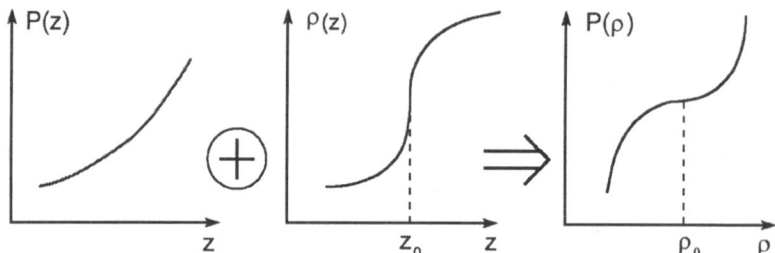

Figure 1b. Illustration of how non-analytic behavior in fugacity produces non-analyticity associated with second-order phase transition.

Yang and Lee tested their results on complex zeros of the grand canonical partition function through equivalence between the lattice hard core gas and the Ising model [3]. According to the equivalence, a state in which a gas atom occupies site j is equivalent to a state in which the spin, S_j, at site j has the value $+1$, and a state in which a gas atom does not occupy site j is equivalent to a state in which the spin at site j has the value 0. The fact that Ising spins only have values $+1$ and 0 implies that a lattice site is either unoccupied or occupied by one gas atom, i.e., a hard core repulsion. The ferromagnetic spin-spin interaction corresponds to an attractive interaction between gas atoms on different sites. The equivalence associates the fugacity z of the gas to a magnetic field h coupling to $\Sigma_{sites\,j}\,S_j$. These equivalences mean that: $Z_{lattice\text{-}gas}(T,\,z)$ $\propto Z_{Ising}(T,\,h)$. Thus, the grand canonical partition function for the lattice hard core gas is proportional to the canonical partition function for the Ising model in a magnetic field.

From this equivalence, the Ising model has a phase transition when its partition function, $Z_{Ising}(T,\,h)$, has complex zeros that approach a real value of $z = \exp(h/kT)$ [3]. Since the zeros of the partition function $Z_{lattice\text{-}gas}(T,\,z)$ are on the unit circle in the fugacity plane [1], the zeros of $Z_{Ising}(T,\,h)$ occur for purely imaginary values of h, i.e., $z = \exp(h/kT)$. Thus, the zeros of $Z_{Ising}(T,\,h)$ have the distribution of either Fig. 2a or Fig. 2b in the thermodynamic limit. In Fig. 2a, the zeros are on the imaginary h-axis, but are separated by a distance Δ from the real axis as $V \to \infty$; i.e., $g(h) = 0$ for $|h| < \Delta$. In Fig. 2b, the zeros are on the imaginary h-axis and have a non-zero density as $V \to \infty$. In the former case, there is no phase transition, which implies that $T > T_c$. In the later case, there is a transition at $h = 0$ implying that $T \leq T_c$. Thus, $\Delta \to 0$ as $T \to T_c$

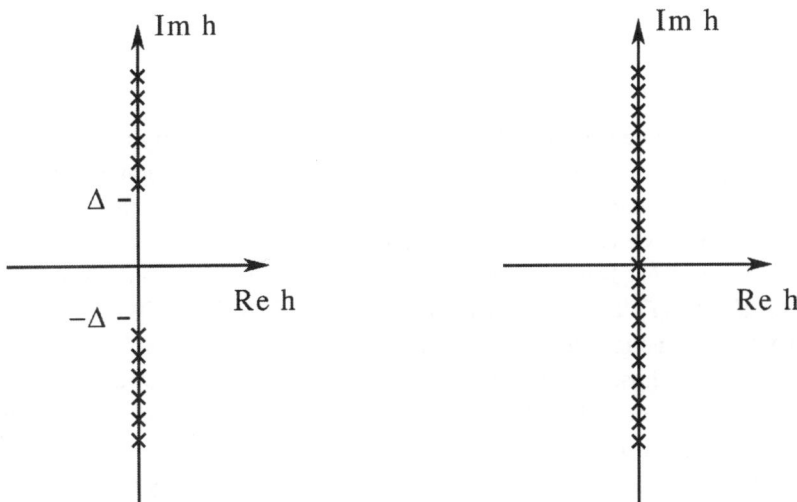

Figure 2a. Zeros of Ising model's partition function above critical temperature.

Figure 2b. Zeros of Ising model's partition function below the critical temperature.

In the high temperature phase where $T > T_c$, points $h = \pm i\Delta$ are Yang-Lee edge singularities. At these special points, the physical variables of the model have properties similar to those at ordinary critical points, e.g., correlation functions have scaling behavior (see below). Since the zeros of $Z_{Ising}(T, h)$ are on the unit "z" circle, each root $z_i = \exp(i\theta_i)$ with $0 \le \theta_i < 2\pi$, the free energy may be written as:

$$F = h - kT \int_0^{2\pi} d\theta\, g(\theta) \ln(z - \exp(i\theta)), \qquad (5)$$

and the spontaneous magnetization, M, defined by

$$M = \frac{z}{kT} \frac{\partial F}{\partial z}, \qquad (6)$$

is:

$$M = 1 - z \int_0^{2\pi} \frac{d\theta\, g(\theta)}{z - \exp(i\theta)} = 1 - i \int_{unit\ circle} \frac{dx\, g(x)}{x - z}. \qquad (7)$$

If the density $g(\theta)$ is nonzero for $\theta \to 0$ or equivalently $h \to 0$, there is a spontaneous magnetization, $<M>$, given by:

$$< M > \propto \lim_{\theta \to 0^+} g(\theta) - \lim_{\theta \to 0^-} g(\theta). \qquad (8)$$

Thus, nonzero density $\theta \to 0$ implies that there is a symmetry breaking phase. This implies that there are two phases, because the high temperature phase does not have a spontaneous magnetization. If there is a gap such that $g(h)$ is zero for $|h| < \Delta(T)$, we can calculate the value of the spontaneous magnetization, M, near $h = \pm i\Delta(T)$. If we assume that $g(h) \approx (h - i\Delta(T))^\sigma$ for $|h| > \Delta(T)$ and $g(h) = 0$ for $|h| < \Delta(T)$, the above equations imply that $M \approx (h - i\Delta(T))^\sigma$, which means that the magnetization diverges with an exponent σ at the Yang-Lee edge singularity. Thus, the Yang-Lee singularity is characterized by an exponent σ. The exponent describes the density of zeros of the partition function and a nonzero spontaneous magnetization, $<M>$.

The exponent σ may be measured numerically either by measurement of the spontaneous magnetization, M near the Yang-Lee singularity or by measurement of the finite-size scaling behavior of $\Delta(T)$.

2.2. Finite-size scaling in lattice models

Finite-size scaling methods use the behavior of mesoscopic-size systems to extract properties of the infinite volume limit. In the mesoscopic-size systems, physical quantities such as $ln(Z)$ are not extensive, i.e., not proportional to V. Rather, extensive scaling behavior receives corrections proportional to inverse powers of V. This non-extensive behavior determines the behavior of the system's infinite volume limit.

Physical quantities of finite-size systems are continuous and have continuous derivatives, because the partition functions of such systems are polynomials without zeros at real values of z. Thus, quantities such as $ln(Z)$; magnetization M; specific heat C_v; correlation length ξ, and magnetic susceptibility χ, do not have discontinuous or divergent behavior in finite-size systems. In comparison, physical properties are discontinuous or divergent at phase transitions. For example, the specific heat C_v, behaves like: $C_v = A|T\text{-}T_c|^{-\alpha}$ at temperatures T, near a critical temperature T_c. Such discontinuous behavior characterizes the $V \to \infty$ limit. In the infinite volume limit, physical quantities can be discontinuous or divergent behavior at a phase transition.

Following Refs. [11], the question arises as to how the smooth properties of finite-size equilibrium systems change as $V \to \infty$. At finite volume, physical quantities, e.g., M, C_v, ξ and/or χ can exhibit rounded peaks at a temperature $T_c(V)$. These rounded peaks are the finite-size analogs of divergences in the $V \to \infty$ limit. As the volume becomes large, the peak heights (PHs) grow higher and diverge at a temperature, $T_c(\infty)$.

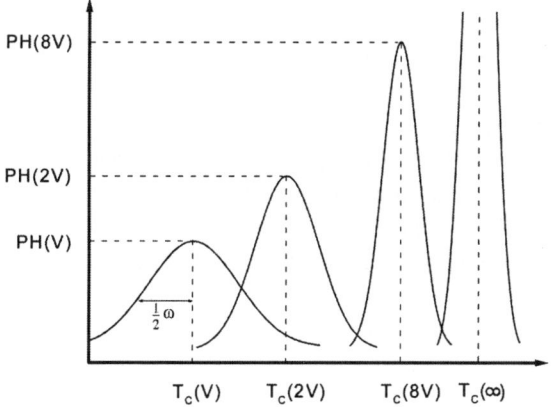

Figure 3. Evolution of peak in the magnetization as a function of system-size at a normal critical point.

Figure 3 illustrates how the rounded peak in C_v changes as $V \to \infty$ for a system near a critical point. For $V \neq \infty$, C_v has a peak of finite height, $PH(V)$, at a position, $T_c(V)$, shifted from the critical temperature $T_c(\infty)$ of the $V \to \infty$ or thermodynamic limit. As the volume V, grows, the peak approaches $T_c(\infty)$ and becomes higher and narrower. Finite-size scaling describes how the rounded peak approaches the divergent peak as $V \rightarrow \infty$. Near a critical point, the peak height, width, and shift have a simple "scaling" dependence on V, because a critical points involves a single dimensionful parameter, i.e., $(T- T_c(\infty))$.

Following Refs. [11], the scaling behavior of the mesoscopic system enables one to extract properties of the $V \rightarrow \infty$ limit. For example, the shift of the peak in C_v is given by: $|T_c(V) - T_c(\infty)|$ where $T_c(V)$ is the temperature of the rounded peak in a mesoscopic D-dimensional system of volume V. As $V \rightarrow \infty$, the peak shift satisfies a scaling law $|T_c(V) - T_c(\infty)| = A V^{-\Theta/D}$ near the critical temperature. The scaling behavior is described by the measurable exponent Θ.

The approach to criticality involves only the dimensionful parameter V. Thus, the correlation length ξ diverges at a critical point like $V^{-1/D}$, because $V^{-1/D}$ is the only parameter with the proper dimension. In the infinite volume system, ξ also scales near $T_c(\infty)$, i.e., $\xi(T) \approx |T - T_c(\infty)|^{-\nu}$. Inserting these relations in the above scaling formula for the peak shift in C_v, one finds that $V^{-1/(Dv)} \approx \xi(T_c(V))^{-1/v} \approx |T_c(V) - T_c(\infty)| \approx V^{-\Theta/D}$. Thus, $v = 1/\Theta$ so that the exponent Θ, which is measurable in mesoscopic systems, determines the scaling form of the correlation length in the thermodynamic limit. The physical exponent v for the thermodynamic limit can be determined by measuring the finite-size scaling behavior of the peak shift for mesoscopic systems. Below, we use such a technique to determine an exponent at the Yang-Lee edge singularity.

We will use hyperscaling laws that relate different exponents at a critical point. Hyperscaling laws come from a dimensional analysis, which provides constraints on the singular part of the free energy, F. The constraints lead to relations among exponents.

In a spin model, the free energy depends on temperature T, magnetic field intensity, H, and D-dimensional volume V, i.e., $F = F(T, H, V)$. In the $V \rightarrow \infty$ limit, the free energy is an extensive quantity, $F(T, H, V) = Vf(T, H)$, which implies the scaling relation $F(T, H, \lambda V) = \lambda F(T, H, V)$.

Constraints on the singular part of F at a critical point come from a renormalization group transformation of the free energy. To make such a transformation, a portion of the spins of the model are integrated out to produce an effective model. Typically, this is done in the partition function Z where:

$$Z = \prod_{i \in \text{lattice}} \left(\sum_{s(i)} \right) e^{\frac{-E(s(i),H)}{kT}} = e^{-\beta Q} \prod_{i \in \text{reduced lattice}} \left(\sum_{s(i)} \right) e^{\frac{-E_{rl}(s(i),H_{rl})}{kT_{rl}}} . \tag{9}$$

Integrating out a portion of the spins replaces couplings by renormalized couplings, i.e., $T \rightarrow T_{rl}$ and $H \rightarrow H_{rl}$ in E_{rl} and introduces additive correction Q

to H. Furthermore, the right hand product has a number of lattice sites i that is equal to a^{-D} times the original number of lattice sites. From the definition, $F = -kT\,ln(Z)$, one can rewrite the above-equation (9) as:

$$kTF(T,H,V) = kTF\left(T_{rl}, H_{rl}, Va^{-D}\right) + Q \text{ or } f(T,H) = a^{-D} f\left(T_{rl}, H_{rl}\right) + \ldots \quad (10)$$

Near a critical point, one is interested in the dependency of the singular part of F on reduced temperature, t, and reduced magnetic field, h. The reduced quantities are defined as:

$$h = H - H_c \text{ and } t = T - T_c \quad (11)$$

Here, H_c and T_c are the critical values of the magnetic field and temperature, respectively. Integrating out spins causes reduced variables t and h to be replaced by reduced lattice variables t_{rl} and h_{rl}. Near a critical point, t and h scale linearly as:

$$t_{rl} = a^y t \text{ and } h_{rl} = a^{y'} h. \quad (12)$$

Here, y and y' are known as the scaling dimensions of the reduced temperature, t, and the reduced magnetic field, h, respectively. In terms of the reduced variables, the renormalization group equation for the free energy becomes:

$$f(t,h) = a^{-D} f\left(t_{rl}, h_{rl}\right) + \ldots \quad (13)$$

Similarly, the correlation function, $G(R, t, h)$, satisfies a renormalization group equation:

$$G(R,t,h) = a^{-x} G\left(\frac{R}{a}, t_{rl}, h_{rl}\right) = a^{-x} G\left(\frac{R}{a}, a^y t, a^{y'} h\right) \quad (14)$$

The exponent x defines the behavior of G at a critical point, $h = t = 0$, because the above-relation implies that $G \rightarrow R^{-x}$.

For the special value of scaling parameter a satisfying $a^{y'} h = 1$, the renormalization group equations simplify. In particular, the equation for free energy density becomes:

$$f(t,h) = a^{-D} f\left(a^y t, a^{y'} h\right) + \ldots = h^{\frac{D}{y'}} f\left(h^{\frac{-y}{y'}} t, 1\right) + \ldots \quad (15)$$

At the Yang-Lee edge singularity, h is the only relevant scaling field. This implies that $f(t, h) \approx h^{D/y'} f(0, 1) + \ldots$. It is customary to define an exponent σ by $\sigma \equiv D/y' - 1$. For the same special value of a, the renormalization group equation for the correlation function becomes:

$$G(R,t,h) = h^{\frac{x}{y'}} G\left(R/h^{\frac{-1}{y'}}, h^{\frac{-y}{y'}}t, 1\right). \tag{16}$$

Since h is the only relevant field, equation (16) provides the scaling law for correlation length, ξ:

$$\xi = h^{\frac{-1}{y'}} \equiv h^{-\nu} \text{ with } \nu = \frac{1}{y'}. \tag{17}$$

The relations defining ν and σ in terms of the same exponent y' also imply that $\sigma = -1 + D\nu$. This equation is a hyperscaling law that relates a finite-size scaling, ν, to the exponent σ defining the Yang-Lee edge singularity.

2.3. *Yang Lee Edge Singularity in 1D*

A simple example of a Yang-Lee singularity is found in the nearest-neighbor 1D Ising model [3, 6]. The 1D Ising model's energy is given by:

$$E_{1D \text{ Ising}} = J\sum_{m=1}^{N}\sigma_m\sigma_{m+1} + h\sigma_m, \tag{18}$$

where the spins $\sigma_m = \pm 1$ and N is the total number of sites, i.e., the 1D volume. The 1D Ising model has a partition function $Z_{1D \text{ Ising}}$, given by:

$$Z_{1D \text{ Ising}} = \sum_{\text{states "s"}} \exp\left[-\beta E_{1D \text{ Ising}}(s)\right] = \sum_{\text{all } \sigma_m = \pm 1} \exp\left[-\beta\sum_{m=1}^{N}\left(J\sigma_m\sigma_{(m+1)} + h\sigma_m\right)\right]. \tag{19}$$

To evaluate $Z_{1D \text{ Ising}}$, one uses the transfer matrix, $T_{1D \text{ Ising} \{\sigma i\}, \{\sigma i+1\}}$ defined by:

$$T_{1D \text{ Ising} \{\sigma_m\},\{\sigma_{m+1}\}} = \exp\left[-\beta\left(J\sigma_m\sigma_{(m+1)} + \frac{1}{2}h\left(\sigma_m\sigma_{(m+1)}\right)\right)\right]. \tag{20}$$

In 1D, the transfer matrix only couples a pair, (σ_m, σ_{m+1}) of adjacent spins. In terms of $T_{1D \text{ Ising}}$, the partition function of the model with periodic boundary conditions is given by the following trace:

$$Z_{1D \text{ Ising}} = \sum_{\text{all } \sigma_i = \pm 1} T_{1D \text{ Ising} \sigma_1,\sigma_2} T_{1D \text{ Ising} \sigma_2,\sigma_3} \dots T_{1D \text{ Ising} \sigma_{N-1},\sigma_N} T_{1D \text{ Ising} \sigma_N,\sigma_1}$$

$$= Trace\left(T_{1D \text{ Ising}}\right)^N = \left(\lambda_+\right)^N + \left(\lambda_-\right)^N. \tag{21}$$

Here, λ_+ and λ_- are the eigenvalues of the 2x2 transfer matrix $T_{1D \text{ Ising}}$. The eigenvalues are easily shown to be:

$$\lambda_i = e^A \cosh(H) \pm \left(e^{2A}\sinh^2(H) + e^{-2A}\right)^{\frac{1}{2}}$$

$$\text{where } A = -\beta J \text{ and } H \equiv \beta h. \tag{22}$$

For H real, $|\lambda_+| > |\lambda_-|$ so that $Z_{1D \text{ Ising}} \neq 0$. Thus, $Z_{1D \text{ Ising}}$ has no zeros for real values of the magnetic field h as anticipated.

Nevertheless, $Z_{1D \text{ Ising}}$ does have N zeros for purely imaginary values of H. From the above form for $Z_{1D \text{ Ising}}$, one sees that zeros occur when $\lambda_+^N = -\lambda_-^N$ or equivalently when $\lambda_+ = e^{i(2p-1)\pi/N}\lambda_-$. Here, p is a positive integer less than or equal to $(N + 1)/2$. Inserting the values λ_+ and λ_- into the equality $(\lambda_+/\lambda_- + \lambda_-/\lambda_+) = 2\cos[(2p-1)\pi/N]$ produces below equation (23) for the magnetic field values, H_p's, at the zeros [3]:

$$\pm\cosh\left(2H_p\right) = -e^{-4A} + \left(1 - e^{-4A}\right)\cos\left[\left(2p-1\right)\frac{\pi}{N}\right]. \tag{23}$$

Equation (23) implies that $\cosh(2H_p)$ is real and has modulus less than 1 at a zero. Thus, the H_p associated with zeros are pure imaginary numbers as anticipated. As $N \to \infty$, the zeros of the partition function densely cover positive and negative segments of the imaginary H – axis. As $N \to \infty$, the closest zero to the real H-axis approaches a point, H_{YL}, where $\pm \cosh(2H_{YL}) = 1 - 2e^{-4A}$. These special points are the Yang-Lee edge singularities.

For $A>0$, the value of the magnetic field at the singularity satisfies $|H_{YL}|>0$. Thus, $A > 0$ implies that no zeros approach the real H – axis as $N \to \infty$. Since $A>0$ implies that the temperature is positive, the absence of zeros that pinch the real H-axis for positive temperature confirms that the 1D Ising model does not have a phase transition.

These results also imply that $\lambda_+ = \lambda_-$ at the Yang-Lee edge singularities, i.e., the eigenvalues of transfer matrix are equal. From $\lambda_+ = \lambda_-$, one sees that the Yang-Lee edge singularities are not zeros of the partition function. Then, $\lambda_+ = e^{i(2p-1)\pi/N}\lambda_-$, shows that zeros only approach Yang-Lee edge singularities as $N \to \infty$. Thus, singularities are accumulation points of zeros of the partition function in the infinite volume limit.

As $N \to \infty$, one defines a variable x by $x = (2p-1)/N$ and treats x as a continuous variable on $(0, 1]$. Then, differentiating equation (23), which defines the zeros produces the following equation for a density function $g(H)$:

$$g\left(H\right) \equiv \frac{dx}{dH} = \frac{\sinh\left(H\right)}{\pi\sqrt{\sinh^2\left(H\right) - e^{-4A}}} \text{ if } |H| > |H_{YL}|$$

$$\text{and } g\left(H\right) = 0 \text{ if } |H| < |H_{YL}|. \tag{24}$$

Near the Yang-Lee edge singularity for positive imaginary magnetic field, the density function $g(H)$ can be expanded as:

$$g(H) \approx \frac{\sqrt{\sin(B_{YL})}}{\pi\sqrt{H - iB_{YL}}} \text{ for } |H| > |H_{YL}| \text{ where } H_{YL} \equiv iB_{YL}. \qquad (25)$$

From section 2.1, we recall that:

$$M = 1 - i \int_{\text{unit circle}} \frac{dx\, g(x)}{x - z}. \qquad (26)$$

Using the relationship between the fugacity and magnetic field, i.e., $z = e^H$, one obtains that:

$$M = 1 - ie^H \int_{B_{YL}}^{\pi} \frac{dB\, g(B)}{e^{iB} - e^H} + B \longleftrightarrow -B. \qquad (27)$$

For $H \to iB_{YL}^+$, the above-described form of the density $g(H)$ implies that:

$$M = 1 - ie^H \int_{\frac{-3\pi}{2}}^{\frac{+\pi}{2}} \frac{dB\, g(B)}{e^{iB} - e^H} \approx \frac{1}{\sqrt{H - iB_{YL}}}. \qquad (28)$$

Thus, explicit calculation shows that the magnetization has a divergent behavior at the Yang-Lee edge singularity and is characterized by an exponent $\sigma = -1/2$. This result suggested to Fisher that the Yang-Lee edge singularity is a critical point [4].

We expand these results by showing that the 1D Ising model exhibits finite-size scaling at the Yang-Lee singularity. A convenient finite-size scaling variable is $|H_1(N) - H_{YL}|$ where $H_1(N) = H_p(N)$ for $p = 1$. This variable is the distance between the Yang-Lee edge singularity and the closest zero of $Z_{1D \text{ Ising}}$. From the equation for $H_1(N)$, one finds that:

$$\cosh(2H_1(N)) - \cosh(2H_{YL}) = (1 - e^{-4A})\left[\cos\left(\frac{\pi}{N}\right) - 1\right]$$

$$\overset{N \to \infty}{=} (1 - e^{-4A})\left[\frac{-1}{2}\left(\frac{\pi}{N}\right)^2\right]. \qquad (29)$$

Thus, we find the finite-size scaling relation: $|H_1(N) - H_{YL}| \approx N^2$, which implies that the exponent Θ defined by: $|H_1 - H_{YL}| \approx N^{-\Theta}$ is equal to 2. Since H plays the role at the Yang-Lee edge singularity that T plays at a normal critical point, $\Theta = 2$ implies that $v = 1/\Theta = \frac{1}{2}$. Thus, the correlation length, ξ satisfies: $\xi \approx |H - H_{YL}|^{-1/2}$.

In section 2.2, we found the hyperscaling relation: $\sigma = -1 + Dv$ for the Yang-Lee edge singularity. From the finite-size scaling result $v = \frac{1}{2}$, the relation: $\sigma = -1 + Dv$ implies that: $\sigma = -\frac{1}{2}$. Thus, the finite-size scaling behavior predicts the same value for σ as the explicit calculation of σ.

2.4. *Finite-size Scaling and Conformal Theory*

Yang-Lee edge singularities of 2D statistical models are believed to be associated with non-unitary conformal field theories [5]. The conformal theories are non-unitary, because Yang-Lee singularities have non-unitary transfer matrices (see below). We further assume that the singularities of the simplest discrete spin models have finite numbers of order parameters. The theories with finite numbers of order parameters are known as minimal models [12] and have central charges [13]:

$$C = 1 - \frac{6(p - p')^2}{pp'}. \tag{30}$$

Here, p and p' coprime integers, $p, p' > 1$ and $p' - p > 1$. The minimal models have order parameters, which are referred to as primary fields. The primary fields conformal dimensions, $h_{r,s}$ and $h_{r',s'}$, given by:

$$h_{r,s} = h_{p'-r,p-s} = \frac{(rp - sp')^2 - (p - p')^2}{4pp'},$$

$$h_{r',s'} = h_{p'-r',p-s'} = \frac{(r'p - s'p')^2 - (p - p')^2}{4pp'} \tag{31}$$

where the integers s, r, s', and r' satisfy: $1 \le r, r' \le p'-1$, $1 \le s, s' \le p - 1$ [13]. The conformal dimensions $h_{r,s}$ and $\underline{h_{r',s'}}$ are associated with transformations on z and z*, respectively [12].

The definition of a conformal model includes a central charge and a set of primary fields, each field being labeled by a set of integers $\{r, s, r', s'\}$. For minimal models, a complete description of the sets of primary fields is given by the ADE classification [14] based on discrete modular symmetries. These symmetries are invariances of the 2D torus that result from periodic boundary conditions in an underlying statistical model. The discrete symmetries constrain the form of the partition functions for minimal models. The ADE classifycation provides four series of modular invariant partition functions. From the ADE partition functions, one can extract the field content, i.e., the sets $\{s, r, s', r'\}$, for any minimal model.

Using the ADE classification, we have found the simplest candidate minimal models for the Yang-Lee edge singularity of the 3-state Potts model. Generally, the constraints on s, r, s', and r' imply that minimal models associated with smaller integers p and p' have fewer primary fields. We enumerated the candidate minimal models with the smallest values of p' for each of the four ADE series. The results are shown in Table 1.

Table 1: Candidates for Yang-Lee Singularity of 3-State Potts models

1st ADE Series ($A_{p'-1}$, A_{p-1}) models			
(p', p)	C	*Unitary*	*Comments*
(3, 2)	0	yes	Trivial → not Yang-Lee
(4, 3)	1/2	yes	Normal Ising critical point
*(5, 2)	-22/5	no!	Ising Yang-Lee singularity!
*(5, 3)	-3/5	no!	(A_4,A_2) – Potts Yang-Lee?
(5, 4)	7/10	yes	Tricritical Ising critical point

2nd ADE Series ($D_{1+p'/2}$, A_{p-1}) models			
(p', p)	C	Unitary	Comments
(6, 5)	4/5	yes	normal 3-Potts critical pt.
*(10, 3)	-44/5	no!	(D_6,A_2) – Potts Yang-Lee?
*(10, 7)	8/35	no!	(D_6,A_6) – Potts Yang-Lee?
(10, 9)	14/15	yes	unitary → not Potts Yang-Lee

3rd ADE Series ($D_{1+p'/2}$, A_{p-1}) models			
(p', p)	C	Unitary	Comments
*(8, 3)	-21/4	no!	(D_5,A_2) – Potts Yang-Lee?
*(8, 5)	-7/20	no!	(D_5,A_4) – Potts Yang-Lee?
(8, 7)	25/28	Yes	unitary → not Potts Yang-Lee

4th Series ($E_{6,7,8}$, A_{p-1}) models
Smallest model has: $p' = 12$, $p = 5$ → 23 fields

In Table 1 known models are identified. The known models include the $C = 1/2$ normal critical point of the 2D Ising model, the $C = 4/5$ normal critical point of the 2D 3-state Potts model, and the $C = -22/5$ Yang-Lee edge singularity of the 2D Ising model. The remaining non-unitary models are candidates for the Yang-Lee edge singularity of the 2D 3-state Potts model. From the ADE partition functions, we found the primary fields of the candidate model. The results are in Table 2.

Table 2: $(r,s|r',s')$ values for the Primary Fields of Candidate
Models for Yang-Lee Edge Singularity of 2D 3-state Potts Model

(A_4,A_2): $(p',p) = (5,3)$ $C = -3/5$ → $C_{effective} = 3/5$			
Field	$(1,2|1,2)$	$(2,1|2,1)$	$(2,2|2,2)$
(h,h')	$(3/4|3/4)$	$(-1/20|-1/20)$	$(1/5|1/5)$

(D_5,A_2): $(p',p)=(8,3)$ $C = -21/4$ → $C_{effective} = 3/4$			
Field	$(1,2|1,2)$	$(3,1|3,1)$	$(3,2|3,2)$
(h,h')	$(3/2|3/2)$	$(-1/4|-1/4)$	$(1/4|1/4)$
Field	$(4,1|4,1)$	$(2,1|6,1)$	$(2,2|6,2)$
(h,h')	$(-3/32|-3/32)$	$(-7/32|25/32)$	$(25/32|-7/32)$

(D_5,A_4): $(p',p) = (8,5)$ $C = -7/20$ → $C_{effective} = 17/20$				
Field	$(1,2|1,2)$	$(1,3|1,3)$	$(1,4|1,4)$	$(3,1|3,1)$	$(3,2|3,2)$
(h,h')	$(7/10|7/10)$	$(11/5|11/5)$	$(9/2|9/2)$	$(1/4|1/4)$	$(-1/20|-1/20)$
Field	$(3,3|3,3)$	$(3,4|3,4)$	$(4,1|4,1)$	$(4,2|4,2)$	$(2,1|6,1)$
(h,h')	$(9/20|9/20)$	$(7/4|7/4)$	$(27/32|27/32)$	$(7/160|7/160)$	$(-1/32|95/32)$
Field	$(2,2|6,2)$	$(2,3|6,3)$	$(2,4|6,4)$		
(h,h')	$(27/160|187/160)$	$(187/160|27/160)$	$(95/32|-1/32)$		

(D_6,A_2): $(p',p) = (10,3)$ $C = -44/5$ → $C_{effective} = 4/5$			
Field	$(1,2|1,2)$	$(3,1|3,1)$	$(3,2|3,2)$	two $(5,1|5,1)$
(h,h')	$(2|2)$	$(-2/5|-2/5)$	$(3/5|3/5)$	$(-1/5|-1/5)$
Field	$(1,1|9,1)$	$(3,1|7,1)$	$(1,2|9,2)$	$(3,2|7,2)$
(h,h')	$(0|2)$	$(-2/5|3/5)$	$(2|0)$	$(3/5|-2/5)$

(D_6,A_6): $(p',p) = (10,7)$ $C = 8/35$ → $C_{effective} = 32/35$				
Field	$(1,2|1,2)$	$(1,3|1,3)$	$(1,4|1,4)$	$(1,5|1,5)$	$(1,6|1,6)$
(h,h')	$(4/7|4/7)$	$(13/7|13/7)$	$(27/7|27/7)$	$(13/2|13/2)$	$(10|10)$
Field	$(3,1|3,1)$	$(3,2|3,2)$	$(3,3|3,3)$	$(3,4|3,4)$	$(3,5|3,5)$
(h,h')	$(2/5|2/5)$	$(-1/35|-1/35)$	$(9/35|9/35)$	$(44/35|44/35)$	$(104/35|104/35)$
Field	$(3,6|3,6)$	two $(5,1|5,1)$	two $(5,2|5,2)$	two $(5,3|5,3)$	$(1,1|9,1)$
(h,h')	$(27/5|27/5)$	$(11/5|11/5)$	$(27/5|27/5)$	$(2/35|2/35)$	$(0|10)$
Field	$(1,2|9,2)$	$(1,3|9,3)$	$(1,4|9,4)$	$(1,5|9,5)$	$(1,6|9,6)$
(h,h')	$(4/7|46/7)$	$(13/7|27/7)$	$(27/7|13/7)$	$(46/7|4/7)$	$(10|0)$
Field	$(3,1|7,1)$	$(3,2|7,2)$	$(3,3|7,3)$	$(3,4|7,4)$	$(3,5|7,5)$
(h,h')	$(2/5|27/5)$	$(-1/35|104/35)$	$(9/35|44/35)$	$(44/35|9/35)$	$(104/35|-1/35)$
Field	$(3,6|7,6)$				
(h,h')	$(27/5|2/5)$				

Table 2 also gives <u>effective</u> central charges and conformal dimensions ($h_{r,s}$, $h_{r',s'}$) of the candidate minimal models.

Finite-size scaling provides a method for determining which candidate model corresponds to the Yang-Lee singularity of the 3-state Potts model. In particular, the finite-size scaling behavior of energy eigenvalues of a critical statistical system is predicted by the associated conformal model. For example, the ground state energy of a critical statistical system on a long cylinder scales as [15]:

$$E_{\text{ground state}}(N) = AN + B - \frac{\xi \pi C_{\text{effective}}}{6N}. \tag{32}$$

Here, N is the width of the cylinder; ξ A and B are non-universal constants; and $C_{\text{effective}}$ is the effective central charge of the conformal model. Below, it will be shown that $E_{\text{ground state}}$ is also the ground state energy of a 1D quantum spin chain associated with the statistical model. In most conformal models, $C_{\text{effective}}$ is the central charge, C. In conformal models whose primary fields having negative scaling dimensions, $C_{\text{effective}} = C - (h + \underline{h})$. Here, h and $\underline{h}$ are the two conformal dimensions of the primary field with the most negative scaling dimension, $(h + \underline{h})$. The energies, E_n, of excited states of the statistical system scale according to [16]:

$$E_n(N) - E_{\text{ground state}}(N) = \frac{2\pi \xi (h_n + \underline{h}_n)}{N}. \tag{33}$$

Here, the $(h_n + \underline{h}_n)$ are scaling dimensions of fields in the associated conformal model. Measuring the scaling of these excited energies also provides the value of the non-universal constant ξ.

The scaling behavior of energy eigenvalues fixes both $C_{\text{effective}}$ and scaling dimensions, $(h_n + \underline{h}_n)$, in the associated conformal field theory. Since the conformal theory is one of the non-unitary minimal models of Table 2, measurements of this scaling behavior at the Yang-Lee edge singularity for the 2D 3-state Potts model should enable an identification of the associated conformal minimal model.

3. II. Quantum chains

3.1. *Lattice Spin Models and Quantum Spin Chains*

In the 2D lattice spin models of interest, the total energy, E_a, for a state a has the form:

$$E_a = \sum_{\langle m,n|k,l \rangle} t_{m,n|k,l} S_{m,n} S_{k,l}^* + H \sum_{m,n} S_{m,n}^* + h.c., \tag{34}$$

where $S_{m,n}$ is the value of the spin at site (m,n). For the Ising and 3-state Potts models, the spins, $S_{m,n}$, are 2nd and 3rd roots of unity, respectively. The $t_{m,n|k,l}$ are nearest neighbor couplings between spins at sites (m,n) and (k,l). For the nonzero couplings, we define $t_{m,n|m,n+1} = t_{horizontal}$ and $t_{m,n|m+1,n} = t_{vertical}$. Here, H is an external magnetic field. On a square lattice, the total energy can be rewritten as:

$$E_a = \left\{ \sum_{m,n} t_{horizontal} S_{m,n} S^*_{m,n+1} + H \sum_{m,n} S_{m,n} \right\} + \left\{ \sum_{m,n} t_{vertical} S_{m,n} S^*_{m+1,n} \right\} + h.c. \quad (35)$$

From this form, one writes the partition function

$$Z = \sum_{states \ "a"} e^{-\beta E_a} \quad (36)$$

as a matrix trace:

$$Z = tr \ T \ x...x \ T, \quad (37)$$

where T is a matrix that couples adjacent rows of spins $\{S_{m,1},..., S_{m,N}\}$ and $\{S_{m+1,1},..., S_{m+1,N}\}$. Products of T's sum over states of a common row of spins. The trace implements periodic boundary conditions. T has the following matrix element:

$$T_{\{row"m"spins\},\{row"m+1"spins\}} = e^{-\beta \sum_n \left(t_{horizontal} S_{m,n} S^*_{m,n+1} + H S_{m,n} \right) - \beta \sum_n \left(t_{vertical} S_{m,n} S^*_{m+1,n} \right) + h.c.} \quad (38)$$

Equilibrium properties of the statistical model are determined by diagonalizing the transfer matrix T. While exact diagonalization is not generally possible, numerical diagonalization of T for small systems is possible at least, to find the largest eigenvalues of T. The numerical diagonalization is much simpler for a special limit known as the highly anisotropic limit. In the highly anisotropic limit, the transfer matrix only includes the diagonal terms and the non-diagonal terms associated with single spin flips. For this reason, the anisotropic limit of the transfer matrix, T, is a sparse matrix. Rapid numerical algorithms are available for determining eigenvalues of sparse matrices. Since the models with isotropic and anisotropic couplings, have the same critical behavior by universality, it is possible to use the more convenient anisotropic couplings for the numerical study of critical properties.

In the highly anisotropic limit, $\beta t_{horizontal} \to 0$ and $\beta t_{vertical} \to \infty$ simultaneously [17]. More precisely, we introduce parameter X and select $t_{horizontal}$ and $t_{vertical}$ such that:

$$\beta t_{horizontal} = tX,$$
$$\beta H = XH \text{ and} \quad (39)$$
$$e^{-\beta t_{vertical}} = X.$$

Here, t is a new parameter that is selected to obtain a linear dispersion relationship between energy and momentum. In this parameterization, the highly anisotropic limit is the $X \to 0$ limit. To obtain the transfer matrix for the Ising or 3-state Potts model in this limit, one expands T in the number of spins flips (denoted SF) between adjacent rows of spins. This expansion gives the following expression for the transfer matrix:

$$T\left(1 - \beta\sum_n\left[t_{horizontal}S_{m,n}S_{m,n+1}^* + HS_{m,n}\right] + O\left[\beta t_{horizontal} + \beta H\right]^2 + ... + c.c.\right)Id +$$
$$+e^{-\beta t_{vertical}}1SF + e^{-2\beta t_{vertical}}2SF + ... + e^{-k\beta t_{vertical}}kSF. \tag{40}$$

In the first term, the identity operator, Id, indicates that the m-th and $(m+1)$-th rows of spins are in the same state, i.e., no spin flips. In the later terms, the operators kSF cause k relative spin flips between the m-th and (m+1)-th rows of spins. The term with k spin flips introduces a factor of $exp(- k\beta t_{vertical})$, i.e., X^k. Rewriting T with the above definitions of $\beta t_{horizontal}$ and $\beta t_{vertical}$ in terms of X gives:

$$T\left(1 - \sum_n\left[XtS_{m,n}S_{m,n+1}^* + HS_{m,n}\right]\right)Id + (X)1SF + O(X^2). \tag{41}$$

As $X \to 0$, the transfer matrix becomes $(1 - XH_{quantum\ chain})$ with $H_{quantum\ chain}$ being defined as the Hamiltonian for a 1D quantum spin chain. In the highly anisotropic limit, the eigenvalues of transfer matrix T are exponents of the energy eigenvalues of a 1D quantum spin chain.

By explicitly taking the above limits, it is straightforward to find the Hamiltonians for the quantum chains of various discrete spin models. In the Ising model, the spin variables have two values. Thus, the $1SF$ operator for a finite width lattice is given by:

$$1SF_{Ising\ chain} = \sigma_x \otimes I \otimes I \otimes...\otimes I + I \otimes \sigma_x \otimes I \otimes I \otimes...\otimes I + ...+ I \otimes I \otimes...\otimes I \otimes \sigma_x, \tag{42}$$

where I is the 2x2 unit matrix. The diagonal part of the Hamiltonian of the Ising quantum spin chain is easily shown to be:

$$Diag\left(H_{Ising\ chain}\right) = t(\sigma_z \otimes \sigma_z \otimes I \otimes...\otimes I + I \otimes \sigma_z \otimes \sigma_z \otimes I \otimes...\otimes I + ...+ \sigma_z \otimes I \otimes...\otimes I \otimes \sigma_z) +$$
$$+H(\sigma_z \otimes I \otimes I \otimes...\otimes I + I \otimes \sigma_z \otimes I \otimes I \otimes...\otimes I + ...+ I \otimes I \otimes...\otimes I \otimes \sigma_z). \tag{43}$$

Adding the diagonal and off-diagonal terms, we obtain the Hamiltonian for an Ising quantum spin chain of length N. The Hamiltonian can be written more simply as:

$$H_{Ising\ chain} = \sum_{n=1}^{N}t\sigma_z(n)\sigma_z(n+1) + H\sigma_z(n) + \sigma_x(n). \tag{44}$$

From the more explicit notation, it is easily seen that the Hamiltonian of the Ising quantum spin chain is a matrix of dimension 2^N. In this matrix, only N entries per row are nonzero. Thus, the Hamiltonian of the quantum spin chain is very sparse.

The highly anisotropic limit of the 3-state Potts model can be derived through a similar calculation [18, 17]. The Hamiltonian for the 3-state Potts quantum spin chain is:

$$H_{3\text{-state Potts chain}} = \sum_{n=1}^{N}\left[tW(n)W(n+1)+H.C.\right]+HD(n)+\Lambda(n). \qquad (45)$$

Here, $W(n)$, $D(n)$, and $\Lambda(n)$ are 3^N x 3^N matrices that are direct products of $(N-1)$ 3x3 identity matrices with a single special 3x3 matrix, i.e., W, D and Λ, respectively. W is the 3x3 matrix whose diagonal entries are the three 3rd roots of unit and whose off-diagonal entries are zero. D is the 3x3 matrix whose (1,1) entry is 1 and whose other entries are zero. Λ is a matrix whose diagonal entries are zero and whose off-diagonal entries are 1, i.e., the 1SF operator. From this form, one sees that $H_{3\text{-state Potts chain}}$ is a matrix of dimension 3^N with of order N nonzero entries per row. The Hamiltonian of the 3-state Potts quantum spin chain is again a very sparse matrix.

3.2. Yang-Lee singularity in 2D Spin Models

The Yang-Lee edge singularity is the endpoint of the curve upon which zeros of the partition function condense as $V \rightarrow \infty$. At this endpoint, the value of the magnetic field is referred to as H_{YL}. Of course, the value of H_{YL} also depends on the coupling t. On a very long cylinder, the transfer matrix's two largest eigenvalues, i.e., λ_1 and λ_2, are equal at the endpoint. The Yang-Lee singularity is associated with a crossing point for the two largest eigenvalues of the transfer matrix. To see this, we write the partition function Z, in terms of the eigenvalues λ_j, of the transfer matrix.

$$Z = \sum_j \left(\lambda_j\right)^N \approx \lambda_1^N + \lambda_2^N. \qquad (46)$$

Keeping only the two largest eigenvalues is not an approximation as $N \rightarrow \infty$, i.e., on a very long cylinder. In this limit, the contribution of smaller eigenvalues, δZ, is negligible, because $|\delta Z|/|\lambda_1^N + \lambda_2^N| \leq \Sigma_{j>2} (\lambda_j/\lambda_1)^N \rightarrow 0$. Thus, on a very long cylinder, Z has a zero if $\lambda_1^N + \lambda_2^N = 0$. The solutions for zeros are: $\lambda_2 = e^{i\pi(2p-1)/N}\lambda_1$ and $0 < p < (N+1)/2$. While this equation implies that the endpoints satisfy $|\lambda_2(H_{YL})| = |\lambda_1(H_{YL})|$, our numerical studies on thin long cylinders show $\lambda_2(H_{YL}) = \lambda_1(H_{YL})$, i.e., both magnitudes and phases are equal at the endpoints. Thus, a Yang-Lee edge singularity is associated with the magnetic field H_{YL}, where the two largest eigenvalues of the transfer matrix actually cross. Since the largest eigenvalues of the transfer matrix correspond to the smallest eigenvalues of the associated quantum spin chain, H_{YL} corresponds to the crossing point for the lowest energy eigenvalues of the associated quantum spin chain.

To characterize Yang-Lee edge singularity as a critical theory and a conformal theory the scaling behavior of various quantities are evaluated at the crossing point. To characterize the critical theory, the exponent σ that defines the magnetization near the singularity is evaluated from the finite-size scaling behavior of $H_{YL}(N)$. To identify the associated conformal field theory, the finite-size scaling behavior is used to determine field scaling dimensions and/or effective central charges.

3.3. Determining Eigenvalues of Quantum Chain Hamiltonians

We will present results for the Yang-Lee singularities of the 2D Ising and 3-state Potts models. The 2D Ising model was our testing ground, because finite-size scaling results exist for the Ising quantum spin chain [10]. The results for the Yang-Lee edge singularity of the 2D 3-state Potts model are new.

To study finite-size scaling behavior, one must determine energy eigenvalues for long quantum spin chains. The associated Hamiltonians are large matrices, e.g., the Hamiltonian of the 10 site Ising quantum chain has dimension 1024. The large dimension of these matrices makes direct diagonalization essentially impossible. The evaluation of some eigenvalues is possible, because the Hamiltonians are sparse matrices for which efficient iterative methods exist. These methods provide the eigenvalues of largest magnitude to a precision sufficient for determining finite-size scaling behavior [19, 20].

Our determination of the lowest energy eigenvalues for quantum spin chains is based on the Lanczos method [19]. From a given matrix H, the Lanczos method iteratively determines a matrix V that is similar to H, i.e., $H = QVQ^T$ with $Q^TQ = 1$. Instead of evaluating the NxN matrix V exactly, the Lanczos method involves evaluating a submatrix of V of dimension MxM. If N is large and sparse, an $M \approx 2\sqrt{N}$ typically suffices to find the largest few eigenvalues of H.

To motivate the Lanczos method, suppose that there is an NxM orthogonal matrix, Q, such that $HQ = QC$ where C is an MxM symmetric matrix with $M < N$. The eigenvectors of C satisfy: $Cy = \lambda_y y$. Then, $HQ = QC$ implies that $Qy = \lambda_y Qy$. Thus, finding the eigenvalues of the smaller matrix C provides some eigenvalues of H.

Since Q is orthogonal, Q can be written as: $Q = [y_1| y_2| ...| y_M]$ where the M vectors $\{y_j\}$ form an orthonormal basis of the invariant subspace, i.e.,

$y_j^T y_k = \delta_{jk}$. The existence of such a subspace is the necessary condition for the existence of a C satisfying: $HQ = QC$. Since the y_j's are orthonormal, $Q^T Q = 1$. Furthermore, $P = QQ^T$ is a projection operator, i.e., $PP = P$, and P projects to the invariant subspace. The existence of a invariant subspace is necessary for Q and C to exist and thus, is necessary to reduce the N-dimensional eigenvalue problem for H to an M-dimensional eigenvalue problem.

The Lanczos method uses an approximate invariant subspace that is known as an M-dimensional Krylov space. The Kyrlov space is generated by the set of vectors $\{y_1, Hy_1, H(Hy_1), H(H(Hy_1)), \dots H^M y_1\}$. For M large, the Krylov space is approximately invariant, because $H^M y_1$ converges to an eigenvector of H. The eigenvector is associated with the largest magnitude eigenvalues λ^{max} : $H^M y_1 \approx \lambda^{max}(H^{M-1} y_1)$ [20]. To see this, we expand $y_1 = \Sigma_{j=1\dots N} a_j X_j$ over the basis of eigenvectors X_j of H. Since each eigenvector X_j satisfies: $HX_j = \lambda_j X_j$, one sees that $H^M y_1 = \Sigma_{j=1\dots N} a_j(\lambda_j)^M X_j$. For M large, the sum is dominated by the term $a_a X_a$ associated with the largest magnitude eigenvalue $|\lambda_a|$, i.e., $|\lambda_a| > |\lambda_b|$ for $b \neq a$. The orthonormal basis vectors of the Krylov space generate a matrix Q satisfying: $Q^T HQ = V$. Eigenvalues of the MxM matrix V are approximate eigenvalues of H.

The Krylov space is useful, because it includes basis vectors, e.g., $H^M y_1$, that have large overlaps with eigenvectors for large magnitude eigenvalues. Thus, the MxM matrix V provides good approximations to the largest eigenvalues of H even for $M \ll N$. Thus, finding H's largest eigenvalues requires diagonalization of the much smaller matrix V.

The matrix V will be a tridiagonal matrix for which simpler diagonalization techniques also exist. The tridiagonal matrix V is defined by:

$$V_{jj} = a_j \text{ for } j = 1, 2, \dots, M \text{ and} \tag{47}$$
$$V_{k,k+1} = V_{k+1,k} = b_k \text{ for } k = 1, 2, \dots, M-1.$$

To find V, one expands the left and right sides of $HQ = QV$ using the definition of Q in terms of the orthonormal column vectors, y_1, y_2, ..., y_M, i.e., $Q = [y_1| y_2| \dots | y_M]$. The expansion produces the following system of equations:

$$Hy_1 = a_1 y_1 + b_1 y_2;$$
$$Hy_2 = b_1 y_1 + a_2 y_2 + b_2 y_3;$$
$$\dots$$
$$Hy_k = b_{k-1} y_{k-1} + a_k y_k + b_{k-1} y_{k+1}; \tag{48}$$
$$\dots$$
$$Hy_M = b_{M-1} y_{M-1} + a_M y_M.$$

The system is solved level-by-level to obtain a_k, b_k, and y_{k+1} at the kth-level. At the 1st level, one selects a starting vector y_1^T, e.g., $y_1^T = (1, 0, 0, ..., 0)$, and then, determines a_1, b_1, and y_2. One multiplies the first equation by y_1^T and uses the orthonormality of the y_k to get $a_1 = y_1^T H y_1$. Multiplying the 1st-level equation by y_2^T and again using orthonormality produces: $b_1 = |H y_1 - a_1 y_1|$. From a_1 and b_1, one finds that $y_2 = b_1^{-1}(H y_1 - a_1 y_1)$. The 2nd-level equation produces: $a_2 = y_2^T H y_2$, $b_2 = |H y_2 - b_1 y_1 - a_2 y_2|$, and $y_3 = b_2^{-1}(H y_2 - b_1 y_1 - a_2 y_2)$. More generally, solving the kth-level produces a_k, b_k, and y_{k+1} from previously found $\{a_j\}$, $\{b_j\}$, and $\{y_{j+1}\}$.

If the method is continued until $M = N$, an exact tridiagonalization of H results. Truncating at the kth-level produces an error proportional to b_k. In particular, if v_s is an eigenvalue of the truncated V, then $| v_s - \lambda_s | \leq b_k$ for some eigenvalue λ_s of H.

In implementing Lanczos tridiagonalization, various techniques enable lowering the dimension, M, of the truncated tridiagonal matrix V needed to obtain largest magnitude eigenvalues to a desired precision.

First, one can add a diagonal contribution to the Hamiltonian, $H \rightarrow H - K_{shift}I$. The diagonal shift increases the rate of convergence for the ground state rather than the highest energy state.

Second, the starting vector y_1 can be selected to produce a more rapid convergence for the largest magnitude eigenvalue. One such starting eigenvector is defined by: $y_1 = (H)^p e_1$ with $e_1 = (1, 0,0,...0)^T$, e.g., for a fixed integer p. Such a starting vector is a combination eigenvectors associated with large eigenvalues and thus, should reduce the size of the truncated V matrix needed to obtain the largest eigenvalues. Such a starting vector should also reduce the need to re-orthogonalize basis vectors $\{y_j\}$, which is a serious problem in standard Lanczos.

To obtain eigenvalues of the tridiagonalized submatrix of V, we look for roots of the characteristic polynomial of V. Since V is tridiagonal, the characteristic polynomials $CP_1(\lambda)$, $CP_2(\lambda)$, $CP_3(\lambda)$, ... $CP_k(\lambda)$ for the respective 1, 2, 3,..., k –dimensional submatrixes form a sequence. The members of the Sturm sequence satisfy:

$$CP_r(\lambda) = (a_r - \lambda) CP_{k-1}(\lambda) - b_{k-1} CP_{k-2}(\lambda),$$
$$CP_1(\lambda) = a_1 - \lambda \text{ and } CP_0(\lambda). \tag{49}$$

This sequence is used to rapidly evaluate the characteristic polynomial of V.

3.4. Yang-Lee singularity of 2D Ising model

Since Yang-Lee edge singularities occur at complex values of the magnetic field H, the associated quantum spin chains have non-hermitian Hamiltonians. Nevertheless, the Ising quantum spin chain has a symmetric Hamiltonian at the Yang-Lee edge singularity, because the singularity corresponds to purely imaginary value of the magnetic field, $H_{YL} = iB_{YL}$:

$$H_{Ising\ chain} = \sum_{n=1}^{N} t\sigma_z(n)\sigma_z(n+1) + iB_{YL}\sigma_z(n) + \sigma_x(n) \text{ with } B_{YL} = B_{YL}^*. \quad (50)$$

The Hamiltonian and its complex conjugate are related by a similarity transforma-tion:

$$\left(H_{Ising\ chain}\right)^* = \sum_{n=1}^{N} t\sigma_z(n)\sigma_z(n+1) - iB_{YL}\sigma_z(n) + \sigma_x(n)$$
$$= \prod_n \sigma_x(n) H_{Ising\ chain} \prod_n \sigma_x(n). \quad (51)$$

The above equivalence means that each eigenstate of $H_{Ising\ chain}$ generates an eigenstate with the same eigenvalue for $H_{Ising\ chain}{}^*$. That is:

$$H_{Ising\ chain}\Psi = E\Psi \rightarrow \left(H_{Ising\ chain}\right)^* \prod_n \sigma_x(n)\Psi = E\prod_n \sigma_x(n)\Psi$$
$$\rightarrow \left(H_{Ising\ chain}\right)\prod_n \sigma_x(n)\Psi^* = E^* \prod_n \sigma_x(n)\Psi^*. \quad (52)$$

Thus, either an eigenvalue of $H_{Ising\ chain}$ is real, or there is a second eigenvalue that is the complex conjugate.

This simplifies the search for the Yang-Lee edge singularity with the Ising quantum spin chain. At $H = 0$, the Hamiltonian is real and the ground state energy of the Ising quantum spin chain is real. For small purely imaginary magnetic fields, H, the ground state eigenvalue remains real by continuity [10]. Otherwise, the eigenvalue would have a complex conjugate. Both the ground state and the first excited state eigenvalues remain real until the first crossing point. This implies that the Yang-Lee edge singularity of the 2D Ising model is associated with an actual crossing point (see section 3.5) where two crossing eigenvalues become a complex conjugate pair.

3.5. Results for Singularity of 2D Ising

Figures 4a-b, 5, and 6 show results for Ising quantum spin chains with 2, 6, and 9 sites an a coupling t equal to 0.4. The 2-site Ising quantum spin chain has 4 energy eigenvalues, and Figures 4a-4b show the motion of these eigenvalues as the magnetic field sweeps along the imaginary axis. The results show the existence of a point, H_{YL}, in the complex magnetic field plane where the two lowest eigenvalues cross. The point H_{YL} is the Yang-Lee edge singularity for the 2-site model. The results also show that the ground state and first excited state have real energies for imaginary magnetic fields smaller than the crossing value and have complex conjugate energies for imaginary magnetic fields greater than the crossing value. At the Yang-Lee edge singularity, the lowest energy levels cross and become complex.

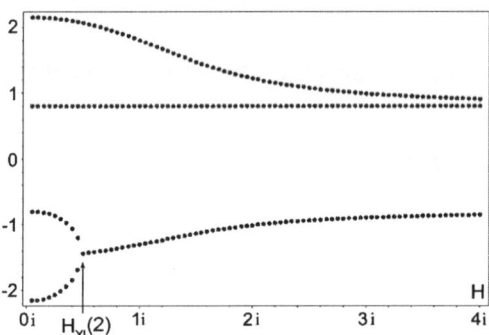

Figure 4a. Real parts of energy eigenvalues of the 2-site Ising quantum spin chain for imaginary values of magnetic field

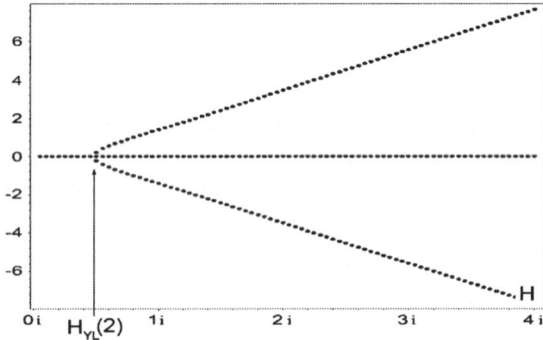

Figure 4b. Imaginary parts of energy eigenvalues of the 2-site Ising quantum spin chain for imaginary values of magnetic field

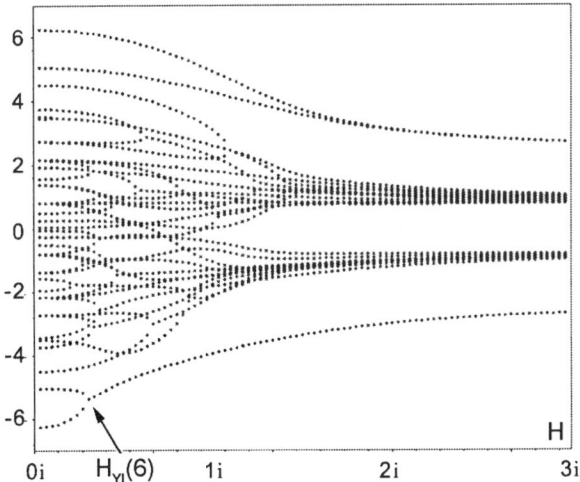

Figure 5. Real parts of energy eigenvalues of the 6-site Ising quantum spin chain for imaginary values of magnetic field

Figure 5 shows the behavior of the 2^6 energy eigenvalues of the 6-site Ising quantum spin chain as the magnetic field is swept along the imaginary axis. The results show the existence of a point H_{YL} where the two lowest eigenvalues cross. The crossing point H_{YL} again corresponds to a purely imaginary value of the magnetic field and is the point where the ground state develops a complex energy.

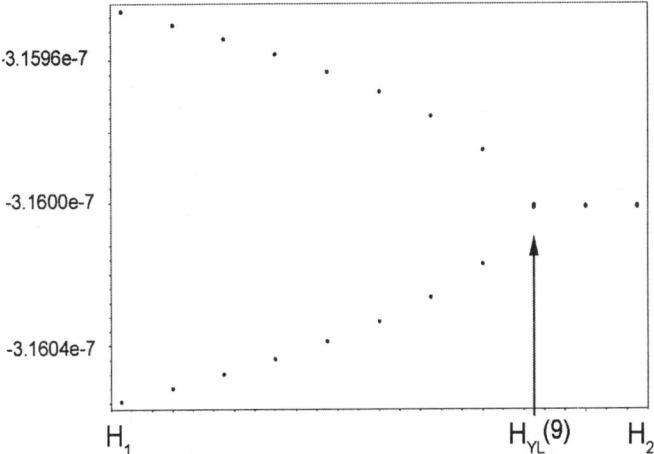

Figure 6a. Real parts of the lowest energy eigenvalues of the 9-site Ising quantum chain showing precise level-crossing point.

34

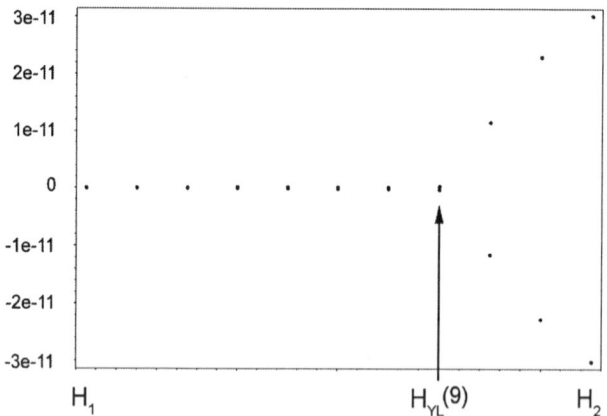

Figure 6b. Imaginary parts of the lowest energy eigenvalues of the 9-site Ising quantum chain showing precise level-crossing point.

Figure 6a and 6b shows results for the crossing point, H_{YL}, of the 9-site Ising quantum spin chain with actual precisions. High precision values of $H_{YL}(N)$ are needed in finite-size scaling studies. Using the Lanczos method, we have found precision values of $H_{YL}(N)$ for Ising quantum spin chains with 2 to 14 sites and $t = 0.4$. Table 3 shows $H_{YL}(N)$ and $E_{ground\ state}(N)$ at $H_{YL}(N)$.

Table 3: Numerical finite-size scaling results for Yang-Lee
edge singularity from Ising quantum spin chain

N	$H_{YL}(N)$	$E_{ground\ state}(N)$
2	0.4281531135 ± 1	
3	0.3209900152 ± 1	-2.4447738 ± 1
4	0.28147048287 ± 1	-3.4359270 ± 1
5	0.26281558764 ± 1	-4.4161731 ± 1
6	0.25270182449 ± 1	-5.3881010 ± 1
7	0.24668840977 ± 1	-6.3537184 ± 1
8	0.24286416531 ± 1	-7.3145422 ± 1
9	0.24030314551 ± 1	-8.2716923 ± 1
10	0.2385161235 ± 1	-9.2259915 ± 1
11	0.2372266713 ± 1	-10.178049 ± 2
12	0.2362699483 ± 1	-11.128317 ± 4
13	$0.23554317 \quad \pm 5$	$-12.07712 \quad \pm 2$
14	$0.2349798 \quad \pm 1$	$-13.0247 \quad \pm 1$

Table 3 shows that $H_{YL}(N)$ is smoothly converging towards a limit at $N = 14$. Assuming that $H_{YL}(N)$ is analytic at $N = \infty$, our results imply that $H_{YL}(\infty) = 0.2320$.

To evaluate σ, the data of table 3 is used to determine Θ, and then, the relations $\Theta = 1/\nu$ and $1 - \sigma = D\nu$ are used to obtain σ. Θ is found from the finite-size scaling relation $|H_{YL}(N) - H_{YL}(\infty)| = f N^{-\Theta}$. Ignoring the dependence of scaling function f on $1/N$ enables determinations of Θ from a triplet of $H_{YL}(N)$ values. Figure 7 plots Θ-values that were determined from adjacent triplets $[H_{YL}(N-1), H_{YL}(N), H_{YL}(N+1)]$. The so-determined Θ-values vary with $1/N$, because the scaling function f varies with $1/N$. The scaling function is analytic at $1/N = 0$ so that $f(1/N) \approx f_0 + f_1/N + f_2/N^2$. Indeed, the determined Θ-values depend nearly linear on $1/N$ for $N > 6$ as would be expected from the dependence on f_1 and f_2. Linearly extrapolating the determined Θ-values to $1/N = 0$ produces a good estimate for the actual value of Θ. From this Θ, we obtain that $\sigma = -1/5.35$. This agrees well the prediction that $\sigma = -1/6$ from the (A_4, A_2) minimal model believed to describe the Yang-Lee edge singularity of the Ising model [4, 5].

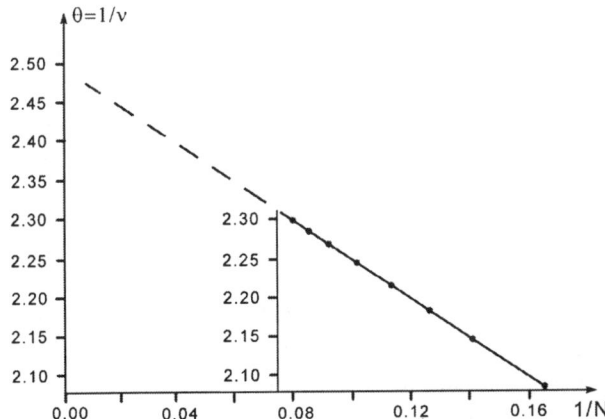

Figure 7. Dots show Θ-values obtained from triplets of crossing points, i.e., $[H_{YL}(N-1), H_{YL}(N), H_{YL}(N+1)]$. Linear extrapolation as $1/N \rightarrow 0$ gives Θ for the 2D Ising model.

We do not report results on magnetization, $<S>$, because the results do not provide values of the exponent σ to as high a precision. For short chains, the results show the correct qualitative comportment of the magnetization in the vicinity of the Yang-Lee edge singularity. Near the singularity, the magnetization has a sharp peak in accordance with expected divergence of $<S>$ due to the negative sign of σ.

3.6. Yang-Lee singularity of 3-state Potts model

Finite-size scaling in the 3-state Potts quantum spin chain is similar to that of the Ising quantum spin chain with a few exceptions. First, zeros of the partition function for the 3-state Potts model are not at purely imaginary values of the magnetic field. Second, the N-site 3-state Potts quantum spin chain has 3^N states whereas the N-site Ising quantum spin chain only has 2^N states. Thus, the spectra of chains are larger in the 3-state Potts model. Third, energy eigenvalues of the 3-state Potts quantum spin chain are not necessarily real or in complex conjugate pairs.

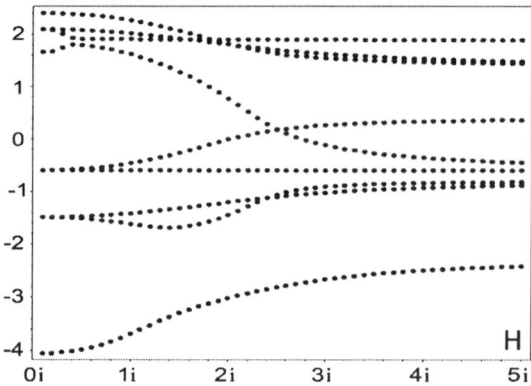

Figure 8. Real parts of energy eigenvalues for 2-site 3-state Potts quantum spin chain for purely imaginary values of the magnetic field.

Figures 8, 9a – 9c show energy eigenvalues for the 2-site 3-state Potts quantum spin chain with coupling $t = 0.4$. Fig. 8 shows real parts of the energy eigenvalues as the magnetic field is swept along the imaginary axis. The absence of a lowest level crossing implies that the Yang-Lee edge singularity is at a complex field value. Figures 9a and 9b show real and imaginary parts of the energy eigenvalues for magnetic field values whose real parts equal 0.5. A near crossing of the lowest energy levels occurs. The energy eigenvalues have equal imaginary parts below the near crossing thereby implying that the Yang-Lee edge singularity is associated with an actual level-crossing. Figure 9c shows real parts of the energy eigenvalues as the magnetic field sweeps through the value for the level-crossing.

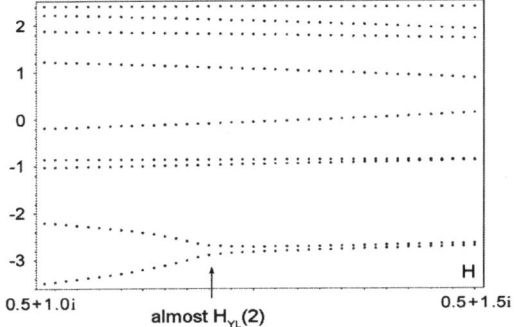

Figure 9a. Real parts of energy eigenvalues for 2-site 3-state Potts quantum spin chain for magnetic fields whose real parts are 0.5.

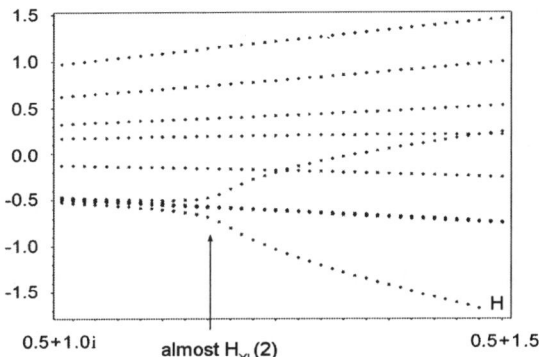

Figure 9b. Imaginary parts of energy eigenvalues for 2-site 3-state Potts quantum spin chain for magnetic fields whose real parts are 0.5.

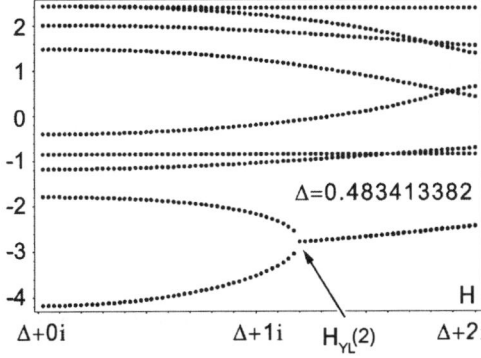

Figure 9c. Real parts of energy eigenvalues for 2-site 3-state Potts quantum spin chain for magnetic fields whose real parts, Δ, are equal to 0.483413382.

Table 4 shows the values of the magnetic field crossings of the lowest energy level-crossings for 3-state Potts quantum spin chains having $t = 0.4$. The results show rapid convergence of $H_{YL}(N)$ to a limit value as $N \to \infty$.

Table 4: Numerical finite-size scaling results for Yang-Lee edge singularity from 3-state Potts quantum chain

N sites	$H_{YL}(N)$	$\Delta H_{YL}(N)$
2	0.483413 + 1.174228i	***
3	0.473980 + 1.059340i	0.009433 + 0.114888i
4	0.473849 + 1.019205i	0.000131 + 0.040135i
5	0.473886 + 1.001125i	-0.000037 + 0.018080i
6	0.473854 + 0.991766i	0.000032 + 0.009359i
7	0.473820 + 0.986407i	0.000034 + 0.005359i
8	0.473805 + 0.983091i	0.000015 + 0.003316i

The exponent σ was evaluated as in the Ising model except that finite-size scaling for a complex valued field, i.e., $[H_{YL}(N) - H_{YL}(\infty)] \approx N^{-\theta}$ was used. Values of exponent Θ obtained from triplets $[H_{YL}(N-1), H_{YL}(N), H_{YL}(N+1)]$ are shown in Figure 10. Linearly extrapolating these values to $1/N \to 0$ provides an estimate of the value of Θ. A more conservative estimate is that $-1/8 > \sigma > -1/4$.

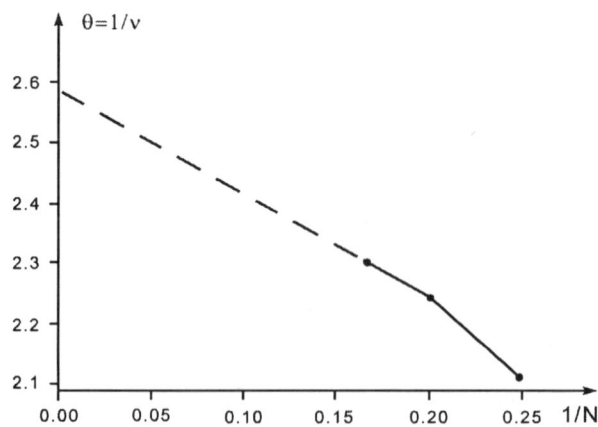

Figure 10. Points show Θ-values obtained from triplets of level crossings $[H_{YL}(N-1), H_{YL}(N), H_{YL}(N+1)$. Linear extrapolation in $1/N$ shows Θ of 2D 3-state Potts model.

Table 5 lists the candidates for the Yang-Lee Singularity of the 3-state Potts model that we obtained from the *ADE* classification. The table also provides values of σ and the conformal dimensions of primary fields associated with the spin.

Table 5: Candidate Conformal Models for the Yang-Lee Edge
Singularity of the 2D 3-state Potts Model

Model	(A_4,A_2)	(D_5,A_2)	(D_5,A_4)	(D_6,A_2)	(D_6,A_6)
(p,p')	(5,3)	(8,3)	(8,5)	(10,3)	(10,7)
Candidate field for spin	(2,1|2,1)	(3,1|3,1)	(3,2|3,2)	(3,1|3,1), (5,1|5,1)	(3,2|3,2)
dimensions	(-1/20|-1/20)	(-1/4|-1/4)	(-1/20|1/20)	(-2/5|-2/5) (-1/5|-1/5)	(-1/35|-1/35)
C_{eff}	3/5	3/4	17/20	4/5	32/35
σ	-1/21	- 1/5	-1/21	-1/7, -1/6	-1/36

Though our results are preliminary, we can almost uniquely identify the conformal minimal model associated with the Yang-Lee singularity of the 2D 3-state Potts model due to the limited number of candidates. The results show that either the *(D5,A2)* minimal model or the *(D6,A2)* minimal model describes the Yang-Lee singularity of the 2D 3-state Potts model. We are analyzing finite-size scaling behavior of excited energy levels to determine which of these two candidate models is the correct one. This further analysis will probably suffice for a unique determination.

References

1. C.N. Yang, T.D. Lee, *Phys. Re.* **87** (1952) 404.
2. K. Uzelac, *Thesis Univ. of Paris XI* (1980) 33-39.
3. C.N. Yang, T.D. Lee, *Phys. Re.* **87** (1952) 410.
4. M.E. Fisher, *Phys. Rev. Lett.* **40** (1978) 1610.
5. J. L. Cardy, *Phys. Rev. Lett.* **54** (1985) 200; C. Itzykson, H. Saleur, and J.-B. Zuber, *Europhys. Lett.* **2** (1986) 91.
6. Z. Glumac and K. Uzelac, *J. Phys. A: Math. Gen.* **27** (1994) 7709; see also B. P. Dolan and D.A. Johnston cond-mat/0010372 (2000).
7. D.A. Kurtze and M.E. Fisher, *Phys. Rev. B* **20** (1979) 2785; P.J. Kortman and R.B. Griffiths, *Phys. Rev. Lett.* **27** (1971) 1439.
8. D. A. Kurtze and M. E. Fisher, *J. Stat. Phys.* **19** (1978) 205; R.B. Griffiths, *J. Math. Phys.* **10** (1969) 1559; B. Simon and R.B. Griffiths, *Comm. Math. Phys.* **33** (1973) 145; C.M. Newman, Commun. *Pure Appl.*

Math. **27** (1974) 143; T. Asano, *Phys. Rev. Lett.* **24** (1970) 1409; ibid, *J. Phys. Soc. Japan,* **29** (1970) 350; A.B. Harris, *Phys. Lett.* **33A** (1970) 161; T. Dunlop and C.M. Newman, *Comm. Math. Phys.* **44** (1975) 223.

9. Ch. Binek, Phys. Rev. Lett. 81 (1998) 5644; Ch. Binek, W. Kleemann, and H. A. Katori, *J. Phys.: Condens. Matter* **14** (2001) L811.

10. G. von Gehlen, *J. Phys. A: Math. Gen.* **24** (1991) 5371.

11. M.E. Fisher and M.N. Barber, *Phys. Rev. Lett.* **28** (1972) 1516; see brief review in M. Henkel *Conformal Invariance and Critical Phenomena* (Springer-Verlag, 1999) chapter 3.

12. A.A. Belavin, A.M. Polyakov, and A.B. Zamolodchikov, *Nucl. Phys. B* **241** (1984) 333.

13. V.G. Kac, *Contravariant forms for infinite dimensional Lie algebras and superalgebras, Lecture Notes in Physics* vol. **94** (Springer-Verlag, Berlin 1979); B.L. Feigin and D. B. Fuchs, *Funct. Anal. and Appl.* **17** (1982) 114; for a review see e.g., C. Itzykson and J.-M. Drouffe, *Statistical Field Theory* (Cambridge University Press, 1989) chapter IX.

14. A. Capelli, C. Itzykson, J.-B. Zuber, *Nucl. Phys.* **B280** (1987) 445; ibid, *Comm. Math. Phys.* **13** (1987) 1; for a review see e.g., C. Itzykson and J.-M. Drouffe, *Statistical Field Theory* (Cambridge University Press, 1989) chapter IX (1989).

15. I. Affleck, *Phys. Rev. Lett.* **56** (1984) 746; H. W. Blöte, J.L. Cardy, and M.P. Nightingale, *Phys. Rev. Lett.* **56** (1984) 742.

16. J.L. Cardy, *J. Phys.* **A17** (1984) L385.

17. E. Fradkin and L. Susskind, *Phys. Rev.* **D17** (1978) 2637; M. Suzuki, *Prog. Theor. Phys.* **46** (1971) 1337; for a review see e.g., M. Henkel in *Conformal Invariance and Critical Phenomena* (Springer-Verlag, 1999) chapters 8-10.

18. G. von Gehlen et al, *J. Phys. A* **20** (1987) 2577- 2591.

19. For a review see e.g.: P.N. Parlett, *The Symmetric Eigenvalue Problem* (Prentice-Hall, 1980); Alan Jennings, *Matrix Computation for Engineers and Scientists* (Wiley, 1977) pages 316-319; W. Kerner, *Journal of Computational Physics* **85** (1989) 1.

20. See e.g., K. E. Atkinson, *An Introduction to Numercial Analysis* (Wiley, 1978) chapter 9.

PATH INTEGRAL REPRESENTATIONS
FOR A CONSTRAINED SYSTEM

Y. OHNUKI

Nagoya Women's University
1302 Takamiya Tempaku Nagoya 468-8507, Japan
E-mail:ohnuki@nagoya-wu.ac.jp

All possible irreducible representations of the Dirac algebra for a system constrained on a manifold diffeomorphic to S^D are explicitly given. It is shown that for the case where $D \geq 2$ the irreducible representation is unique, whereas for $D = 1$ there exist an infinite number of unitary-inequivalent irreducible representations specified by a parameter $\alpha \in [0, 1)$. Applying them we derive some rigorous expressions of the path integral, which automatically involve quantum corrections not existing in the Faddeev-Senjanovic formula. Especially, for $D = 1$ the terms with the parameter α in the path integral are shown to be equivalent to those given by the Aharonov-Bohm gauge potential produced by the magnetic flux $\Phi = -2\pi\alpha\hbar c/e$.

1. Preliminaries

We denote a D-dimensional manifold embedded in the flat space $\mathbb{R}^{D+1}$ as $f(x) = 0$, where $f(x)$ is a real function and x stands for the coordinates x_α ($\alpha = 1, 2, \cdots, D + 1$) in $\mathbb{R}^{D+1}$. The manifold is assumed to be diffeomorphic to S^{D+1}.

Then let us consider a quantum-mechanical system constrained to move on

$$f(\hat{x}) = 0 \tag{1.1}$$

with the Hamiltonian

$$\hat{H} = \frac{1}{2}\hat{p}_\alpha\hat{p}_\alpha + V(\hat{x}), \tag{1.2}$$

where and in what follows the repeated use of the same Greek indices in a single term means the summation over $1, 2, \cdots, D + 1$, and the symbol ^ indicates operators. In addition to (1.1) the algebra describing this system is known to be given[1,2] by

$$\{\hat{p}_\alpha, f_{,\alpha}(\hat{x})\} = 0, \tag{1.3}$$

$$[\hat{x}_\alpha, \, \hat{x}_\beta] = 0, \tag{1.4}$$

$$[\hat{x}_\alpha, \, \hat{p}_\beta] = i\hbar \Lambda_{\alpha\beta}(\hat{x}), \tag{1.5}$$

$$[\hat{p}_\alpha, \, \hat{p}_\beta] = -i\hbar \left\{ \frac{f_{,\alpha}(\hat{x})f_{,\beta\gamma}(\hat{x}) - f_{,\beta}(\hat{x})f_{,\alpha\gamma}(\hat{x})}{2R^2(\hat{x})}, \, \hat{p}_\gamma \right\} \tag{1.6}$$

$$(\alpha, \, \beta, \, \gamma = 1, \, 2, \, \cdots, \, D + 1),$$

where $R(x) \, (> 0)$ and the $(D + 1) \times (D + 1)$ matrix $\Lambda(x) (\equiv \| \Lambda_{\alpha\beta}(x) \|)$ are defined by

$$R^2(x) = f_{,\alpha}(x)f_{,\alpha}(x), \qquad \Lambda_{\alpha\beta}(x) = \delta_{\alpha\beta} - \frac{f_{,\alpha}(x)f_{,\beta}(x)}{R^2(x)} \tag{1.7}$$

with

$$f_{,\beta_1\beta_2\cdots\beta_s}(x) \equiv \partial_{\beta_1}\partial_{\beta_2} \cdots \partial_{\beta_s} f(x). \tag{1.8}$$

We will call the set of the equations (1.1) and (1.3)–(1.6) the Dirac algebra on $f(x) = 0$.

All possible irreducible representations of this algebra have recently been determined[2] with the aid of the canonical variables $\hat{x}_\alpha$, and $\hat{\pi}_\alpha$ in $(D + 1)$-dimension, which obey

$$[\hat{x}_\alpha, \, \hat{x}_\beta] = [\hat{\pi}_\alpha, \, \hat{\pi}_\beta] = 0, \qquad [\hat{x}_\alpha, \, \hat{\pi}_\beta] = i\hbar\delta_{\alpha\beta}. \tag{1.9}$$

Then the operators $\hat{p}_\beta$ in the irreducible representation of the Dirac algebra are expressed as follows:

$D = 1$.

$$\hat{p}_\beta = \frac{1}{2} \{ \Lambda_{\beta\gamma}(\hat{x}), \, \hat{\pi}_\gamma \} - \alpha\hbar \frac{\Lambda_{\beta\gamma}(\hat{x})f_{,\gamma\rho}(\hat{x})f_{,\sigma}(\hat{x})\epsilon_{\rho\sigma}}{R^2(\hat{x})}, \tag{1.10}$$

where α is a real parameter which uniquely specifies the irreducible representation, and $\epsilon_{\rho\sigma}$ stands for the two-dimensional Levi-Civita symbol defined by $\epsilon_{\rho\sigma} = -\epsilon_{\sigma\rho}$ and $\epsilon_{12} = 1$. It was shown[2] that (i) no other irreducible representation exists than the above for $D = 1$ and (ii) two irreducible representations specified respectively by α and α' are unitarily equivalent to each other if and only if $\alpha' = \alpha + integer$. Hence without loss of generality we may restrict ourselves to the cases where $0 \le \alpha < 1$.

$D \ge 2$.

$$\hat{p}_\beta = \frac{1}{2} \{ \Lambda_{\beta\gamma}(\hat{x}), \, \hat{\pi}_\gamma \}. \tag{1.11}$$

This expression for $\hat{p}_\beta$ is unique up to unitary-equivalent transformations.

We will apply these results to obtain path integral representations for the system constrained on $f(x) = 0$.

Now given an irreducible representation we denote the representation space as $\underline{\mathcal{H}}$ and state vectors belonging to it as $|\underline{\psi}\rangle$, $|\underline{\chi}\rangle$, $\cdots$. We further write the point on $f(x) = 0$ as $\underline{x}$. Since the eigenstates $|\underline{x}\rangle$ of the position operators on the manifold form an ortho-complete system in $\underline{\mathcal{H}}$ we may write the completeness condition as

$$\int_{\Sigma_f} d^D\sigma(\underline{x}) \, |\underline{x}\rangle(\underline{x}| = \hat{\underline{1}}, \tag{1.12}$$

where $\hat{\underline{1}}$ is the unit operator on $\underline{\mathcal{H}}$. The infinitesimal volume element $d^D\sigma(\underline{x})$ of the manifold is defined referring to the measure of the flat space $\mathbb{R}^{D+1}$, and Σ_f denotes the whole domain of the manifold. In the position diagonal representation we write the wave functions corresponding to $|\underline{\psi}\rangle$ and $|\underline{\chi}\rangle$ as $\underline{\psi}(\underline{x}) \equiv (\underline{x}|\underline{\psi})$ and $\underline{\chi}(\underline{x}) \equiv (\underline{x}|\underline{\chi})$, respectively.

On the other hand we will denote the representation space of the canonical commutation relations (1.9) as $\mathcal{H}$, which is spanned by the eigenstates $|x\rangle$ of the operators $\hat{x}_\beta$. They of course obey

$$\hat{x}_\beta|x\rangle = x_\beta|x\rangle, \qquad \langle x|x'\rangle = \delta^{D+1}(x - x'), \tag{1.13}$$

$$\int d^{D+1}x \, |x\rangle\langle x| = \hat{1} \tag{1.14}$$

with $\hat{1}$ being the unit operator on $\mathcal{H}$. Corresponding to $\underline{\psi}(\underline{x})$ and $\underline{\chi}(\underline{x})$ we introduce the 'wave functions' $\psi(x) \equiv \langle x|\psi\rangle$ and $\chi(x) \equiv \langle x|\chi\rangle$ for $|\psi\rangle$, $|\chi\rangle \in \mathcal{H}$. They are assumed to satisfy the conditions

$$\underline{\psi}(\underline{x}) = \psi(x)\Big|_{x=\underline{x}}, \qquad \underline{\chi}(\underline{x}) = \chi(x)\Big|_{x=\underline{x}}. \tag{1.15}$$

Thus for a normalized $f(x)$, which is defined[a] by

$$R^2(\underline{x}) = f_{,\alpha}(\underline{x})f_{,\alpha}(\underline{x}) = 1 \qquad (\forall \underline{x} \in \Sigma_f), \tag{1.16}$$

we have[2]

$$(\underline{\psi}, \underline{\chi}) = \int_{\Sigma_f} d^D\sigma(\underline{x}) \, \underline{\psi}^*(\underline{x})\underline{\chi}(\underline{x}) = \int d^{D+1}x \, \delta(f(x))\psi^*(x)\chi(x). \tag{1.17}$$

If use is made of (1.15) the above relation is immediately generalized to

$$(\underline{\psi}|O(\hat{x}, \hat{p})|\underline{\chi}) = \int d^{D+1}x \, d^{D+1}x' \delta(f(x))\psi^*(x)\langle x|O(\hat{x}, \hat{p})|x'\rangle\chi(x') \tag{1.18}$$

since the operator $O(\hat{x}, \hat{p})$ on $\underline{\mathcal{H}}$ can also act on $\mathcal{H}$ by virtue of (1.10) and (1.11).

---

[a]There exists an ambiguity of multiplying $f(x)$ by a non-vanishing factor. Utilizing this we can always convert $f(x)$ into the normalized form.

Then by means of (1.18) we can express the transition amplitude T_{FI} from the initial state $|\psi_I\rangle$ at $t = t_I$ to the final state $|\psi_F\rangle$ at $t = t_F$ as

$$T_{FI} = (\psi_F|e^{-\frac{i}{\hbar}\hat{H}(t_F-t_I)}|\psi_I)$$

$$= \int d^{D+1}x_F \, d^{D+1}x_I \, \delta(f(x_F))$$

$$\times \psi_F^*(x_F)\langle x_F|e^{-\frac{i}{\hbar}\hat{H}(t_F-t_I)}|x_I\rangle\psi_I(x_I), \qquad (1.19)$$

which enables us to obtain T_{FI} by calculating the propagation function $\langle x_F|e^{-\frac{i}{\hbar}\hat{H}(t_F-t_I)}|x_I\rangle$ on $\mathcal{H}$. Thus in the next section on the basis of (1.19) we will try to derive a rigorous form of the path integral for the constrained system under consideration.

Before closing this section we insert here a short remark. If we employ the function

$$f_c(x) \equiv f(x) - c \qquad (c : \text{real}) \qquad (1.20)$$

in lieu of $f(x)$ in the Dirac algebra, then the primary constraint will become $f_c(x) = 0$, which for small $|c|$ provides us with a D-dimensional manifold in the neighborhood of $f(x) = 0$. Since $f_{c,\beta}(x) = f_{,\beta}(x)$ in this domain, one finds that the secondary constraint $\{\hat{p}_\alpha, f_{c,\alpha}(\hat{x})\} = 0$ takes the same form as (1.3) and the commutators (1.4)–(1.6) also remain unchanged under $f(x) \rightarrow f_c(x)$. Especially, the respective $\hat{p}_\beta$'s of (1.10) and (1.11) form irreducible representations of the Dirac algebra on $f_c(x) = 0$ together with $\hat{x}_\alpha$'s that satisfy $f_c(\hat{x}) = 0$.

2. Path Integral Representations

With these preparations we will, first of all, rewrite the Hamiltonian (1.2) in terms of $\hat{x}_\beta$'s and $\hat{\pi}_\beta$'s by applying (1.10) and (1.11). Then after some calculations we find

$$\hat{H} = \frac{1}{2}\hat{\pi}_\beta \Lambda_{\beta\gamma}(\hat{x})\hat{\pi}_\gamma + K(\hat{x}) + V(\hat{x})$$

$$+ \frac{\alpha\hbar}{2}\epsilon_{\beta\tau}\left\{\frac{f_{,\beta}(\hat{x})f_{,\tau\sigma}(\hat{x})\Lambda_{\sigma\rho}(\hat{x})}{R^2(\hat{x})}, \, \hat{\pi}_\rho\right\}$$

$$+ \frac{\alpha^2\hbar^2}{2}\frac{\Lambda_{\beta\sigma}(\hat{x})f_{,\sigma\tau}(\hat{x})\Lambda_{\tau\rho}(\hat{x})f_{,\rho\beta}(\hat{x})}{R^2(\hat{x})} \qquad (D=1) \qquad (2.1)$$

and

$$\hat{H} = \frac{1}{2}\hat{\pi}_\beta \Lambda_{\beta\gamma}(\hat{x})\hat{\pi}_\gamma + K(\hat{x}) + V(\hat{x}) \qquad (D \geq 2), \qquad (2.2)$$

where

$$K(x) = -\frac{\hbar^2}{8} \{\partial_\beta \partial_\sigma \Lambda_{\beta\sigma}(x) - \partial_\beta \Lambda_{\sigma\tau}(x) \partial_\sigma \Lambda_{\tau\beta}(x)\}. \tag{2.3}$$

Using these expressions we now evaluate $\langle x^{(k)} | \hat{H} | x^{(k-1)} \rangle$ $(k = 1, 2, \cdots, N)$ to obtain

$$\langle x^{(k)} | \hat{H} | x^{(k-1)} \rangle = \int \frac{d^{D+1} p^{(k)}}{(2\pi\hbar)^{D+1}} \exp\left[\frac{i}{\hbar} p^{(k)} \Delta x^{(k)}\right] H(p^{(k)\perp}, \bar{x}^{(k)}), \tag{2.4}$$

with

$$H(p^{(k)\perp}, \bar{x}^{(k)}) = \frac{1}{2}(p^{(k)\perp})^2 + V_{\text{eff}}(\bar{x}^{(k)})$$

$$+ \alpha\hbar \frac{\epsilon_{\beta\tau} f_{,\beta x}(\bar{x}^{(k)}) f_{,\tau\sigma}(\bar{x}^{(k)}) p_\sigma^{(k)\perp}}{R^2(\bar{x}^{(k)})}$$

$$+ \alpha^2\hbar^2 \frac{\Lambda_{\beta\sigma}(\bar{x}^{(k)}) f_{,\sigma\tau}(\bar{x}^{(k)}) \Lambda_{\tau\rho}(\bar{x}^{(k)}) f_{,\rho\beta}(\bar{x}^{(k)})}{2R^2(\bar{x}^{(k)})}$$

$$(D = 1) \tag{2.5}$$

and

$$H(p^{(k)\perp}, \bar{x}^{(k)}) = \frac{1}{2}(p^{(k)\perp})^2 + V_{\text{eff}}(\bar{x}^{(k)}) \qquad (D \geq 2), \tag{2.6}$$

where

$$p_\beta^{(k)\perp} = \Lambda_{\beta\gamma}(\bar{x}^{(k)}) p_\gamma^{(k)}, \qquad \bar{x}_\beta^{(k)} = \frac{x_\beta^{(k)} + x_\beta^{(k-1)}}{2},$$

$$\Delta x_\beta^{(k)} = x_\beta^{(k)} - x_\beta^{(k-1)},$$

$$V_{\text{eff}}(\bar{x}^{(k)}) = \frac{\hbar^2}{8} \partial_\beta \Lambda_{\sigma\tau}(\bar{x}^{(k)}) \partial_\sigma \Lambda_{\beta\tau}(\bar{x}^{(k)}) + V(\bar{x}^{(k)}). \tag{2.7}$$

In calculating the above we have, for the sake of simplicity, employed the Weyl ordering for the product of canonical operators.

Thus applying the well-known technique in path integral we are now led to

$$\langle x_F | e^{-\frac{i}{\hbar}(t_F - t_I)\hat{H}} | x_I \rangle$$

$$= \lim_{N \to \infty} \int \prod_{k=1}^{N-1} d^{D+1} x^{(k)} \int \prod_{k=1}^{N} \frac{d^{D+1} p^{(k)}}{(2\pi\hbar)^{(D+1)}}$$

$$\times \exp\left[\frac{i}{\hbar} \sum_{k=1}^{N} \{p^{(k)} \cdot \Delta x^{(k)} - H(p^{(k)\perp}, \bar{x}^{(k)}) \Delta t\}\right] \tag{2.8}$$

with $t^{(N)} \equiv t_F$ and $t^{(0)} \equiv t_I$. Then applying (2.8) to the right hand side of (1.19) we arrive at

$$T_{FI} = \lim_{N \to \infty} \int \delta(f(x^{(N)})) \prod_{k=0}^{N} d^{D+1} x^{(k)} \int \prod_{k=1}^{N} \frac{d^{D+1} p^{(k)}}{(2\pi\hbar)^{(D+1)}}$$

$$\times \exp\left[\frac{i}{\hbar} \sum_{k=1}^{N} \{p^{(k)} \cdot \Delta x^{(k)} - H(p^{(k)\perp}, \bar{x}^{(k)}) \Delta t\}\right]$$

$$\times \psi_F^*(x^{(N)}) \psi_I(x^{(0)}), \quad (2.9)$$

which provides a rigorous expression of the path integral for the constrained system under consideration. It apparently differs from the Faddeev-Senjanovic (F-S)[3,4] formula derived by a semi-classical approach. The relation of (2.9) with the F-S formula will be discussed in Section 3.

We are now ready to perform the p-integration in (2.9) using (2.5) for $D = 1$ and (2.6) for $D \geq 2$. As a result we obtain

$$T_{FI} = \lim_{N \to \infty} \frac{1}{(2\pi i \hbar \Delta t)^{DN/2}} \prod_{k=0}^{N} \int d^{D+1} x^{(k)} \cdot \delta(f(x^{(N)}))$$

$$\times \prod_{k=1}^{N} \delta(\Delta x_\beta^{(k)} f_{,\beta}(\bar{x}^{(k)})/R(\bar{x}^{(k)}))$$

$$\times \exp\left[\frac{i\Delta t}{\hbar} \sum_{k=1}^{N} L_{k,\,\text{eff}}\right] \psi_F^*(x^{(N)}) \psi_I(x^{(0)}), \quad (2.10)$$

where the effective Lagrangian $L_{k,\,\text{eff}}$ is given by

$$L_{k,\,\text{eff}} = \frac{1}{2}\left(\frac{\Delta x_\beta^{(k)}}{\Delta t}\right)^2 + A_\beta(\bar{x}^{(k)}) \frac{\Delta x_\beta^{(k)}}{\Delta t} - V_{\text{eff}}(\bar{x}^{(k)}) \qquad (D = 1) \quad (2.11)$$

with

$$A_\beta(x) \equiv -\alpha\hbar \frac{\epsilon_{\sigma\tau} f_{,\sigma}(x) f_{,\tau\beta}(x)}{R^2(x)} \qquad (2.12)$$

and

$$L_{k,\,\text{eff}} = \frac{1}{2}\left(\frac{\Delta x_\beta^{(k)}}{\Delta t}\right)^2 - V_{\text{eff}}(\bar{x}^{(k)}) \qquad (D \geq 2). \quad (2.13)$$

Emergence of the factor $\delta(\Delta x_\beta^{(k)} f_{,\beta}(\bar{x}^{(k)})/R(\bar{x}^{(k)}))$ is quite reasonable since it requires that in large N limit the vector $\Delta x^{(k)}$ should be perpendicular to the normal standing on the manifold at $\bar{x}^{(k)}$. Especially in this limit the Lagrangian (2.13) for $D \geq 2$ becomes of the standard form with the effective potential V_{eff}, in which $\hbar^2$-terms have been brought through the process for the Weyl ordering.

In contrast to the above, for $D = 1$ the Lagrangian has a kind of gauge interaction, which reduces to the form $A_\beta(x)\dot{x}_\beta$ in the continuous limit. The situation is very characteristic of $D = 1$. Hence in the following we will examine properties of the gauge potential (2.12), which will be denoted in the unit of $e/c = 1$ for simplicity. Since the components of the gauge potential are of the form

$$\begin{cases} A_1(x) = \alpha\hbar\dfrac{f_{,2}(x)f_{,11}(x) - f_{,1}(x)f_{,12}(x)}{R^2(x)} \\[2ex] A_2(x) = \alpha\hbar\dfrac{f_{,2}(x)f_{,12}(x) - f_{,1}(x)f_{,22}(x)}{R^2(x)}, \end{cases} \tag{2.14}$$

they are seen to satisfy

$$\frac{\partial A_1(x)}{\partial x_2} - \frac{\partial A_2(x)}{\partial x_1} = 0 \qquad (R(x) \neq 0), \tag{2.15}$$

that is, no magnetic flux exists in the neighborhood of the manifold $f(x) = 0$ which is now a closed loop diffeomorphic to the circle. Related to this we introduce the quantity

$$\xi(x) \equiv \tan^{-1}\left(\frac{f_{,2}(x)}{f_{,1}(x)}\right). \tag{2.16}$$

For x on $f(x) = 0$ it stands for the angle between the x_1-axis and the normal to the closed loop at x (Figure 1). Then the gauge potential (2.14) is expressed as

$$A_\beta(x) = -\alpha\hbar\frac{\partial\xi(x)}{\partial x_\beta} \qquad (R(x) \neq 0). \tag{2.17}$$

Thus the contour integral $\oint A_\beta(x)dx_\beta$ evaluated along the closed loop is found to be $-2\pi\alpha\hbar$, which tells us that the gauge potential is non-trivial for $\alpha \neq 0$. Now let us consider an arbitrary point $q = (q_1, q_2)$ fixed inside the domain $\mathbb{D}$ that is surrounded by the closed loop $f(x) = 0$ on $\mathbb{R}^2$. Then we introduce the potential such that

$$\mathfrak{A}_\beta(x) = -\frac{1}{2\pi}\epsilon_{\beta\tau}\frac{(x - q)_\tau}{(x_\sigma - q_\sigma)^2}$$

$$= \frac{1}{2\pi}\frac{\partial\theta(x)}{\partial x_\beta} \tag{2.18}$$

with

$$\theta(x) = \tan^{-1}\left(\frac{x_2 - q_2}{x_1 - q_1}\right) \qquad (x \neq q). \tag{2.19}$$

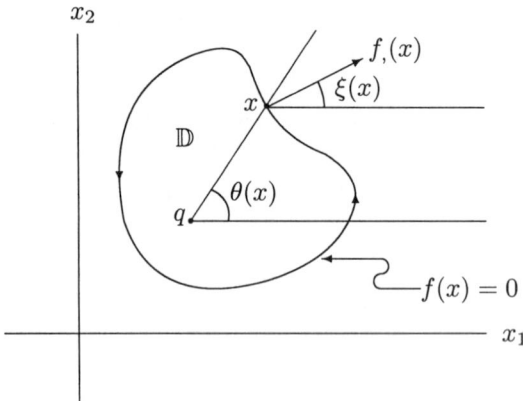

Figure 1. The domain $\mathbb{D}$ is surrounded by the closed loop $f(x) = 0$. The vector $f_,(x) = (f_{,1}(x), f_{,2}(x))$ is a normal to the closed loop at x.

The potential $\mathfrak{A}_\beta(x)$ is nothing but the Aharonov-Bohm gauge potential[5] produced by the unit magnetic flux confined in an extremely thin solenoid perpendicular to $\mathbb{R}^2$ at q. As the point x circles once along the closed loop, the angles $\xi(x)$ and $\theta(x)$ are changed to be

$$\xi(x) \rightarrow \xi(x) + 2\pi, \qquad \theta(x) \rightarrow \theta(x) + 2\pi. \tag{2.20}$$

Consequently the function

$$U(x) = \exp\left[\frac{i}{\hbar}F(x)\right] \tag{2.21}$$

defined with

$$F(x) = -\alpha\hbar(\xi(x) - \theta(x)) \tag{2.22}$$

is a single-valued function in the neighborhood of the closed loop. Then it enables us to make the gauge transformation such that

$$A_\beta(x) \rightarrow A_\beta(x) - \frac{\hbar}{i}U^*(x)\frac{\partial U(x)}{\partial x_\beta} = -2\pi\alpha\hbar\,\mathfrak{A}_\beta(x). \tag{2.23}$$

As mentioned in Section 1 the argument based on the closed loop $f(x) = 0$ can be generalized to that on $f_c(x) = 0$. Thus we may conclude that the potential $A_\beta(x)$ is definable in the neighborhood of $f(x) = 0$ and is gauge-equivalent to the Aharonov-Bohm potential produced by the magnetic flux $\Phi = -2\pi\alpha\hbar$ that perpendicularly crosses the domain $\mathbb{D}$ at q.

Furthermore it is shown that the parameter α is essentially the same as that introduced by Schulman[6] in his path integral formalism given without recourse to the operator formalism. It originates in the multiply connected structure of the closed loop.

3. Relation with the Faddeev-Senjanovic Formula

The Faddeev-Senjanovic path integral formula[3,4] for a system constrained on the D-dimensional manifold $f(x) = 0$ is written as

$$T_{FI} = \int D\mu \, \psi_F^*(x_F) \exp\left[\frac{i}{\hbar} \int_{T_I}^{T_F} dt \, (p\dot{x} - H(p,x))\right] \psi_I(x_I), \qquad (3.1)$$

with $H = \frac{1}{2}p^2 + V(x)$, where the integral measure $D\mu$ is given by

$$D\mu = \delta(f(x))\delta(p_\alpha f_{,\alpha}(x))R^2(x) \, DpDx. \qquad (3.2)$$

The amplitude (3.1) was derived by applying a semi-classical treatment based on the Dirac formalism[1] for the constrained system, since at that time nothing was known on the irreducible representation of the Dirac algebra. Accordingly there may arise a question about quantum effects which may emerge in the path integral representation and also about the concrete definition of the measure (3.2). In this connection Kashiwa and Fukudaka[7] carefully examined the formula (3.1) and inductively determined the correct form of the path integral in some cases without use of irreducible representations.

In the following, starting from (2.9) we will derive a rigorous expression for the path integral that may correspond to (3.1). For this purpose we will rewrite the quantity

$$I(x, x') = \int d^{D+1}p \, \exp\left[\frac{i}{\hbar}\{p \cdot \Delta x - H(p^\perp, \bar{x})\Delta t\}\right] \qquad (3.3)$$

$$(\Delta x \equiv x - x', \quad \bar{x} \equiv (x + x')/2),$$

which appears in the right hand side of (2.9) as a product of the form $\prod_{k=1}^{N}\{I(x^{(k)}, x^{(k-1)})/(2\pi\hbar)^{D+1}\}$. We then introduce an orthogonal transformation represented by the matrix $\|a_{\beta\gamma}(\bar{x})\| \in SO(D+1)$, which rotates the vector $(f_{,1}(\bar{x}), f_{,2}(\bar{x}), \ldots, f_{,D+1}(\bar{x}))$ to the direction of the $(D+1)$-th axis, i.e., $a_{\beta\gamma}(\bar{x})f_{,\gamma}(\bar{x}) = \delta_{\beta \, D+1}R(\bar{x})$. By means of it we define the quantities such that

$$X_\beta = a_{\beta\gamma}(\bar{x})\, x_\gamma, \qquad X'_\beta = a_{\beta\gamma}(\bar{x})\, x'_\gamma,$$

$$P_\beta = a_{\beta\gamma}(\bar{x})\, p_\gamma, \qquad P_\beta^\perp = a_{\beta\gamma}(\bar{x})\, p_\gamma^\perp. \qquad (3.4)$$

We will use the boldface letter for a D-dimensional vector obtained by dropping out the $(D+1)$-th component from a vector in $\mathbb{R}^{D+1}$, and hence, for example, we will denote it as $\boldsymbol{X} \equiv (X_1, X_2, \ldots, X_D)$ for $X = (X_1, X_2, \ldots, X_{D+1})$. In this connection it is noted that there holds the relation $P^{\perp} = (\boldsymbol{P}, 0)$.

Then we find that (3.3) is rewritten as

$$I(x, x')$$

$$= \int d^D\boldsymbol{P}\, dP_{D+1} \exp\left[\frac{i}{\hbar}\{\boldsymbol{P}\cdot\Delta\boldsymbol{X} + P_{D+1}\Delta X_{D+1} - \boldsymbol{H}(P^{\perp}, \bar{X})\Delta t\}\right]$$

$$= 2\pi\hbar\,\delta(\Delta X_{D+1}) \int d^D\boldsymbol{P} \exp\left[\frac{i}{\hbar}\{\boldsymbol{P}\cdot\Delta\boldsymbol{X} - \boldsymbol{H}(P^{\perp}, \bar{X})\Delta t\}\right]$$

$$= 2\pi\hbar\,\delta(\Delta X_{D+1}) \int d^{D+1}P\, \delta(P_{D+1}R(\bar{x}))R(\bar{x})$$

$$\times \exp\left[\frac{i}{\hbar}\{P\cdot\Delta X - \boldsymbol{H}(P, \bar{X})\Delta t\}\right]$$

$$= 2\pi\hbar R(\bar{x})\,\delta(\Delta X_{D+1}) \int d^{D+1}p\, \delta(p_\beta f_{,\beta}(\bar{x}))$$

$$\times \exp\left[\frac{i}{\hbar}\{p\cdot(x - x') - H(p, \bar{x})\Delta t\}\right], \quad (3.5)$$

where

$$\Delta X_\beta = X_\beta - X'_\beta, \qquad \bar{X}_\beta = \frac{X_\beta + X'_\beta}{2},$$

$$\boldsymbol{H}(P^{\perp}, \bar{X}) = H(p^{\perp}, \bar{x}) \qquad (3.6)$$

and $H(p, \bar{x})$ is obtained by simply replacing $p_\beta^{\perp}$ with p_β in $H(p^{\perp}, \bar{x})$.

To rewrite the factor $\delta(\Delta X_{D+1})$ we will proceed in the following way:

$$f(x) - f(x')$$

$$= f(\bar{x} + \frac{\Delta x}{2}) - f(\bar{x} - \frac{\Delta x}{2})$$

$$= \sum_{n=0}^{\infty} \frac{\Delta x_{\beta_1}\Delta x_{\beta_2}\cdots\Delta x_{\beta_{2n+1}}}{2^{2n}\cdot(2n+1)!} f_{,\beta_1\beta_2\cdots\beta_{2n+1}}(\bar{x})$$

$$= \sum_{n=0}^{\infty} \frac{\Delta X_{\beta_1}\Delta X_{\beta_2}\cdots\Delta X_{\beta_{2n+1}}}{2^{2n}\cdot(2n+1)!} F_{\beta_1\beta_2\cdots\beta_{2n+1}}(\bar{x})$$

$$= \Delta X_{D+1}R(\bar{x}) + \sum_{n=1}^{\infty} \frac{\Delta X_{\beta_1}\Delta X_{\beta_2}\cdots\Delta X_{\beta_{2n+1}}}{2^{2n}\cdot(2n+1)!} F_{\beta_1\beta_2\cdots\beta_{2n+1}}(\bar{x}),$$

$$(3.7)$$

where

$$F_{\beta_1\beta_2\cdots\beta_{2n+1}}(\bar{x}) = a_{\beta_1\gamma_1}(\bar{x})a_{\beta_2\gamma_2}(\bar{x})\cdots a_{\beta_{2n+1}\gamma_{2n+1}}(\bar{x})\, f_{,\gamma_1\gamma_2\cdots\gamma_{2n+1}}(\bar{x}).$$

We divide the sum in the right hand side of (3.7) into two parts, the one of which is the sum of those terms which involve factors ΔX_{D+1} and the other consists of the rest terms. Since $|\Delta X_j| \sim \sqrt{\hbar\Delta t}$ in the path integral for the Hamiltonian $\hat{H} = \frac{1}{2}\hat{p}^2 + V(\hat{x})$, the latter is of the order of magnitude $O(|\Delta t|^{3/2})$ for small Δt.

Thus we can write (3.7) as

$$f(x) - f(x') = \Delta X_{D+1}\{R(\bar{x}) + A(x, x')\} + O(|\Delta t|^{3/2}), \tag{3.8}$$

where $A(x, x')$ takes a form such that

$$
\begin{aligned}
&A(x, x')\\
&= \frac{1}{8}\sum_{j,l=1}^{D} \Delta X_j \Delta X_l\, F_{D+1\,j\,l}(\bar{x}) + \frac{1}{8}\Delta X_{D+1}\sum_{j=1}^{D}\Delta X_j\, F_{D+1\,D+1\,j}(\bar{x})\\
&\quad + \frac{1}{24}(\Delta X_{D+1})^2\, F_{D+1\,D+1\,D+1}(\bar{x}) + \text{higher terms in } \Delta. \tag{3.9}
\end{aligned}
$$

We will assume that $R(\bar{x}) + A(x, x')$ is non-vanishing for small Δt when regarded as a function of ΔX_{D+1}. Thus we are led to

$$\delta(f(x) - f(x')) = \delta(\Delta X_{D+1})\Big\{R(\bar{x}) + \frac{Q(x, x')}{R(\bar{x})}\Big\}^{-1} + O(|\Delta t|^{3/2}), \tag{3.10}$$

where

$$
\begin{aligned}
Q(x, x') &= \frac{R(\bar{x})}{8}\sum_{j,l=1}^{D}\Delta X_j \Delta X_l\, F_{D+1\,j\,l}(\bar{x})\\
&= \frac{1}{8}\Delta x_\beta \Delta x_\gamma \Lambda_{\beta\rho}(\bar{x})\Lambda_{\gamma\sigma}(\bar{x})f_{,\rho\sigma\tau}(\bar{x})f_{,\tau}(\bar{x}). \tag{3.11}
\end{aligned}
$$

Hence we have

$$R(\bar{x})\delta(\Delta X_{D+1}) = \delta(f(x) - f(x'))\{R^2(\bar{x}) + Q(x, x')\} + O(|\Delta t|^{3/2}). \tag{3.12}$$

Combining this relation with (3.5) we obtain

$$
\begin{aligned}
&\delta(f(x))I(x, x')\\
&\quad = 2\pi\hbar\int d^{D+1}p\, \delta(f(x))\delta(p_\beta f_{,\beta}(\bar{x}))\{R^2(\bar{x}) + Q(x, x')\}\\
&\qquad \times \exp\Big[\frac{i}{\hbar}\{p\cdot(x - x') - H(p, \bar{x})\Delta t\}\Big]\delta(f(x')) + O(|\Delta t|^{3/2})
\end{aligned}
$$

and consequently

$$\delta(f(x^{(k)})) \int d^{D+1}p^{(k)} \exp\left[\frac{i}{\hbar}\{p^{(k)} \cdot \Delta x^{(k)} - H(p^{(k)\perp}, \bar{x}^{(k)})\Delta t\}\right]$$

$$= 2\pi\hbar \int d^{D+1}p\, \delta(f(x^{(k)}))\delta(p_\beta^{(k)} f_{,\beta}(\bar{x}^{(k)}))$$

$$\times \{R^2(\bar{x}^{(k)}) + Q(x^{(k)}, x^{(k-1)})\}$$

$$\times \exp\left[\frac{i}{\hbar}\{p^{(k)} \cdot \Delta x^{(k)} - H(p^{(k)}, \bar{x}^{(k)})\Delta t\}\right]$$

$$\times \delta(f(x^{(k-1)})) + O(|\Delta t|^{3/2}). \tag{3.13}$$

We now apply the equation (3.13) to the right hand side of (2.9) by putting k equal to N, $N-1$, ..., 2, 1 successively. Then we finally arrive at

$$T_{FI} = \lim_{N\to\infty} \int \prod_{k=0}^{N} d^{D+1}x^{(k)}\delta(f(x^{(k)}))$$

$$\times \int \prod_{k=1}^{N} \frac{d^{D+1}p^{(k)}}{(2\pi\hbar)^D}\, \delta(p_\beta^{(k)} f_{,\beta}(\bar{x}^{(k)}))R^2(\bar{x}^{(k)})\{1+Q(x^{(k)}, x^{(k-1)})\}$$

$$\times \exp\left[\frac{i}{\hbar}\sum_{k=1}^{N}\{p^{(k)} \cdot \Delta x^{(k)} - H(p^{(k)}, \bar{x}^{(k)})\Delta t\}\right]\psi_F^*(x^{(N)})\,\psi_I(x^{(0)}),$$

$$\tag{3.14}$$

which provides us with a rigorous form of path integral representation of the Faddeev-Senjanovic type. In the above calculation the term $O(|\Delta t|^{3/2})$ appearing in (3.13) has been omitted because of its vanishing contribution in the limit of $\Delta t \to 0$. On the other hand, since $Q(x^{(k)}, x^{(k-1)})$ is a quantity of the order of magnitude $\hbar\Delta t$, we expect that unless $Q = 0$ it will produce a finite effect proportional to $\hbar$.

4. Concluding Remarks

We have formulated rigorous expressions for the path integral, which describes the transition amplitude for the system constrained on $f(x) = 0$. The basic equation for this is Eq.(1.19), in which the constraint function $f(x)$ is required to be normalized like (1.16) and the wave functions defined on $\mathcal{H}$ are assumed to obey the relation (1.15). The result then obtained has been found to differ from the classical work by Faddeev and Senjanovic in some respects.

In closing this note we will add a few remarks.

The simplest example of our argument is given by the manifold S^D, where the normalized $f(x)$ is written as

$$f(x) = \frac{(x^2 - a^2)}{2a} \qquad (a > 0, \ x^2 \equiv x_\alpha x_\alpha) \tag{4.1}$$

and hence

$$\Lambda_{\beta\gamma}(x) = \delta_{\beta\gamma} - \frac{x_\beta x_\gamma}{x^2}, \qquad Q(x) = 0. \tag{4.2}$$

The Hamiltonian $H(p, x)$ in the right hand side of (3.5) is found to take the form

$$H(p, x) = \begin{cases} \dfrac{1}{2}\left\{(p_1 - \alpha\hbar\dfrac{x_2}{x^2})^2 + (p_2 + \alpha\hbar\dfrac{x_1}{x^2})^2\right\} + V_{\text{eff}}(x) & (D = 1), \\ \dfrac{1}{2}p^2 + V_{\text{eff}}(x) & (D \geq 2). \end{cases} \tag{4.3}$$

As expected from the argument in section 3, the first line in the above is just the Aharonov-Bohm Hamiltonian with effective potential $V_{\text{eff}}(x)$.

With a slight modification our formalism is also applicable to the case where the manifold $f(x) = 0$ is diffeomorphic to $\mathbb{R}^D$ which is embedded in $\mathbb{R}^{D+1}$. The Dirac algebra in this case takes the same form as (1.1) and (1.3)–(1.6). Thus the operators $\hat{p}_\beta$ of (1.10) for $D = 1$ and of (1.11) for $D \geq 2$ satisfy this algebra. It is noted, however, that p_β's specified with $\alpha \neq 0$ in (1.10) are unitarily equivalent to those with $\alpha = 0$ because of the simply-connected structure of the manifold. Thus, irrespective of the value of D we may write $\hat{p}_\beta$ as

$$\hat{p}_\beta = \frac{1}{2}\{\Lambda_{\beta\gamma}(\hat{x}), \ \hat{\pi}_\gamma\}. \tag{4.4}$$

in any irreducible representation. Accordingly the path integral representation is given by (2.10) with $L_{k,\text{eff}}$ of (2.13) for any D.

Finally we remark that we are unable to represent the trace of $e^{-\beta\hat{H}}$ in a form of path integral with the manner stated in the present note. The reason is as follows:

Since the trace is to be taken on the physical Hilbert space $\mathcal{H}$, it is written as

$$\text{Tr}e^{-\beta\hat{H}} = \sum_{n=1,2,3,\ldots} (\underline{\psi}_n|e^{-\beta\hat{H}}|\underline{\psi}_n) \tag{4.5}$$

with the complete set of ortho-normalized vectors $|\underline{\psi}_n)$ $(n = 1, 2, 3, \ldots)$ on $\mathcal{H}$. Then owing to (1.18) we obtain

$$\text{Tr}e^{-\beta\hat{H}}$$
$$= \sum_{n=1,2,3,\ldots} \int d^{D+1}x\, d^{D+1}x'\, \delta(f(x))\langle x|\text{Tr}e^{-\beta\hat{H}}|x'\rangle\, \psi_n^*(x)\psi_n(x'), \tag{4.6}$$

in which square integrable functions $\psi_n(x) \in \mathcal{H}$ are related to $\underline{\psi}_n(\underline{x})$ by

$$\psi_n(x)|_{x=\underline{x}} = \underline{\psi}_n(\underline{x}). \qquad (4.7)$$

Therefore, if $\psi_n(x)$ were made to satisfy the ortho-completeness conditions

$$\int dx^{D+1} \, \psi_n^*(x)\psi_{n'}(x) = \delta_{nn'},$$

$$\sum_{n=1,2,3,\ldots} \psi_n(x)\psi_n^*(x') = \delta^{D+1}(x - x') \qquad (4.8)$$

in $\mathcal{H}$, the right hand side of (4.5) could be written as $\int d^{D+1}x \, \delta(f(x))$ $\times \langle x|\mathrm{Tr}e^{-\beta\hat{H}}|x\rangle$, which would provide us with a path integral representation for the trace. However, the situation is not so simple. For instance let us consider the case where the manifold is S^2 with the central potential $V_{\text{eff}}(r)$. Then $\underline{\psi}_n(\underline{x})$ is represented by the spherical harmonics $Y_l^m(\theta, \varphi)$ with $n = (l, m)$ and $\underline{x} = (\theta, \varphi)$. Accordingly, $\psi_n(x)$ may be written as $Y_l^m(\theta, \varphi)F_{lm}(r)$ from the symmetric reason. It should be noted here that no quantum number exists corresponding to the radial degree of freedom, and thereby $Y_l^m(\theta, \varphi)F_{lm}(r)$'s as whole cannot span an ortho-complete system on $\mathcal{H}$. This clearly contradicts the condition (4.8) which would enables us to realize the path integral representation for the trace.

Accordingly a study of alternative approaches to attack this problem would highly be desired.

References

1. P. A. M. Dirac, *Can. J. Math.* **2**, 129 (1950); *Lectures on Quantum Mechanics* (Belfer Graduate School of Science, Yeshiva University, New York, 1964).
2. Y. Ohnuki, *Proc. 6th Int. School of Theoretical Physics –SSPCM–*, 73 (eds T. Lulek et al., World Scientific 2001), and for details, to be published.
3. L. D. Faddeev, *Theor. Math. Phys.* **1**, 1(1970)
4. P. Senjanovic, *Ann. of Phys.* **100**, 227 (1976).
5. Y. Aharonov and D. Bohm, *Phys. Rev.* **115**, 485 (1959).
6. L. S. Schulman, *J. Math. Phys.* **12**, 305 (1971).
7. H. Fukudaka and T. Kashiwa, *Ann. Phys*, **175** 301 (1987); T. Kashiwa, *Prog. Theor. Phys.* **95** (1996), 431.

FINITE QUANTUM SYSTEMS AND THEIR APPLICATIONS TO QUANTUM INFORMATION PROCESSING

A. VOURDAS

Department of Computing,
University of Bradford,
Bradford BD7 1DP, United Kingdom

Quantum mechanics for finite systems is briefly reviewed and used in the context of quantum information processing. In an angular momentum Hilbert space, the displacement operators play the role of $SU(2j + 1)$ generators. Unitary transformations are expressed as a finite sum of the displacement operators with the Weyl function as coefficients. A factorization of large qudits in terms of smaller qudits is studied. All unitary transformations on large qudits can be performed through appropriate unitary transformations on the smaller qudits. Coding with these states is also considered. A concatenated code that introduces redundancy in both amplitude and phase, is studied.

1. Introduction

Finite quantum systems and the relevant Heisenberg-Weyl group have been studied for a long time [1-5]. It is an important topic in its own right and it has several applications in condensed matter in areas like the quantum Hall effect, two-dimensional electron gas in external magnetostatic fields, the magnetic translation group, etc.

In the last few years two-dimensional quantum systems (qubits) amd more recently multi-dimensional quantum systems (qudits [6]) have been used in the context of quantum information processing. Many condensed matter systems are potential candidates for the practical implementation of qubits and more generally qudits.

In this paper we study finite quantum systems with emphasis on those aspects which are important for quantum information processing. We consider unitary transformations (which play a central role in quantum computation) and show that they can be expressed as a finite sum of the displacement operators with the Weyl function as coefficients.

We also study a factorization of large qudits in terms of smaller qudits. Smaller qudits are easier to implement in a practical physical system, and at the same time large qudits might have some advantages in terms of computational power. The factorization considered, combines the merits of both because we perform quantum computation with large qudits which are really composed from smaller qudits.

An important area within the general subject of quantum computation is quantum coding. Here we generalize Shor's coding scheme for qubits [7] into qudits [8]. The results are based heavily on the theory of finite quantum systems and demonstrate its importance for quantum information processing with qudits.

2. Finite quantum systems

2.1. *Displacements in the $\theta_z - J_z$ phase space*

Let $|J, jm>$ be the usual angular momentum states where m belongs to $\mathcal{Z}(2j + 1)$ (the integers modulo $2j + 1$). We denote as $H(2j + 1)$ the corresponding Hilbert space . The finite Fourier transform is defined as:

$$F = (2j + 1)^{-1/2} \sum_{m,n} \omega(mn)|J; j\ m\rangle\langle J; j\ n| \tag{1}$$

$$\omega(\alpha) = \exp\left[i\frac{2\pi\alpha}{2j + 1}\right]; \quad FF^\dagger = F^\dagger F = 1; \quad F^4 = 1 \tag{2}$$

Using this Fourier transform we have introduced [4] another orthonormal basis. It is the θ-basis of angle states $|\theta; j\ m>$ defined as:

$$|\theta; j\ m\rangle = F|J; j\ m\rangle = (2j + 1)^{-1/2} \sum_n \omega(mn)|J; j\ n\rangle \tag{3}$$

We have also introduced the angle operators θ_+, θ_-, θ_z

$$\theta_z = FJ_zF^\dagger; \quad \theta_+ = FJ_+F^\dagger; \quad \theta_- = FJ_-F^\dagger \tag{4}$$

which obey the $SU(2)$ algebra. The θ-operators act on the θ-states in an analogous way to the J-operators acting on the J-states .

We next consider the $\theta_z - J_z$ phase space which is analogous to the $x - p$ phase space of the harmonic oscillator. The $\theta_z - J_z$ phase space is $\mathcal{Z}(2j+1) \times \mathcal{Z}(2j+1)$ (toroidal lattice). The displacement operators in this phase space are :

$$X = \exp\left[-i\frac{2\pi}{2j + 1}\theta_z\right]; \quad Z = \exp\left[i\frac{2\pi}{2j + 1}J_z\right] \tag{5}$$

$$X^{2j+1} = Z^{2j+1} = 1; \quad X^\beta Z^\alpha = Z^\alpha X^\beta \omega(-\alpha\beta) \tag{6}$$

where α, β are integers in $\mathcal{Z}(2j+1)$. They act on the angle and angular momentum states as follows:

$$X^\beta |J; j\ m\rangle = |J; j\ m+\beta\rangle; \qquad X^\beta |\theta; j\ m\rangle = \omega(-\beta m)|\theta; j\ m\rangle \qquad (7)$$

$$Z^\alpha |J; j\ m\rangle = \omega(m\alpha)|J; j\ m\rangle; \qquad Z^\alpha |\theta; j\ m\rangle = |\theta; j\ m+\alpha\rangle \qquad (8)$$

The general displacement operators are defined as:

$$D(\alpha, \beta) = Z^\alpha X^\beta \omega(-2^{-1}\alpha\beta); \qquad [D(\alpha, \beta)]^\dagger = D(-\alpha, -\beta) \qquad (9)$$

where 2^{-1} is the inverse of 2 within $\mathcal{Z}(2j+1)$ (which in fact is $j+1$, but for similarity with the harmonic oscillator case we prefer to keep the notation 2^{-1}).

The Weyl function corresponding to an operator U, is defined in terms of the displaced operator as:

$$\tilde{W}(U; \alpha, \beta) = Tr[UD(\alpha, \beta)] \qquad (10)$$

The tilde in the notation indicates that it is related to the Wigner function (which in the present context has been discussed in [4]) through the Fourier transform:

$$\tilde{W}(U; \alpha, \beta) = (2j+1)^{-1}W(U; \gamma, \delta)\omega(\alpha\delta - \beta\gamma) \qquad (11)$$

We can prove that

$$U = (2j+1)^{-1}\sum_{\alpha,\beta} \tilde{W}(U; -\alpha, -\beta)D(\alpha, \beta) \qquad (12)$$

This is proved if we take the matrix elements of both sides with regard to the states $\langle J; j\ n|$ and $|J; j\ m\rangle$ and use the relations:

$$\langle J; j\ n|D(\alpha, \beta)|J; j\ m\rangle = \delta(n, m+\beta)\omega(2^{-1}\alpha\beta + \alpha m)$$

$$\sum_{m=-j}^{j} \omega(m\alpha) = (2j+1)\delta(\alpha, 0) \qquad (13)$$

where $\delta(n, m)$ is the Kronecker delta which is equal to 1 when $n = m(mod(2j+1))$.

2.2. $SU(2j+1)$ transformations

The displacement operators $D(\alpha, \beta)$ are generators of the $SU(2j+1)$ transformations in the Hilbert space $H(2j+1)$ [9]. Their commutator is

$$[D(\alpha_1, \beta_1), D(\alpha_2, \beta_2)] \equiv D(\alpha_1, \beta_1)D(\alpha_2, \beta_2) - D(\alpha_2, \beta_2)D(\alpha_1, \beta_1)$$

$$= 2i\sin\left[\frac{2\pi}{2j+1}2^{-1}(\alpha_1\beta_2 - \alpha_2\beta_1)\right] \times \qquad (14)$$

$$D(\alpha_1 + \alpha_2, \beta_1 + \beta_2)$$

Finite $SU(2j + 1)$ transformations involve the exponentials of the generators which are here finite matrices. The exponential of a finite matrix is a polynomial and we can write an arbitrary unitary operator U as:

$$U = \sum_{\alpha,\beta} \mu(\alpha, \beta) D(\alpha, \beta); \quad \mu(\alpha, \beta) = (2j + 1)^{-1} \tilde{W}(U; -\alpha, -\beta) \quad (15)$$

where from Eq(12) we see that the coefficients are the Weyl functions. Eq(15) shows that the displacement operators are the basic 'building blocks' for general unitary transformations of qudits.

3. Factorization

We consider the case where $2j + 1$ can be factorized as $2j + 1 = \prod_{i=1}^{N}(2j_i + 1)$, where any two of the factors $2j_i + 1$ are coprime. In this case we introduce an isomporphism between the Hilbert space $H(2j + 1)$, and a product of Hilbert spaces $\prod_{i=1}^{N} H(2j_i + 1)$. This isomorphism is based on the Chinese remainder theorem and it is similar to the one used in a clasical context in 'Fast Fourier transform' in order to reduce the computation time. The Chinese remainder theorem has also been used in quantum Fourier transforms (e.g., ref. [10]).

Using the same notation as in ref [5], we introduce the integers:

$$r_i = \frac{2j + 1}{2j_i + 1}; \quad t_i r_i = 1 \ (mod \ 2j_i + 1). \quad (16)$$

t_i always exists because the r_i and $2j_i + 1$ are coprime. We also define the $s_i = t_i r_i$ in $Z(2j + 1)$. We note that since t_i is the inverse of r_i in $Z(2j_i + 1)$, the $s_i = t_i r_i$ defined in $Z(2j + 1)$ is an integer multiple of $(2j_i + 1)$ plus 1. We next introduce two isomorphisms between $Z(2j + 1)$ and $\prod_{i=1}^{N} Z(2j_i + 1)$. For a given m in $Z(2j + 1)$ we define the corresponding m_i and $\bar{m}_i$ in $Z(2j_i + 1)$ as:

$$m_i = m(mod \ 2j_i + 1); \quad \bar{m}_i = mt_i(mod \ 2j_i + 1);$$
$$m = \sum_i m_i s_i = \sum_i \bar{m}_i r_i(mod \ 2j + 1) \quad (17)$$

We then have the one-to-one mappings $m \leftrightarrow \{m_i\} \leftrightarrow \{\bar{m}_i\}$. Using this we define a unitary isomorphism between the Hilbert space $H(2j + 1)$ and the product $\prod H(2j_i + 1)$ with:

$$|J; j \ m\rangle \leftrightarrow \prod_{i=1}^{N} |J; j_i \ \bar{m}_i\rangle; \quad |\theta; j \ m\rangle \leftrightarrow \prod_{i=1}^{N} |\theta; j_i \ m_i\rangle. \quad (18)$$

The displacement operators (generators of $SU(2j+1)$ transformations) in $H(2j+1)$ can be expressed as products of the displacement operators (generators of $SU(2j_i+1)$ transformations) in the various $H(2j_i+1)$:

$$D(\alpha,\beta) = \prod_{i=1}^{N} D_i(\alpha_i, \bar{\beta}_i) \tag{19}$$

where $\alpha_i = \alpha(mod\ 2j_i+1)$ and $\bar{\beta}_i = \beta t_i(mod\ 2j_i+1)$. Infinitesimal $SU(2j+1)$ transformations can be written as

$$g = 1 + \sum_{\alpha,\beta} \lambda(\alpha,\beta)\, D(\alpha,\beta) = 1 + \sum_{\{\alpha_i\},\{\beta_i\}} \lambda\left(\{\alpha_i\},\{\bar{\beta}_i\}\right) \prod_{i=1}^{N} D_i(\alpha,\bar{\beta}_i), \tag{20}$$

where $\lambda(\alpha,\beta)$ are infinitesimal coefficients. All unitary transformations on the large qudits can be constructed as combinations of unitary transformations on the smaller qudits.

We next consider finite $SU(2j+1)$ transformations, and using Eqs(12),(19) we get

$$U = \sum_{\{\alpha_i\},\{\beta_i\}} \tilde{W}\left(\{-\alpha_i\},\{-\bar{\beta}_i\}\right) \prod_{i=1}^{N} D_i(\alpha_i, \bar{\beta}_i)$$

$$\tilde{W}\left(\{\alpha_i\},\{\bar{\beta}_i\}\right) = (2j+1)^{-1} Tr\left[U \prod_{i=1}^{N} D_i(\alpha_i, \bar{\beta}_i)\right] \tag{21}$$

This shows again that the displacement operators $D_i(\alpha_i, \bar{\beta}_i)$ are the basic 'building blocks' for general unitary transformations on the large qudits in $H(2j+1)$ The required coefficients are the Weyl functions given above.

4. Quantum coding for qudits

In this section we generalize Shor's coding method for qudits [8]. The idea is to construct a code in two steps. In the first step we introduce amplitude redundancy by considering the subspace H_A of H^N spanned by the direct products of N angular momentum states with the same m. In the second step we introduce phase redundancy by considering the subspace H_B of H_A^M spanned by the direct products of M angle states with the same m. The combined effect of amplitude and phase redundancy is general redundancy with regard to general transformations.

4.1. *Amplitude redundancy*

We consider the $(2j+1)^N$-dimensional space $[H(2j+1)]^N$ and its $(2j+1)$-dimensional subspace spanned by the vectors

$$H_A = \{|J_A; j\ m\rangle \equiv |J; j\ m\rangle \otimes ... \otimes |J; j\ m\rangle, m = -j, ..., j\} \tag{22}$$

The Hilbert space H_A is isomorphic to the Hilbert space $H(2j+1)$ through the mapping $|J_A; j\ m\rangle \leftrightarrow |J; j\ m\rangle$. Given some states and operators in $H(2j+1)$, we use the same notation for their counterparts in H_A with an additional index A. We call Π_A the projection operators in H_A and we use the notation $W_i \equiv 1 \otimes ... \otimes W \otimes ... \otimes 1$ $(i = 1, ..., N)$ for operators acting on H^N, with the operator W acting on the i Hilbert space H.

We next define angle states in the space H_A:

$$F_A = (2j+1)^{-1/2} \sum_{m,n} \omega(mn)|J; j\ m\rangle\langle J; j\ n| \otimes ... \otimes |J; j\ m\rangle\langle J; j\ n|$$

$$|\theta_A; j\ m\rangle \equiv F_A|J_A; j\ m\rangle \tag{23}$$

We can prove that

$$Z_A = Z_i\Pi_A; \qquad X_A = X_1...X_N\Pi_A \tag{24}$$

The states $|J_A; j\ m\rangle$ are far from each other in the sense that the operator X_A that performs shifts among them, is equal to the product of all X_i. Therefore, it is unlikely that small noise (acting on a one qudit) will perform such transformations, causing errors. In contrast, the states $|\theta_A, j\ m\rangle$ are near each other in the sense that the operator $Z_A = Z_i\Pi_A$ that performs shifts between them requires the action of only one Z_i. Therefore, it is very easy for noise to perform such transformations causing errors.

Infinitesimal $SU(2j+1)$ transformations on the states in the Hilbert space H_A can be written as:

$$g_A = 1 + \sum_{\alpha,\beta} \lambda_{\alpha\beta}X_A^\alpha Z_A^\beta = \Pi_A[1 + \sum_{\alpha,\beta} \lambda_{\alpha\beta}(X_1^\alpha...X_N^\alpha)Z_i^\beta] \tag{25}$$

It is seen that transformations that contain X_A^α are performed with the $X_1^\alpha...X_N^\alpha$ while transformations that contain Z_A^β are performed with the Z_i^β. So we have amplitude redundancy (i.e., redundancy in the J-direction) only.

4.2. *General redundancy*

We consider the space $(H_A)^M$ which is clearly a $(2j+1)^M$-dimensional subspace of the space H^{NM}. The operator $P_A = \Pi_A \otimes ... \otimes \Pi_A$ projects the space H^{NM} to the space $(H_A)^M$. We use the notation $W_{A\mu} \equiv \Pi_A \otimes$

$\ldots \otimes W_A \otimes \ldots \otimes \Pi_A$ ($\mu = 1, \ldots, M$) for operators acting on $(H_A)^M$, with the operator W_A acting on the μ Hilbert space H_A. Clearly $\Pi_{A\mu} = P_A$ for any μ. For a product of two operators $W_A V_A$ it is easily seen that $(W_A V_A)_\mu = W_{A\mu} V_{A\mu}$. Using this notation we can write $Z_{A\mu} = (Z_i \Pi_A)_\mu = Z_{i\mu} P_A$. In $Z_{i\mu}$ the indices i and μ refer to the positions in the words considered at the first and second step of the concatenated code, correspondingly.

We consider the $(2j + 1)$-dimensional subspace spanned by the vectors

$$H_B = \{|\theta_B; j\ m\rangle \equiv |\theta_A; j\ m\rangle \otimes \ldots \otimes |\theta_A; j\ m\rangle, m = -j, \ldots, j\} \qquad (26)$$

The Hilbert space H_B, is isomorphic to the Hilbert space H_A and also to the Hilbert space H, through the mapping: $|\theta_B; j\ m\rangle \leftrightarrow |\theta_A; j\ m\rangle \leftrightarrow |\theta; j\ m\rangle$. We call P_B the projection operator in H_B. We can show that

$$X_B = X_{A\mu} P_B = X_{1\mu} \ldots X_{N\mu} P_B; \qquad Z_B = Z_{i1} \ldots Z_{iM} P_B \qquad (27)$$

It is seen that we have general redundancy in all directions. In the previous subsection we have introduced amplitude redundancy and here we have introduced extra phase redundancy. The operator X_B that performs shifts among the states $|J_B; j\ m\rangle$ is equal to the product of all $X_{i\mu}$ and the operator Z_B that performs shifts among the states $|\theta_B; j\ m\rangle$, is equal to the product of all $Z_{i\mu}$. Therefore, it is unlikely that small noise will perform such transformations causing errors. It is important that the combined effect of amplitude and phase redundancy is general redundancy in the sense that all infinitesimal $SU(2j + 1)$ transformations can be written as:

$$g_B = 1 + \sum_{\alpha,\beta} \lambda_{\alpha\beta} X_B^\alpha Z_B^\beta = P_B[1 + \sum_{\alpha,\beta} \lambda_{\alpha\beta}(X_{1\mu}^\alpha \ldots X_{N\mu}^\alpha)(Z_{i1}^\beta \ldots Z_{iM}^\beta)] \quad (28)$$

It is seen that any transformation requires the action of many operators.

Noise transformations can be written as

$$\rho' = \sum_l E_\ell \rho E_\ell^\dagger$$

$$(29)$$

$$E_\ell = \sum \lambda_{\ell;\alpha_{11}\beta_{11}\ldots\alpha_{NM}\beta_{NM}}(X_{11}^{\alpha_{11}} Z_{11}^{\beta_{11}})\ldots(X_{NM}^{\alpha_{NM}} Z_{NM}^{\beta_{NM}})$$

Our coding provides protection against noise in any direction provided that we have 'small noise' which acts only on one qudit.

5. Discussion

We have studied several aspects of the theory of finite quantum systems which are important in the context of quantum information processing. We have considered angular momentum systems and studied the displacement operators which are generators of the $SU(2j + 1)$ group. Finite $SU(2j + 1)$

62

transformations are a finite sum of the displacement operators with the Weyl function as coefficients. Consequently the displacement operators are the 'building blocks' for general unitary transformations of qudits.

The Chinese remainder theorem has been used to factorize large qudits in terms of smaller qudits. All unitary transformations on large qudits can be performed through appropriate unitary transformations on the smaller qudits. Smaller qudits might be easier to implement physically, but large qudits might be more powerful computationally. The factorization that we have studied uses large qudits for quantum computation and performs the necessary unitary transformations, through appropriate unitary transformations in smaller qudits, which can be implemented physically more easily.

Another important field within the general area of quantum computation, is quantum coding. We have considered a concatenated code that introduces amplitude and phase redundancy in two steps. We have explained that this introduces general redundancy with respect to any transformation.

The work provides the theoretical background for qudit quantum computation and its implementation with condensed matter systems described by the theory of finite quantum systems.

References

1. H. Weyl, *Theory of Groups and Quantum Mechanics* (Dover, New York, 1950);
   J. Schwinger, *Proc. Nat. Acad. Sci. U.S.A.* **46**, 570 (1960); *Quantum Kinematics and Dynamics* (Benjamin, New York, 1970).
2. L. Auslander, R. Tolimieri *Bull. Am. Math.Soc.* **1**, 847 9 (1979);
   R. Balian and C. Itzykson, *C.R. Acad. Sci.* **303**, 773 (1986);
   W.K. Wootters and B.D. Fields, *Ann. Phys (N.Y)* **191**, 363 (1989);
   M.L. Mehta, *J.Math. Phys.* **28**, 781 (1987);
   V.S. Varadarajan, *Lett. Math. Phys.* **34**, 319 (1995);
   G.Chadzitaskos, J.Tolar, *Intern. J . Theo. Phys* **32**, 517 (1993).
3. T. Lulek, *Acta Phys. Polon.* **A82**, 377 (1992); *Rep. Math.Phys.* **34**, 71 (1994);
   W. Florek, *Rep. Math. Phys.* **34**, 81 (1994);
   D. Lipinski, *Rep. Math. Phys.* **34**, 97 (1994);
   S. Walcerz, *Rep. Math.Phys.* **34**, 107 (1994).
4. A. Vourdas, *Phys. Rev.* **A41**, 1653 (1990); **A43**;
   A. Vourdas, *J.Phys.A* **29**, 4275 (1996);
   A. Vourdas in *Proc. 4th Intern. School Theo. Phys.*, Poland 1996, Ed. T. Lulek, W. Florek, B. Lulek, (World Sci. Singapore 1997), pp.310-317.
5. A. Vourdas, C. Bendjaballah, *Phys.Rev.* **A47**, 3523 (1993).
6. D. Gottesman, A. Kitaev, J. Preskill, *Phys. Rev.* **A64**, 012310 (2001);
   D. Gottesman, *Lecture Notes Computer Science* **1509**, 302 (1999);
   S.D. Bartlett, H. de Guise, B.C. Sanders, *Phys. Rev.* **A65**, 052316 (2002).

7. P. Shor, *Phys. Rev.* **A52**, 2493 (1995).
8. A. Vourdas, *Phys. Rev.* **A65**, 042321 (2002).
9. D.B. Fairlie, P. Fletcher, C.K. Zachos, *J. Math. Phys.* **31**, 1088 (1990).
10. A. Ekert, R. Jozsa, *Rev. Mod. Phys.* **68**, 733 (1996).

COHERENT STATE MEASURES AND THE EXTENDED DOBIŃSKI RELATIONS

KAROL A. PENSON AND ALLAN I. SOLOMON

Laboratoire de Physique Théorique des Liquides,
Université Paris VI, 75252 Paris Cedex 05, France
E-mail:penson@lptl.jussieu.fr
E-mail:a.i.solomon@open.ac.uk

Conventional Bell and Stirling numbers arise naturally in the normal ordering of simple monomials in boson operators. By extending this process we obtain generalizations of these combinatorial numbers, defined as coherent state matrix elements of arbitrary monomials, as well as the associated Dobiński relations. These Bell-type numbers may be considered as power moments and give rise to positive measures which allow the explicit construction of new classes of coherent states.

1. Introduction

A defining characteristic of coherent states is the *resolution of unity* property, which is another expression of the classical Stieltjes moment problem. For the conventional coherent states of quantum optics the moments are simply $n!$, in some sense the simplest combinatorial numbers, leading to a simple exponential weight function. Not many other solutions to the moment problem are known, and it is in a sense not surprising that one should seek for solutions in the realm of other combinatorial integers. One such set is the Bell and Stirling numbers, which we define explicitly later, and which arise naturally in the normal ordering properties of boson operators.

The standard boson commutation relation $[a, a^\dagger] = 1$ can be *formally* realised by identifying $a = \frac{d}{dx}$ and $a^\dagger = x$, since $[\frac{d}{dx}, x] = 1$. In the present note we shall use both of the above forms. Integer sequences arise naturally when considering the action of $(x\,d/dx)^n$ on $f(x)$ as the following low order examples reveal:

$$(x\frac{d}{dx})f(x) = xf'(x) \tag{1}$$

$$(x\frac{d}{dx})^2 f(x) = xf'(x) + x^2 f''(x) \tag{2}$$

$$(x\frac{d}{dx})^3 f(x) = xf'(x) + 3x^2 f''(x) + x^3 f'''(x) \quad \text{etc.,} \tag{3}$$

which in general can be written as

$$(x\frac{d}{dx})^n f(x) = \sum_{k=1}^{n} S(n,k)x^k(d/dx)^k f(x) \tag{4}$$

or, alternatively[1]

$$(a^\dagger a)^n = \sum_{k=1}^{n} S(n,k)(a^\dagger)^k a^k. \tag{5}$$

The *Stirling numbers of the second kind* $S(n,k)$ appearing in Eqs.(4) and (5) have been known for over 250 years.[2] Eq.(5) exemplifies the *normal ordering problem*, that is, finding the form of $(a^\dagger a)^n$ with the powers of a on the right. Although explicit expressions for $S(n,k)$ are known,[3] of particular interest here are the *Bell numbers* $B(n)$ given by the sums

$$B(n) = \sum_{k=1}^{n} S(n,k), \qquad n = 1, 2, \ldots \tag{6}$$

with $B(0) = 1$ by convention $(S(n,0) = \delta_{n,0})$. A closed-form expression for $B(n)$ can be found by considering the action of $(xd/dx)^n$ on a function $f(x)$ having a Taylor expansion around $x=0$, i.e. $f(x) = \sum_{k=0}^{\infty} c_k x^k$.

Applying Eq.(4) to $f(x)$ gives:

$$(x\frac{d}{dx})^n f(x) = \sum_{k=0}^{\infty} c_k k^n x^k. \tag{7}$$

Now specify $f(x) = e^x$ in Eq.(7) so that $c_k = 1/k!$ and deduce from Eq.(4) that

$$(1/e^x)\sum_{k=0}^{\infty} \frac{k^n}{k!}x^k = \sum_{k=1}^{n} S(n,k)x^k, \tag{8}$$

which for $x = 1$ reduces to

$$(1/e)\sum_{k=0}^{\infty} \frac{k^n}{k!} = \sum_{k=1}^{n} S(n,k) = B(n). \tag{9}$$

Equations (8) and (9) are the celebrated Dobiński formulas[2,3,4,5] which have been the subject of much combinatorial interest. For completeness, we recall the combinatorial definitions of $B(n)$ and $S(n,k)$: $B(n)$ counts the number of partitions of a set of n distinguishable elements; $S(n,k)$ counts the number of partitions of a set of n distinguishable elements into k *non-empty* sets. Eq.(9) represents the integer $B(n)$ as an infinite series, which is however *not* a power series in n. An immediate consequence of Eqs.(8) and (9) is that $B(n)$ is the n-th moment of a (singular) probability distribution, consisting of weighted Dirac delta functions located at the positive integers (the so-called Dirac comb) :

$$B(n) = \int_0^\infty x^n W(x) dx, \quad n = 0, 1, \ldots \tag{10}$$

where

$$W(x) = (1/e) \sum_{k=1}^{\infty} \frac{\delta(x-k)}{k!}. \tag{11}$$

The discrete measure $W(x)$ serves as a weight function for a family of orthogonal polynomials $C_n^{(1)}(x)$, the Charlier polynomials.[6] The exponential generating function (EGF) of the sequence $B(n)$ can be obtained from Eq.(9) as

$$e^{(e^\lambda - 1)} = \sum_{n=0}^{\infty} B(n) \frac{\lambda^n}{n!}. \tag{12}$$

This equation is related via Eq.(5) to a formula giving the normal ordered form[7,8] of $e^{\lambda a^\dagger a}$

$$e^{\lambda a^\dagger a} = \mathcal{N}(e^{\lambda a^\dagger a}) =: e^{a^\dagger a(e^\lambda - 1)} : \tag{13}$$

The symbol $\mathcal{N}$ denotes normal ordering and $: A(a^\dagger, a) :$ means expand A in a Taylor series and normally order *without* taking account of the commutation relation $[a, a^\dagger] = 1$. We stress that in the derivation[7,8] of Eq.(13) no use has been made of the Stirling and Bell numbers. It may readily be seen that Eq.(12) is the expectation value of Eq.(13) in the coherent state $|z\rangle$ defined by $a|z\rangle = z|z\rangle$ at the value $|z| = 1$. This circumstance has been used recently to re-establish the link between the matrix element $\langle z|e^{\lambda a^\dagger a}|z\rangle$ and the properties of Stirling and Bell numbers.[9]

2. Extending Dobiński formulas

The purpose of this note is to show that the above results on functions of $a^\dagger a$ may be extended to functions of $(a^\dagger)^r a^s$, $(r, s = 1, 2, \ldots)$ with $r \geq s$.

Specifically, we pose the following questions:

(1) What extensions of the conventional Stirling and Bell numbers occur in the normal ordering of $[(a^\dagger)^r a^s]^n$?
(2) Can the generalised Bell numbers $B_{r,s}(n)$ so defined be represented by an infinite series of the type of Eq.(9) - that is, do they satisfy a generalised Dobiński formula ?

(3) May one consider the $B_{r,s}(n)$ as the n-th moments of a positive weight function $W_{r,s}(x)$ on the positive half-axis, and may this latter be explicitly obtained ?

(4) Can one attach a combinatorial significance to the sequences $\{B_{r,s}(n)\}$?

In this note we indicate affirmative answers to the first three questions and a partial answer to the fourth.

To this end we generalize Eq.(5) by defining for $r \geq s$:

$$[(a^\dagger)^r a^s]^n = (a^\dagger)^{n(r-s)} \sum_{k=s}^{ns} S_{r,s}(n,k)(a^\dagger)^k a^k \tag{14}$$

or, alternatively,

$$[x^r (d/dx)^s]^n = x^{n(r-s)} \sum_{k=s}^{ns} S_{r,s}(n,k) x^k (d/dx)^k. \tag{15}$$

Eqs.(14) and (15) introduce generalized Stirling numbers $S_{r,s}(n,k)$ which imply an extended definition of generalized Bell numbers:

$$B_{r,s}(n) \equiv \sum_{k=s}^{ns} S_{r,s}(n,k). \tag{16}$$

Note that the Stirling numbers $S_{r,1}(n,k)$ have been previously studied.[10] Also $B_{1,1}(n) = B(n)$ of Eq.(6).

We have found a representation of the numbers $B_{r,s}(n)$ as an infinite series, which is a generalization of the Dobiński formula Eq.(9). For $r = s$ one obtains:

$$B_{r,r}(n) = (1/e) \sum_{k=0}^{\infty} \frac{1}{k!} \left[\frac{(k+r)!}{k!} \right]^{n-1} , \quad n = 1,2\ldots \tag{17}$$

with $B_{r,r}(0) = 1$ by convention. For $r > s$ the corresponding formula is:

$$B_{r,s}(n) = [(r-s)^{s(n-1)}/e] \sum_{k=0}^{\infty} \left[\prod_{j=1}^{s} \frac{\Gamma(n + \frac{k+j}{r-s})}{\Gamma(1 + \frac{k+j}{r-s})} \right], \quad B_{r,s}(0) = 1 \tag{18}$$

The formula Eq.(9) and its extensions Eqs.(17) and (18) share a common feature, namely, the fact that they give rise to a series of integers is by no means evident.

Choosing various pairs (r,s) gives alternative representations of many integer sequences. Some examples are:

(1) $(r > 1, s = 1)$
$B_{r,1}(n) = [(r-1)^n/e] \sum_{k=1}^{\infty} \Gamma(n + \frac{k}{r-1})/[k!\,\Gamma(\frac{k}{r-1})]$

(2) The pair $(r+1, r)$ leads to a hypergeometric function $_pF_q$:

$$B_{r+1,r}(n) = (1/e)[\prod_{j=1}^{r} \frac{(n-1+j)!}{j!}] \times$$

$$\times {}_rF_r(n+1, n+2, \ldots, n+r; 2, 3, \ldots, r+1; 1)$$

(3) as does $(2r, r)$

$$B_{2r,r}(n) = [(rn)!/(e\, r!)]\, {}_1F_1(rn+1; r+1; 1).$$

A still more general family of sequences arising from Eq.(18) has the form $(p, r = 1, 2, \ldots)$:

$$B_{pr+p,pr}(n) = (1/e) \left[\prod_{j=1}^{r} \frac{(p(n-1)+j)!}{(pj)!} \right] \times$$

$$\times {}_rF_r(pn+1, \ldots, pn+1+p(r-1); 1+p, \ldots, 1+p+p(r-1); 1).$$

For example, for $p = 3, r = 2$, this reduces to

$$B_{9,6}(n) = (1/e) \frac{(3(n-1)+1)!(3(n-1)+2)!}{3!6!}\, {}_2F_2(3n+1, 3n+4; 4, 7; 1) \quad (19)$$

whose first four terms are $1, 207775, 566828686621, 9011375448568566265$.

Knowledge of the generalized Stirling numbers in Eq.(14) solves the normal ordering problem for $[(a^\dagger)^r a^s]^n$. We are able to give the appropriate generating functions for the sequences $B_{r,s}(n)$. It then follows that, at least formally, we can furnish the generating functions for $S_{r,s}(n, k)$ as well.[11] Additionally, it turns out that in certain circumstances one may obtain explicit expressions for them. We quote two such cases:

$$S_{r,r}(n, k) = \sum_{p=0}^{k-r} \frac{(-1)^p[\frac{(k-p)!}{(k-p-r)!}]^n}{(k-p)!p!}, \quad (r \le k \le rn) \quad (20)$$

and

$$S_{2,1}(n, k) = \frac{n!}{k!}\binom{n-1}{k-1}, \quad (1 \le k \le n) \quad (21)$$

which are the so-called unsigned Lah numbers.[3,10]

For those pairs (r, s) for which we have an explicit expression for $S_{r,s}(n)$ we may generalize Eq.(13) to obtain the normal ordered form of $e^{\lambda(a^\dagger)^r a^s}$. For example, the matrix element $\langle z|e^{\lambda(a^\dagger)^r a}|z\rangle$ leads to the following normally ordered expression:

$$e^{\lambda(a^\dagger)^r a} = \mathcal{N}(e^{\lambda(a^\dagger)^r a}) =: \exp\{[(1-\lambda(a^\dagger)^{r-1}(r-1))^{\frac{1}{r-1}} - 1]a^\dagger a\}: \quad (22)$$

We apply to Eq.(22) the same method whereby we obtained the EGF of Eq.(12) by taking the expectation of Eq.(13) in the coherent state $|z\rangle$, to get:

$$\langle z|e^{\lambda(a^\dagger)^r a}|z\rangle =: \exp\{[(1 - \lambda(z^*)^{r-1}(r-1))^{\frac{1}{r-1}} - 1]|z|^2\} : \qquad (23)$$

which evaluates at $z = 1$ to give

$$\langle z|e^{\lambda(a^\dagger)^r a}|z\rangle_{z=1} =: \exp\{(1 - \lambda(r-1))^{\frac{1}{r-1}} - 1\} : \qquad (24)$$

which is precisely[10] the EGF for the numbers $B_{r,1}(n)$.

For general $r > s$ the corresponding $B_{r,s}(n)$ grow much more rapidly than $n!$ and thus may not be obtained via the usual form of EGF. One instead defines the EGF in terms of $B_{r,s}(n)/(n!)^t$ where t is an integer chosen to ensure that $\sum_{n=0}^{\infty} B_{r,s}(n)/(n!)^{t+1}$ has a non-zero radius of convergence. As a result one obtains variants of Eq.(23) involving different hypergeometric functions.[11]

3. Generalized Bell numbers as moments of positive measures

The formulae Eq.(17) and (18) can be used to demonstrate what we consider to be the main thrust of our results, namely that $B_{r,s}(n)$ is the n-th moment of a positive probability measure. For $r = s$ this takes the form of a sort of Dirac comb, while for $r > s$ gives a continuous distribution. Consider the case $r = s$ first. We rewrite Eq.(17) as

$$B_{r,r}(n) = (1/e)\sum_{k=0}^{\infty}\frac{1}{(k+r-1)!}[k(k+1)\ldots(k+r-1)]^n \quad n = 1, 2\ldots(25)$$

which immediately indicates that $B_{r,s}(n)$ is the n-th moment of

$$W_{r,r}(x) = (1/e)\sum_{k=0}^{\infty}\frac{\delta(x - k(k+1)\ldots(k+r-1))}{(k+r-1)!} \qquad (26)$$

which is a "rarefied" Dirac comb whose delta spikes are situated on the x-axis at $x = k(k+1)\ldots(k+r-1)$ $k = 0, 1, \ldots$.

For $r > s$ it is necessary to excise the part of formula Eq.(18) which contains n, that is $\prod_{j=1}^{s}\Gamma(n+\frac{k+j}{r-s})$, for fixed k. We reparametrize this with $n = \sigma - 1$ and consider it as a Mellin transform; that is, we seek $w_{r,s}^{(k)}(x)$ such that when extended to complex σ

$$\int_0^{\infty} x^{\sigma-1}w_{r,s}^{(k)}(x)dx = \prod_{j=1}^{s}\Gamma(\sigma - 1 + \frac{k+j}{r-s}). \qquad (27)$$

The positivity of $\omega_{r,s}^{(k)}(x)$ follows from the fact that it may be considered as resulting from an s-fold Mellin convolution of shifted Gamma-functions[12,13] which does preserve positivity. We conclude that $\omega_{r,s}^{(k)}(x)$ is positive and so is $W_{r,s}(x)$, where

$$\int_0^\infty x^n W_{r,s}(x)dx = B_{r,s}(n). \tag{28}$$

Use of the inverse Mellin transform yields many analytic forms of $W_{r,s}(x)$. We exhibit some solutions of the Stieltjes moment problem Eq.(28):

(1) For $r = 2, 3, \ldots$

$$W_{r,1}(x) = \frac{1}{e(r-1)} \left(\frac{x}{r-1}\right)^{\frac{2-r}{r-1}} \exp\left(-\frac{x}{r-1}\right) \times$$

$$\times \sum_{k=0}^\infty \frac{1}{k!} \left(\frac{x}{r-1}\right)^{\frac{k}{r-1}} / \Gamma\left(\frac{r+k}{r-1}\right);$$

(2) In particular

$$W_{2,1}(x) = e^{-x} I_1(2\sqrt{x})/(e\sqrt{x}),$$

where $I_1(y)$ is the modified Bessel function of the first kind.

(3) Taking $r = 3$ and $s = 1$ we have

$$W_{3,1}(x) = \frac{1}{e\sqrt{8x}} e^{-\frac{x}{2}} \left[\frac{2}{\sqrt{\pi}} \, {}_0F_2\left(\frac{1}{2}, \frac{3}{2}; \frac{x}{8}\right) + \frac{x}{\sqrt{2}} \, {}_0F_2\left(\frac{3}{2}, 2; \frac{x}{8}\right)\right]$$

(4) while for $(r, s) = (5, 2)$ we obtain

$$W_{5,2}(x) = \frac{2}{27e} \frac{1}{\sqrt{x}} K_{\frac{1}{3}}\left(\frac{2\sqrt{x}}{3}\right) \cdot u_{5,2}(x), \text{ where}$$

$$u_{5,2}(x = \frac{3}{32\pi} \left[24\sqrt{3} \, {}_0F_4\left(\frac{1}{3}, \frac{2}{3}, \frac{4}{3}, \frac{5}{3}; \frac{x}{243}\right) + \right.$$

$$+ 8 \cdot 3^{\frac{5}{6}} x^{\frac{1}{3}} \, {}_0F_4\left(\frac{2}{3}, \frac{4}{3}, \frac{5}{3}, \frac{6}{3}; \frac{x}{243}\right) +$$

$$\left. + 3 \cdot 3^{\frac{1}{6}} x^{\frac{2}{3}} \, {}_0F_4\left(\frac{4}{3}, \frac{5}{3}, \frac{6}{3}, \frac{7}{3}; \frac{x}{243}\right)\right]$$

and $K_\nu(y)$ is the modified Bessel function of the second kind.

(5) Finally

$$W_{2r,r}(x) = \frac{1}{er} x^{\frac{2-3r}{2r}} e^{-x^{\frac{1}{r}}} I_r(2x^{\frac{1}{2r}}).$$

Our proof of the existence and positivity of $W_{r,s}(x)$ implies the following *Theorem*: For all the sequences defined by Eqs. (17) and (18) the following inequalities are satisfied:

$$\det\left[B_{r,s}(i+j-2)_{1\leq i,j\leq n}\right] > 0 \tag{29}$$

$$\det\left[B_{r,s}(i+j-1)_{1\leq i,j\leq n}\right] > 0 \quad \text{for all } n = 1, 2 \dots . \tag{30}$$

The above expresses the positivity of the two series of Hankel-Hadamard determinants spanned by the moment sequences. This positivity is the necessary and sufficient condition for the existence of a positive measure $W_{r,s}(x)$ resulting from the given sequence of moments.[14]

For completeness we give here the two asymptotic expansions as $n \to \infty$ for the series $B_{2,1}(n)$ and $B_{3,1}(n)$, obtained by the method of Hayman.[15] $B_{2,1}(n) \equiv \langle z|[(a^\dagger)^2 a]^n|z\rangle_{z=1}$ has the EGF

$$e^{\frac{x}{1-x}} = \sum_{n=0}^{\infty} \frac{B_{2,1}(n)}{n!} x^n \tag{31}$$

and has the following asymptotics:

$$B_{2,1}(n) \overset{n\to\infty}{\longrightarrow} \frac{1}{\sqrt{2e}} (n^{-\frac{1}{4}} + \frac{1}{12} n^{-\frac{3}{4}} + O(n^{-\frac{5}{4}})) \cdot n^n \exp(-n + 2\sqrt{n}). \tag{32}$$

Similarly, for $B_{3,1}(n) \equiv \langle z|[(a^\dagger)^3 a]^n|z\rangle_{z=1}$

$$e^{\frac{1-\sqrt{1-2x}}{\sqrt{1-2x}}} = \sum_{n=0}^{\infty} \frac{B_{3,1}(n)}{n!} x^n \tag{33}$$

$$B_{3,1}(n) \overset{n\to\infty}{\longrightarrow} \frac{2^{\frac{1}{6}}}{\sqrt{3e}} (n^{-\frac{1}{3}} + 2^{-\frac{3}{7}} n^{-\frac{2}{3}} + O(n^{-1})) \cdot (2n)^n \exp(-n + (3/2)(2n)^{\frac{1}{3}}). \tag{34}$$

The role played by these approximations is analogous to that of the Stirling approximation to n-factorial. The growth of other $B_{r,s}(n)$ with n is in general even more rapid. However, we do not have the asymptotic expansions for them at the present time.

4. Postcript

Our interest in integer combinatorial sequences (see also papers[11,12,17,18]) is a consequence of our parallel work on the construction of complete sets of coherent states (see paper [16] and references therein). Completeness - or, equivalently, the resolution of unity property - is an essential ingredient in such construction, and can be shown to be satisfied if one can furnish an appropriate positive measure $W(x)$. Thus consider a state $|z\rangle$ constructed from an arbitrary discrete set of states $|m\rangle$ which are both orthonormal ($\langle m|m'\rangle = \delta_{m,m'}$) and complete:

$$|z\rangle = N^{-\frac{1}{2}}(|z|^2) \sum_{m=0}^{\infty} \frac{z^m}{\sqrt{\rho(m)}} |m\rangle, \quad \rho(m) > 0. \tag{35}$$

Resolution of unity is achieved via

$$\iint d^2z \, |z\rangle W(|z|^2) \langle z| = I \equiv \sum_{m=0}^{\infty} |m\rangle\langle m|. \tag{36}$$

The conditions Eqs.(35) and (36) boil down to a relation closely resembling Eq.(28), namely

$$\pi \int_0^{\infty} x^n \left[\frac{W(x)}{N(x)} \right] dx = \rho(n), \quad x \equiv |z|^2 \quad n = 0, 1 \ldots \tag{37}$$

where it is assumed that the normalization of $N(x)$ defined by

$$N(x) = \sum_{m=0}^{\infty} \frac{x^n}{\rho(n)} \tag{38}$$

has a *finite* radius of convergence. It is clear from Eq.(28) that when identifying $W_{r,s}(x)$ with $\pi W(x)/N(x)$ we can use the $B_{r,s}(n)$ as $\rho(n)$ to construct states of the form

$$|z\rangle_{r,s} = N_{r,s}^{-\frac{1}{2}}(|z|^2) \sum_{n=0}^{\infty} \frac{z^n}{\sqrt{B_{r,s}(n)}} |n\rangle \tag{39}$$

where

$$N_{r,s}(x) = \sum_{n=0}^{\infty} \frac{x^n}{B_{r,s}(n)}. \tag{40}$$

As pointed out above, $B_{r,s}(n)$ grows much more quickly than $n!$ so the series Eq.(40) is very rapidly convergent for all r, s. This ensures that the problem defined by Eq.(37) is well-defined for all $x > 0$.

The properties of the coherent states defined by Eq.(39) remain to be investigated but they automatically satisfy the resolution of unity, due to Eq.(28).

We are currently attempting to unravel the combinatorial meaning of some of the $B_{r,s}(n)$. It appears that this can be done systematically for $B_{r,1}(n)$ at least, which will be the subject of a forthcoming publication.[19]

The integral representations of combinatorial numbers, used by us for the construction of coherent states, have triggered off applications for hyperdeterminants and Selberg integrals.[20] Although of evident utility, Stirling and Bell numbers have not yet found their way into most textbooks on mathematical physics, a notable exception being that of Aldrovandi.[21]

Acknowledgments

We thank P. Blasiak, G. Duchamp, M. Mendez and J.Y. Thibon for enlightening discussions.

References

1. J. Katriel and G. Duchamp, *J. Phys. A* **28**, 7209 (1995).
2. S.V. Yablonsky, *Introduction to Discrete Mathematics*, (Moscow: Mir Publishers) (1989).
3. L. Comtet, *Advanced Combinatorics*,(Dordrecht: Reidel) (1974).
4. G.M. Constantine and T.H. Savits, *SIAM J. Discrete Math.* **7** 194 (1994).
5. H. Wilf, *Generatingfunctionology*, (New York: Academic) (1994).
6. R. Koekoek and R.F. Swarttouv, *The Askey scheme of hypergeometric polynomials and its q-analogue*, Dept. of Technical Mathematics and Informatics, Report No. 98-17 Delft University of Technology (1998).
7. J.R. Klauder and E.C.G. Sudarshan, *Fundamentals of Quantum Optics*, (New York: Benjamin) (1968).
8. W.H. Louisell,*Radiation and Noise in Quantum Electronics*, (Florida: Krieger) (1977).
9. J. Katriel, *Phys. Lett.* **A273**, 159 (2000).
10. W. Lang, *J. Int. Seqs.* **12** Article 00.2.4(2000).
11. K.A. Penson and A.I. Solomon, *to be published* (2002).
12. K.A. Penson and A.I. Solomon, *Coherent states from combinatorial sequences* arXiv: quant-ph/0111151. Proceedings of the 2nd International Symposium on Quantum Theory and Symmetries, E. Kapuscik and A. Horzela, Editors (World Scientific, Singapore, p. 527) (2001).
13. O.I. Marichev, *Handbook of integral transforms of higher transcendental functions* (Chichester:Ellis Horwood) (1983).
14. N.I. Akhiezer, *The classical moment problem and some related questions in analysis*, (London: Oliver and Boyd) (1965).
15. W.K. Hayman, *J. Reine Angew. Math.* **196** 6795 (1956).
16. J.R. Klauder, K.A. Penson and J.M. Sixdeniers, *Phys.Rev.* **A64**, 013817 (2001).
17. K.A. Penson and J.M. Sixdeniers, *J. Int. Seqs.* Article 01.2.5 (2001).
18. J.M. Sixdeniers, K.A. Penson and A.I. Solomon, *J. Int. Seqs. Article 01.1.4* (2001).
19. M. Mendez and K.A. Penson, *(in preparation)* (2002).
20. J.G. Luque and J.Y. Thibon, *(preprint)*(2002).
21. R. Aldrovandi, *Special Matrices of Mathematical Physics* (Singapore: World Scientific) (2001).

THE VANDERMONDE DETERMINANT REVISITED

BRIAN G WYBOURNE

Instytut Fizyki,
Uniwersytet Mikołaja Kopernika,
87-100 Toruń,
Poland
E-mail: bgw@phys.uni.torun.pl
WEB: http://www.phys.uni.torun.pl/~bgw

Dedicated to the memory of Claude Itzykson (1938-1995)

The expansion of the even powers of the Vandermonde determinant in terms of signed sequences of Schur functions is considered. The q–discriminant is introduced to provide an explanation for the vanishing of certain expansion coefficients. A number of compelling conjectures are given. Finally connections with Hankel hyperdeterminants are noted.

In most sciences one generation tears down what another has built, and what one has established, another undoes. In mathematics alone each generation adds a new storey to the old structure. (Hermann Hankel 1839-1873)

1. Introduction

The Vandermonde determinant plays a crucial role in the description of the Quantum Hall Effect (QHE) via the Laughlin wavefunction ansatz [1] and in the description of One Component Plasmas [2] (OCP). There has been considerable interest in the expansion of the Laughlin wavefunction as a linear combination of Slater determinantal wavefunctions for N particles. [3,4]

The *even* powers of the Vandermonde alternating function play a key role in determining the coefficients of the expansion of the Laughlin wavefunction as a linear combination of Slater determinantal wavefunctions. Indeed, the relevant coefficients are directly related to the signed integer coefficients that arise in the expansion of the even powers of the Vandermonde alternating function into Schur functions. The problem of determining the expansion coefficients for increasingly large values of N is a combinatorially explosive problem.

The primary problem is to determine the signed integer coefficients that arise in the Schur function expansion of the second power of the Vandermonde alternating function, higher powers following by application of the Littlewood-Richardson rule [5]. The Schur functions that arise in the expansion of the second power of the Vandermonde are indexed by ordered partitions, (λ), of the integer $n = N(N - 1)$. Di Francesco *etal* [3] defined a class of *admissible partitions*, being those partitions of n associated with non-zero expansion coefficients, c^λ, and determined their number $A(N)$ for up to $N = 29$. They conjectured that these numbers would be exact for every value of N *provided none of the coefficients accidently vanished.* At the time of their conjecture numerical calculations [3,4] supported their conjecture for $N \leq 6$. Scharf *etal* [6] developed algorithms for calculating the expansion coefficents and computed the coefficients for $N = 7, 8, 9$ finding agreement for the conjecture for $N = 7$ but disagreement for $N = 8, 9$. I have recently extended the calculations to $N = 10$.

In this paper I first outline the relationship of the Laughlin wavefunction to the problem of its expansion in terms of signed integers of the even powers of the Vandermonde and define the admissible partitions. In searching for an explanation for the vanishing coefficients I introduce the $q-$discriminant and show that some of the $q-$polynomials have factors that vanish when $q = 1$. Explicit $q-$polynomials are given for the vanishing coefficients when $N = 8$. We then consider properties of the $q-$polynomials and the various possible sums of coefficients which leads to several surprising, though compelling conjectures. Finally we consider, briefly, the relationship of the higher even powers of the Vandermonde and their relationship to Hankel hyperdeterminants.

2. The Laughlin Wavefunction

Laughlin [1] has described the fractional quantum Hall effect in terms of a wavefunction

$$\Psi^m_{Laughlin}(z_1, \ldots, z_N) = \prod_{i<j}^{N}(z_i - z_j)^{2m+1} \exp\left(-\frac{1}{2}\sum_{i=1}^{N}|z_i|^2\right) \qquad (1)$$

The Vandermonde alternating function in N variables is defined as

$$V(z_1, \ldots, z_N) = \prod_{i<j}^{N}(z_i - z_j) \qquad (2)$$

In terms of the Vandermonde alternating function in N variables, to within an overall normalisation, the Laughlin wavefunction may be written as

$$\frac{\Psi^m_{Laughlin}}{V} = V^{2m} = \sum_{\lambda \vdash n} c^\lambda s_\lambda \quad n = mN(N-1) \tag{3}$$

where the s_λ are Schur functions and the c^λ are signed integers. In most of the following we will discuss the case of $m = 1$. The partitions (λ) indexing the Schur functions are of weight $N(N-1)$. For a given N the partitions are bounded by a highest partition $(2N-2, 2N-4, \ldots, 0)$ and a lowest partition $((N-1)^{N-1})$ with the partitions being of length N and $N-1$.

3. Admissible Partitions

Let

$$n_k = \sum_{i=0}^{k} \lambda_{N-i} - k(k+1) \quad k = 0, 1, \ldots, N-1 \tag{4}$$

Following Di Francesco et al [3], we define *Admissible partitions* as satisfying Eq(4) with *all* $n_k \geq 0$. Di Francesco et al conjectured that the number of admissible partitions, A_N, was the number of distinct partitions arising in the expansion, Eq(3), *provided none of the coefficients vanished*. The conjecture fails [6] for $N \geq 8$. We find the number of admissible partitions associated with vanishing coefficients as

$$(N = 8) \quad 8, \ (N = 9) \quad 66, \ (N = 10) \quad 389$$

The coefficients of s_λ and s_{λ_r} are equal if [3]

$$(\lambda_r) = (2(N-1) - \lambda_N, \ldots, 2(N-1) - \lambda_1) \tag{5}$$

Such pairs of partitions are said to exhibit *reversal symmetry*. We list the partitions for $N = 8, 9, 10$, having vanishing coefficients, in tables I,II and III respectively.

4. The q-discriminant

Let $q\mathbf{x} = (qx_1, qx_2, \ldots, qx_N)$ and the q-discriminant of $\mathbf{x}$ be

$$D_N(q; \mathbf{x}) = \prod_{1 \leq i \neq j \leq N} (x_i - qx_j) \tag{6}$$

and

$$R_N(q; \mathbf{x}) = \prod_{1 \leq i \neq j \leq N} (x_i - qx_j)(qx_i - x_j) = \sum_\lambda c^\lambda(q) s_\lambda(\mathbf{x}) \tag{7}$$

So that

$$V_N^2(\mathbf{x}) = \prod_{1 \leq i \neq j \leq N} (x_i - x_j)^2 = R_N(1; \mathbf{x}) \tag{8}$$

Introduce q-polynomials such that

$$R_N(q; \mathbf{x}) = \sum_\lambda c^\lambda(q) s_\lambda(\mathbf{x}) \tag{9}$$

$$R_N(q; \mathbf{x}) = \frac{(-1)^{N(N-1)/2}}{(1-q)^N} \sum_{\nu \subseteq (N-1)^N} ((-q)^{|\nu|} + (-q)^{N^2-|\nu|}) s_{(N-1)^N/\nu}(\mathbf{x}) s_{\nu'}(\mathbf{x}) \tag{10}$$

Such expansions have been evaluated as polynomials in q for all admissible partitions for $N = 2, \ldots, 6$ with many examples for $N = 7, 8, 9$. Below we give in Table IV the q−polynomials for $N = 2, 3, 4$.

Note that at $N = 4$ some q−polynomial factors with some negative coefficients start to appear. As N increases the number of these types of factors grows making possible the "accidental" vanishing of some q−polynomials when $q = 1$. This first happens at $N = 8$.

5. The Zero Coefficients and q−polynomials

The q−polynomials associated with the 8 vanishing coefficients for $N = 8$ have been constructed and are given in Table V. The polynomials are given for each reversal pair of Schur functions. Each q−polynomial has a factor $(q - 1)^4$ which vanishes for $q = 1$.

6. Sums of Coefficients and Conjectures

Each q−polynomial, $P(N, q)$ is of the form

$$P(N, q) = (-1)^\phi q^p Q(q) \tag{11}$$

where ϕ is a phase, p is a positive integer and $Q(q)$ is a polynomial in q. Inspection of Table IV, and many more complex examples, suggests the following conjecture:-

If $N \to N + 1$ then

$$\phi \to \phi, \quad p \to p + N, \quad Q(q) \to Q(q), \quad \{\lambda\} \to \{2N - 2, \lambda\} \tag{12}$$

Di Francesco et al [3] give the remarkable result that if

$$V^2(N) = \sum_\lambda c^\lambda s_\lambda$$

then

$$\sum_\lambda |c^\lambda|^2 = \frac{(3N)!}{N!(3!)^N} \tag{13}$$

Can one write a similar expression for

$$\sum_\lambda |c(q)|^2? \tag{14}$$

We give below the sums of squares for $N = 2, \ldots, 4$

$N = 2$ $\qquad q^4 + 2q^3 + 4q^2 + 2q + 1$

$N = 3$ $\qquad q^{12} + 4q^{11} + 11q^{10} + 20q^9 + 34q^8 + 44q^7$

$\qquad\qquad +52q^6 + 44q^5 + 34q^4 + 20q^3 + 11q^2 + 4q + 1$

$N = 4$ $\qquad q^{24} + 6q^{23} + 22q^{22} + 58q^{21} + 128q^{20} + 242q^{19}$

$\qquad\qquad +418q^{18} + 646q^{17} + 929q^{16} + 1210q^{15} + 1490q^{14}$

$\qquad\qquad +1670q^{13} + 1760q^{12} + 1670q^{11} + 1490q^{10} + 1210q^9$

$\qquad\qquad +646q^8 + 418q^6 + 242q^5 + 128q^4 + 58q^3 + 22q^2 + 6q + 1$

$N = 5$ $\quad q^{40} + 8q^{39} + 37q^{38} + 124q^{37} + 339q^{36} + 796q^{35} + 1671q^{34}$

$\qquad\qquad +3192q^{33} + 5662q^{32} + 9392q^{31} + 14755q^{30} + 21946q^{29}$

$\qquad\qquad +31190q^{28} + 42202q^{27} + 54902q^{26} + 68238q^{25} + 81835q^{24}$

$\qquad\qquad +93846q^{23} + 104006q^{22} + 110180q^{21} + 112756q^{20}$

$\qquad\qquad +110180q^{19} + 104006q^{18} + 93846q^{17} + 81835q^{16}$

$\qquad\qquad +68238q^{15} + 54902q^{14} + 42202q^{13} + 31190q^{12}$

$\qquad\qquad +21946q^{11} + 14773q^{10} + 9392q^9 + 5662q^8 + 3192q^7$

$\qquad\qquad +1671q^6 + 796q^5 + 339q^4 + 124q^3 + 37q^2 + 8q + 1$

We note that in each case the coefficient distribution is symmetric and unimodal.

Consider the sum $CS(N) = \sum_\lambda c^\lambda(1)$

N	No. Partitions	CS(N)
2	2	-2
3	5	-14
4	16	70
5	59	910
6	247	-7280
7	1111	-138320
8	5294	1521520
9	26310	38038000
10	135281	-532532000

From the results tabulated for $N = 2, \ldots, 10$ we conjecture that

$$CS(N) = \prod_{x=0}^{[N/2]} (-3x + 1) \prod_{x=0}^{[(N-1)/2]} (6x + 1) \tag{15}$$

Again one would like to obtain a q−dependent extension of this conjecture.

7. Hankel Determinants and Vandermonde Expansions

The Hankel matrix of order $n + 1$ of a sequence $c_0, c_1, \ldots$ is the $n + 1$ by $n + 1$ matrix whose (i, j) element is c_{i+j} with $0 \leq i, j, \leq n$.

The Hankel determinant of order $n + 1$ is the determinant of the corresponding Hankel matrix,

$$det|c_{i+j}|_{0 \leq i,j, \leq n} = det \begin{vmatrix} c_0 & c_1 & \cdots & c_n \\ c_1 & c_2 & \cdots & c_{n+1} \\ \vdots & \vdots & \vdots & \\ c_n & c_{n+1} & \cdots & c_{2n} \end{vmatrix} \tag{16}$$

Consider the sequence where $c_n = n!$. Then for $n = 3$ we have the sequence $1, 1, 2, 6, \ldots$

$$D_n^1(c) = det|c_{i+j}|_{0 \leq i,j, \leq 3} = det \begin{vmatrix} 1 & 1 & 2 & 6 \\ 1 & 2 & 6 & 24 \\ 2 & 6 & 24 & 120 \\ 6 & 24 & 120 & 720 \end{vmatrix} = 144 \tag{17}$$

We may define a Hankel hyperdeterminant as

$$D_n^k(c) = det_{2k}|c_{i_1 + \ldots + i_{2k}}|_{0 \leq i_p \leq n-1} \tag{18}$$

NB. The sequence c need not be restricted to just integers but may involve sequences of polynomials etc. Thus Luque and Thibon [7] consider the case where $c_n = h_n(X)$, the $n-th$ complete homogeneous symmetric function of some auxiliary variables $X = \{x_i\}$ and show that $D_n^{(k)}(h)$ may be expressed in terms of Schur functions $s_\lambda(X)$ This problem is equivalent to determining the Schur function expansion of the even powers, Δ^{2k}, of the Vandermonde determinant, Δ!

8. Concluding Remarks

My interest in this subject was stimulated by a meeting with Claude Itzykson at the Formal Power Series and Algebraic Combinatorics meeting held in Florence June 1993. In early 1994 I communicated to him the $N = 8$ counter examples to his conjecture on admissible tableaux. On 1 March 1994 he wrote to me an enthusiastic letter raising many questions for future work. He died on 22 May 1995. This paper discusses some of the questions he raised - but not all - the area seems fruitful with many possible directions remaining to be explored.

Acknowledgments

I have benefited from detailed discussions with Prof. R C King (Southampton University) and Prof. J-Y Thibon (Université de Marne-la-Vallée). It is a real pleasure to take the opportunity to thank Professors Barbara and Tadeusz Lulek and their splendid team for making such a stimulating meeting possible. This research has been performed under a grant from the Polish KBN (Contract No. 5P03B 057 21).

References

1. R. B. Laughlin, *Phys. Rev. Lett.* **50**, 1395 (1983).
2. G. Tellez and P. J. Forrester, *J. Stat. Phys.* **97**, 489 (1999).
3. G. Dunne, *Int. J. Mod. Phys.* **B7**, 4783 (1993).
4. P. Di Francesco, M. Gaudin, C. Itzykson and F. Lesage, *Int. J. Mod. Phys.* **A9**, 4257 (1994)
5. I. G. Macdonald, *Symmetric Functions and Hall Polynomials* Oxford: Clarendon Press (1979).
6. T. Scharf, J-Y. Thibon and B. G. Wybourne, *J. Phys. A: Math. Gen.* **27**, 4211 (1994).
7. Jean-Gabriel Luque and Jean-Yves Thibon, *Hankel hyperdeterminants and Selberg Integrals*(In preparation August 2002).

Table I. The partitions (λ) associated with vanishing c^λ coefficients for $N = 8$.

(13 11 $985^2$41)	(13 11 $9854^2$2)	(13 11 976541)	(13 10 $9^2$6531)
(13 10 987531)	(12 11 $97^24^2$2)	(12 $10^2$96531)	(12 $10^27^2$532)

Table II. The partitions (λ) associated with vanishing c^λ coefficients for $N = 9$.

(16 13 11 $985^2$41)	(16 13 11 $9854^2$2)	(16 13 11 976541)	(16 13 10 $9^2$6531)
(16 13 10 987531)	(16 12 11 $97^24^2$2)	(16 12 $10^2$96531)	(16 12 $10^27^2$532)
(15 14 11 $985^2$41)	(15 14 11 $9854^2$2)	(15 14 11 976541)	(15 14 10 $9^2$6531)
(15 14 10 987531)	(15 13 11 10 $7^2$63)	(15 13 11 10 $7^2$621)	(15 13 11 10 7^252^2)
(15 13 11 10 $76^2$4)	(15 13 11 9^2652^2)	(15 13 11 $9^264^2$1)	(15 13 11 $9^2$6432)
(15 13 11 98763)	(15 13 11 987621)	(15 13 11 $976^2$32)	(15 13 11 976542)
(15 13 $10^285^2$42)	(15 13 $9^27^2$543)	(15 $12^2$10 $7^2$531)	(15 $12^29^25^2$41)
(15 $12^29^254^2$2)	(15 12 $11^2$8753)	(15 12 $11^2$87521)	(15 12 $11^27^24^2$1)
(15 12 $11^27^2$432)	(15 12 11 10 9753)	(15 12 11 10 97521)	(15 12 11 10 $76^2$41)
(15 12 $10^2$96541)	(15 10 9^35^4)	($14^2$11 10 $7^2$531)	($14^2$11 $9^2$6531)
(14 13 12 10 $7^2$531)	(14 13 12 $9^25^2$41)	(14 13 12 $9^254^2$2)	(14 13 11 $9^26^2$4)
(14 13 $10^2$97531)	(14 $12^2$11 8753)	(14 $12^2$11 87521)	(14 $12^2$11 $7^24^2$1)
(14 $12^2$11 $7^2$432)	(14 $12^29^2$754)	(14 12 $11^286^2$31)	(14 12 11 10 97531)
($13^3$10 7643^2)	($13^2$12 10 963^3)	(13 12 11 $9^27^2$31)	(13 12 $10^295^3$3)
(13 $11^376^2$43)	($12^2987^3$64)	(12 $10^376^3$5)	(12 10 9^3874^2)
(12 10 9^385^3)	($11^47^3$61)	($11^3976^3$5)	($11^387^3$64)
(11 10^3975^3)	(11 $10^396^3$4)		

Table III. The partitions (λ) associated with vanishing c^λ coefficients for $N = 10$.

$(18\ 16\ 13\ 11\ 985^241)$ $(18\ 16\ 13\ 11\ 9854^22)$ $(18\ 16\ 13\ 11\ 976541)$ $(18\ 16\ 13\ 10\ 9^26531)$

$(18\ 16\ 13\ 10\ 987531)$ $(18\ 16\ 12\ 11\ 97^24^22)$ $(18\ 16\ 12\ 10^296531)$ $(18\ 16\ 12\ 10^27^2532)$

$(18\ 15\ 14\ 11\ 985^241)$ $(18\ 15\ 14\ 11\ 9854^22)$ $(18\ 15\ 14\ 11\ 976541)$ $(18\ 15\ 14\ 10\ 9^26531)$

$(18\ 15\ 14\ 10\ 987531)$ $(18\ 15\ 13\ 11\ 10\ 7^263)$ $(18\ 15\ 13\ 11\ 10\ 7^2621)$ $(18\ 15\ 13\ 11\ 10\ 7^252^2)$

$(18\ 15\ 13\ 11\ 10\ 76^24)$ $(18\ 15\ 13\ 11\ 9^2652^2)$ $(18\ 15\ 13\ 11\ 9^264^21)$ $(18\ 15\ 13\ 11\ 9^26432)$

$(18\ 15\ 13\ 11\ 98763)$ $(18\ 15\ 13\ 11\ 987621)$ $(18\ 15\ 13\ 11\ 976^232)$ $(18\ 15\ 13\ 11\ 976542)$

$(18\ 15\ 13\ 10^285^242)$ $(18\ 15\ 13\ 9^27^2543)$ $(18\ 15\ 12^210\ 7^2531)$ $(18\ 15\ 12^29^25^241)$

$(18\ 15\ 12^29^254^22)$ $(18\ 15\ 12\ 11^28753)$ $(18\ 15\ 12\ 11^287521)$ $(18\ 15\ 12\ 11^27^24^21)$

$(18\ 15\ 12\ 11^27^2432)$ $(18\ 15\ 12\ 11\ 10\ 9753)$ $(18\ 15\ 12\ 11\ 10\ 97521)$ $(18\ 15\ 12\ 11\ 10\ 76^241)$

$(18\ 15\ 12\ 10^296541)$ $(18\ 15\ 10\ 9^35^4)$ $(18\ 14^211\ 10\ 7^2531)$ $(18\ 14^211\ 9^26531)$

$(18\ 14\ 13\ 12\ 10\ 7^2531)$ $(18\ 14\ 13\ 12\ 9^25^241)$ $(18\ 14\ 13\ 12\ 9^254^22)$ $(18\ 14\ 13\ 11\ 9^26^24)$

$(18\ 14\ 13\ 10^297531)$ $(18\ 14\ 12^211\ 8753)$ $(18\ 14\ 12^211\ 87521)$ $(18\ 14\ 12^211\ 7^24^21)$

$(18\ 14\ 12^211\ 7^2432)$ $(18\ 14\ 12^29^2754)$ $(18\ 14\ 12\ 11^286^231)$ $(18\ 14\ 12\ 11\ 10\ 97531)$

$(18\ 13^310\ 7643^2)$ $(18\ 13^212\ 10\ 963^3)$ $(18\ 13\ 12\ 11\ 9^27^231)$ $(18\ 13\ 12\ 10^295^33)$

$(18\ 13\ 11^376^243)$ $(18\ 12^2987^364)$ $(18\ 12\ 10^376^35)$ $(18\ 12\ 10\ 9^3874^2)$

$(18\ 12\ 10\ 9^385^3)$ $(18\ 11^47^361)$ $(18\ 11^3976^35)$ $(18\ 11^387^364)$

$(18\ 11\ 10^3975^3)$ $(18\ 11\ 10^396^34)$ $(17^213\ 11\ 985^241)$ $(17^213\ 11\ 9854^22)$

$(17^213\ 11\ 976541)$ $(17^213\ 10\ 9^26531)$ $(17^213\ 10\ 987531)$ $(17^212\ 11\ 97^24^22)$

$(17^212\ 10^296531)$ $(17^212\ 10^27^2532)$ $(17\ 16\ 14\ 11\ 985^241)$ $(17\ 16\ 14\ 11\ 9854^22)$

$(17\ 16\ 14\ 11\ 976541)$ $(17\ 16\ 14\ 10\ 9^26531)$ $(17\ 16\ 14\ 10\ 987531)$ $(17\ 16\ 13\ 11\ 10\ 7^263)$

$(17\ 16\ 13\ 11\ 10\ 7^2621)$ $(17\ 16\ 13\ 11\ 10\ 7^252^2)$ $(17\ 16\ 13\ 11\ 10\ 76^24)$ $(17\ 16\ 13\ 11\ 9^2652^2)$

$(17\ 16\ 13\ 11\ 9^264^21)$ $(17\ 16\ 13\ 11\ 9^26432)$ $(17\ 16\ 13\ 11\ 98763)$ $(17\ 16\ 13\ 11\ 987621)$

$(17\ 16\ 13\ 11\ 976^232)$ $(17\ 16\ 13\ 11\ 976542)$ $(17\ 16\ 13\ 10^285^242)$ $(17\ 16\ 13\ 9^27^2543)$

$(17\ 16\ 12^210\ 7^2531)$ $(17\ 16\ 12^29^25^241)$ $(17\ 16\ 12^29^254^22)$ $(17\ 16\ 12\ 11^28753)$

$(17\ 16\ 12\ 11^287521)$ $(17\ 16\ 12\ 11^27^24^21)$ $(17\ 16\ 12\ 11^27^2432)$ $(17\ 16\ 12\ 11\ 10\ 9753)$

$(17\ 16\ 12\ 11\ 10\ 97521)$ $(17\ 16\ 12\ 11\ 10\ 76^241)$ $(17\ 16\ 12\ 10^296541)$ $(17\ 16\ 10\ 9^35^4)$

$(17\ 15^211\ 985^241)$ $(17\ 15^211\ 9854^22)$ $(17\ 15^211\ 976541)$ $(17\ 15^210\ 9^26531)$

$(17\ 15^210\ 987531)$ $(17\ 15\ 14\ 11^2763^3)$ $(17\ 15\ 14\ 11\ 7^3642)$ $(17\ 15\ 14\ 11\ 7^26^243)$

$(17\ 15\ 14\ 8^37652)$ $(17\ 15\ 14\ 8^27^343)$ $(17\ 15\ 13\ 12\ 11\ 763^3)$ $(17\ 15\ 13\ 12\ 9^2852)$

$(17\ 15\ 13\ 12\ 9^2851^2)$ $(17\ 15\ 13\ 12\ 9^2843)$ $(17\ 15\ 13\ 12\ 9^28421)$ $(17\ 15\ 13\ 12\ 9^283^21)$

$(17\ 15\ 13\ 12\ 9^2832^2)$ $(17\ 15\ 13\ 12\ 9^274^2)$ $(17\ 15\ 13\ 12\ 9^273^22)$ $(17\ 15\ 13\ 12\ 9^25^241)$

$(17\ 15\ 13\ 12\ 9^254^22)$ $(17\ 15\ 13\ 12\ 98^262)$ $(17\ 15\ 13\ 12\ 98^261^2)$ $(17\ 15\ 13\ 12\ 8^2743^2)$

$(17\ 15\ 13\ 12\ 876543)$ $(17\ 15\ 13\ 11^2874^2)$ $(17\ 15\ 13\ 11^2873^22)$ $(17\ 15\ 13\ 11^286^23)$

$(17\ 15\ 13\ 11^286^221)$ $(17\ 15\ 13\ 11^28654)$ $(17\ 15\ 13\ 11\ 10\ 9852)$ $(17\ 15\ 13\ 11\ 10\ 9851^2)$

$(17\ 15\ 13\ 11\ 10\ 9843)$ $(17\ 15\ 13\ 11\ 10\ 98421)$ $(17\ 15\ 13\ 11\ 10\ 983^21)$ $(17\ 15\ 13\ 11\ 10\ 9832^2)$

$(17\ 15\ 13\ 11\ 10\ 75^32)$ $(17\ 15\ 13\ 11\ 10\ 75^243)$ $(17\ 15\ 13\ 11\ 98^254)$ $(17\ 15\ 13\ 11\ 98764)$

$(17\ 15\ 13\ 11\ 8^362^2)$ $(17\ 15\ 13\ 11\ 876^243)$ $(17\ 15\ 13\ 10^295^31)$ $(17\ 15\ 13\ 10^2954^23)$

Table III. The partitions (λ) associated with vanishing c^λ coefficients for $N = 10$.(Contd)

$(17\,15\,13\,10\,9^2 5^3 2)$ $\quad$ $(17\,15\,13\,10\,9^2 5^2 43)$ $\quad$ $(17\,15\,13\,10\,97^2 642)$ $\quad$ $(17\,15\,13\,9^3 7632)$

$(17\,15\,13\,9^2 875^2 2)$ $\quad$ $(17\,15\,12^3 7652^2)$ $\quad$ $(17\,15\,12^3 764^2 1)$ $\quad$ $(17\,15\,12^3 76432)$

$(17\,15\,12^3 75^2 32)$ $\quad$ $(17\,15\,12^2 10\,7^2 64)$ $\quad$ $(17\,15\,12\,11^2 75^3 2)$ $\quad$ $(17\,15\,12\,11^2 75^2 43)$

$(17\,15\,12\,11\,97^2 543)$ $\quad$ $(17\,15\,12\,10^2 8^2 631)$ $\quad$ $(17\,15\,11^2 9^2 765)$ $\quad$ $(17\,14^2 13\,87^2 43^2)$

$(17\,14^2 12\,11\,6^3 31)$ $\quad$ $(17\,14^2 12\,11\,6^2 541)$ $\quad$ $(17\,14^2 12\,11\,65^2 3^2)$ $\quad$ $(17\,14^2 12\,9^2 753)$

$(17\,14^2 12\,9^2 7521)$ $\quad$ $(17\,14^2 12\,9^2 5^2 41)$ $\quad$ $(17\,14^2 12\,9^2 54^2 2)$ $\quad$ $(17\,14^2\,11^2 7^2 63)$

$(17\,14^2\,11^2 7^2 621)$ $\quad$ $(17\,14^2\,11^2 76^2 4)$ $\quad$ $(17\,14^2 10\,9^2 7541)$ $\quad$ $(17\,14\,13^2 10\,9752)$

$(17\,14\,13^2 10\,9751^2)$ $\quad$ $(17\,14\,13^2 10\,9743)$ $\quad$ $(17\,14\,13^2 10\,97421)$ $\quad$ $(17\,14\,13^2 10\,973^2 1)$

$(17\,14\,13^2 10\,9732^2)$ $\quad$ $(17\,14\,13^2 9^2 623)$ $\quad$ $(17\,14\,13^2 9^2 6^2 21)$ $\quad$ $(17\,14\,13^2 9^2 654)$

$(17\,14\,13^2 9^2 6531)$ $\quad$ $(17\,14\,13^2 9^2 652^2)$ $\quad$ $(17\,14\,13^2 9^2 64^2 1)$ $\quad$ $(17\,14\,13^2 9^2 6432)$

$(17\,14\,13^2 9^2 63^3)$ $\quad$ $(17\,14\,13^2 8^2 7541)$ $\quad$ $(17\,14\,13\,12^2 7652^2)$ $\quad$ $(17\,14\,13\,12^2 764^2 1)$

$(17\,14\,13\,12^2 76432)$ $\quad$ $(17\,14\,13\,12^2 75^2 32)$ $\quad$ $(17\,14\,13\,12\,11\,9752)$ $\quad$ $(17\,14\,13\,12\,11\,9751^2)$

$(17\,14\,13\,12\,11\,9743)$ $\quad$ $(17\,14\,13\,12\,11\,97421)$ $\quad$ $(17\,14\,13\,12\,11\,973^2 1)$ $\quad$ $(17\,14\,13\,12\,11\,9732^2)$

$(17\,14\,13\,12\,11\,76^2 2^2)$ $\quad$ $(17\,14\,13\,12\,11\,75^2 42)$ $\quad$ $(17\,14\,13\,12\,98^2 63)$ $\quad$ $(17\,14\,13\,12\,98^2 621)$

$(17\,14\,13\,12\,8^3 62^2)$ $\quad$ $(17\,14\,13\,12\,8^2 6^2 51)$ $\quad$ $(17\,14\,13\,11\,10^2 5^2 41)$ $\quad$ $(17\,14\,13\,11\,10^2 54^2 2)$

$(17\,14\,13\,11\,9^2 84^2 1)$ $\quad$ $(17\,14\,13\,11\,9^2 8432)$ $\quad$ $(17\,14\,13\,9^3 84^2 3)$ $\quad$ $(17\,14\,12^2 11\,8763)$

$(17\,14\,12^2 11\,87621)$ $\quad$ $(17\,14\,12^2 11\,765^2 1)$ $\quad$ $(17\,14\,12\,10^3 5^3 2)$ $\quad$ $(17\,14\,12\,10^3 5^2 43)$

$(17\,14\,12\,10\,9876^2 1)$ $\quad$ $(17\,14\,11\,10^3 843^2)$ $\quad$ $(17\,14\,11\,10^3 6^2 42)$ $\quad$ $(17\,14\,11\,10^3 5^3 3)$

$(17\,13^3 98^2 531)$ $\quad$ $(17\,13^2\,12^2 75^2 3^2)$ $\quad$ $(17\,13^2 12\,11\,76^2 41)$ $\quad$ $(17\,13\,12^2 10^2 6541)$

$(17\,13\,12\,11\,97^3 43)$ $\quad$ $(17\,12^2 11\,10\,98641)$ $\quad$ $(17\,12\,11^3 7^4)$ $\quad$ $(17\,12\,10^5 4^2 3)$

$(17\,12\,10^4 94^3)$ $\quad$ $(17\,12\,10^4 65^3)$ $\quad$ $(17\,11^2 10\,9^3 83^2)$ $\quad$ $(16^2 15\,11\,985^2 41)$

$(16^2 15\,11\,9854^2 2)$ $\quad$ $(16^2 15\,11\,976541)$ $\quad$ $(16^2 15\,10\,9^2 6531)$ $\quad$ $(16^2 15\,10\,987531)$

$(16^2 13\,12\,11\,6^3 31)$ $\quad$ $(16^2 13\,12\,11\,6^2 541)$ $\quad$ $(16^2 13\,12\,11\,65^2 3^2)$ $\quad$ $(16^2 13\,12\,9^2 753)$

$(16^2 13\,12\,9^2 7521)$ $\quad$ $(16^2 13\,12\,9^2 5^2 41)$ $\quad$ $(16^2 13\,12\,9^2 54^2 2)$ $\quad$ $(16^2 13\,11^2 8753)$

$(16^2 13\,11^2 87521)$ $\quad$ $(16^2\,12^2 11\,76541)$ $\quad$ $(16^2 12\,10^3 7531)$ $\quad$ $(16^2 12\,10^3 6541)$

$(16^2 11\,10^3 5^3 2)$ $\quad$ $(16^2 11\,10^3 5^2 43)$ $\quad$ $(16\,15^2 11\,10\,7^2 531)$ $\quad$ $(16\,15^2 11\,9^2 6531)$

$(16\,15\,14\,13\,87^2 43^2)$ $\quad$ $(16\,15\,14\,12\,11\,6^3 31)$ $\quad$ $(16\,15\,14\,12\,11\,6^2 541)$ $\quad$ $(16\,15\,14\,12\,11\,65^2 3^2)$

$(16\,15\,14\,12\,9^2 753)$ $\quad$ $(16\,15\,14\,12\,9^2 7521)$ $\quad$ $(16\,15\,14\,12\,9^2 5^2 41)$ $\quad$ $(16\,15\,14\,12\,9^2 54^2 2)$

$(16\,15\,14\,11^2 7^2 63)$ $\quad$ $(16\,15\,14\,11^2 7^2 621)$ $\quad$ $(16\,15\,14\,11^2 76^2 4)$ $\quad$ $(16\,15\,14\,10\,9^2 7541)$

$(16\,15\,13\,11\,6^3 31)$ $\quad$ $(16\,15\,13^2 11\,6^2 541)$ $\quad$ $(16\,15\,13^2 11\,65^2 3^2)$ $\quad$ $(16\,15\,13\,11^2 8^2 62)$

$(16\,15\,13\,11^2 8^2 61^2)$ $\quad$ $(16\,15\,12^2 11\,9753)$ $\quad$ $(16\,15\,12^2 11\,97521)$ $\quad$ $(16\,15\,12\,11\,9^3 531)$

$(16\,15\,12\,10^2 8763^2)$ $\quad$ $(16\,14^2 13\,10\,9752)$ $\quad$ $(16\,14^2 13\,10\,9751^2)$ $\quad$ $(16\,14^2 13\,10\,9743)$

$(16\,14^2 13\,10\,97421)$ $\quad$ $(16\,14^2 13\,10\,973^2 1)$ $\quad$ $(16\,14^2 13\,10\,9732^2)$ $\quad$ $(16\,14^2 13\,9^2 623)$

$(16\,14^2 13\,9^2 6^2 21)$ $\quad$ $(16\,14^2 13\,9^2 654)$ $\quad$ $(16\,14^2 13\,9^2 6531)$ $\quad$ $(16\,14^2 13\,9^2 652^2)$

Table III. The partitions (λ) associated with vanishing c^λ coefficients for $N = 10$.(Contd)

$(16\ 14^2 13\ 9^2 64^2 1)$ $(16\ 14^2 13\ 9^2 6432)$ $(16\ 14^2 13\ 9^2 63^3)$ $(16\ 14^2 13\ 8^2 7541)$

$(16\ 14^2\ 11^2 9762)$ $(16\ 14^2\ 11^2 9761^2)$ $(16\ 14^2 11\ 987542)$ $(16\ 14\ 13^2 11\ 76541)$

$(16\ 14\ 13^2 10\ 8^2 53)$ $(16\ 14\ 13^2 10\ 8^2 521)$ $(16\ 14\ 13\ 12\ 11\ 9753)$ $(16\ 14\ 13\ 12\ 11\ 97521)$

$(16\ 14\ 13\ 11\ 10\ 974^2 2)$ $(16\ 14\ 12^2 8^3 741)$ $(16\ 14\ 12\ 11^3 7431)$ $(16\ 14\ 12\ 11^2 98531)$

$(16\ 14\ 12\ 11\ 987652)$ $(16\ 13^3 12\ 75^2 3^2)$ $(16\ 13^3 11\ 87531)$ $(16\ 13^3 11\ 7^2 631)$

$(16\ 13^3 9^2 8531)$ $(16\ 13^3 8^3 72^2)$ $(16\ 13^3 8^3 641)$ $(16\ 13^2 11\ 10\ 9^2 531)$

$(16\ 13\ 12^2 11\ 76^2 43)$ $(16\ 13\ 12\ 11\ 10^3 431)$ $(16\ 13\ 12\ 11\ 10\ 97642)$ $(16\ 11^2 10\ 987^3 4)$

$(16\ 11^2 987^5)$ $(16\ 11\ 10^5 54^2)$ $(15^3 12\ 11\ 7^2 431)$ $(15^3 12\ 11\ 76531)$

$(15^3 12\ 9^2 5^2 41)$ $(15^3 12\ 9^2 54^2 2)$ $(15^3 12\ 9865^2)$ $(15^2 14\ 12\ 11\ 85^3)$

$(15^2 14\ 12\ 11\ 84^3 3)$ $(15^2 14\ 11^2 10\ 54^2 1)$ $(15^2 14\ 11^2 10\ 5432)$ $(15^2 14\ 11\ 10^2 6531)$

$(15^2 14\ 10\ 8^3 741)$ $(15^2\ 13^2 12\ 7652^2)$ $(15^2\ 13^2 12\ 764^2 1)$ $(15^2\ 13^2 12\ 76432)$

$(15^2\ 13^2 12\ 75^2 32)$ $(15^2\ 13^2 11\ 6^2 5^2 1)$ $(15^2\ 13^2 11\ 65^3 2)$ $(15^2\ 13^2 11\ 65^2 43)$

$(15^2 12\ 11\ 10\ 8^2 632)$ $(15^2 10\ 9^3 87^2 1)$ $(15\ 14^3 10\ 7643^2)$ $(15\ 14^2 13\ 98^2 531)$

$(15\ 14^2 10\ 9^3 541)$ $(15\ 14^2 8^5 61)$ $(15\ 14\ 13^2 12\ 75^2 3^2)$ $(15\ 14\ 13^2 11\ 87531)$

$(15\ 14\ 13^2 11\ 7^2 631)$ $(15\ 14\ 13^2 9^2 8531)$ $(15\ 14\ 13^2 8^3 72^2)$ $(15\ 14\ 13^2 8^3 641)$

$(15\ 14\ 13\ 12\ 11\ 10\ 6531)$ $(15\ 14\ 13\ 11^2 9^2 53)$ $(15\ 14\ 13\ 11^2 9^2 521)$ $(15\ 14\ 13\ 11^2 97631)$

$(15\ 14\ 12^2\ 11^2 7431)$ $(15\ 14\ 12^2 11\ 10\ 7531)$ $(15\ 14\ 12^2 11\ 7^3 5)$ $(15\ 14\ 12^2 11\ 76^2 52)$

$(15\ 14\ 12^2 11\ 765^2 3)$ $(15\ 14\ 11^3\ 10^2 431)$ $(15\ 14\ 11^3 97651)$ $(15\ 13^3 98^2 65)$

$(15\ 13^3 8^3 741)$ $(15\ 13^2 12\ 11\ 76^2 43)$ $(15\ 12^2 10\ 8^2 7^2 65)$ $(15\ 12\ 11\ 9^2 8^2 7^2 4)$

$(15\ 11^2 10\ 9^2 7^3 4)$ $(15\ 11\ 10^3 98764)$ $(14^3 98^4 61)$ $(14^2 13\ 8^2 7^4 5)$

$(14^2 13\ 8^2 7^4 5)$ $(14^2 11\ 10\ 9^3 86)$ $(14\ 12^3 98^3 7)$ $(14\ 12\ 11^3 10\ 96^2)$

$(14\ 12\ 11^3 10\ 7^3)$ $(14\ 12\ 11\ 10\ 98^3 73)$ $(14\ 11^3 10\ 987^2 2)$ $(14\ 11^3 9^2 87^2 3)$

$(14\ 11^2\ 10^2 9^2 763)$ $(14\ 11^2 10\ 9^2 7^3 5)$ $(14\ 11\ 10^2 8^3 7^3)$ $(13^4 9^3 83)$

$(13^4 9^3 821)$ $(13^3 12\ 8^4 61)$ $(13^3 11\ 98^3 7)$ $(13^3 10\ 9^3 86)$

$(13\ 12^3 11\ 97^3)$ $(13\ 12^3 11\ 8^3 6)$ $(13\ 12\ 11^2\ 10^2 86^2 3)$ $(13\ 12\ 11\ 8^5 7^2)$

$(13\ 11^4\ 10^2 54^2)$ $(13\ 11^3 9^2 87^2 4)$ $(11^5 10\ 97^2 2)$ $(11^3\ 10^3 8^2 74)$

$(11^2\ 10^5 765)$

Table IV. The q–polynomials for $N = 2, 3, 4$. The square brackets encase the signed coefficients, $c^\lambda(1)$, for $q = 1$, the next column the q–polynomial and the last column the associated Schur functions, $\{\lambda\}$.

N = 2

[1]	q	$\{2\}$
[−3]	$-(q^2 + q + 1)$	$\{1^2\}$

N = 3

[1]	q^3	$\{42\}$
[−3]	$-q^2(q^2 + q + 1)$	$\{41^2\} + \{3^2\}$
[6]	$+q(q^2 + q + 1)(q^2 + 1)$	$\{321\}$
[−15]	$-(q^2 + q + 1)(q^4 + q^3 + q^2 + q + 1)$	$\{2^3\}$

N = 4

[1]	q^6	$\{642\}$
[−3]	$-q^5(q^2 + q + 1)$	$\{641^2\} + \{63^2\} + \{5^22\}$
[6]	$+q^4(q^2 + 1)(q^2 + q + 1)$	$\{6321\} + \{543\}$
[9]	$+q^4(q^2 + q + 1)^2$	$\{5^21^2\}$
[−15]	$-q^3(q^2 + q + 1)(q^4 + q^3 + q^2 + q + 1)$	$\{62^3\} + \{4^3\}$
[−12]	$-q^3(q^2 + q + 1)(q^2 + 1)^2$	$\{5421\}$
[−9]	$-q^3(q^2 - q + 1)(q^2 + q + 1)^2$	$\{53^21\}$
[−6]	$-q^3(q^2 + q + 1)(q^4 + 1)$	$\{4^22^2\}$
[27]	$+q^2(q^2 - q + 1)(q^2 + q + 1)^3$	$\{532^2\} + \{4^231\}$
[−45]	$-q(q^2 - q + 1)(q^4 + q^3 + q^2 + q + 1)(q^2 + q + 1)^2$	$\{43^22\}$
[105]	$+(q^2 + q + 1)(q^4 + q^3 + q^2 + q + 1)$ $\times(q^6 + q^5 + q^4 + q^3 + q^2 + q + 1)$	$\{3^4\}$

Table IV. The q–polynomials for $N = 2, 3, 4$. The square brackets encase the signed coefficients, $c^\lambda(1)$, for $q = 1$, the next column the q–polynomial and the last column the associated Schur functions, $\{\lambda\}$. (Contd)

N = 5

[1]	q^{10}	$\{8642\}$
[−3]	$-q^9(q^2 + q + 1)$	$\{8641^2\} + \{863^2\} + \{85^22\}+$ $+\{7^242\}$
[6]	$+q^8(q^2 + 1)(q^2 + q + 1)$	$\{86321\} + \{8543\} + \{7652\}$
[9]	$+q^8(q^2 + q + 1)^2$	$\{85^21^2\} + \{7^241^2\} + \{7^23^2\}$
[−12]	$-q^7(q^2 + q + 1)(q^2 + 1)^2$	$\{85421\} + \{7643\}$
[−9]	$-q^7(q^2 - q + 1)(q^2 + q + 1)^2$	$\{853^21\} + \{75^23\}$
[−6]	$-q^7(q^2 + q + 1)(q^4 + 1)$	$\{84^22^2\} + \{6^24^2\}$
[−15]	$-q^7(q^2 + q + 1)(q^4 + q^3 + q^2 + q + 1)$	$\{862^3\} + \{84^3\} + \{6^32\}$
[−18]	$-q^7(q^2 + 1)(q^2 + q + 1)^2$	$\{7^2321\} + \{7651^2\}$
[27]	$+q^6(q^2 - q + 1)(q^2 + q + 1)^3$	$\{8532^2\} + \{84^231\} + \{754^2\}+$ $+\{6^253\}$
[24]	$+q^6(q^2 + q + 1)(q^2 + 1)^3$	$\{76421\}$
[18]	$+q^6(q^2 + 1)(q^2 - q + 1)(q^2 + q + 1)^2$	$\{763^21\} + \{75^221\}$
[45]	$+q^6(q^4 + q^3 + q^2 + q + 1)(q^2 + q + 1)^2$	$\{7^22^3\} + \{6^31^2\}$
[−45]	$-q^5(q^2 - q + 1)(q^4 + q^3 + q^2 + q + 1)$ $\times(q^2 + q + 1)^2$	$\{843^22\} + \{65^24\}$
[−54]	$-q^5(q^2 + 1)(q^2 - q + 1)(q^2 + q + 1)^3$	$\{7632^2\} + \{6^2521\}$
[−36]	$-q^5(q^2 - q + 1)(q^2 + q + 1)^2(q^2 + 1)^2$	$\{75431\} + \{74^31\}$
[−27]	$-q^5(q^2 - q + 1)^2(q^2 + q + 1)^3$	$\{7542^2\} + \{6^2431\}$
[−18]	$-q^5(q^2 - q + 1)(q^4 + 1)(q^2 + q + 1)^2$	$\{6^23^22\} + \{65^22^2\}$
[105]	$+q^4(q^2 + q + 1)(q^4 + q^3 + q^2 + q + 1)$ $\times(q^6 + q^5 + q^4 + q^3 + q^2 + q + 1)$	$\{83^4\} + \{5^4\}$
[81]	$+q^4(q^2 - q + 1)^2(q^2 + q + 1)^4$	$(\{753^22\} + \{65^231\}$
[72]	$+q^4(q^4 + 1)(q^2 + 1)^2(q^2 + q + 1)^2$	$(\{74^232\} + \{654^21\}$
[111]	$+q^4(q^2 + q + 1)(q^{10} + 2q^9 + 4q^8 + 3q^7$ $+6q^6 + 5q^5 + 6q^4 + 3q^3 + 4q^2 + 2q + 1)$	$\{6^242^2\}$
[45]	$+q^4(q^4 - q^3 + q^2 - q + 1)(q^4 + q^3 + q^2 + q + 1)$ $\times(q^2 + q + 1)^2$	$\{653^3\} + \{5^332\}$
[−180]	$-q^3(q^2 + 1)(q^4 + q^3 + q^2 + q + 1)(q^4 + 1)$ $\times(q^2 + q + 1)^2$	$\{743^3\} + \{5^341\}$
[−144]	$-q^3(q^4 + 1)(q^2 + q + 1)^2(q^2 + 1)^3$	$\{65432\}$
[−90]	$-q^3(q^4 + q^3 + q^2 + q + 1)(q^2 - q + 1)(q^4 + 1)$ $(q^2 + q + 1)^2$	$\{64^32\}$
[−75]	$-q^3(q^2 + q + 1)(q^4 - q^3 + q^2 - q + 1)$ $\times(q^4 + q^3 + q^2 + q + 1)^2$	$\{5^243^2\}$
[270]	$+q^2(q^4 + q^3 + q^2 + q + 1)(q^2 - q + 1)(q^4 + 1)$ $(q^2 + q + 1)^3$	$\{64^23^2\} + \{5^24^22\}$
[−420]	$-q(q^2 + q + 1)(q^2 + 1)(q^4 + q^3 + q^2 + q + 1)$ $\times(q^4 + 1)(q^6 + q^5 + q^4 + q^3 + q^2 + q + 1)$	$\{54^33\}$
[945]	$+(q^4 + q^3 + q^2 + q + 1)(q^6 + q^3 + 1)$ $\times(q^2 + q + 1)^2(q^6 + q^5 + q^4 + q^3 + q^2 + q + 1)$	$\{4^5\}$

Table V. The q–polynomials for the eight vanishing coefficients, $c^\lambda(1)$, for $N = 8$.

$$-q^{17}(q^2 - q + 1)^2(q^2 + 1)^2(q - 1)^4(q^2 + q + 1)^5$$
$$+q^{16}(q^2 + 1)(q^2 - q + 1)^3(q - 1)^4(q^2 + q + 1)^6$$
$$+q^{16}(q^2 - q + 1)^2(q^2 + 1)^3(q - 1)^4(q^2 + q + 1)^5$$
$$+q^{14}(q^{10} + q^9 + 3q^8 + 4q^6 + q^5 + 4q^4 + 3q^2 + q + 1)$$
$$\times(q^2 - q + 1)^2(q - 1)^4(q^2 + q + 1)^5$$

$\{13\ 11985^241\}, \{13\ 10\ 9^26531\}$
$\{13\ 11\ 9854^22\}, \{13\ 10\ 987531\}$
$\{13\ 11\ 976541\}, \{12\ 10^2\ 96531\}$
$\{12\ 11\ 97^24^22\}, \{12\ 10^2\ 7^2532\}$

FASCINATING PROPERTIES OF CARBON NANOTUBES

M.CZECHOWSKA, M. LISOWSKI, M.MARGANSKA, M.SZOPA, E.ZIPPER

University of Silesia, Institute of Physics
ul. Uniwersytecka 4, 40-007 Katowice, Poland

A brief review of the remarkable properties of carbon nanotubes is presented. The manifestation of mesoscopic coherent transport are persistent currents flowing in carbon nanotubes in the presence of static magnetic field. Persistent currents as a function of hole doping are investigated. The possibility of flux expulsion and flux trapping in a set of concentric mesoscopic cylinders nested inside each other is also discussed.

1. Introduction

Carbon nanotubes (CN) are currently the focus of intense interest worldwide. This attention to CN is not suprising in light of their possibilities to exhibit unique physical properties that cover broad areas of science and technology ranging from super strong composites to nanoelectronics. We would like to emphasize the very rich physics encountered in the study of CN.

The structure of the paper is as follows.

First, we review the remarkable properties of CN and their possible applications.

Second, we discuss persistent currents which flow in CN in the presence of static magnetic field parallel to the tube axis.

Lastly, we discuss the possibility of flux expulsion and flux trapping in a set of concentric mesoscopic cylinders. This considerations could have potential applications in Multi-Walled Carbon Nanotubes if all the sheets forming the nanotube would exhibit the same (metallic or semiconducting) behaviour.

2. Electric, magnetic and mechanical properties of Carbon Nanotubes and their applications

The remarkable properties of carbon nanotubes[1] (CN) stem from the unusual electronic structure of graphite - a material which has a structure of a set of parallel sheets. In graphite adjacent sheets are only weakly bounded by van der Waals forces, that's why layers of it can be easily peeled apart. Carbon nanotube can be considered as a single sheet of graphite that has been rolled up into a tube.

In CN each carbon atom has four valence electron which are hybrydized in sp^3 form. In a single sheet each carbon atom is strongly bounded to three others atoms (σ bonding), the fourth valence electron associated with each carbon atom is delocalized (π bonding) and is free to move. This electron is responsible for the electrical conductivity.

The electronic properties of CN depend on the direction in which the sheet of graphite was rolled up and its size. They change from metallic with high electrical conductivity to semiconducting with relatively large band gaps (fig.1). If the graphene (one sheet of graphite) is rolled up around the y axis, the nanotube is a metal, but if it is rolled up around the x axis, the nanotube is a semiconductor.

A nanotube is specified by a chiral vector

$$\mathbf{C_h} = n\mathbf{a_1} + m\mathbf{a_2} = (n, m) \tag{1}$$

where $\mathbf{a_1}$, $\mathbf{a_2}$ are the basic vectors of the graphite lattice, the pair of integers (n, m) are crucially important for the electronic properties of CN; tubes with $n - m = 3i$ (i is integer) are metallic, all other are semiconducting (fig.2). If the chiral angle θ is $0°$ CN are called zigzag nanotubes, if $\theta = 30°$ they are known as armchair nanotubes. When the chiral angle lies between $0°$ and $30°$ we get a chiral nanotube.

The circumference of CN is expressed in terms of the chiral vector C_h which connects two crystalographically equivalent sites. Tubes with $m = 0$ are zigzag CN and tubes with $n = m$ are armchair CN. CN with other (n, m) indices are chiral (fig.3). It is amazing that similarly shaped molecules consisting of only one element (carbon) may have very different electronic behaviour. Slight differences in parameters (tube diameter, chirality) cause a shift from metallic to semiconducting state.

CN can be grown by the catalytic decomposition of a reaction gas that contains carbon, at temperatures of the order of $T \sim 500° - 700°C$, with Fe, Co, Ni often used as the catalyst. CN can be grown in various shapes such as single-wall (SWNT) with diameters $4 - 18$Å, multi-wall (MWNT) tubes consisting of several concentric tubes with external diameters $100 - 500$Å, tori or springs.

CN produced nowadays are almost perfect strucures; their mean free path is of the order of $L_e > 10\mu m$ or more, which suggests ballistic transport. The measured conductance exhibits two types of behaviour corresponding

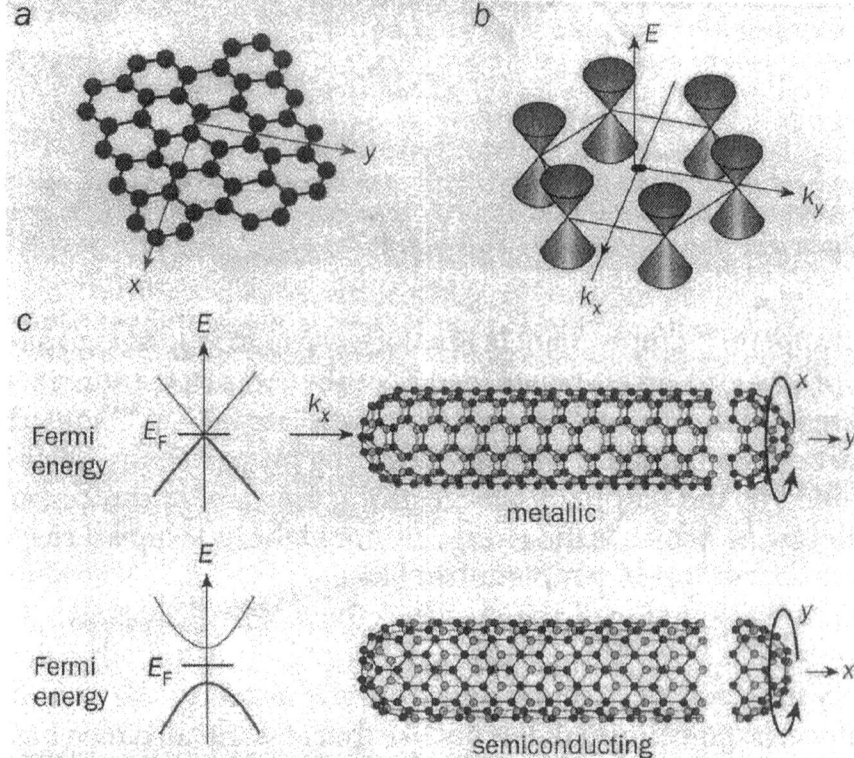

(a) The lattice structure of graphene – the two-dimensional material that is rolled up to form a nanotube. The lattice is made up of a honeycomb of carbon atoms. (b) The energy of the conducting states in graphene as a function of the wavevector, k, of the electrons. The material does not conduct, except along certain, special directions where "cones" of states exist. (c) If the graphene is rolled up around the y axis, the nanotube is a metal (upper figure), but if it is rolled up around the x axis, the nanotube is a semiconductor (lower figure). The band structure of the nanotube is then given by one-dimensional slices through the two-dimensional band structure shown in (b). The permitted wavevectors are quantized along the axis of the tube.

Figure 1. Curling up with a nanotube. From P.L.McEuen, *Phys. World*, June, 31 (2000).

to metal or semiconductor.

C.T.White and T.N.Todorov calculated[4] that conduction electrons in armchair SWNT experience an effective disorder averaged over the tube's circumference, leading to electron mean free paths that increase with nanotube

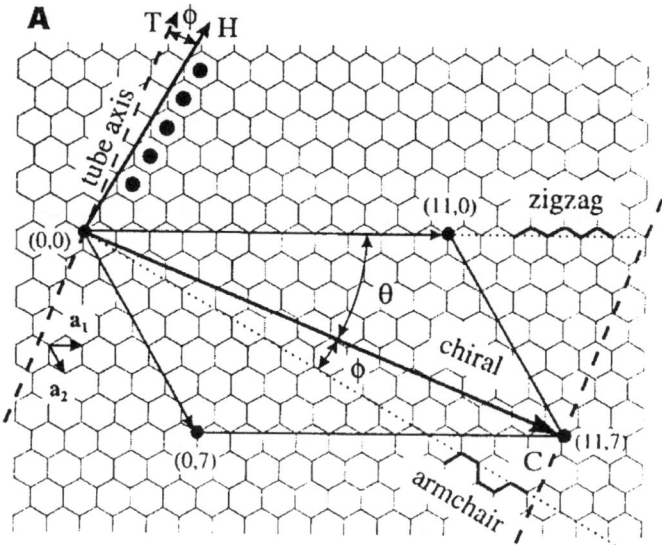

Figure 2. Relation between the hexagonal carbon lattice and the chirality of carbon nanotubes. From J.W.G.Wildöer, L.C.Venema, A.G.Rinzler, R.E.Smalley, C.Dekker, *Nature* **391**, 59 (1998).

diameter. This results in exeptional ballistic transport properties.

A ballistic conductor has some resistance and this resistance is independent of its length, which means that Ohm's law does not apply. In 1998 Walt de Heer and co-workers[5] found that all MWNT's have nearly the same conductance, $G_0 = \frac{2e^2}{h}$ (the inverse of resistance) and that dependence of the resistance on the length was very weak. It is a proof that MWNT's are ballistic conductors.

Semiconducting nanotubes can work as transistors. A negative bias applied to the gate electrode induces holes on the tube and makes it conduct. Positive biases deplete the holes and decrease the conductance. The resistance of the off state can be more than a milion times greater than the on state. This behaviour is analogous to that of a p-type metal-oxide-silicon field effect transistor (MOSFET), exept that nanotube replaces silicon as the material that hosts the charge carriers.

What is even more exciting, first reports on the superconductivity in CN have recently been published[6,7,8].

In MWNT the Bohm-Aharonov effect has been observed in a magnetic field parallel to the tube axis - it is a manifestation of coherent transport. In the standard Aharonov-Bohm effect the magnetic flux through the solenoid

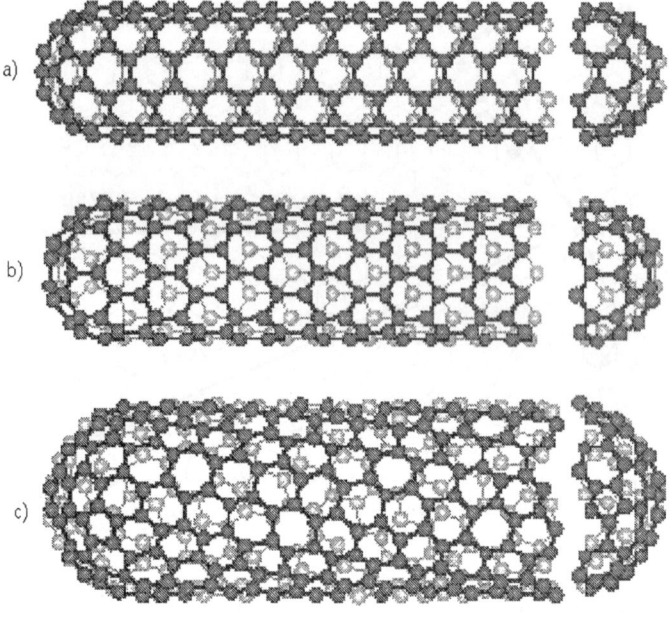

Figure 3. Different types of carbon nanotubes: a) armchair, b) zigzag, c) chiral.

changes the relative phase of the electron waves in paths 1 and 2, leading to a shift of the interference pattern on the screen. In carbon nanotube[9], the two paths are clockwise and anticlockwise around the nanotube, and the shift in the interference pattern manifests itself as a change in the electrical resistance along the nanotube as function of the magnetic field (fig.4). In the remaining part of this chapter we describe some of the remarkable material properties of CN and interesting applications which follow from them.

Carbon nanotubes due to their low intrinsic resistance can transport high current densities up to 10^9 A/cm^2 without any damage. They can find applications in non-silicon electronics as interconnects between electronic elements, logic and storage devices, diodes, rectifiers or transistors.

Conductivity of CN is sensitive to the presence of various gases, such as ammonia, oxygen, etc. so they can be used also as novel chemical sensors. High thermal conductivity (ca. 6000 W/Km at 300 K) can help to remove heat from electronic elements.

High chemical inertness and hollow interior are properties that make the

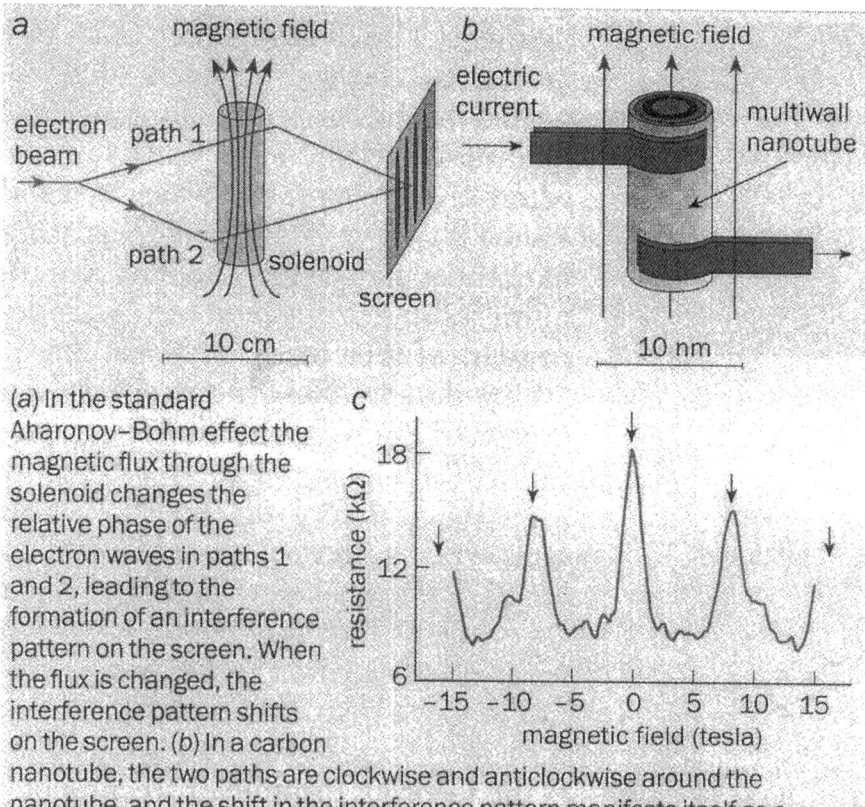

(a) In the standard Aharonov–Bohm effect the magnetic flux through the solenoid changes the relative phase of the electron waves in paths 1 and 2, leading to the formation of an interference pattern on the screen. When the flux is changed, the interference pattern shifts on the screen. (b) In a carbon nanotube, the two paths are clockwise and anticlockwise around the nanotube, and the shift in the interference pattern manifests itself as a change in the electrical resistance along the nanotube as a function of magnetic field (c). The magnetic field at the peaks can be related to the quantum of magnetic flux, $h/2e$, and the cross-section of the nanotube

Figure 4. The Bohm-Aharonov effect in multiwall carbon nanotubes. From A.Bachtold, C.Strunk, J.P.Salvetat, J.M.Bohard, L.Forro, T.Nussbaumer, C.Schonenberger, *Nature* **397**, 673 (1999).

CN useful as containers of gases and chemicals and can be used in fuel cells and batteries.

Small diameter of CN is favorable for field emissions and can have an application in field emitters such as lamps and flat-panel displays.

The very strong bonds between C atoms in the graphite planes gives CN exeptional strength. This bond is also responsible for the hardness of diamond. The very high Young's modulus ca. $10^{12} \ N/m^2$, which is 5 times the value for steel, makes that deformations (bending, squeezing) are elastic, i.e. they disappear when the load is removed. These properties will probably find an application in new composite materials with high strength and elasticity.

Nanotweezers made of CN are used to manipulate objects 500 nm across. CN tips for scanning microscope increase the resolution 10 times and allow to investigate microobjects like proteins and DNA spirals.

The remarkable properties of CN may allow them to play a crucial role in the drive towards miniaturization at the nanometre scale.

3. Persistent currents in SWCN

In mesoscopic objects with periodic boundary conditions (like rings and cylinders) the external magnetic field induces, via Aharonov-Bohm effect, a peculiar sort of current. It was first investigated in a small metallic ring by Cheung et al.[10] and can be understood in the following way.

In a circular system there exists a symmetry between states with momentum $\mathbf{k}$ and $-\mathbf{k}$. The currents they carry cancel out. This symmetry, however, is destroyed by the introduction of the magnetic field perpendicular to the direction of the periodic boundary conditions. Since the magnetic flux through the ring changes the phase of the wavefunction, it modifies the boundary condition, effectively adding a term proportional to ϕ/ϕ_0 to the momentum of a state ($\phi_0 = hc/e$ is the flux quantum). The current carried by the electron in a state $-\mathbf{k}$ decreases by an amount determined by the value of ϕ/ϕ_0, while that carried by the $\mathbf{k}$ increases by the same amount. They do not cancel out any more. When we sum the contributions from all the states below the Fermi level, we obtain an overall, non-vanishing current, persistent at temperatures much lower than quantum size energy gaps.

The formula for the current carried by an individual state $\mathbf{k}$ is:

$$I_\mathbf{k}(\phi) = -c\frac{\partial E_\mathbf{k}}{\partial \phi}, \tag{2}$$

with total current given by the formula

$$I(\phi) = \sum_{\mathbf{k}\ occupied} I_\mathbf{k}(\phi). \tag{3}$$

In nanotubes there are two ways in which these currents can be induced. We can, for instance, curl the nanotube into a torus, analogous to the metallic ring considered above. However, the persistent currents induced in such structures are very small[11] and we shall not dwell upon them here. We can also align a straight nanotube with the magnetic field, thereby obtaining an analogue of a metallic cylinder[12]. Here the persistent currents are considerably greater, because of the coherence of states over the whole length of the tube.

The dispersion relation is that of graphene:

$$E_{\mathbf{k}} = -\gamma\sqrt{1 + 4\cos^2(\frac{\sqrt{3}}{2}k_x) + 4\cos(\frac{\sqrt{3}}{2}k_x)\cos(\frac{3}{2}k_y)}, \qquad (4)$$

where $\gamma \simeq 3.3\text{eV}$. The boundary conditions in this case have the following form:

$$\mathbf{k} \cdot \mathbf{L}_t = 2\pi(l_t + \phi/\phi_0), \quad l_t = 0, \pm 1, \pm 2... \qquad (5)$$

$$\mathbf{k} \cdot \mathbf{L}_l = \pi l_l, \quad l_l = 0, 1, 2... \qquad (6)$$

where $\mathbf{L}_t$ is the circumference of the nanotube, determined by the n, m parameters. The length of the nanotube, $\mathbf{L}_l$ is defined in the same way by the parameters p, q. The first is the usual cyclic boundary condition, the second can be either a cyclic one[14] if we want to maintain the original Brillouin zone of the graphene, or an open one if we identify the states with $-k_l$ and k_l, thereby halving the Brillouin zone (see Fig. 5). The currents in both cases are practically the same, within the numerical error. Here we have chosen the open boundary condition[15]. We see that the dispersion relation is modified by the flux (eq.5) and therefore leads to finite persistent current. The amplitude and shape of the persistent currents depend on

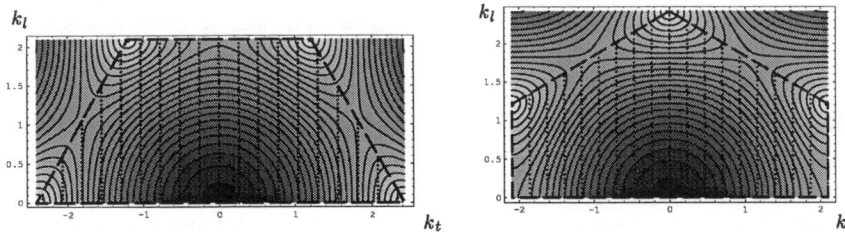

Figure 5. The background is the density plot of dispersion relation for graphene. The points represent allowed momentum states in a zigzag $(14, 0) \times (-60, 120)$ nanotube (left), momentum states in an armchair $(9, 9) \times (-90, 90)$ of similar size (right). For better display the Brillouin zone of the armchair nanotube is turned by $30°$.

the size and geometry of the nanotube. Since the individual currents are proportional to $1/L_t$, the overall current decreases with the tube radius. It generally increases with the nanotube's length, but the strength of this dependence varies with the geometry of the tube.

By the geometry of the tube we understand its chirality and twist angle (a twist is introduced when $\mathbf{L}_t$ and $\mathbf{L}_l$ are not perpendicular). We have not yet investigated directly the dependence of the persistent currents in SWCN on the twist, but it did not change significantly the currents in tori, so we suppose the twist is not a significant factor for nanotubes either. This, however, has to be checked precisely.

Chirality is another matter by far. The persistent currents in armchair and zigzag nanotubes of similar size are very different. In every mesoscopic system at sufficiently large ϕ the momentum of the most energetic states increases so much that they leave the first Brillouin zone (their energy exceeds the Fermi energy) while the states with opposite momentum are boosted just enough to enter it. This results in a sudden jump in the overall current, whose amplitude depends on the amount of states thus displaced. Here comes into play the position of the states in the Brillouin zone. In

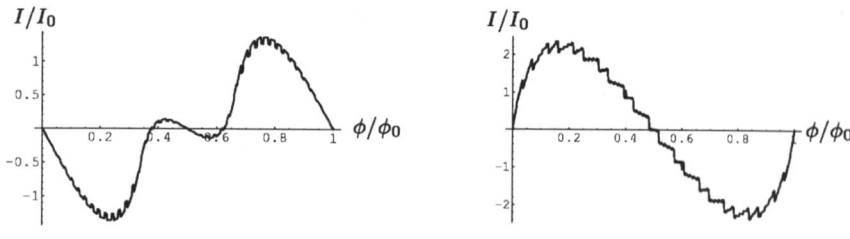

Figure 6. Persistent currents vs ϕ/ϕ_0 in undoped zigzag $(14, 0) \times (-100, 200)$ (left) and armchair $(9, 9) \times (-150, 150)$ (right) nanotubes. The unit on the y axis is $I_0 = 1.27 \cdot 10^{-4}$A.

undoped nanotubes of different chiralities the currents are similar, both in amplitude and shape (see Fig. 6); but in hole-doped nanotubes the Fermi energy lowers and instead of consisting of just six points, it changes into a set of lines. At such doping that $E_F = -\gamma$ ($\simeq 3$ eV) the Fermi surface becomes a smaller hexagone inside the original Brillouin zone. For zigzag nanotubes the momentum lines are parallel to the sides of the Fermi surface and when ϕ is sufficiently large to push some states above the Fermi level, it pushes the whole line of them and the current is strongly enhanced in comparison with the undoped case; for our nanotube it increases almost 20 times (note the almost vertical jumps at $\phi = 0$ and $\phi = \phi_0$). In the contrary, in armchair tubes any correlations there were are now lost and the current diminishes (see Fig. 7).

Theoretically it is possible, in systems with self-inductance, to obtain the so called spontaneous flux. This mechanism will be explained in the next

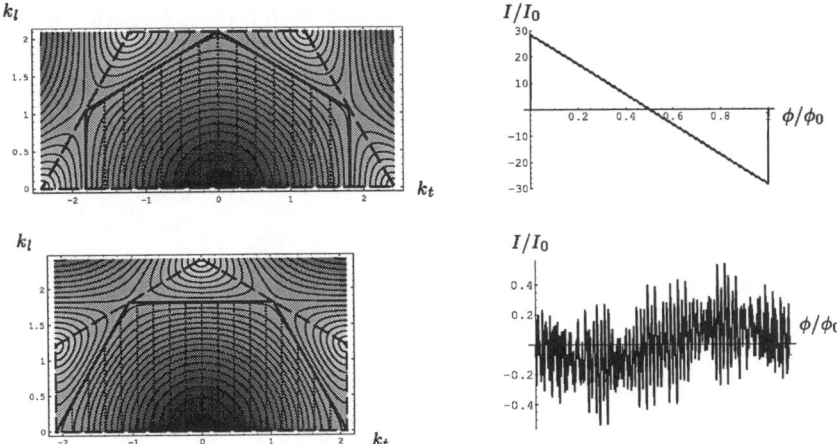

Figure 7. Upper left: the Brillouin zone of a $(14, 0) \times (-100, 200)$ zigzag nanotube with $E_F = -\gamma$, bold line is the Fermi level. Upper right: the persistent current for this tube and this value of doping. Lower left: Brillouin zone for a $(9, 9) \times (-150, 150)$ armchair nanotube of similar size, $E_F = -\gamma$, bold line is the Fermi level. Lower right: the persistent current vs ϕ/ϕ_0 for this tube and this value of doping.

section; we shall only mention here that these spontaneous fluxes are rather unlikely to appear in single-wall nanotubes, since even the enhanced currents are not very large. However, in multiwall nanotubes the situation may be different. It might happen that all tube layers have similar chirality and that the currents from all sheets superpose, increasing the total current and favouring the appearance of spontaneous flux.

4. The possibility of flux trapping and flux quantization in a set of concentric cylinders

It is well known that flux expulsion and flux trapping are one of the main features characterizing the superconducting state being a hallmark of a phase coherence.

There is a question whether these phenomena can be obtained in non-superconducting structures such as metallic or semiconducting mesoscopic systems.

It is well known that in mesoscopic cylinders orbital currents[10] can be induced by the external magnetic field applied parallel to the cylinder axis. The induced currents are persistent at low T[12,16,17] and if their amplitude is high there exists a possibility of creating self sustaining currents which run even if we switch the external field off. For simplicity we consider a system of spinless fermions and we assume $T = 0$.

Let us consider a 3D cylinder made of material with layered structure under the influence of a static magnetic field H. On the thickness d of the cylinder we have M_d coaxial closely packed layers ($d = M_d b$, b is the distance between layers) and the conduction takes place within the layers.

The field is partially of external origin and partially due to the average field produced by all electrons. It is described by a vector potential $\mathbf{A}$,

$$\mathbf{A} = \mathbf{A_e} + \mathbf{A_i}, \qquad (7)$$

where $\mathbf{A_e}$ is the vector potential of an external magnetic field $\mathbf{H_e}$ parallel the to z axis, $\mathbf{A_i}$ is the vector potential due to the currents in the system. Let ϕ_s be the flux inside the s-th cylinder (s=1,...,M_d), s=1 for the cylinder at $r = R_1$, $s = M_d$ at $r = R_2$. The flux[18] through the area between the $(s+1)$-th and s-th cylinder is created by $\mathbf{H_e}$ and $\mathbf{H_i}$ where $\mathbf{H_i}$ is the field created by the currents $I(\phi)$ in the outer cylinders. We get

$$\phi_{s+1} - \phi_s = 2\pi r_s b[H_e + \frac{4\pi}{cL} \sum_{j=s+1}^{M_d} I(\phi_s)] \equiv 2\pi r_s b H_s \qquad (8)$$

Let us assume that ϕ_s is a smooth function of s. One can then transform Eq.(8) into a differential equation

$$H(r) = \frac{1}{2\pi r} \frac{d\phi(r)}{dr}, \qquad (9)$$

where $[\phi(r) = 2\pi r A(r)]$,

$$\phi = \phi_e + \phi_i, \qquad (10)$$

ϕ_e, ϕ_i are the external, internal fluxes respectively.

The formula for the total energy takes the form[19]

$$E = \frac{\pi L \hbar^2 n}{m} \int_{R_1}^{R_2} r dr \sum_{k_z l} \frac{N_{k_z l}}{N} [k_z^2 + \frac{(l - \phi')^2}{r^2}] + \frac{L}{4} (\frac{\hbar c}{e})^2 \int_{R_1}^{R_2} \frac{1}{r} (\frac{d\phi_i'}{dr})^2 dr, (11)$$

where N is the number of electrons in a single layer, l is the orbital quantum number, k_z is a wave vector in the z direction, $\phi' = \frac{\phi}{\phi_0}$, $\phi_0 = \frac{hc}{e}$, ϕ_o is the flux unit, $n = \frac{1}{a^2 b}$ is the electron density, a is the lattice constant in the conducting planes. The occupation numbers $N_{k_z l}$ are choosen to be

$$N_{k_z l} = \begin{cases} 1 \text{ for } (k_z l) \text{ below the Fermi surface (FS)} \equiv (k_z l)_{occ} \\ 0 \text{ otherwise} \end{cases} \qquad (12)$$

Notice that the flux ϕ' shifts the energy levels connected with the angular momentum leading to unequal (asymmetric) occupation of $\pm l$ states. Eq. (11) can now be used to obtain the formula for the selfconsistent flux.

Demanding that the energy E has an extremum against variation of the function $\phi_i'(r)$ and using the standard Euler-Lagrange procedure we get the following equation for $\phi'(r)$

$$r\frac{d}{dr}\left[\frac{1}{r}\frac{d\phi'(r)}{dr}\right] = -\frac{1}{\lambda^2}[i_p(\phi') - \phi'(r)] \tag{13}$$

where

$$\frac{1}{\lambda^2} = \frac{4\pi e^2 n}{mc^2} \tag{14}$$

λ is the parameter containing only the material constants, known as a penetration depth,

$$i_p(\phi') = \frac{\sum_{(k_z l)_{occ}} l}{N}. \tag{15}$$

It is easy to see that the rhs of Eq. (13) is proportional to the current I in a single cylindrical layer where

$$I = \frac{e\hbar N}{4\pi^2 mr^2}[i_p(\phi') - \phi']. \tag{16}$$

It is well known[13,17] that persistent currents in mesoscopic 2D cylinders depend strongly on the correlation of currents from different channels labelled by k_z i.e. on the shape of the Fermi Surface (FS). The most favorable situation is when the FS is flat, perpendicular to the angular momentum l in the reciprocal space. The currents then add coherently and the resulting current is the strongest. With decreasing curvature of the FS the current becomes smaller.

In this review we consider only the most favourable situation corresponding to *rectangular FS perpendicular to l and k_z direction*.

For such a Fermi surface the sums over k_z and l in Eq. (15) are independent and

$$i_p(\phi') = \frac{1}{N}\sum_{k_z=-k_{zF}}^{k_{zF}}\sum_{l=-l_F+l'}^{l_F+l'} l = \frac{(2k_{zF}+1)(2l_F+1)}{N}l' = l'. \tag{17}$$

Eq. (13) takes now the form

$$r\frac{d}{dr}(\frac{1}{r}\frac{d\phi'}{dr}) = -\frac{1}{\lambda^2}(l' - \phi'). \tag{18}$$

To solve Eq. (18) we denote $\beta(r) \equiv l' - \phi'(r)$, note that $\beta(r)$ is proportional to the current I (16). Further we assume that in the considered range of r the value of l' is constant. In this case the equation reads

$$r\frac{d}{dr}\left[\frac{1}{r}\frac{d\beta(r)}{dr}\right] = \frac{1}{\lambda^2}\beta(r). \tag{19}$$

By introducing a new variable x, $r = \lambda x$ and a function $y(x)$, where $\beta(r) = \lambda x \cdot y(x)$ the equation (19) can be reduced to a form

$$x^2\frac{d^2y}{dx^2} + x\frac{dy}{dx} - (x^2+1)y = 0. \tag{20}$$

It has well known solutions $y(x) = c_1 I_1(x) + c_2 K_1(x)$ in terms of modified Bessel functions of the first kind $I_1(x)$ and second kind $K_1(x)$, c_1 and c_2 are constants, to be found from the boundary conditions given by the magnitude of the flux at the inner and outer cylinder wall. The general solution of (19) is thus of the form

$$\beta(r) = r[c_1 I_1(\frac{r}{\lambda}) + c_2 K_1(\frac{r}{\lambda})]. \tag{21}$$

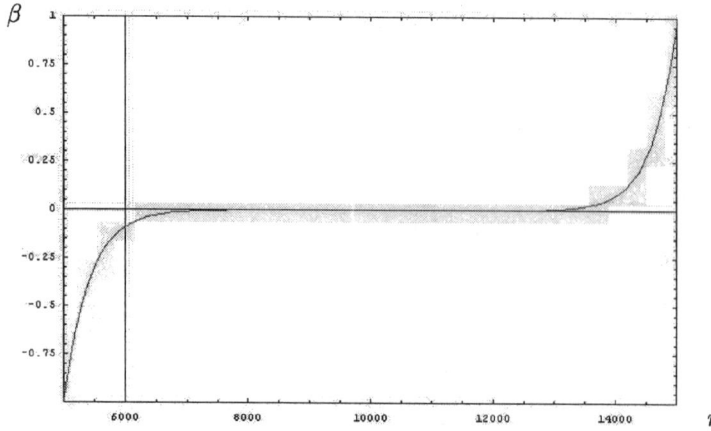

Figure 8. β as a function of r for $\lambda = 400$Å, $R_1 = 5000$Åand $R_2 = 15000$Å.

Figs. 8 and 9 show this function for two different sets of λ, R_1, R_2, and $\beta(R_1) = -\beta(R_2) = -1$. Note that by definition $\beta(r) \equiv l' - \phi'(r)$ and hence $\beta(r) = 0$ means that the flux inside the cylinder of the radius r is constant $\phi'(r) = l'$.

Thus for thick ($d \gg \lambda$) cylinder (fig. 8) we get $\phi' = l'$ i.e. the quantization of the flux trapped in ϕ_o units.

For $l' = 0$ this solution corresponds to flux expulsion i.e. to the Meissner-like effect.

For systems with smaller thickness (fig. 9) we get only partial flux expulsion and trapping of the quantized flux, but the effective magnitude of the flux quantum is less then its full value[19].

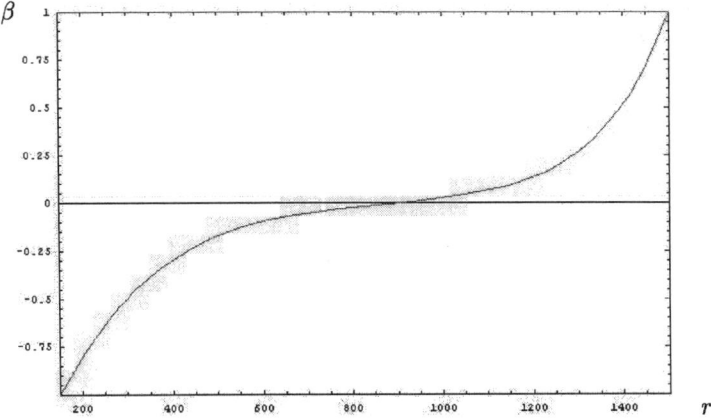

Figure 9. β as a function of r for $\lambda = 160\text{Å}$, $R_1 = 150\text{Å}$ and $R_2 = 1500\text{Å}$.

Multiwall Carbon Nanotubes (MWNT) consisting of a set concentric sheets nested inside each other have also the desired, cylindrical structure. However in MWNT produced nowadays different sheets may have in general different chiralities and thus different electric and magnetic behaviour. Thus the assumption, which the present chapter is based on seems not to be fulfilled. However if in the future it becomes possible to produce MWNT consisting of sheets having the same (metallic or semiconducting) properties then such a MWNT should also exhibit flux expulsion and flux quantization.

5. Conclusions

This paper gives a short review of highly unusual electronic, magnetic and mechanical properties of CN and their possible applications.In particular we discussed the possibility of creating in CN persistent currents which run in the presence of the static magnetic field. We have shown that by hole doping the shape of the FS is changed and it strongly influences the magnitude of persistent currents and hence of the related magnetic moments.We have also presented the considerations showing that a set of concentric mesoscopic cylinders exhibiting the same metallic or semiconducting behaviour can show flux expulsion and trapping of the quantized flux. This analysis could have possible applications to MWNT if the nanotubes of sheets showing the same (metallic or semiconducting) properties could be been synthetized.

The carbon nanotubes are a fascinating new class of materials with many unique and desirable properties which offer great intellectual challenges and the potential for novel applications.

6. Aknowledgements

Work supported by KBN grant 5P03B0320.

References

1. for a review see e.g. *Carbon Nanotubes: Synthesis, Structure, Properties and Applications*, M.S.Dresselhaus, G.Dresselhaus, Ph.Avouris (eds.), Springer-Verlag (2001)
2. P.L.McEuen, *Phys. World*, June, 31 (2000),
3. J.W.G.Wildöer, L.C.Venema, A.G.Rinzler, R.E.Smalley, C.Dekker, *Nature* **391**, 59 (1998)
4. C.T.White, T.N.Todorov, *Nature* **393**, (1998)
5. W.A.de Heer, A.Chatelain, D.Ugarte, *Science* **270**, 1179 (1995) ibid p. 1119
6. G.Zhao, Y.S.Wang, *Cond. Matt.* **2**, 0111268 (2001)
7. M.Kociak, A.Yu.Kasumov, S.Gueron, B.Reulet, I.I.Khodos, Yu.B.Gorbatov, V.T.Volkov, L.Vaccarini, H.Bouchiat, *Phys. Rev. Lett.* **86**, No.11, 2416 (2001)
8. Z.K.Tang, L.Zhang, N.Wang, X.X.Zhang, G.H.Wen, G.D.Li, N.Wang, C.T.Chan, P.Sheng, *Science* **292**, 2462 (2001)
9. A.Bachtold, C.Strunk, J.P.Salvetat, J.M.Bohard, L.Forro, T.Nussbaumer, C.Schonenberger, *Nature* **397**, 673 (1999)
10. H.Cheung, Y.Gefen, E.K.Riedel, IBM, *J. Res. Rev.* **32**, 359 (1988)
11. M.Margańska, M.Szopa, *Acta Phys. Pol.* **32** (2001) 427
12. D.Wohlleben, M.Esser, P.Freche, E.Zipper, M.Szopa, *hys Rev Lett*, **66** (1991) 3191
13. M. Stebelski, M. Szopa, E. Zipper, *Z. Phys.* **B103**, 79 (1997)
14. F. Bloch, H.E. Rorschach, Phys Rev **128**, 1697 (1962)
15. M.Szopa, M.Margańska, E.Zipper, *Phys. Lett. A* **299** (2002) 593
16. M.Buttiker, Y.Imry, R.Landauer, *Phys. Rev. Lett.* **A96**, 365 (1969)
17. M.Lisowski, M.Stebelski, E.Zipper, *Phys. Rev* **B59**, 8305 (1999)
18. E. V. Tsiper, A. L. Efros, *J. Phys. Condens. Matter*, **10**, 1053 (1998).
19. M.Czechowska, M.Lisowski, M.Szopa, E.Zipper, to be published

JUSTIFICATION FOR THE
COMPOSITE FERMION PICTURE

A. WÓJS,[1,2] J. J. QUINN,[1] AND L. JACAK[2]

[1] *University of Tennessee, Knoxville, Tennessee 37996, USA*
[2] *Wroclaw University of Technology, Wroclaw 50-370, Poland*
E-mail: wojs@if.pwr.wroc.pl

The mean field (MF) composite Fermion (CF) picture successfully predicts the low-lying bands of states of fractional quantum Hall systems. This success cannot be attributed to the originally proposed cancellation between Coulomb and Chern–Simons interactions beyond the mean field and solely depends on the short range of the repulsive Coulomb pseudopotential in the lowest Landau level (LL). The class of pseudopotentials is defined for which the MFCF picture can be applied. The success or failure of the MFCF picture in various systems (electrons in the lowest and excited LL's, Laughlin quasiparticles) is explained.

1. Introduction

The quantum Hall effect (QHE)[1,2] is the quantization of Hall conductance of a two-dimensional electron gas (2DEG) in high magnetic fields that occurs at certain fillings of the macroscopically degenerate single-electron Landau levels (LL's). The LL filling is defined by the filling factor ν equal to the number of electrons divided by the LL degeneracy, which is proportional to the magnetic field and the physical area occupied by the 2DEG. The series of values of ν at which the QHE is observed contains small integers and simple, almost exclusively odd-denominator fractions such as $\nu = \frac{1}{3}$, $\frac{2}{5}$, etc. The series is universal for all samples, which means that the occurrence of QHE at a particular value of ν depends on the quality of the sample, temperature, or other similar conditions, but not on the material parameters such as lattice composition. Moreover, the observed values of ν are "exact" in a sense that poor sample quality (e.g., lattice imperfections) can destroy QHE at some or all of the values of ν, but cannot shift these values.

The quantization of Hall conductance is always accompanied by a rapid drop of longitudinal conductance, and both effects signal the appearance of (incompressible) nondegenerate many-body ground states (GS's) in the spectrum of the 2DEG, separated from the continuum of excited states by a finite gap. At integer $\nu = 1, 2 \ldots$ (IQHE), the origin of incompress-ibility is the single particle cyclotron gap between the LL's. On the other hand, at fractional $\nu = 1/3, 1/5, 2/5 \ldots$ (FQHE) electrons partially fill a degenerate (lowest) LL and the formation of incompressible GS's is a com-plicated many-body phenomenon. The gap and resulting incompressibility are entirely due to electron-electron (Coulomb) interactions and reveal the unique properties of this interaction within the lowest LL.[3,4]

In this note we shall concentrate on the latter effect. We will apply the pseudopotential formalism[5,6] to the FQH systems, and show that the form of the pseudopotential $V(L')$ [pair energy vs. pair angular momentum] in the lowest LL rather than of the interaction potential $V(r)$, is responsible for incompressibility of the FQH states. The idea of fractional parentage[7] will be used to characterize many-body states by the ability of electrons to avoid pair states with largest repulsion. The condition on the form of $V(L')$ necessary for the occurrence of FQH states will be given, which defines the short-range repulsive (SRR) pseudopotentials to which MFCF picture can be applied. As an example, we explain the success or failure of MFCF predictions for the electrons in the lowest and excited LL's and for Laughlin quasiparticles (QP's) in hierarchy picture of FQH states.[8,9]

2. Theoretical concepts for FQHE

2.1. Single-electron states and Laughlin wavefunction

The Hamiltonian for an electron confined to the x–y plane in the presence of a perpendicular magnetic field B is

$$H_0 = \frac{1}{2\mu} \left(\vec{p} + \frac{e}{c}\vec{A} \right)^2. \tag{1}$$

Here μ is the effective mass, $\vec{p} = (p_x, p_y, 0)$ is the momentum operator and $\vec{A}(x, y)$ is the vector potential (whose curl gives B). For the "symmetric gauge," $\vec{A} = \frac{1}{2}B(-y, x, 0)$, the single particle eigenfunctions[10] are of the form $\psi_{nm}(r, \theta) = e^{-im\theta}u_{nm}(r)$, and the eigenvalues are given by

$$E_{nm} = \frac{1}{2}\hbar\omega_c(2n + 1 + |m| - m). \tag{2}$$

In these equations, $n = 0, 1, 2, \ldots$, and $m = 0, \pm 1, \pm 2, \ldots$. The lowest energy states (lowest Landau level) have $n = 0$, $m = 0, 1, 2, \ldots$, and energy $E_{0m} = \hbar\omega_c/2$. It is convenient to introduce a complex coordinate $z = re^{-i\theta} = x - iy$, and to write the lowest Landau level wavefunctions as $\psi_{0,-m} = N_m z^m \exp(-|z|^2/4)$, where N_m is a normalization constant, and m can take on any non-negative integral value. In this expression we have used the magnetic length $\lambda = (\hbar c/eB)^{1/2}$ as the unit of length. The function $|\psi_{0,-m}|^2$ has its maximum at a radius r_m which is proportional to $m^{1/2}$. All single particle states from a given Landau level are degenerate, and separated in energy from neighboring levels by $\hbar\omega_c$.

If m is restricted to being less than some maximum value, N_L, chosen so that the system has a "finite radial range," then the alowed m values are 0, 1, 2, ..., $N_L - 1$. The value of N_L is equal to the flux through the sample, $B \cdot A$ (where A is the area), divided by the quantum of flux $\phi_0 = hc/e$. The filling factor ν is defined as the ratio of the number of electrons, N, to N_L. An infinitesimal decrease in the area A when ν has an integral value requires promotion of an electron across the gap $\hbar\omega_c$ to the first unoccupied level, making the system incompressible.

In order to construct a many electron wavefunction corresponding to filling factor $\nu = 1$, the product function which places one electron in each of the N_L orbitals $\psi_{0,-m}$ ($m = 0, 1, \ldots, N_L - 1$) must be antisymmetrized. This can be done with the aid of a Slater determinant

$$\Psi(z_1, z_2, \ldots, z_N) \propto \begin{vmatrix} 1 & 1 & \cdots & 1 \\ z_1 & z_2 & \cdots & z_N \\ z_1^2 & z_2^2 & \cdots & z_N^2 \\ \vdots & \vdots & & \vdots \\ z_1^{N_L-1} & z_2^{N_L-1} & \cdots & z_N^{N_L-1} \end{vmatrix} e^{-\sum_k |z_k|^2/4}. \tag{3}$$

The determinant in Eq. (3) is the well-known Vandemonde determinant. It is not difficult to show that it is equal to $\prod_{i<j}(z_i - z_j)$. Of course, $N_L = N$ (since each of the N_L orbitals is occupied by one electron) and $\nu = 1$.

Laughlin noticed that if the factor $(z_i - z_j)$ of Vandemonde determinant was replaced by $(z_i - z_j)^m$, where m was an odd integer, the wavefunction

$$\Psi_m(z_1, z_2, \ldots, z_N) \propto \prod_{i<j}(z_i - z_j)^m e^{-\sum_k |z_k|^2/4} \tag{4}$$

would be antisymmetric, keep the electrons further apart (and therefore reduce repulsion), and correspond to a filling factor $\nu = m^{-1}$. This results because the highest power of the orbital index entering Ψ_m is $N_L - 1 = m(N - 1)$ giving $\nu = N/N_L = m^{-1}$ in the limit of large systems. The additional factor $\prod_{i<j}(z_i - z_j)^{m-1}$ multiplying $\Psi_{m=1}(z_1, z_2, \ldots, z_N)$ is the Jastrow factor which accounts for correlations.

2.2. Laughlin quasiparticles and Haldane hierarchy

The elementary charged excitations of the Laughlin $\nu = (2p+1)^{-1}$ GS are Laughlin quasiparticles (QP's) corresponding to a vortex (for the quasihole,

QH) or anti-vortex (for the quasielectron, QE) at an arbitrary point z_0, and described by the following wavefunctions,

$$\Psi_{\mathrm{QH}} \propto (z - z_0)\Psi_m \ \ \text{and} \ \ \Psi_{\mathrm{QE}} \propto \frac{\partial}{\partial(z - z_0)}\Psi_m. \tag{5}$$

Laughlin QP's carry fractional electric charge $\pm e/(2p + 1)$ and obey fractional statistics (although in some situations they can be conveniently treated as either fermions or bosons thanks to a statistics transformation valid in two dimensions). Being charged particles (of the finite size of the order of the magnetic length λ) moving in the magnetic field, Laughlin QP form (quasi)LL's similar to those of electrons, except for the m-times lower degeneracy due to theor reduced charge. They are also naturally expected to interact with one another via normal, charge-charge Coulomb forces.

As an extension of Laughlin's idea, Haldane,[8] and others[9,11] proposed that in analogy to electrons, Laughlin QP's must form Laughlin-like incompressible states of their own. According to this idea, each Laughlin state of electrons would stand atop entire family of so-called "daughter" Laughlin state of its QE's and QH's. And on the following level, each daughter state would have its own family of daughter states, and so on. This construction results in entire family of incompressible states that corresponds to many more fractional filling factors ν at which incompressibility and in result also the FQHE are expected in the underlying 2DEG. For example, the $\nu = \frac{2}{5}$ state can be interpreted as $\nu = \frac{1}{3}$ QE daugter state of the parent $\nu = \frac{1}{3}$ electron state. In fact, all odd-denominator fractions $\nu = p/q$ can be generated in this way, which brings us to the major problem of the (original) concept of hierarchy. On one hand, the hierarchy predicts too many fractions (only a finite number of fractions are observed experimentally, and it seems evident that FQHE will not occur at most of the other predicted fractions regarless of the experimental conditions[2,12,13]). On the other hand, the hierarchy model gives no apparent connection between the stability of a given state and its position in the hierarchy (explanation of some of the easily experimentally observed FQH states requires introducing many generations of QP's). As we shall explain later, these problems of the hierarchy model resulted from an erroneous assumption that, being charged particles, Laughlin QP's will form Laughlin states at each Laughlin filling factor, $\nu = (2p + 1)^{-1}$. With the knowledge of the form of QP–QP interactions, one can eliminate "false" daughter states from the hierarchy and reach an agreement with the experimental observation. This makes the Laughlin–Haldane theory the only microscopic theory of the FQHE.

2.3. *Jain composite Fermion model*

Independently of the Laughlin–Haldane model, from the similar energy spectra of the FQH and IQH systems one can expect that some kind of effective, charged particle-like excitations may form in the interacting 2DEG. These excitations would be the relevant charge carriers near $\nu = \frac{1}{3}$ and they would fill exactly their quasi-LL's at precisely this value, giving rise to incompressibility. This idea leads to the composite Fermion (CF) picture.[14,15]

In the mean field (MF) CF picture, in a 2DEG of density n at a strong magnetic field B, each electron is assumed to bind an even number $2p$ of magnetic flux quanta $\phi_0 = hc/e$ (in form of an infinitely thin flux tube) forming a CF. Because of the Pauli exclusion principle, the magnetic field confined into a flux tube within one CF has no effect on the motion of other CF's, and the average effective magnetic field B^* seen by CF's is reduced, $B^* = B - 2p\phi_0 n$. Because $B^*\nu^* = B\nu = n\phi_0$, the relation between the electron and CF filling factors is

$$(\nu^*)^{-1} = \nu^{-1} - 2p. \tag{6}$$

Since the low band of energy levels of the original (interacting) 2DEG has similar structure to that of the noninteracting CF's in a uniform effective field B^*, it was proposed[14] that the Coulomb charge-charge and Chern–Simons (CS) charge-flux interactions beyond the MF largely cancel one another, and the original strongly interacting system of electrons is converted into one of weakly interacting CF's. Consequently, the FQHE of electrons was interpreted as the IQHE of CF's.

Although the MFCF picture correctly predicts the structure of low-energy spectra of FQH systems, the energy scale it uses (the CF cyclotron energy $\hbar\omega_c^*$) is totally irrelevant. Moreover, since the characteristic energies of CS ($\hbar\omega_c^* \propto B$) and Coulomb ($e^2/\lambda \propto \sqrt{B}$, where λ is the magnetic length) interactions between fluctuations beyond MF scale differently with the magnetic field, the reason for its success cannot be found in originally suggested cancellation between those interactions. Since the MFCF picture is commonly used to interpret various numerical and experimental results, it is important to understand why and under what conditions it is correct.

3. Numerical exact diagonalization studies

Because of the LL degeneracy, the electron-electron interaction in the FQH states cannot be treated perturbatively, and the exact (numerical) diagonalization techniques have been commonly used in their study. In order to model an infinite 2DEG by a finite (small) system that can be handled numerically, it is very convenient to confine N electrons to a surface of a (Haldane) sphere of radius R, with the normal magnetic field B produced by a magnetic monopole of integer strength $2S$ (total flux of $4\pi BR^2 = 2S\phi_0$) in the center.[8] The obvious advantages of such geometry is the absence of an edge and preserving full 2D symmetry of a 2DEG (good quantum numbers are the total angular momentum L and its projection M). The numerical experiments in this geometry have shown that even relatively small systems that can be solved exactly on a computer behave in many ways like an infinite 2DEG, and a number of parameters of a 2DEG (e.g. excitation energies) can be obtained from such small scale calculations.

The single particle states on a Haldane sphere (monopole harmonics) are labeled by angular momentum l and its projection m.[16] The energies, $\varepsilon_l = \hbar\omega_c[l(l+1) - S^2]/2S$, fall into degenerate shells and the nth shell ($n = l - |S| = 0, 1, \ldots$) corresponds to the nth LL. For the FQH states at filling factor $\nu < 1$, only the lowest, spin polarized LL need be considered.

The object of numerical studies is to diagonalize the electron-electron interaction Hamiltonian H in the space of degenerate antisymmetric N electron states of a given (lowest) LL. Although matrix H is easily block diagonalized into blocks with specified M, the exact diagonalization becomes difficult (matrix dimension over 10^6) for $N > 10$ and $2S > 27$ ($\nu = 1/3$).[6] Typical results for ten electrons at filling factors near $\nu = 1/3$ are presented in Fig. 1. Energy E, plotted as a function of L in the magnetic units, includes shift $-(Ne)^2/2R$ due to charge compensating background. There is always one or more L multiplets (marked with open circles) forming a low-energy band separated from the continuum by a gap. If the lowest band consists of a single $L = 0$ GS (Fig. 1d), it is expected to be incompressible in the thermodynamic limit (for $N \to \infty$ at the same ν) and an infinite 2DEG at this filling factor is expected to exhibit the FQHE.

The MFCF interpretation of the spectra in Fig. 1 is the following. The effective magnetic monopole strength seen by CF's is[14,6]

$$2S^* = 2S - 2p(N-1), \tag{7}$$

and the angular momenta of lowest CF shells (CF LL's) are $l_n^* = |S^*| + n$.[17] At $2S = 27$, $l_0^* = 9/2$ and ten CF's fill completely the lowest CF shell ($L = 0$ and $\nu^* = 1$). The excitations of the $\nu^* = 1$ CF GS involve an excitation of at least one CF to a higher CF LL, and thus (if the CF-CF interaction is weak on the scale of $\hbar\omega_c^*$) the $\nu^* = 1$ GS is incompressible and so is Laughlin[3] $\nu = 1/3$ GS of underlying electrons. The lowest lying excited states contain a pair of QP's: a quasihole (QH) with $l_{QH} = l_0^* = 9/2$ in the lowest CF LL and a quasielectron (QE) with $l_{QE} = l_1^* = 11/2$ in the first

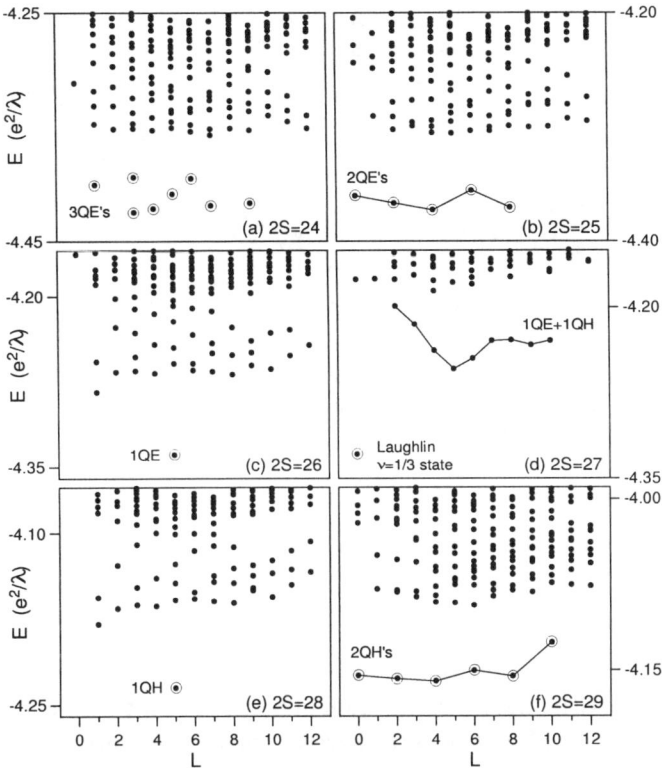

Figure 1. Energy spectra of ten electrons in the lowest LL at the monopole strength $2S$ between 24 and 29. Open circles mark lowest energy bands with fewest CF QP's.

excited one. The allowed angular momenta of such pair are $L = 1, 2, \ldots,$ 10. The $L = 1$ state usually has high energy and the states with $L \geq 2$ form a well defined band with a magnetoroton minimum at a finite value of L. The lowest CF states at $2S = 26$ and 28 contain a single QE and a single QH, respectively (in the $\nu^* = 1$ CF state, i.e. the $\nu = 1/3$ electron state), both with $l_{QP} = 5$, and the excited states will contain additional QE-QH pairs. At $2S = 25$ and 29 the lowest bands correspond to a pair of QP's, and the values of energy within those bands define the QP-QP interaction pseudopotential V_{QP}. At $2S = 25$ there are two QE's each with $l_{QE} = 9/2$ and the allowed angular momenta (of two identical Fermions) are $L = 0$, 2, 4, 6, and 8, while at $2S = 29$ there are two QH's each with $l_{QH} = 11/2$ and $L = 0, 2, 4, 6, 8,$ and 10. Finally, at $2S = 24$, the lowest band contains three QE's each with $l_{QE} = 4$ and $L = 1, 3^2, 4, 5, 6, 7,$ and 9.

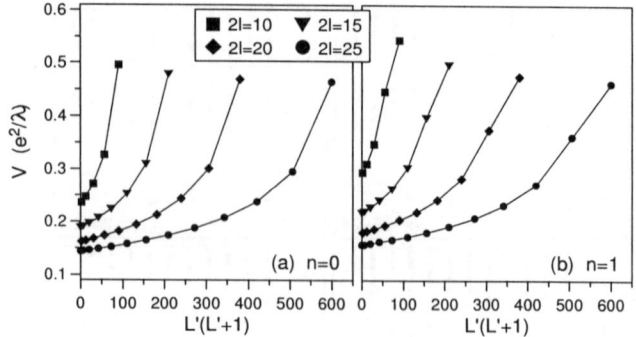

Figure 2. Pseudopotentials V of the Coulomb interaction in the lowest (a), and first excited LL (b) as a function of squared pair angular momentum $L'(L'+1)$. Different symbols mark data for different $S = l + n$.

4. Pseudopotential and fractional grandparentage

The two body interaction Hamiltonian H can be expressed as

$$\hat{H} = \sum_{i<j} \sum_{L'} V(L')\, \hat{\mathbf{P}}_{ij}(L'),\tag{8}$$

where $V(L')$ is the interaction pseudopotential[5] and $\hat{\mathbf{P}}_{ij}(L')$ projects onto the subspace with angular momentum of pair ij equal to L'. For electrons confined to a LL, L' measures the average squared distance d^2,[6]

$$\frac{\hat{d}^2}{R^2} = 2 + \frac{S^2}{l(l+1)}\left(2 - \frac{\hat{L}'^2}{l(l+1)}\right),\tag{9}$$

and larger L' corresponds to smaller separation. Due to the confinement of electrons to one (lowest) LL, interaction potential $V(r)$ enters Hamiltonian H only through a small number of pseudopotential parameters $V(2l - \mathbf{R})$, where $\mathbf{R}$, relative pair angular momentum, is an odd integer.

In Fig. 2 we compare Coulomb pseudopotentials $V(L')$ calculated for a pair of electrons on the Haldane sphere each with $l = 5$, $15/2$, 10, and $25/2$, in the lowest and first excited LL. For the reason that will become clear later, $V(L')$ is plotted as a function of $L'(L'+1)$. All pseudopotentials in Fig. 2 increase with increasing L'. If $V(L')$ increased very quickly with increasing L' (we define ideal short-range repulsion, SRR, as: $dV_{\mathrm{SR}}/dL' \gg 0$ and $d^2V_{\mathrm{SR}}/dL'^2 \gg 0$), the low-lying many-body states would be the ones maximally avoiding pair states with largest L'.[5,6] At filling factor $\nu = 1/m$ (m is odd) the many-body Hilbert space contains exactly one multiplet in which all pairs completely avoid states with $L' > 2l - m$. This multiplet is the $L = 0$ incompressible Laughlin state[3] and it is an exact GS of V_{SRR}.

The ability of electrons in a given many-body state to avoid strongly repulsive pair states can be conveniently described using the idea of fractional parentage.[6,7] An antisymmetric state $|l^N, L\alpha\rangle$ of N electrons each with angular momentum l that are combined to give total angular momentum L can be written as

$$|l^N, L\alpha\rangle = \sum_{L'} \sum_{L''\alpha''} G_{L\alpha}^{L''\alpha''}(L') \, |l^2, L'; l^{N-2}, L''\alpha''; L\rangle. \tag{10}$$

Here, $|l^2, L'; l^{N-2}, L''\alpha''; L\rangle$ denote product states in which $l_1 = l_2 = l$ are added to obtain L', $l_3 = l_4 = \ldots = l_N = l$ are added to obtain L'' (different L'' multiplets are distinguished by a label α''), and finally L' is added to L'' to obtain L. The state $|l^N, L\alpha\rangle$ is totally antisymmetric, and states $|l^2, L'; l^{N-2}, L''\alpha''; L\rangle$ are antisymmetric under interchange of particles 1 and 2, and under interchange of any pair of particles 3, 4, ... N. The factor $G_{L\alpha}^{L''\alpha''}(L')$ is called the coefficient of fractional grandparentage (CFGP). The two-body interaction matrix element is expressed as

$$\langle l^N, L\alpha | V | l^N, L\beta\rangle = \frac{N(N-1)}{2} \sum_{L'; L''\alpha''} G_{L\alpha}^{L''\alpha''}(L') \, G_{L\beta}^{L''\alpha''}(L') \, V(L'), \tag{11}$$

and expectation value of energy is

$$E_\alpha(L) = \frac{N(N-1)}{2} \sum_{L'} \mathbf{G}_{L\alpha}(L') \, V(L'), \tag{12}$$

where the coefficient

$$\mathbf{G}_{L\alpha}(L') = \sum_{L''\alpha''} \left| G_{L\alpha}^{L''\alpha''}(L') \right|^2 \tag{13}$$

gives the probability that pair ij is in the state with L'.

5. Energy spectra of short-range repulsive pseudopotentials

The very good description of actual GS's of a 2DEG at fillings $\nu = 1/m$ by the Laughlin wavefunction (overlaps typically larger that 0.99) and the success of the MFCF picture at $\nu < 1$ both rely on the fact that pseudopotential of Coulomb repulsion in the lowest LL falls into the same class of SRR pseudopotentials as V_{SRR}. Due to a huge difference between all parameters $V_{SRR}(L')$, the corresponding many-body Hamiltonian has the following hidden symmetry: the Hilbert space $\mathbf{H}$ contains eigensubspaces $\mathbf{H}_p$ of states with $\mathbf{G}(L') = 0$ for $L' > 2(l - p)$, i.e. with $L' < 2(l - p)$. Hence, $\mathbf{H}$ splits into subspaces $\tilde{\mathbf{H}}_p = \mathbf{H}_p \setminus \mathbf{H}_{p+1}$, containing states that do

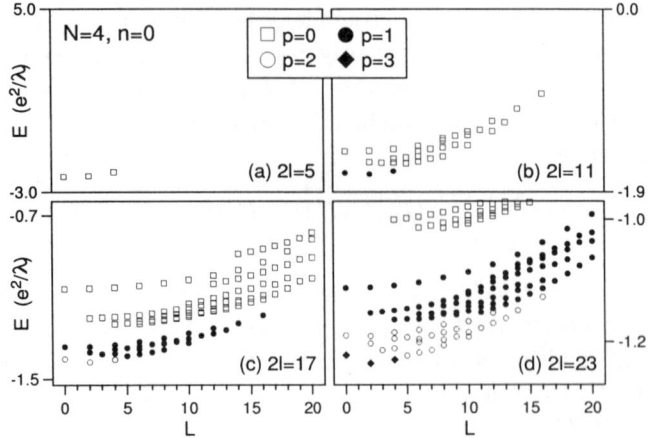

Figure 3. Energy spectra of four electrons in the lowest LL each with angular momentum $l = 5/2$ (a), $11/2$ (b), $17/2$ (c), and $23/2$ (d). Different subspaces $\mathbf{H}_p$ are marked with squares ($p = 0$), full ($p = 1$) and open circles ($p = 2$), and diamonds ($p = 3$).

not have grandparentage from $L' > 2(l - p)$, but have some grandparentage from $L' = 2(l - p) - 1$,

$$\mathbf{H} = \tilde{\mathbf{H}}_0 \oplus \tilde{\mathbf{H}}_1 \oplus \tilde{\mathbf{H}}_2 \oplus \dots \qquad (14)$$

The subspace $\tilde{\mathbf{H}}_p$ is not empty (some states with $L' < 2(l - p)$ can be constructed) at filling factors $\nu \le (2p + 1)^{-1}$. Since the energy of states from each subspace $\tilde{\mathbf{H}}_p$ is measured on a different scale of $V(2(l - p) - 1)$, the energy spectrum splits into bands corresponding to those subspaces. The energy gap between the pth and $(p + 1)$st bands is of the order of $V(2(l - p) - 1) - V(2(l - p - 1) - 1)$, and hence the largest gap is that between the 0th band and the 1st band, the next largest is that between the 1st band and 2nd band, etc.

Fig. 3 demonstrates on the example of four electrons to what extent this hidden symmetry holds for the Coulomb pseudopotential in the lowest LL. The subspaces $\mathbf{H}_p$ are identified by calculating CFGP's of all states. They are not exact eigenspaces of the Coulomb interaction, but the mixing between different $\mathbf{H}_p$ is weak and the coefficients $\mathbf{G}(L')$ for $L' > 2(l - p)$ are indeed much smaller in states marked with a given p than in all other states. For example, for $2l = 11$, $\mathbf{G}(10) < 0.003$ for states marked with full circles, and $\mathbf{G}(10) > 0.1$ for all other states (squares).

Note that the set of angular momentum multiplets which form subspace $\tilde{\mathbf{H}}_p$ of N electrons each with angular momentum l is always the same as the set of multiplets in subspace $\tilde{\mathbf{H}}_{p+1}$ of N electrons each with angular momentum $l + (N - 1)$. When l is increased by $N - 1$, an additional band appears at high energy, but the structure of the low-energy part of the spectrum is completely unchanged. For example, all three allowed multiplets for $l = 5/2$ ($L = 0$, 2, and 4) form the lowest energy band for $l = 11/2$, $17/2$, and $23/2$, where they span the $\tilde{\mathbf{H}}_1$, $\tilde{\mathbf{H}}_2$ and $\tilde{\mathbf{H}}_3$ subspace, respectively. Similarly, the first excited band for $l = 11/2$ is repeated for $l = 17/2$ and $23/2$, where it corresponds to $\tilde{\mathbf{H}}_1$ and $\tilde{\mathbf{H}}_2$ subspace.

Let us stress that the fact that identical sets of multiplets occur in subspace $\tilde{\mathbf{H}}_p$ for a given l and in subspace $\tilde{\mathbf{H}}_{q+1}$ for l replaced by $l + (N - 1)$, does not depend on the form of interaction, and follows solely from the rules of addition of angular momenta of identical Fermions. However, if the interaction pseudopotential has SRR, then: (i) $\tilde{\mathbf{H}}_p$ are interaction eigensubspaces; (ii) energy bands corresponding to $\tilde{\mathbf{H}}_p$ with higher p lie below those of lower p; (iii) spacing between neighboring bands is governed by a difference between appropriate pseudopotential coefficients; and (iv) wavefunctions and structure of energy levels within each band are insensitive to the details of interaction. Replacing V_{SRR} by a pseudopotential that increases more slowly with increasing L' leads to: (v) coupling between subspaces $\tilde{\mathbf{H}}_p$; (vi) mixing, overlap, or even order reversal of bands; (vii) deviation of wavefunctions and the structure of energy levels within bands from those of the hard core repulsion (and thus their dependence on details of the interaction pseudopotential). The numerical calculations for the Coulomb pseudopotential in the lowest LL show (to a large extent) all SRR properties (i)–(iv), and virtually no effects (v)–(vii), characteristic of 'non-SRR' pseudopotentials.

The reoccurrence of L multiplets forming the low-energy band when l is replaced by $l \pm (N - 1)$ has the following crucial implication. In the lowest LL, the lowest energy (pth) band of the N electron spectrum at the monopole strength $2S$ contains L multiplets which are all the allowed N electron multiplets at $2S - 2p(N - 1)$. But $2S - 2p(N - 1)$ is just $2S^*$, the effective monopole strength of CF's! The MFCS transformation which binds $2p$ fluxes (vortices) to each electron selects the same L multiplets from the entire spectrum as does the introduction of a hard core, which forbids a pair of electrons to be in a state with $L' > 2(l - p)$.

6. Definition of short-range repulsive pseudopotential

A useful operator identity relates total (L) and pair ($\hat{L}_{ij}$) angular momenta[6]

$$\sum_{i<j} \hat{L}_{ij}^2 = \hat{L}^2 + N(N - 2)\, \hat{l}^2. \tag{15}$$

It implies that interaction given by a pseudopotential $V(L')$ that is linear in $\hat{L}'^2$ (e.g. the harmonic repulsion within each LL) is degenerate in each L subspace and its energy is a linear function of $L(L+1)$. The many-body GS has the lowest available L while the maximum L corresponds to the largest energy. Note that this result is opposite to the Hund rule valid for spherical harmonics, due to the opposite behavior of $V(L')$ for the FQH ($n = 0$ and $l = S$) and atomic ($S = 0$ and $l = n$) systems.

Deviations of $V(L')$ from a linear function of $L'(L'+1)$ lead to the level repulsion within each L subspace, and the GS is no longer necessarily the state with minimum L. Rather, it is the state at a low L whose multiplicity N_L (number of different L multiplets) is large. It interesting to observe that the L subspaces with relatively high N_L coincide with the MFCF prediction. In particular, for a given N, they reoccur at the same L's when l is replaced by $l \pm (N-1)$, and the set of allowed L's at a given l is always a subset of the set at $l + (N-1)$.

As we said earlier, if $V(L')$ has short range, the lowest energy states within each L subspace are those maximally avoiding large L', and the lowest band (separated from higher states by a gap) contains states in which a number of largest values of L' is avoided altogether. This property is valid for all pseudopotentials which increase more quickly than linearly as a function of $L'(L'+1)$. For $V_\beta(L') = [L'(L'+1)]^\beta$, exponent $\beta > 1$ defines the class of SRR pseudopotentials, to which the MFCF picture can be applied. Within this class, the structure of low-lying energy spectrum and the corresponding wavefunctions very weakly depend on β and converge to those of V_{SRR} for $\beta \to \infty$.

The extension of the SRR definition to $V(L')$ that are not strictly in the form of $V_\beta(L')$ is straightforward. If $V(L') > V(2l-m)$ for $L' > 2l-m$ and $V(L') < V(2l-m)$ for $L' < 2l-m$ and $V(L')$ increases more quickly than linearly as a function of $L'(L'+1)$ in the vicinity of $L' = 2l-m$, then pseudopotential $V(L')$ behaves like SRR at filling factors near $\nu = 1/m$.

7. Application to various interacting systems

It follows from Fig. 2a that the Coulomb pseudopotential in the lowest LL satisfies the SRR condition in the entire range of L'; this is what validates the MFCF picture for filling factors $\nu \leq 1$. However, in a higher, nth LL this is only true for $L' < 2(l-n)-1$ (see Fig. 2b for $n = 1$) and the MFCF picture is valid only for ν_n (filling factor in the nth LL) around and below $(2n+3)^{-1}$. Indeed, the MFCF features in the ten electron energy spectra around $\nu = 1/3$ (in Fig. 1) are absent for the same fillings of the $n = 1$ LL.[6]

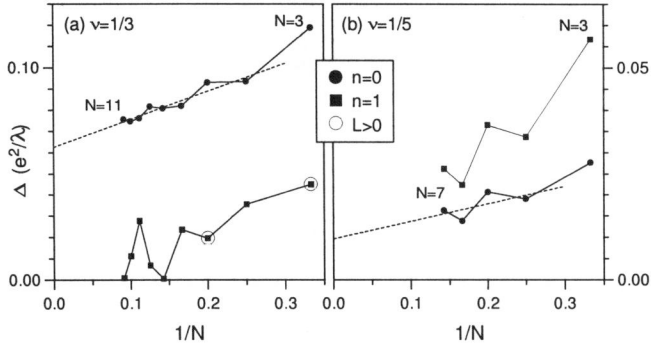

Figure 4. Excitation gap Δ as a function of inverse electron number $1/N$ for filling factors $\nu = 1/3$ (a) and $1/5$ (b) in the $n = 0$ (dots) and $n = 1$ (squares) LL's. Open circles mark degenerate ground states $(L > 0)$.

One consequence of this is that the MFCF picture or Laughlin-like wave-function cannot be used to describe the reported[12] incompressible state at $\nu = 2 + 1/3 = 7/3$ $(\nu_1 = 1/3)$. The correlations in the $\nu = 7/3$ GS are different than at $\nu = 1/3$; the origin of (apparent) incompressibility cannot be attributed to the formation of a Laughlin-like $\nu_1 = 1/3$ state on top of the $\nu = 2$ state and connection between the excitation gap and the pseudopotential parameters is different. This is clearly visible in the dependence of the excitation gap Δ on the electron number N, plotted in Fig. 4 for $\nu = 1/3$ and $1/5$ fillings of the lowest and first excited LL. The gaps for $\nu = 1/5$ behave similarly as a function of N in both LL's, while it is not even possible to make a conclusive statement about degeneracy or incompressibility of the $\nu = 7/3$ state based on data for up to 11 electrons.

The SRR criterion can be applied to the QP pseudopotentials to understand why QP's do not form incompressible states at all Laughlin filling factors $\nu_{QP} = 1/m$ in the hierarchy picture[8,9] of FQH states. Lines in Fig. 1b and 1f mark V_{QE} and V_{QH} for the Laughlin $\nu = 1/3$ state of ten electrons. From similar calculations for diffent N one can deduce the behavior of the QP pseudopotentials in the $N \to \infty$ limit. Such analysis leads to a surprising conclusion that the SRR character characteristic of the electron–electron pseudopotential in the lowest LL does not generally hold for the QP pseudopotentials.[18,19] In Fig. 5 we show the data for QE's and QH's in Laughlin $\nu = 1/3$ (data for $N \leq 8$ was published before[19]) and $\nu = 1/5$ states. The plotted energy is given in units of e^2/λ where λ is the magnetic length in the parent state. Different symbols mark pseudopotentials obtained in diagonalization of N electron systems with different N

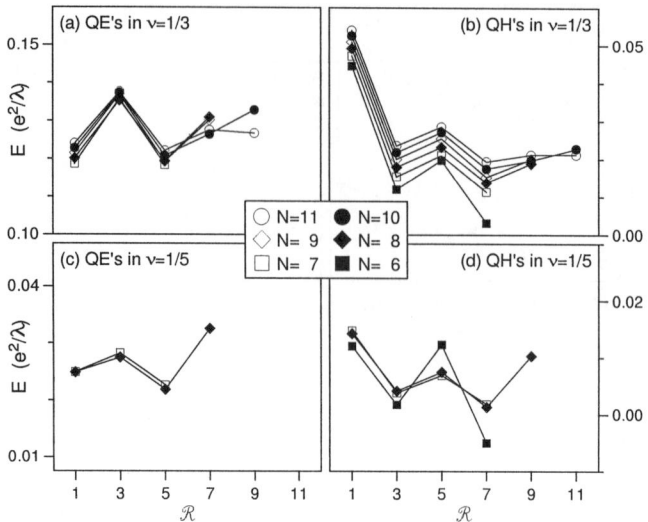

Figure 5. Energies of a pair of quasielectrons (left) and quasiholes (right) in Laughlin $\nu = 1/3$ (top) and $\nu = 1/5$ (bottom) states, as a function of relative pair angular momentum $\mathbf{R}$, obtained in diagonalization of N electrons.

and thus with different l_{QP}). Clearly, the QE and QH pseudopotentials are quite different and neither one decreases monotonically with increasing $\mathbf{R}$. On the other hand, the corresponding pseudopotentials in $\nu = 1/3$ and $1/5$ states look similar, only the energy scale is different. The convergence of energies at small $\mathbf{R}$ obtained for larger N suggests that the maxima at $\mathbf{R} = 3$ for QE's and at $\mathbf{R} = 1$ and 5 for QH's, as well as the minima at $\mathbf{R} = 1$ and 5 for QE's and at $\mathbf{R} = 3$ and 7 for QH's, persist in the limit of large N (i.e. for an infinite system on a plane). Consequently, the only incompressible daughter states of Laughlin $\nu = 1/3$ and $1/5$ states are those with $\nu_{QE} = 1$ or $\nu_{QH} = 1/3$ (asterisks in Fig. 1) and (maybe) $\nu_{QE} = 1/5$ and $\nu_{QH} = 1/7$ (question marks in Fig. 1). It is also clear that no incompressible daughter states will form at e.g. $\nu = 4/11$ or $4/13$. Taking into account the behavior of involved QP pseudopotentials on all levels of hierarchy explains all observed odd-denominator FQH states and allows prediction of their relative stability (without using trial wavefunctions involving multiple LL's and projections onto the lowest LL needed in Jain's CF picture).

8. Conclusion

Using the pseudopotential formalism, we have described the FQH states in terms of the ability of electrons to avoid strongly repulsive pair states. We have defined the class of SRR pseudopotentials leading to the formation of incompressible FQH states. We argue that the MFCF picture is justified for the SRR interactions and fails for others. The pseudopotentials of the Coulomb interaction in excited LL's and of Laughlin QP's in the $\nu = 1/3$ state are shown to belong to the SRR class only at certain filling factors.

Acknowledgment

AW and JJQ acknowledge partial support by the Materials Research Program of Basic Energy Sciences, US Department of Energy.

References

1. K. von Klitzing, G. Dorda, and M. Pepper, *Phys. Rev. Lett.* **45**, 494 (1980).
2. D. C. Tsui, H. L. Störmer, A. C. Gossard, *Phys. Rev. Lett.* **48**, 1559 (1982).
3. R. Laughlin, *Phys. Rev. Lett.* **50**, 1395 (1983).
4. *The Quantum Hall Effect*, edited by R. E. Prange and S. M. Girvin, Springer-Verlag, New York (1987).
5. F. D. M. Haldane and E. H. Rezayi, *Phys. Rev. Lett.* **60**, 956 (1988).
6. A. Wójs and J. J. Quinn, *Solid State Commun.* **108**, 493 (1998); ibid. **110**, 45 (1999); *Philos. Mag.* **B80**, 1405 (2000); J. J. Quinn, A. Wójs, *J. Phys.: Cond. Mat.* **12**, R265 (2000); A. Wójs, *Phys. Rev.* **B63**, 125312 (2001).
7. A. de Shalit and I. Talmi, *Nuclear Shell Theory*, Academic Press, New York (1963); R. D. Cowan, *The Theory of Atomic Structure and Spectra*, University of California Press, Berkeley (1981).
8. F. D. M. Haldane, *Phys. Rev. Lett.* **51**, 605 (1983).
9. P. Sitko, K.-S. Yi, and J. J. Quinn, *Phys. Rev.* B **56**, 12417 (1997).
10. S. Gasiorowicz, *Quantum Physics*, John Wiley and Sons, New York (1974).
11. R. B. Laughlin, *Surf. Sci.* **142**, 163 (1984); B. I. Halperin, *Phys. Rev. Lett.* **52**, 1583 (1984); J. K. Jain and V. J. Goldman, *Phys. Rev.* **B45**, 1255 (1992).
12. R. Willet, J. P. Eisenstein, H. L. Störmer, D. C. Tsui, A. C. Gossard, and J. H. English, *Phys. Rev. Lett.* **59**, 1776 (1987).
13. J. R. Mallet, R. G. Clark, R. J. Nicholas, R. L. Willet, J. J. Harris, and C. T. Foxon, *Phys. Rev.* **B38**, 2200 (1988); T. Sajoto, Y. W. Suen, L. W. Engel, M. B. Santos, and M. Shayegan, *Phys. Rev.* **B41**, 8449 (1990).
14. J. Jain, *Phys. Rev. Lett.* **63**, 199 (1989).
15. A. Lopez and E. Fradkin, *Phys. Rev.* **B44**, 5246 (1991).
16. T. T. Wu and C. N. Yang, *Nucl. Phys.* **B107**, 365 (1976).
17. X. M. Chen and J. J. Quinn, *Solid State Commun.* **92**, 865 (1996).
18. P. Béran and R. Morf, *Phys. Rev.* **B43**, 12654 (1991).
19. S. N. Yi, X. M. Chen, and J. J. Quinn, *Phys. Rev.* **B53**, 9599 (1996).

CORRELATION PROPERTIES OF INTERFERING ELECTRONS IN A MESOSCOPIC RING UNDER NONCLASSICAL MICROWAVE RADIATION

D. I. TSOMOKOS, C. C. CHONG AND A. VOURDAS

Department of Computing, School of Informatics,
University of Bradford,
Bradford BD7 1DP, United Kingdom

Interfering electrons in a mesoscopic ring are irradiated with both classical and nonclassical microwaves. The average intensity of the charges is calculated as a function of time and it is found that it depends on the nature of the irradiating electromagnetic field. For various quantum states of the microwaves, the electron autocorrelation function is calculated and it shows that the quantum noise of the external field affects the interference of the charges. Two-mode entangled microwaves are also considered and the results for electron average intensity and autocorrelation are compared with those of the corresponding separable state. In both cases, the results depend on whether the ratio of the two frequencies is rational or irrational.

1. Introduction

The Aharonov-Bohm effect [1] manifests itself as a nontrivial quantum phase, whenever electric charges travel in a field-free region enclosing a magnetostatic flux. This 'geometrical phase' has been generalized [2] and the original results have found applications in various contexts, for example in conductance oscillations in mesoscopic rings [3] and 'which-path' experiments that use novel solid-state devices [4].

A recent development of these ideas has been to replace the magnetostatic flux by an electromagnetic field [5]. The objective here is very different, since this 'ac Aharonov-Bohm experiment' constitutes a nonlinear device where the interaction between the interfering electrons and the photons leads to interesting nonlinear phenomena [6]. For an overview of related studies on the interaction of mesoscopic devices with microwaves we refer the reader to [7].

It is interesting to investigate the same phenomena with quantized electromagnetic fields. This 'quantum ac Aharonov-Bohm experiment' with nonclassical microwaves, has been studied [5] and one can quantify how the quantum noise destroys slightly the electron interference [8]. The aim is to investigate how various quantum phenomena and the quantum statistics of the nonclassical microwaves link to corresponding quantum phenomena on the electrons.

In what follows we study the interference of the electrons by calculating their intensity, while they are being irradiated with classical or nonclassical microwaves. The correlation properties of electron interference are then studied by calculating the autocorrelation function of the electron intensity (Sec. 2). We also consider two-mode microwaves with frequencies ω_1 and ω_2 (Sec. 3). Two-mode microwaves can be factorizable, separable or entangled [9] and since the problem of entanglement is generally complex, we approached it using an example. In particular, we assumed that the two modes of the microwave field form a Bell state and calculated its effect on electron interference. We found that the result is very different from that of the corresponding separable case. We conclude in Sec. 4 with a discussion of our results.

2. One-mode microwaves

2.1. Classical microwaves

The following system is considered: a beam of electric charges splits into two possible paths C_0 and C_1. The charges enter a region that is irradiated with microwaves (using a suitable waveguide). The microwaves propagate in the waveguide with the time-dependent magnetic field perpendicular to the plane of the two paths and the electric field parallel to it. Let ψ_0, ψ_1 be the electron wavefunctions with total winding equal to 1, in the absence of magnetic field. The effect of the electromagnetic field is the phase factor $\exp[ie\phi(t)]$ and the intensity is

$$I(t) = |\psi_0 + \psi_1 \exp[ie\phi(t)]|^2 = |\psi_0|^2 + |\psi_1|^2 + 2|\psi_0||\psi_1|\Re\{\exp[i(\sigma + e\phi(t))]\} \quad (1)$$

where $\sigma = \arg(\psi_1) - \arg(\psi_0)$. Units in which $k_B = \hbar = c = 1$ are used throughout. For simplicity we consider the case of equal splitting, in which $|\psi_0|^2 = |\psi_1|^2 = 1/2$ and let $\sigma = 0$. In this case we get

$$I(t) = 1 + \cos[e\phi(t)]. \quad (2)$$

We calculate the autocorrelation function of the electron intensity:

$$\Gamma(\tau) = \lim_{T \to \infty} \frac{1}{2T} \int_{-T}^{T} R(t,\tau)dt; \qquad R(t,\tau) \equiv I(t)I(t+\tau). \quad (3)$$

An expansion of $\Gamma(\tau)$ into a Fourier series gives the spectral density S_K:

$$S_K = \frac{\Omega}{2\pi} \int_0^{2\pi/\Omega} \Gamma(\tau) \exp(-iK\Omega\tau)d\tau$$

$$\Gamma(\tau) = \sum_{K=-\infty}^{\infty} S_K \exp(iK\Omega\tau). \tag{4}$$

Firstly, we consider the case where the classical time-dependent flux is given by

$$\phi(t) = \phi_1 \sin(\omega t) \tag{5}$$

and using Eqs. (2) and (3) we find the autocorrelation function:

$$\Gamma_{cl}(\tau) = [1 + J_0(e\phi_1)]^2 + 2\sum_{K=1}^{\infty} [J_{2K}(e\phi_1)]^2 \cos(2K\omega\tau), \tag{6}$$

where J_K are Bessel functions. Comparison of Eqs. (4) and (6) shows that $\Omega = 2\omega$ and

$$S_0 = [1 + J_0(e\phi_1)]^2; \qquad S_K = [J_{2K}(e\phi_1)]^2. \tag{7}$$

2.2. Nonclassical microwaves

A monochromatic electromagnetic field of frequency ω is considered, at temperatures $k_B T \ll \hbar\omega$. We quantize the electromagnetic field by considering the vector potential A_i and the electric field E_i as dual quantum variables. The loop $C = C_0 - C_1$ is small in comparison to the wavelength of the microwaves, hence the A_i and the E_i can be integrated around it and yield the magnetic flux ϕ and the electromotive force V_{EMF}, respectively, as dual quantum variables. The annihilation operator can be introduced as $a = 2^{-\frac{1}{2}}\xi^{-1}\left(\phi + i\omega^{-1}V_{\text{EMF}}\right)$, and similarly the creation operator, where ξ is a constant proportional to the area enclosed by C. The flux operator is consequently written as $\phi(t) = \exp(itH)\phi(0)\exp(-itH)$, where H is the Hamiltonian that contains the $\omega a^\dagger a$ term and an interaction term. This interaction term can be neglected for small currents.

Under these conditions the magnetic flux, which defines the phase factor, becomes the operator $\hat{\phi}(t) = (\xi/\sqrt{2})\left[\exp(i\omega t)a^\dagger + \exp(-i\omega t)a\right]$. Hence this phase factor $\exp(ie\phi)$ now is

$$\exp\left[ie\hat{\phi}(t)\right] = D\left[iq\exp(i\omega t)\right], \quad q = \frac{\xi e}{\sqrt{2}} \tag{8}$$

where $D(\lambda)$ is the displacement operator $D(\lambda) = \exp(\lambda a^\dagger - \lambda^* a)$. The interference between the two electron beams is described by the intensity operator

$$\hat{I}(t) = 1 + \cos\left[e\hat{\phi}(t)\right] = 1 + \frac{1}{2}D\left[iq\exp(i\omega t)\right] + \frac{1}{2}D\left[-iq\exp(i\omega t)\right]. \quad (9)$$

Let ρ be the density matrix describing the external nonclassical microwaves. The expectation value of the electron intensity is

$$\langle I(t)\rangle \equiv \mathrm{Tr}\left[\rho\hat{I}(t)\right] = 1 + \frac{1}{2}\tilde{W}(\lambda) + \frac{1}{2}\tilde{W}(-\lambda); \quad \lambda = iq\exp(i\omega t), \quad (10)$$

where $\mathrm{Tr}\left[\rho D(\lambda)\right] \equiv \tilde{W}(\lambda)$ is the Weyl (or characteristic) function which has been studied by various authors including ourselves (e.g. [10] and references therein).

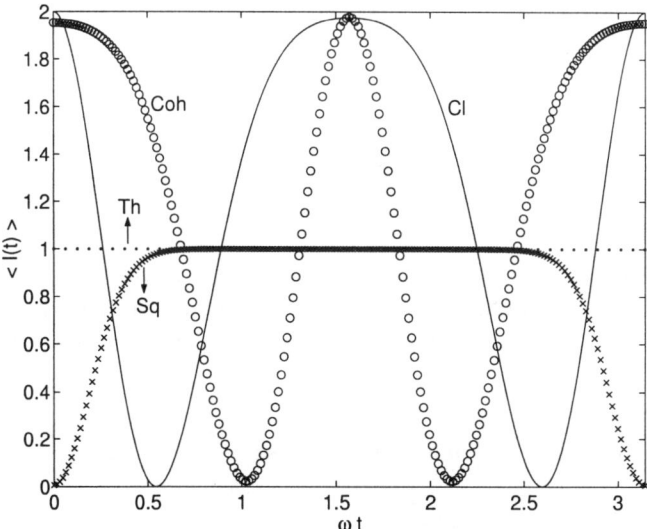

Figure 1. $\langle I(t)\rangle$ as a function of ωt for $\omega = 10^{-4}$, $\langle N \rangle = 200$, $r = 6.4$. We use units where $\hbar = k_B = c = 1$. Continuous line represents the case of irradiation with classical microwaves; line of circles, coherent states; line of crosses, squeezed states; and dotted line, thermal states.

We have calculated $\langle I(t)\rangle$ for various quantum states of the microwaves (using results for $\tilde{W}(\lambda)$ in Ref. [11]). In order to find the $\Gamma(\tau)$ from Eq. (3), one needs to calculate the quantity

$$R(t,\tau) \equiv \mathrm{Tr}\left[\rho\hat{I}^\dagger(t)\hat{I}(t+\tau)\right] \quad (11)$$

Numerical results are presented for different quantum states that we calculated. In particular, we plot four cases: classical microwaves and

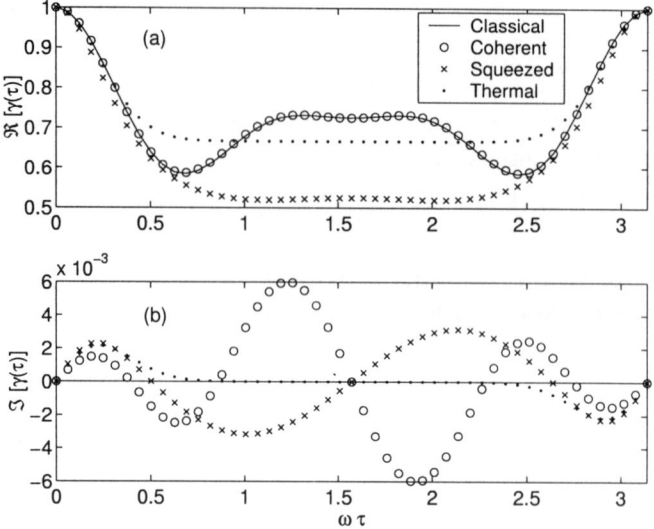

Figure 2. $\gamma(\tau)$ as a function of $\omega\tau$ for $\omega = 10^{-4}$, $\langle N \rangle = 200$, $r = 6.4$. Part (a) shows the real part of $\gamma(\tau)$; part (b) shows the imaginary part. We use units where $\hbar = k_B = c = 1$.

nonclassical microwaves in coherent, squeezed, and thermal states. For a meaningful comparison, we consider the case where the average number of photons $\langle N \rangle$ in coherent, squeezed, and thermal states is the same:

$$\langle N \rangle = |A|^2 = \left[\sinh\left(\frac{r}{2}\right)\right]^2 + \left[\cosh\left(\frac{r}{2}\right) - \sinh\left(\frac{r}{2}\right)\right]^2 B^2$$

$$= \frac{1}{\exp(\beta\omega) - 1}. \qquad (12)$$

For the classical case we took $\phi_1^2 = 2|A|^2 = 2\langle N \rangle$. In all results of Figs. 1 to 3, $\omega = 10^{-4}$ (which in our units is eV), $\langle N \rangle = 200$, $r = 6.4$.

The results show that the quantum noise in the irradiating microwaves affects the electron interference. All microwaves that we have considered have the **same average number of photons**, but differ in the quantum noise. These four types of microwaves lead to different electron interference results and different autocorrelation functions. Irradiation of the electrons by nonclassical microwaves leads to nonzero value of the imaginary part of the electron autocorrelation function. This is not so (i.e. the imaginary part of $\Gamma(\tau)$ vanishes) when the ring is irradiated with classical microwaves.

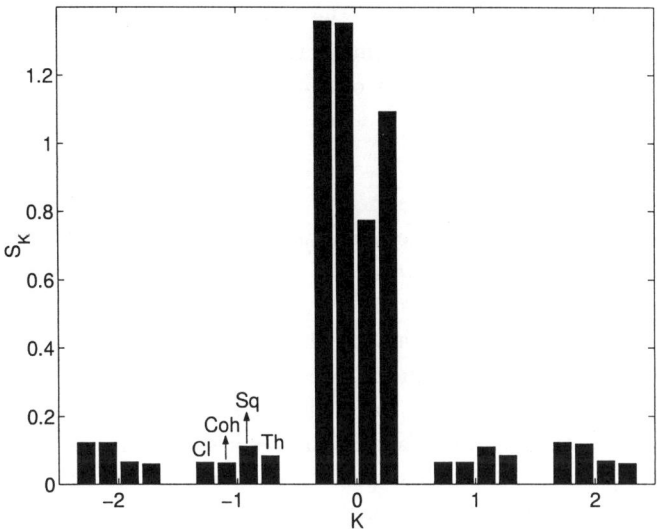

Figure 3. S_K coefficients for the electrons with $\langle N \rangle = 200$, $r = 6.4$. We use units where $\hbar = k_B = c = 1$. The bars correspond to irradiation of the ring by (from left to right): classical, coherent, squeezed, and thermal states.

3. Two-mode nonclassical microwaves

We consider two-mode nonclassical microwaves. We are particularly interested to study how entangled two-mode microwaves affect the electron interference. For this reason we consider a Bell state $|s\rangle = 2^{-1/2}(|01\rangle + |10\rangle)$ where $|01\rangle$, $|10\rangle$ are two mode number eigenstates. For comparison we also consider the separable (disentangled) state

$$\rho_{sep} = \frac{1}{2}(|01\rangle\langle 01| + |10\rangle\langle 10|). \tag{13}$$

Clearly, the density matrix of the entangled state $\rho_{ent} = |s\rangle\langle s|$ can be written as

$$\rho_{ent} = \rho_{sep} + \frac{1}{2}(|01\rangle\langle 10| + |10\rangle\langle 01|). \tag{14}$$

In this case the phase factor $\exp[ie\phi(t)]$ becomes the product of two displacement operators and, consequently, the intensity becomes

$$\hat{I}(t) = 1 + \frac{1}{2}D_1(\lambda_1)D_2(\lambda_2) + \frac{1}{2}D_1(-\lambda_1)D_2(-\lambda_2); \quad \lambda_j = iq\exp(i\omega_j t) \tag{15}$$

for two modes $(j = 1, 2)$. Therefore, we find that

$$\langle I(t) \rangle_{sep} = 1 + \left(1 - q^2\right)\exp\left(-q^2\right), \tag{16}$$

$$\langle I(t) \rangle_{ent} = \langle I(t) \rangle_{sep} - q^2 \exp\left(-q^2\right)\cos[(\omega_1 - \omega_2)t]. \tag{17}$$

It is seen that for this example, the $\langle I(t) \rangle_{sep}$ is constant in time, while the $\langle I(t) \rangle_{ent}$ is an oscillatory function of time. Clearly, different correlations among the two irradiating modes of the microwaves may lead to different average electron intensities.

4. Discussion

The subject of mesoscopic devices interacting with microwaves has received attention in the last few years (e.g., Ref. [7]). Our contribution has been to consider that these microwaves are prepared in various nonclassical states [5,8,12]. Here we have quantified the effect of the quantum noise on electron interference. More specifically we have calculated both the electron average intensity and the spectral density for several types of nonclassical microwaves and a comparison of the results with the case of classical microwaves (Figs. 1-3), demonstrates clearly that the presence of both classical and quantum noise in the nonclassical microwaves affects the electron intensity. What is more, when the ring is irradiated with two-mode microwaves, then entanglement among these two modes (i.e., the formation of a Bell state) leads to a time-dependent expectation value of the electron intensity.

References

1. M. Peshkin and A. Tonomura, *The Aharonov-Bohm effect*, Lecture notes in Physics Vol. 340, Berlin: Springer (1989).
2. A. Shapere and F. Wilczek (ed), *Geometric Phases in Physics*, Singapore: World Scientific (1989).
3. S. Washburn and R.A. Webb, *Adv. Phys.* **35**, 375 (1986)
   A.G. Aronov and Y.V. Sharvin, *Rev. Mod. Phys.* **59**, 755 (1987).
4. G. Hackenbroich, *Phys. Rep.* **343**, 464 (2001).
5. A. Vourdas, *Phys. Rev.* **B54**, 13175 (1996)
   A. Vourdas and B.C. Sanders, *Europhys. Lett.* **43**, 659 (1998).
6. M.P. Silverman, *Nuovo Cimento* **B97**, 200 (1987)
   M. Buttiker, *Phys. Rev.* **B46**, 12485 (1992).
7. M. Buttiker, *J. Low Temp. Phys.* **118**, 519 (2000)
   R. Deblock *et. al.*, *Phys. Rev.* **B65**, 075301 (2002).
8. P. Cedraschi, V.V. Ponomarenko, M. Buttiker, *Phys. Rev. Lett.* **84**, 346 (2000)
   A. Vourdas, *Phys. Rev.* **A64**, 053814 (2001).
9. R.F. Werner, *Phys. Rev.* **A40**, 4277 (1989); A. Peres, *Phys. Rev. Lett.* **77**, 1413 (1996); R. Horodecki and M. Horodecki, *Phys. Rev.* **A54**, 1838 (1996); V. Vedral *et. al.*, *Phys. Rev. Lett.* **78**, 2275 (1997).
10. S. Chountasis and A. Vourdas, *Phys. Rev.* **A58**, 848 (1998).
11. A. Vourdas, *Phys. Rev.* **B49**, 12040 (1994).
12. C.C. Chong, D.I. Tsomokos, A. Vourdas, *Phys. Rev.* **A66**, 33813 (2002).

COHERENT SPIN TRANSPORT AND QUANTUM INTERFERENCE IN MESOSCOPIC LOOP STRUCTURES

I.TRALLE AND W.PAŚKO

Institute of Physics, University of Rzeszów
Al.Rejtana 16A, 35-310 Rzeszów, Poland
e-mail: tralle@univ.rzeszow.pl

The quantum interference in a loop structure caused by spin coherent transport and Larmor precession of the electron spin is presented. A 'spin ballistic' regime is assumed, where the phase relaxation length for the spin part of the wave function $(L_\varphi^{(s)})$ is much greater than the relaxation length for the 'orbital part' $(L_\varphi^{(e)})$. In the presence of an additional magnetic field, the spin part of the electron wave function (WF) acquires a phase shift due to additional spin precession about that field. If the structure length L is chosen to be $L_\varphi^{(s)} > L > L_\varphi^{(e)}$, it is possible to 'wash out' the quantum interference related to the phase coherence of the 'orbital part' of the WF, retaining at the same time that related to the phase coherence of the spin part and hence, to reveal corresponding conductance oscillations. The additional mechanism of spin relaxation, the scattering by the surface of the structure and its influence on the spin coherent transport is also considered.

1. Introduction

There was a time of great excitement, scientific and engineering activities when vacuum tubes were replaced by solid state devices. Nowadays, an approach to electronics is emerging that is based on controlling the electron spin rather than its charge. Many propositions for substituting the charge of carriers (electrons) by their spins have been made already [1-3], so even the name of this emerging field, 'spintronics' was coined.

Up to now most of attention was paid to the manipulation of electron spins in micro- and nanostructures by means, for instance, spin injection [4,5], while another possible non-classical devices based on quantum interference of spins attracted less attention.

One of the main ideas which underpins different possible applications of 'spin transport', including information storage and computation, is that the spins of electrons in semiconductors may have very long quantum coherence times [3,6], or in other words, electrons can travel a long way without flipping their spins. But this also gives the possibility to observe quantum effects which involve the interference of electron waves. In the classical picture of transport phenomena, the total probability for a particle to transfer from one point to another is the sum of the probabilities for such transfer over all possible trajectories. In the quantum description, this result corresponds to neglecting the interference of scattered electron waves propagating along different paths. The destruction of quantum coherence is controlled by the phase relaxation time or phase relaxation length. Since for the electron spin this length may be very long, it is naturally to expect that the spin interference can reveal itself in the conductance oscillations similar to that ones which are due to Aharonov-Bohm effect [7]. Most of the researchers who dealt with the Aharonov-Bohm effect considered mainly the Hamiltonian $\hat{H} = (\mathbf{p} - (e/c)\mathbf{A})^2/2m^* + U(y)$, where $U(y)$ is the energy corresponding to the transverse motion, and almost nobody takes into account the spin-part $\mu\hat{\sigma}_B\mathbf{B}$ of the Hamiltonian (μ is the Bohr magneton, $\hat{\sigma}$ is the electron spin operator, $\mathbf{B}$ - magnetic field). However, if the quantum interference is concerned, the quantity of main importance is the coherence length. If one considers the total Hamiltonian which includes Pauli term, one can write down the electron wave function in factorized form as the direct product: $\Psi(\mathbf{r}, s) = \varphi(\mathbf{r}) \otimes \chi(s)$ and consider the coherence of each part separately as it was done in the previous papers [8,9] of one of us. As a result, it is possible to introduce two phase relaxation lengths, the first one for the 'orbital part' of the electron wave function, $L_\varphi^{(e)}$, and the second one, $L_\varphi^{(s)}$ for the spin part of the wave function. It turns out [8,9] that $L_\varphi^{(s)} >> L_\varphi^{(e)}$ which is in total agreement with the experiment [3,6,10]. The physics which is behind that is the following. An electron during its transfer along some path in the solid (semiconductor, for definiteness) interacts all the time with the environment. As a rule rigid scatters such as impurities and other defects of crystalline structure do not contribute to the phase relaxation; only dynamical scatters like phonons do. On the other hand, the electron scattering by phonons is mainly inelastic, while impurity scattering is mainly elastic, so we can say that only inelastic scattering contributes to the phase relaxation. But what does it mean inelastic scattering in case of spin ? It means spin flips which are the consequences of scattering by phonons accompanied by spin-orbit interaction. This interaction is very weak and that is why the spin flips are rare events and the phase relaxation length for the spin part of the electron wave function is very long. But now, if the structure length L is chosen to be $L_\varphi^{(s)} > L > L_\varphi^{(e)}$, it is possible to 'wash out' the quantum interference related to phase coherence of the 'orbital part' of the wave function retaining at the same time the phase coherence of the spin part one and hence, to reveal the corresponding conductance oscillations of the microstructure. Such model was considered in the papers [8,9] where the simple theory of the quantum interference in a loop structure due to Larmor precession of electron spin in semiconductor microstructure was presented for the first time. The aim of this paper is to develop the approach further, discussing other aspects of the problem such as, for instance, electron scattering by the edges of the structure and its influence on the quantum interference of spins.

2. The model and necessary preliminaries

Let us start with a generic microstructure with two end regions ($x < 0$ and $x > L$) and a middle region $0 \leq x \leq L$ consisting of two channels (Figure 1) similar to that one considered in [9].

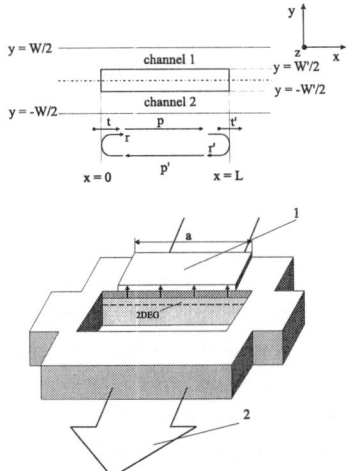

Figure 1. A sketch of a two-channel semiconductor mesoscopic structure with an additional magnetic field (1) across one of the channels. On the upper panel t, t', r, r' stand for the transmission and reflection matrices at the two junctions; P, P' stand for the propagation matrices in the middle region; 2 - the external magnetic field B_0.

The main difference however is the following. In the Ref [9], apart an external magnetic field $\mathbf{B_0}$ in the plane of the microstructure, on the upper surface of one of the channels there was a regular periodic array of micromagnets which created and additional magnetic field. As it was mentioned in [9], the periodic magnetic field was not obligatory and that choice was motivated by the current interest in the study of electron motion in inhomogeneous magnetic fields on the nanometer scale. Here instead, even more simpler case of uniform magnetic field is considered, since the only thing which is really needed is that the magnetic fields to be different in the two arms of the loop. An additional advantage of the uniform magnetic field is that it is more appropriate from the experimentalist's point of view.

Consider an electron enters the domain occupied by the magnetic field, say, from the left-end region. The electron's spin wave function is a coherent superposition of the spin-up and spin-down eigenstates, which are split in the magnetic field by the Zeeman energy $\Delta E = g\mu_B B$, (g is the Lande factor). Coherent evolution under the spin Hamiltonian results in oscillations between the two eigenstates; classically this oscillation corresponds to the precession of spin vector at the Larmor frequency $\Delta\varepsilon/\hbar$. In other words, we consider the non-relativistic electon motion in the magnetic field as the motion of a classical top which precesses about the magnetic field. Since the magnetic fields are different in the two arms of the structure, the phase shifts acquired by the spin wave functions are also different and if one of the field (say, B_0) can be altered, it should lead to the specific conductance oscillations periodic in B.

So, the magnetic field $\mathbf{B}$ which affects the electron in the first arm of the structure is equal to $\mathbf{B} = \mathbf{B_0}$, while in the other one it is equal to $\mathbf{B} = \mathbf{B_0} + \mathbf{B_1}$. Suppose the Hamiltonian of the electron is $H = H_0 + H_1$, where

$$H_0 = 1/2m^* \left(\mathbf{p} - (e/c)\,\mathbf{A}\right)^2 + U(\mathbf{r}) \ , H_1 = -\mu_B \hat{\sigma} \mathbf{B} \ , \qquad (1)$$

where m^* is the electron effective mass, $\mathbf{A}$ is vector potential corresponding to the magnetic field $\mathbf{B}$, μ_B and $\hat{\sigma}$ are Bohr magneton and the spin operator respectively. We also assume that $U(\mathbf{r})$ describes conduction bands bending due to space charge and discontinuities of any band. Since H_0 does not depend on spin, the wave function is the direct product: $\Psi(\mathbf{r}, s) = \varphi(\mathbf{r}) \otimes \chi(s)$. Ever since for convenience we shall refer to $\varphi(\mathbf{r})$ as the 'orbital part' of the total wave function, keeping in mind that it corresponds to H_0 describing the charge-field interaction, and we will refer to $\chi(s)$ as the spin-part of the wave function related to H_1, the spin part of the Hamiltonian H in (1).

Let us now introduce the phase-relaxation length $L_\varphi^{(s)}$ for the spin part of the wave function, in just the same way as the one usually introduced for the 'orbital part', $L_\varphi^{(e)}$. As it was argued in the Introduction, the phase relaxation length $L_\varphi^{(s)}$ is much greater than $L_\varphi^{(e)}$. The proof of the statement can be found in Ref.[9]; however, for the reader's convenience we give here its brief discussion.

First of all, as it already mentioned above, as a rule rigid scatters such as impurities and other defects of crystalline structure do not contribute to phase-relaxation; only dynamical scatters like phonons do. But impurity scattering also can be phase-randomizing if the impurity has an internal degree of freedom with the result that it can change its state. For example if magnetic impurities have an internal spin that fluctuates with time, the collisions with such impurities cause phase relaxation. So, we suppose there are no such impurities here. Let us also add some more comments concerning the phase-relaxation length $L_\varphi^{(s)}$ which is connected to the corresponding phase-relaxation time $\tau_\varphi^{(s)}$. The formula for $L_\varphi^{(s)}$ was obtained in Ref.[9], but in order to make estimates, it is sufficient to use its simplified version: $L_\varphi^{(s)} \sim v_F \tau_\varphi^{(s)}$.

In order to estimate the phase-relaxation time $\tau_\varphi^{(s)}$ consider a simple model. Let us take a two-state quantum system (which we shall refer to as a sub-system $\mathcal{A}$) with excitation energy ε interacting with a phonon bath, and identify the two states with 'spin-up'($|\uparrow>$) and 'spin-down' ($|\downarrow>$) states of a spin in an external magnetic field. For simplicity we suppose the subsystem $\mathcal{A}$ interaction with the phonons to be resonant; this means that only those modes of the phonon bath whose energy is equal to ε interact with the two-level subsystem. Other modes are taken into account indirectly by choosing all mean values of phonon bath parameters to be equal to their statistical average at given temperature T. As a result, for the model of phonon bath we can take a great number ($N \gg 1$) of identical non-interacting subsystems $\mathcal{B}_n$ with excitation energy ε.

Thus, the Hamiltonian of the entire system (subsystem $\mathcal{A}$ + phonon bath) is

$$H = \varepsilon \left(a^\dagger a + \sum_{n=1}^{N} b_n^\dagger b_n \right), \tag{2}$$

where $a^\dagger, a$ are the Fermi creation and anihilation operators related to the excitations of subsystem $\mathcal{A}$, while $b_n^\dagger, b_n$ are Bose creation and anihilation operators related to the excitations of the n-th subsystem of phonon bath.

So, for the subsystems of the equidistant spectra, we have

$$< b_n^\dagger b_n >= Sp(\rho_n b_n^\dagger b_n) = (\exp(\beta\varepsilon) - 1)^{-1}, \beta = 1/k_B T,$$

where ρ_n is the statistical operator for the $\mathcal{B}$ subsystems, k_B is the Boltzmann constant, T is the temperature, $Sp(...)$ is the trace operator.

The interaction of two-level systems with phonons can be described by the term $\mathcal{H}_{int}(t)$:

$$\mathcal{H}_{int}(t) = \sum_{n=1}^{N} (\theta(t - \tau(n - 1)) - \theta(t - \tau n)) H_n, \tag{3}$$

where

$$\theta = \begin{cases} 1 & \text{if } t > 0 \\ 0 & \text{if } t \le 0, \end{cases}, H_n = \varepsilon_{int}(a^\dagger b_n + b_n^\dagger a)$$

and ε_{int} is the interaction energy.

The physical meaning of (3) is that the subsystem $\mathcal{A}$ interacts each time during interval τ with those subsystem $\mathcal{B}_n$ which did not interact with $\mathcal{A}$ during previous time interval, or, in other words, τ is the 'electron-phonon collision time'.

Introduce now two probabilities $p_1(t)$ and $p_2(t)$ for the subsystem $\mathcal{A}$ to be at a time t in excited state and unexcited one, respectively. It is well-known that the steady state for the two-level system corresponds to $p_1 = p_2 = 1/2$. As it can be shown (see Ref. [9]), for the model described above, the time t which is needed for the subsystem $\mathcal{A}$ to achieve the state with $p_1 = p_2 = 1/2$ is equal to:

$$t = (\hbar^2/\tau\varepsilon_{int}^2) \ln 2 \tanh(\beta\varepsilon/2). \tag{4}$$

Since the two levels of the subsystem $\mathcal{A}$ correspond to the 'spin-up' and 'spin-down' states, the steady state corresponds to the redistribution of initially non-equilibrium spin distribution due to spin-flips and hence, to the total destruction of spin coherence. Thus, time t is the spin relaxation time, which can be identified with $\tau_{\varphi,ph}^{(s)}$, because it relates to the phase coherence destruction caused by phonon inelastic scattering accompanied by the spin flips.

For the magnetic field about $1T$ and the temperature about $5K$, finally we have for the time $t \sim \tau_{\varphi,ph}^{(s)}$ the value $\sim 2.2 \times 10^{-10}s$.

The values of spin relaxation time measured for the A_3B_5 semiconductors are widely ranged from 10^{-12} to 10^{-7} s [12] and hence we can conclude that our estimation for $\tau_\varphi^{(s)}$ is quite reasonable. Anyway, we can take it for certain that $\tau_{\varphi,ph}^{(s)} \gg \tau_\varphi^{(e)}$. Indeed, the experiments show [13] that at 5 K the phase relaxation time $\tau_\varphi^{(e)}$ is about 1.6×10^{-12} s and as a result, $L_\varphi^{(s)} \gg L_\varphi^{(e)}$ and hence, we can suppose that the structure length L can be chosen to be $L_\varphi^{(s)} > L > L_\varphi^{(e)}$.

However, there si another mechanism which also, in principle, can contribute to the spin relaxation. This mechanism is the scattering by the surface and the edges of the structure. The role of this mechanism in spin relaxation is the topic of the next section.

3. Scattering by the edges of the structure

Now our aim is to calculate the probability of the spin-flips caused by the electron scattering by the edges of the structure in question, since such spin-flips results in the 'washing up' of 'phase memory' of the spin-part of the electron wave function.

It is commonly used to treat the interaction of electrons with the surface of the sample in terms of a phenomenological parameter ε_0 introduced for the first time by F. Dyson [14]. This quantity can be defined as the mean probability of spin-flip of the conduction electrons having the energy $\mathcal{E} = \mathcal{E}_F$ at their collision with the surface of the sample, averaged over the incident angles. Following [15], define ε_0 as

$$\varepsilon_0 = \hbar^{-1} \int d\Omega (\mathbf{JS})^{-1} \int d\mathbf{k'} \rho(\mathbf{k'}) \delta[\mathcal{E}(\mathbf{k}, \sigma) - \mathcal{E}(\mathbf{k'}, \sigma')] |V_{\mathbf{k'}, \sigma'; \mathbf{k}, \sigma}|^2,$$

where Ω is the incident solid angle, σ, σ' are the spin variables corresponding to the states before and after the collision with the surface, $\mathbf{k}, \mathbf{k'}$ are the electron wave vectors before and after collision; $\mathbf{J}$ is the flux density of electrons incident on the surface from within; S is the surface area, $\rho(\mathbf{k'})$ is the density of states in the $\mathbf{k}$ - space over the single spin; $V_{\mathbf{k'}, \sigma'; \mathbf{k}, \sigma}$ is the matrix element of the perturbation operator responsible for the spin-flips; $\mathcal{E}(\mathbf{k}, \sigma), \mathcal{E}(\mathbf{k'}, \sigma')$ are the electron energies in the corresponding quantum states.

It is quite obvious that the crystal potential which is supposed to be periodic deep inside the sample, changes abruptly at the edges and vicinity of the surface where it is not periodic altogether. These abrupt changes of the potential lead to the emerging of an electric field in a thin layer near the surface of the sample. The thickness of the layer can be estimated as to be 2-4 monolayers. Hence, if the electrons are moving in this layer, one can treat it as the quasi-two dimensional electron gas (2DEG).

Introduce now the space variables x, y, z where z-axis is normal to the 2DEG. Suppose the external magnetic field $\mathbf{B_0}$ is directed along y-axis, $\mathbf{B} = (0, B, 0)$ and let the vector potential $\mathbf{A}$ to be: $\mathbf{A} = (B, 0, 0)$. Now the Hamiltonian of the electron moving in our 2DEG is of the form:

$$H = \frac{1}{2m^*} \left[\left(p_x - \frac{eBz}{c} \right)^2 + p_y^2 + p_z^2 \right] + U(z) + \mu_B \hat{\sigma} B_0. \tag{5}$$

Here $U(z)$ is the potential responsible for the space quantization in 2DEG in z-direction. The electron functions corresponding to such Hamiltonian can be written down as

$$\Psi_{\mathbf{k}\sigma}(\mathbf{r}) = C \exp(ik_x + ik_y) \varphi(k_x, z) \exp(-i\mathcal{E}_k/\hbar) \otimes \chi_\sigma \tag{6}$$

and the normalizing conditions defined as

$$\int dx \psi^*_{k'_x k'_y \sigma'}(x,y)\psi_{k_x k_y \sigma}(x,y) = \delta(\sigma,\sigma')\delta(\mathbf{k}-\mathbf{k}')$$

$$C = ((L_x L_y)\int_{-\infty}^{\infty}|\varphi(k_x,z)|^2 dz)^{-1/2}.$$

Here L_x, L_y are the structure sizes in x, y-directions.

As it was mentioned above, due to the potential $U(z)$, the strong electric field emerges at the surface of the structure . Obviously, the field is equal to $E = -e^{-1}\frac{\partial}{\partial z}U(z)$. If, just like in the Sect. 2 we suppose the absence of magnetic impurities in semiconductor, the only mechanism responsible for the spin-flips is the spin-orbit interaction. In the refrence frame of moving electron, the electric field generates an effective magnetic field of the form $\sim [\mathbf{E} \times (d\mathbf{r}/dt)c^{-1}]$, which causes spin flips. The operator of spin-orbit interaction has the form (see Ref. [16]):

$$V^{so} = \frac{\hbar^2}{4m^{*2}c^2}\frac{\partial}{\partial z}U(z)\left(\sigma_y \frac{1}{i}\frac{\partial}{\partial x} - \sigma_x \frac{1}{i}\frac{\partial}{\partial y}\right), \tag{7}$$

where σ_y, σ_x are the corresponding Pauli matrices and this is the perturbation mentioned above.

After some manipulations carried out in accordance with procedure developed in Ref. [15], the probability of the spin flips caused by the potential (7) and averaged over the incident angles, can be written as:

$$\varepsilon_0 = \frac{1}{60}(mc^2)^{-2}(cos^2(\hat{\vec{B},\vec{n}}) + 1)I,$$

where $(\hat{\vec{B},\vec{n}})$ is the angle between $\mathbf{B}$ and the normal to the surface and

$$I = \left|\int_{-\infty}^{\infty}dz\varphi^*(-k_x,z)\frac{\partial U}{\partial z}\varphi(k_x,z)\right|^2. \tag{8}$$

Assuming, as it was done in [15], that the parameter $\alpha = g\mu_B B/\mathcal{E}_F \ll 1$ (g is the Lande factor), it is possible to calculate integral in (8) even without having known the wave functions $\varphi_{k_x z}$ and the form of the potential $U(z)$. The authors of [15] considered the electron scattering by the surface in a metal where the Fermi energy is high enough and the parameter α is small. However, in our case this assumption is no longer valid, since in 2DEG Fermi energy is not so high. So, we calculate the integral in (8) using the simple model of 'triangular potential well'. To this end, we write down the Schrödinger equation for the 'orbital part' of the wave functions (6) (it corresponds to the Hamiltonian (5) without Pauli's term) in the form:

$$-\frac{\hbar^2}{2m^*}\frac{\partial^2\varphi(k_x,z)}{\partial z^2}+$$

$$+\left[\frac{\hbar^2 k_x^2}{2m^*}+\frac{\hbar eBk_x z}{m^*c}+\frac{(eBz)^2}{2m^*c^2}+U(z)-\mathcal{E}(k_x,k_y)\right]\varphi(k_x,z)=0, \quad (9)$$

where $\mathcal{E}(k_x,k_y)=\mathcal{E}_{k_x}-\hbar^2 k_y^2/2m^*$.

Now suppose the potential $U(z)$ has the form of triangular potential well, that is

$$U(z)=\begin{cases}\infty, & z<0\\ eEz, & z\geq 0\end{cases}$$

Suppose also the so called quantum limit to occur, when only the lowest bands are occupied by the electrons. In the absence of a magnetic field, electrons in lower subbands (small subband index n) have a large mean free path, because the amplitude near the boundary is smaller and the scattering is less frequent. Therefore, the current is carried mostly by electrons in lower subband. In a magnetic field, electrons are pushed toward the boundaries by a Lorentz force and the scattering is enhanced. From now on, we assume also the condition $(d_0/l_B)^4 \ll 1$ is fulfilled. In the last relation $d_0=(3\hbar^2\pi^2/16m^*eE)^{1/3}$ is the thickness of the 2DEG layer and $l_B=(\hbar c/eB)^{1/2}$ is the so called magnetic length. Then, the solution of the eigenvalue problem for Eq.(7) in the quantum limit approximation is of the form (see [17]):

$$\varphi(k_x,z)=\begin{cases}Ai\left(\left[\frac{2m^*eE}{\hbar^2}+\frac{2eBk_x}{\hbar c}\right]^{1/3}\left[z-\frac{\mathcal{E}(k_x,k_y)-\hbar^2 k_x^2/2m^*}{eE+\hbar eBk_x/m^*c}\right]\right), & z\geq 0\\ 0, & z<0\end{cases}$$

$$\mathcal{E}(k_x,k_y)-\hbar^2 k_x^2/2m^*=(\hbar^2/2m^*)^{1/3}\left[\frac{9\pi}{8}\left(eE+\frac{\hbar eBk_x}{m^*c}\right)\right]^{2/3},$$

where $Ai(...)$ means Airy function (see, for instance, [18]). By means of the last formulae and numerical evaluation of the integral in (8), we estimated the spin flip probability due to edge scattering. We used also the next values of the parameters: $B\sim 0.1T, k_x\sim k_F=2\pi/\lambda_F, \lambda_F\sim 4\times 10^{-6}cm, E\sim 10^3 Vcm^{-1}, m^*\sim 0.1m_e$; then, according to our estimates, the dimensionless probability of spin flip is equal approximately $\sim 3.7\times 10^{-17}$. Now the estimates of the time of spin flips due to scattering by the edges of the structure can be done as follows. It is clear that the considered mechanism can cause the spin flip only if the electrons are in the domain where the potential $U(z)$ is essential. So, the spin flip time is just approximately equal to the product of the time which electrons spend in this domain multiplied by the inverse probability of the spin flip ε_0. The time which electrons spend in this domain can be estimated from as the distance where electron wave function in the potential $U(z)$ is essentially non-zero, divided by v_F. According to our estimates, the dimensionless probability of spin flip is equal approximately $\sim 3.7 \times 10^{-17}$, while the time of spin flip due to the interaction with the surface is about $\sim 0.11 sec$. Considering two mechanisms of spin flips (the scattering by phonons and by surface or the edges) as independent, we can calculate the total probability of spin flip as the sum of the probabilities of these two events. As a result, the inverse spin relaxation time is the sum of the inverse relaxation times for two scattering mechanisms: $(\tau_\varphi^{(s)})^{-1} = (\tau_{\varphi,ph}^{(s)})^{-1} + (\tau_{\varphi,s}^{(s)})^{-1}$, where $\tau_{\varphi,ph}^{(s)}, \tau_{\varphi,s}^{(s)}$ stand for the phase relaxation times due to scattering by the phonons (and which was calculated in Sec.2) and by edges (surface), respectively. Since $\tau_{\varphi,ph}^{(s)} \gg \tau_{\varphi,s}^{(s)}$, the total spin relaxation time is practically equal to $\tau_{\varphi,ph}^{(s)}$. We can conclude that the scattering by the surface (edges) of the structure does not contribute essentially to the spin relaxation time and can be neglected. Therefore, $L_\varphi^{(s)}$ is indeed much greater than $L_\varphi^{(e)}$ and hence, the structure length can be chosen to be $L_\varphi^{(s)} > L > L_\varphi^{(e)}$.

4. Calculation of the transmission coefficient

The current I through the structure considered in the previous section, for the small applied potential V, can be written as [11,19]:

$$I = \frac{2e}{h} \int d\mathcal{E} \int (w_z dk_z / 2\pi)[f(\mathcal{E}) - f(\mathcal{E} + eV) \sum_{n',n''} |T_{n',n''}|^2. \qquad (10)$$

Here w_z is the width of the structure in the z-direction, $T_{n',n''}$ is the transmission coefficient from the state n' in the left-hand end to the state n'' in the right-hand end, $\mathcal{E}$ and k_z are the energy and the transverse wave vector of the electrons as they enter from the left-hand end.
Taking into account the relation $\Psi(\mathbf{r}, s) = \varphi(\mathbf{r}) \otimes \chi(s)$ and using the next property of the direct product $(A \otimes B)(C \otimes D) = AC \otimes BD$, one can demonstrate that $T_{n',n''} = T_{k',k''} \otimes T_{\sigma'\sigma''}$ where subscripts k', k'' relate to the states of H_0, while σ', σ'' relate to the states of H_1. Since $L_\varphi^{(s)} >> L_\varphi^{(e)}$, we can demonstrate (see [9]) that $T_{n',n''} = <T> T_{\sigma'\sigma''}$, where $<T>$ is the averaged transition coefficient which does not depend on the phase relation between the states of H_0 in the left-hand end and in the rigth-hand end of the structure. So, in accordance with the assumptions above, there are two states ('spin up' and 'spin down') to consider in the end regions, while in

the middle region there are four states corresponding to the channels 1 and 2. Dropping the subscripts σ', σ'', one can write down the next expression for the transmission coefficient T [20]:

$$T = t' [I - PrP'r']^{-1} Pt. \tag{11}$$

Here I is the unit matrix, t is 4×1 matrix describing the transmission from the left-hand end into the two channels, while t' is 1×4 matrix describing the transmission from the channels into the right-hand end. Similarly, r and r' are 4×4 matrices describing the reflections at the two junctions of the channels back into the channels. Matrices P and P' describe forward and reverse propagation of the electron wave through the channels 1 and 2, respectively. In order to construct matrices r and r', let us suppose $i, j = 1, ..., 4$ each stand for one of four states: 'spin up' or 'spin down' in the channels 1 or 2. Then the r_{ii} stand for the scattering from a state of definite spin ('spin up' or spin down') to the same state (in other words, for 'self-scattering')in the channels 1 or 2 at the first junction; say, r_{11} means $| \uparrow > \rightarrow | \uparrow >$ scattering in the first channel, r_{33} means $| \uparrow > \rightarrow | \uparrow >$ and so on. The same is true for r'_{ii} but at the second junction, while r_{ij}, r'_{ij} stand for the scattering from the 'spin-up' state to the 'spin-down' state and vice versa or for the scattering between the same spin states but of different channels at the first second junction, respectively.

In accordance with the consideration given in section 2, there are no spin flips ($| \uparrow > \rightarrow | \downarrow >$) in two channels considered, and only the following matrix elements of r are non-zero: $r_{11}, r_{13}, r_{22}, r_{24}, r_{31}, r_{33}, r_{42}, r_{44}$ (the same is true for the matrix elements of r'). Hence, the matrices r, r' are of the form:

$$r = \begin{pmatrix} r_{11} & 0 & r_{13} & 0 \\ 0 & r_{22} & 0 & r_{24} \\ r_{31} & 0 & r_{33} & 0 \\ 0 & r_{42} & 0 & r_{44} \end{pmatrix} \quad r' = \begin{pmatrix} r'_{11} & 0 & r'_{13} & 0 \\ 0 & r'_{22} & 0 & r'_{24} \\ r'_{31} & 0 & r'_{33} & 0 \\ 0 & r'_{42} & 0 & r'_{44} \end{pmatrix}$$

In order to construct P and P', it is necessary to note that the spin parts of the wave functions acquire the phase factors due to Larmor spin precession about $\mathbf{B}$ - axis. Since magnetic field in the channels are different, these phase factors are also different.

One can treat the states 'spin up' and 'spin down' as the two opposite points on a unit sphere S^2 which can be transformed one into another under rotation by an angle $\varphi = \pm \pi$ about some axis $\mathbf{a}$. Introduce also formally $\mathbf{b}$ - axis which is a unit vector of the precession axis: $+\mathbf{b}$ corresponds to the electron propagation from $x = 0$ to $x = L$ while $-\mathbf{b}$ corresponds to reverse propagation, and θ_1 and θ_2 are the phase acquired by spin part of the wave functions in the channels 1 and 2, respectively. Then the matrix elements describing the phase shifts in the two channels can be written as:

$$P_{\pm 1} = exp(\pm i\varphi_a)exp(i\theta_{1,b}), \quad P'_{\pm 1} = exp(\pm i\varphi_a)exp(-i\theta_{1,b}), \quad (12)$$
$$P_{\pm 2} = exp(\pm i\varphi_a)exp(i\theta_{2,b}), \quad P'_{\pm 2} = exp(\pm i\varphi_a)exp(-i\theta_{2,b}). \quad (13)$$

The idea of (12)-(13) is to express the elements of the matrices P, P' as the two rotations about two independent axis. Then, these objects are nothing else but the unitary quaternions [21]. As is known [21], any quaternion can be written in the form $q = c_0 + i_1 c_1 + i_2 c_2 + i_3 c_3 = \sum_{\alpha=0}^{3} i_\alpha c_\alpha$, where $i_0 = 1$ and $i_1^2 = i_2^2 = i_3^2 = i_1 i_2 i_3 = -1$. However, it is possible also to define, for instance, i_1, i_2 as

$$i_1 = \begin{pmatrix} 0 & 1 \\ -1 & 0 \end{pmatrix}, i_2 = \begin{pmatrix} 0 & i \\ i & 0 \end{pmatrix},$$

where i is the ordinary complex square root of -1, thus forcing

$$i_3 = i_1 i_2 = \begin{pmatrix} i & 0 \\ 0 & -i \end{pmatrix}.$$

If these three matrices are multiplied by $-i$, one obtains Pauli spin matrices. Thus, the quaternion q could have been identified with the complex 2-by-2 matrix

$$\begin{pmatrix} c_0 + ic_3 & c_1 + ic_2 \\ -c_1 + ic_2 & c_0 - ic_3 \end{pmatrix} = \begin{pmatrix} u & v \\ -v^* & u^* \end{pmatrix},$$

where u and v are complex numbers with complex conjugates u^* and v^*. Replacing $0, 1$ and i in these complex matrices by

$$\begin{pmatrix} 0 & 0 \\ 0 & 0 \end{pmatrix}, \begin{pmatrix} 1 & 0 \\ 0 & 1 \end{pmatrix}, \begin{pmatrix} 0 & 1 \\ -1 & 0 \end{pmatrix},$$

respectively, one can obtain a representation of quaternions as 4-by-4 matrices.

Since two channels 1 and 2 are supposed to be isolated, in this way the matrices P and P' can be represented as the diagonal 4×4-matrices with the diagonal elements defined by (12)-(13).

After a great deal of algebra (see Ref.[9]), we have:

$$|T|^2 = |a_1|^2 + |a_2|^2 + |a_3|^2 + |a_4|^2 + (a_1^* a_3 + a_1 a_3^* + a_2^* a_4 + a_2 a_4^*) + (a_1^* a_2$$
$$+ a_1 a_2^* + a_2^* a_3 + a_2 a_3^* + a_3^* a_4 + a_3 a_4^*) cos\Delta\theta, \Delta\theta = \theta_1 - \theta_2,$$

where $a_i, (i = 1, 2, 3, 4)$ do not depend on θ_1, θ_2 and are the complicated functions of $r_{ij}, r'_{ij}, t_i, t'_i$.

So, the problem now is to calculate the additional phase shift $\Delta\theta = \theta_1 - \theta_2$ which arises due to the precession of electron spin in the magnetic fields of two arms of the structure.

5. Calculation of the phase shift

Consider now the non-relativistic motion of the particle (electron) with the spin $s = 1/2$ in a two- component magnetic field: $\mathbf{B} = \mathbf{B_0} + \mathbf{B_1}$, $\mathbf{B_0} = (0, B_0, 0)$, and $\mathbf{B_1} = (0, 0, B_1)$, where B_1 is an additional uniform magnetic field in one of the channels of the structure. The spin part of electron wave function can be considered as a two-component vector defined by the pair of functions $\chi(|\uparrow>)$ and $\chi(|\downarrow>)$ which stand for the probability amplitudes of the two possible orientations of spin. The spin operator $\hat{\sigma}(\sigma_x, \sigma_y, \sigma_z)$ is defined in terms of Pauli matrices:

$$\sigma_x = \begin{pmatrix} 0 & 1 \\ 1 & 0 \end{pmatrix}, \sigma_y = \begin{pmatrix} 0 & -i \\ i & 0 \end{pmatrix}, \sigma_z = \begin{pmatrix} 1 & 0 \\ 0 & -1 \end{pmatrix}$$

Thus,we can treat the mean value of the magnetic moment of the electron moving within the channels of microstructure as the classical quantity $\mathbf{P} = <\sigma>$, its evolution under magnetic field being defined by the equation:

$$\frac{d\mathbf{P}}{dt} = \gamma\left[\mathbf{P}, \mathbf{B}\right],$$

where $\gamma = e/mc$ is the electron gyromagnetic constant.

In other words, the vector $\mathbf{P}$ can be treated as classical magnetic top and, if this classical top having the initial orientation $\mathbf{P_0} = (P_x^0, P_y^0, P_z^0)$ enters magnetic field $\mathbf{B} = (B_x, B_y, B_z)$, it begins to precess about magnetic field with the frequency $\Omega = \gamma B$, where $B = \sqrt{B_x^2 + B_y^2 + B_z^2}$.

It is interesting to note that despite its purely quantum character, the spin of the particle during its movement in external fields often can be treated classically. The accuracy of such treatment can be estimated by means of Heisenberg uncertainty relation, since classical treatment is possible if one can neglect the commutator $[\mathbf{r}, \mathbf{p}]$ where $\mathbf{p}$ is the particle momentum operator. So, the measure of accuracy of the classical approximation is $|\Delta p|/p$. Δp in our case can be estimated as $\sim m\Delta v = m(v^2/l_B)\Delta t$, where $l_B = \sqrt{\hbar c/|e|B}$ is the magnetic length and $\Delta t \sim 2\pi/\omega_c, \omega_c = |e|B/mc$ is the cyclotron frequency, while $p \sim mv_F$. As a result, $|\Delta p|/p \sim 2\pi m v_F \sqrt{mc}/\sqrt{\hbar|e|B}$. Assuming $v_F \sim 3 \times 10^7 cms^{-1}$ and $B \sim 0.1T$, we have $|\Delta p|/p \approx 2 \times 10^{-8}$. Therefore, indeed to a good approximation we can treat the evolution of vector $\mathbf{P}$ as the evolution of the classical magnetic top under external magnetic field.

Let us introduce now the phase of precessing spin by means of the formula

$$\theta(v,x) = \mu_B/\hbar \int_0^x B(v,x)dt = \gamma \int_0^x B(v,x)dt,$$

and take into account that the fields $\mathbf{B_0}$, $\mathbf{B_1}$ are uniform. Then the phase of precessing spin depends on t linearly: $\theta(v,t) = \gamma Bt$. Now the calculation of the phase shift $\Delta\theta$ can easily be done. Moreover, it is clear that under certain conditions including appropriate structure length L, electron velocity and the values of magnetic fields B_0, B_1, the phase shift $\Delta\theta = \theta_2 - \theta_1$ can be multiple of $\pi/2$. Indeed,

$$\Delta\theta = \theta_2 - \theta_1 = (n+1/2)\pi = (\gamma L/v)(\sqrt{B_0^2 + B_1^2} - B_0) \ , n = 0,1,2...$$

If the values of B_1, L, v, n are given, the value of B_0 which is needed for the $\Delta\theta$ to be equal of multiple of $\pi/2$ can be easily calculated:

$$B_0 = \frac{\gamma L}{\pi v}B_1^2 - \frac{\pi v n^2}{4\gamma L}.$$

Hence, changing the external magnetic field B_0, one can change the phase shift and the quantum interference from constructive to destructive one and back. Also it is seen that $\Delta\theta = \theta_2 - \theta_1 = f(B_0, B_1, v)$ is the function of B_0, B_1, v. That is, the phase shift generally speaking is different for the electrons with different velocities. At first sight, this makes matters worse, because it means that the 'interference pattern 'should be blurred. One should remember, however, that the temperature is considered to be sufficiently low. That is, the electron distribution fuction $f(\mathcal{E}) = \chi(\mathcal{E}_F - \mathcal{E})$ and $v = v_F$, where $\chi(...)$ is the Heaviside step-like function, $\mathcal{E}_F, v_F$ are the Fermi energy and Fermi velocity, respectively. So, the calculation by means of (10) now can easily be done and we have:

$$I = (2e/h)\,K(A + Dcos\Delta\theta(v_F)),$$

where K, A, D are the coefficients dependent on the peculiarities of the structure. Now it is clear that changing B_0 one can approach very deep modulation of the conductance and since $A \sim D$, the 'contrast' of the 'interference pattern' is defined only by the ratio $\sqrt{\frac{\mathcal{E}_F - k_B T}{\mathcal{E}_F}}$ which at the temperature of about 40 K is of the order of 90 per cent.

6. Discussion and conclusion

A simple theory of the quantum interference due to Larmor precession of an electron spin in a loop structure is presented in this paper. Also, we assumed here 'ballistic spin transport' - that is, the phase-relaxation length $L_\varphi^{(s)}$ of the spin part of the electron wave function is assumed to be greater than the microstructure length L. If in one of the arms of microstructure there is an additional magnetic field, the spin wave function acquires a phase shift due to additional spin precession about that field. In addition to the phonon scattering, here we considered another possible mechanism of spin relaxation, namely scattering by the edges or the surface of the structure. It turns out, that the spin relaxation time due to this mechanism is even greater than that of caused by phonon scattering. Hence, the probability of spin flips is determined mainly by the phonon scatterring, while spin scattering by the edges or by the surface of the structure can be neglected. That the spin flips are more frequent under the phonon scattering, can be clearly understood, if one remember, as it mentioned above, the rigid scatterers do not contribute to the phase relaxation; only the dynamical - that is, time-dependent - scatterers, do. In our model the spin flips, as inelastic scattering events, are caused by the *explicitly time-dependent interaction* (see Eq (3)), which above all, is resonant. Certainly, among the great many modes of phonon bath the resonant ones are always present. On the other hand, the operator of spin-orbit interaction in case of edge (or surface) scattering is *explicitly time-independent* (Eq (7)). However, in most of the cases this interaction is far to be resonant and that is why the spin flips under surface scattering are much more rare events than even under phonon scattering. Now if we suppose the microstructure length is chosen to be greater than the $L_\varphi^{(e)}$, it is possible to 'wash out' the quantum interference related to phase coherence of the 'orbital' part of the wave function retaining at the same time that related to the phase coherence of the spin part and hence, reveal the corresponding conductance oscillations.

In conclusion, it is worth emphasizing that extremely long spin coherence times were indeed observed in recent experiments done by J. Kikkawa and D. Awshalom [3,6].

References

1. B. Doudin, J-Ph. Ansermet, Europhys. News 28 (1997) 14
2. G.A. Prinz, Science 282 (1998) 1660
3. J.M. Kikkawa, J.A. Gupta, I. Malajovich, D.D. Awschalom, Physica E 9 (2001) 194
4. G. Burkard, D. Loss, Physica E 9 (2001) 175

5. G. Schmdt, L.W. Molenkamp, Physica E 9 (2001) 202
6. D.D. Awschalom, J.M. Kikkawa, Physics Today 52 (1999) 33
7. S. Washburn, R.A. Webb, Adv.in Phys. 35 (1986) 375
8. I.Tralle, Acta phys. polonica 94 (1998) 603
9. I.Tralle, J Phys: Condens. Matter 11 (1999) 8239
10. D. Hägle, M. Oestreich, W.W. Rüle, N. Nestle, K.Eberl Appl Phys Lett 73 (1998) 1580
11. S. Datta, Electronic Transport in Mesoscopic Systems, Cambridge: Cambr, Univ. Press 1995
12. R.P. Parsons, Can. J Phys 49 (1971) 1850
13. K.K. Choi, D.C. Tsui and K. Alavi, Phys Rev B 36 (1987) 7751
14. F.J. Dyson, Phys Rev 98 (1955) 349
15. V.N. Lisin, B.M. Khabibulin, Fiz Tverd Tel (Sov. Solid State Phys) 17 (1975) 1600
16. D. Bohm, Quantum Theory, NY: Prentice-Hall Inc. 1952
17. O.V. Kibis, Pis'ma JETF (Sov. JETPH Lett) 66 (1997) 551
18. M. Abramowitz, I.A. Stegun, eds. Handbook of Mathematical Functions, Dover Publ. Inc., 1965
19. S. Datta and S. Bandyopadhyay, Phys Rev Lett 58 (1987) 717
20. P.W. Andreson, Phys Rev B 23 (1981) 4828
21. G. Casanova, L'algebre vectorielle, Paris: Pesses Universitaires de France 1976
22. M. Peshkin, in: Fundamental Problems in Quantum Theory: a Conference Held in Honor of Professor John A. Wheeler (Annals of New York Academy of Sciences, vol 755) eds. D.M. Greenberg and A. Zeilinger, NY, New York Acad Sci, 1995

THE POLYHEDRAL REPRESENTATION OF HYPERSURFACES FOR INTRAMOLECULAR REARRANGEMENTS: APPLICATION TO THE ICOSAHEDRAL H ⊗ 2H JAHN-TELLER SURFACE

E. LIJNEN AND A. CEULEMANS

Division of Quantum Chemistry, Katholieke Universiteit Leuven
Celestijnenlaan 200F,
B-3001 Leuven, Belgium
E-mail: erwin.lijnen@chem.kuleuven.ac.be

The polyhedral representation is introduced as a new way to depict intramolecular rearrangements. In this representation the reaction graph is embedded on the appropriate 2D closed manifold. The resulting structure is capable of visualizing spatial relationships. As an illustration we apply this technique to obtain 2D representations of the 5D hypersurface, which describes the Jahn-Teller distortions of a fivefold degeneracy.

1. Introduction

In most studies of chemical reactions or intramolecular rearrangements, one has to deal with complex multidimensional PES. Because the human mind is not trained to visualize objects in more than three dimensions, one has to search for simplified representations that still retain the most essential features of the surface. Already in 1968 Muetterties [1] proposed to visualize these surfaces by means of a graph where the vertices denote the minima on the PES and the edges the possible interconversion paths. Although this representation is extremely useful in most cases, one has to realize that a graph is a purely combinatorial object and is not capable of describing spatial relationships. So there is a clear need for a more elaborated representation that also takes into account spatial aspects. One way to do this is by mapping the graph on a 2D closed manifold to give a *polyhedral* representation. [2] In this representation the underlying manifold is divided into polygons, called faces, which form a new entity that was not apparent in the graph-like representation.

In the following we shortly review the classification of the 2D closed surfaces and describe how to extend a graph to its polyhedral representation. As an example we derive the polyhedral representations of the graphs corresponding to the linear H $\otimes$ 2h Jahn-Teller problem.

2. From Graph to Polyhedron

The main properties of a polyhedral surface are undoubtedly related to the topology of the underlying manifold which it decorates. This topology is completely described by means of two properties: the Euler characteristic χ and the orientability of the surface. The former can be calculated easily from the celebrated Euler theorem [3] which states that the number of vertices, edges and faces, denoted as V, E, and F, obey the following rule:

$$V - E + F = \chi \qquad (1)$$

Here χ is a fixed integer, the Euler characteristic, which marks the particular topology of the surface on which the polyhedron is embedded. However, a surface is only completely characterized when also its orientability is given. A surface is called orientable if there exists no path on the surface that would take you from the outside to the inside, otherwise it is said to be non-orientable. Taking into account this orientability, by a theorem of Brahana,[4] all 2D closed surfaces can be divided into two subclasses:

(1) The sphere with handles: Take the sphere and attach p handles to form the orientable surface S_p. The corresponding Euler characteristic is given by $\chi(S_p) = 2 - 2p$. For example the torus is homeomorphic to a sphere with one handle and has $\chi = 0$.

(2) The sphere with crosscaps: Take the sphere and attach q crosscaps to form the non-orientable surfaces N_q which have the Euler characteristic $\chi(N_q) = 2 - q$. For example the projective plane is homeomorphic to the sphere with one crosscap and has $\chi = -1$

Connected with every surface is also its Euler genus which is defined as $\bar{\gamma} = 2 - \chi$, and can therefore be calculated directly from the Euler equation. This brings us to the formulation of the problem that is at the basis of our discussion, namely:

Given a certain graph: what is the surface with the lowest Euler genus on which this graph can be embedded without edges crossing each other?

In general this problem is known to be NP-complete.[5] However, in some highly symmetrical cases a solution to the problem can be found. Among these are the complete graphs for which in the orientable case:

$$\bar{\gamma}(K_n) = 2 \left\{ \frac{(n-3)(n-4)}{12} \right\} \; for \; n \geq 3 \qquad (2)$$

and in the non-orientable case:

$$\bar{\gamma}(K_n) = \{\frac{(n-3)(n-4)}{6}\} \; for \; n \neq 7 \; and \; n \geq 3, \; \gamma(K_7) = 3 \qquad (3)$$

where the notation $\{x\}$ is the ceiling function, i.e. it corresponds to the smallest integer not less than x. This result is related to the long standing map coloring problem and was successfully solved by G. Ringel in 1968.[6]

3. The H ⊗ 2h Jahn-Teller Case

The linear H ⊗ 2h Jahn-Teller problem describes the instability of icosahedral molecules in five-fold degenerate states. Previous studies have revealed the possibility of two cases, depending on the coupling regime, which can both be described in a five dimensional coordinate space.[7] In the first regime one finds six equivalent pentagonal minima of D_{5d} symmetry, while in the other there are ten equivalent trigonal minima of D_{3d} symmetry. Several attempts have been made in the literature to represent the essential geometric structure of this surface. These studies were motivated by the observation of a ground state crossover under certain coupling conditions, accompanied by a disappearance of the Berry phase for the smallest tunneling cycles.[8,9] While phase tracking along the interconversion paths has yielded an explanation of the crossover, the general structure of the surface remains unknown. In the following we will investigate this problem using the polyhedral representation.

3.1. *The Pentagonal Case*

In the pentagonal case one has six equivalent minima which are all equidistant in coordinate space. It is obvious that the corresponding graph will be the complete graph K_6 on six vertices, in which the vertices are the stable minima and the edges the possible tunneling paths between these minima

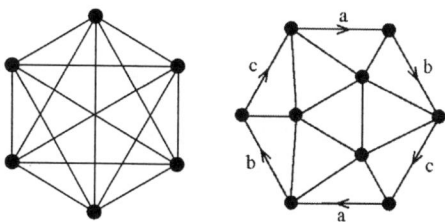

Figure 1. The complete graph K_6(left) and its embedding in the projective plane(right). Sides labeled with the same index have to be identified in such a way that the orientation is preserved

which in these case all have the same probability. Following formulas 2 and 3, one can deduce that it is possible to embed this graph on the sphere with one crosscap, also known as the projective plane. In Figure 1 this embedding is shown to be a triangulation of the projective plane. This means that the shortest closed tunneling paths (cycles of length three) are realized as faces of the embedding. Notice that the mapping of the graph lowers the symmetry from S_6 to I. This is due to the fact that in the polyhedral representation, one has the extra condition that automorphisms also have to map faces to faces.

3.2. The Trigonal Case

Although we have seen that in the trigonal case one has ten equivalent minima, these minima are no longer equidistant in coordinate space. Every minimum has three neighboring minima at distance r_A and six minima at distance r_B. Depending on the magnitude of the coupling-parameters these distances will vary and give rise to three possible cases (Figure 2). One can intuitively see that tunneling paths between nearest neighbors have the greatest probability to occur, so one finds the corresponding graphs by drawing edges between nearest neighbors only. The intermediate case where both distances r_A and r_B are equal corresponds to the complete graph K_{10} on ten vertices. According to equations 2 and 3 this graph can be embedded on the non-orientable surface with seven crosscaps. Starting from this embedding the induced embedding of both the Petersen graph and its complement can easily be deduced.

In Figure 3 we give a planar representation of the minimal genus embedding of K_{10}, derived from the proof of Ringel. The original surface with seven crosscaps can be reconstructed from this representation by gluing appropriate segments on the outer boundary to each other. Numbers on the outer boundary denote which vertex will be the destination of the edge

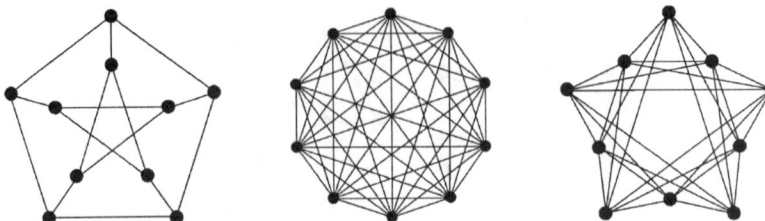

Figure 2. The three possible cases of tunneling paths in the trigonal case. With increasing r_A the Petersen-graph(left) with $r_A < r_b$ transforms into its complement(right) with $r_A > r_b$ via the intermediate K_{10} graph for which $r_a = r_b$

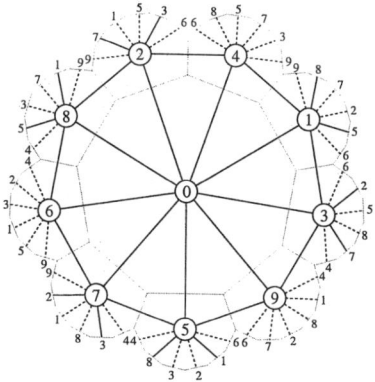

Figure 3. Embedding of K_{10} on the surface with seven crosscaps. Dashed lines represent edges that run over a crosscap

leaving the boundary there. One sees for instance that the edge from 8 to 1 runs from vertex 8 till the boundary and reappears on the right side of the figure to complete its course to vertex 1. Notice that there are two different types of edges, full lines and dashed lines, which respectively denote normal edges and edges that run over a crosscap. This means that along a dashed line from one vertex to another we will change our relative position (inside or outside) to the surface. A closer examination shows that K_{10} divides the surface into thirty triangles. Using the Euler equation, one can therefore simply show that the underlying surface is indeed the sphere with seven crosscaps. One remarkable fact is that by embedding K_{10} on this surface, its symmetry is lowered from the total symmetric group S_{10} to the icosahedral group I, which is exactly the symmetry of our initial undistorted system. Notice that this result was obtained with only one restriction, namely that the genus of the surface should be minimal.

Figure 4 represents the induced embedding of both the Petersen graph (left) and its complement (right), obtained directly from the embedding of K_{10}. In case of the Petersen graph one finds six faces of length five, meaning (by application of the Euler formula) it is an embedding on the projective plane. Its symmetry is lowered from the total symmetric group S_5 to the icosahedral group I. For the complement one finds fifteen square faces, giving an embedding on the surface with seven crosscaps. In this case the symmetry of the graph and its embedding is in both cases S_5. Looking at the Petersen-graph, one sees that the smallest closed tunneling paths are of length five, and are realized as faces of the embedding. For its complement however the smallest closed paths are of length three and are not faces of the

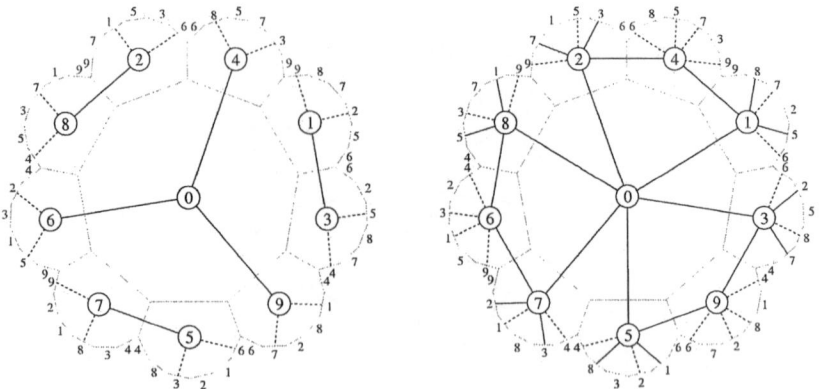

Figure 4. Induced embedding of the Petersen graph(left) and its complement(right), obtained from the embedding of K_{10}(Figure 3)

embedding. In fact, all cycles of length three are non-contractible, meaning that they encircle a point that is not part of the surface and therefore are not homotopic to a point. This observation matches the result of phase tracking calculations on the H $\otimes$ 2h JT problem.[8] The Berry phase for pentagonal loops, corresponding to faces on the Petersen graph, was found to vanish, while in contrast the non-orientable triangular loops on the complement were characterized by a Berry phase of π. This indicates that the flux lines which give rise to the topological phase factors leave the surface through the non-contractible cycles on the K_{10} polyhedron. In this way we obtain a transparent pictorial representation of the entire JT surface.

Acknowledgments

Financial support from the Fund for Scientific Research - Flanders and the Flemish Government (Concerted Action Scheme) is gratefully acknowledged.

References

1. E.L. Muetterties, *J. Am. Chem. Soc.* **90**, 5097 (1968).
2. A. Ceulemans and E. Lijnen, *Eur. J. Inorg. Chem.* **7**, 1571 (2002).
3. H.S.M. Coxeter, *Convex Polytopes*, 3rd *Edition* (Dover, New York, 1973).
4. H.R. Brahana, *Ann. of Math.* **30**, 234 (1923).
5. C. Thomassen, *Journal of Algorithms* **10**, 568 (1989).
6. G. Ringel, *Map Color Theorem* (Springer, Berlin, 1974).
7. A. Ceulemans and P.W. Fowler, *J. Chem. Phys.* **93**, 1221 (1990).
8. C.P. Moate et al., *Phys. Rev. Lett.* **77**, 4362 (1996).
9. N. Manini and P. De Los Rios, *Phys. Rev.* **B62**, 29 (2000).

SCALE INVARIANT BEHAVIOUR IN GROWING AND COALESCING DROPLETS

J. A. BLACKMAN

Department of Physics, University of Reading,
Whiteknights, P.O. Box 220,
Reading RG6 6AF, UK
E-mail: j.a.blackman@reading.ac.uk

J. POULTER

Department of Mathematics, Faculty of Science,
Mahidol University, Rama 6 Road,
Bangkok 10400, Thailand
E-mail: scjpt@mahidol.ac.th

Aggregation phenomena occur in many branches of physics. We give a review of scaling behaviour that is general to a wide range of systems, and then focus on a particular model of droplet growth and coalescence that has certain unusual features. The problem is related to that of a dense random packing of hard discs or spheres for which a fractal description is useful. We discuss an analytic solution within mean field theory, and examine the limits of validity.

1. Introduction

1.1. *Background*

Non-equilibrium growth processes are basic to a wide range of physical phenomena in fields as diverse as nanotechnology, atmospheric science and cosmology [1-5]. We envisage some basic entities (atoms, droplets, etc) being introduced into a system (eg onto a surface) at a constant rate. Aggregation occurs and a random array of clusters with a range of sizes is formed. Our concern here will be the deposition of vapour droplets onto a surface and the way in which their growth and size distribution evolves with time. Common to all growth phenomena is the concept of scaling, with exponents that reflect the microscopic details of the aggregation process. We will review in this introduction the basic ideas of scaling, and introduce the issues involved in developing a theory for various types of phenomena. In subsequent sections we will introduce a particular model for the vapour deposition problem, explore the results of computer simulations using this model, and discuss attempts to develop a consistent theory.

1.2. *Scaling*

The key feature of the scaling hypothesis is the observation that the configuration of cluster sizes at one time looks the same as that at another time apart from a rescaling of lengths. The distribution of clusters of sizes s at time t can be written in the following form (or several equivalent ones) for a wide range of systems that exhibit scaling

$$N_s(t) = s^{-\theta} f(s/S(t)), \tag{1}$$

where $f(x)$ is some function of the scaled size and $S(t)$, the mean cluster size, acts as the characteristic scaling variable. The exponent θ depends on the particular physical system. An overview of scaling theory in terms of generalised homogeneous functions is given by Stanley [6].

It is useful to introduce moments of the distribution function, which with a trivial manipulation can be shown to scale with some power of S

$$M_n = \sum_s s^n N_s(t) = S^{n-\theta+1} \int x^{n-\theta} f(x) dx. \tag{2}$$

A second exponent, z, is introduced to describe the time evolution of the mean cluster size

$$S \sim t^z, \tag{3}$$

and so the explicit time dependence of the moments takes the form

$$M_n \sim t^{z(n-\theta+1)}. \tag{4}$$

For most systems of interest, new material is being deposited at a constant rate, and so the first moment scales like t. From Eq. (4), we obtain a relation between the two exponents

$$z(2-\theta) = 1. \tag{5}$$

Eq.(5) has very general applicability. We have thus reduced the problem, at least at this stage, to one of a single independent exponent.

1.3. *Types of processes*

To evaluate this exponent, we have to include some information about the microscopic processes underlying the aggregation and growth. The way to proceed also depends on the form of the distribution function. Two generic types of distribution are illustrated in Figure 1. The one on the left (a) represents the commonest situation, with a function having a single peak and going to zero (or to a finite value) for small clusters and exponentially to zero for large clusters. The one on the right (b) is bimodal and describes a system containing a small number of very large clusters embedded in a sea comprising a large number of small clusters. The former is typical of systems in which atoms are deposited on a surface. The atoms are highly mobile, but are quickly captured by immobile islands that have already formed on the surface. If, on the other hand, new material is injected into the system as vapour droplets, these can remain where they land, and the alternative distribution may result.

(a)   (b)

Figure 1. Schematic illustrations of two types of distribution functions: $N_s(t)$ against s for a particular instant of time. Plot (b) is understood to be on a logarithmic scale.

Eq. (5) applies for both types of distribution. For distributions of type (a), the theory is usually developed via the Smoluchowski equation, which describes the time evolution of the distribution function

$$dN_s/dt = \sum_{i+j=s} K_{ij} N_i N_j - N_s \sum_j K_{sj} N_j. \tag{6}$$

There is an extensive body of literature on the solutions of Eq. (6) and the resulting exponents for a wide variety of reaction kernels, K_{ij}. Over the last few years, however, it has been realised that one has to exercise considerable care in applying Eq. (6) to actual systems, because it basically represents a mean field theory and ignores spatial fluctuations in the distribution function. Despite the mean field nature, it does provide reliable values of exponents in most cases, but generally does a poor job on the distribution function itself. Much of the recent work in this field has been an attempt to go beyond mean field theory. The work of Stroscio and Pierce [7] contains a clear experimental demonstration of scaling for type (a) systems. References [8-12] provide an introduction to the issues mentioned and contain references to the extensive literature on the subject.

There has also been some work [13, 14] done on applying Eq. (6) to systems of type (b). However, there is an additional complication in the case of the bimodal distribution. The predominant (in number) small clusters largely determine the cluster density, whereas it is the large clusters that contribute most to the mean cluster size. Mathematically this means that the lower cut-off to the integral in Eq. (2) is important in the calculation of the lower order moments, whereas it is irrelevant (at least for large times) for the higher moments. For the bimodal distribution, therefore, Eq. (4) only applies for n larger than a certain value.

Although less common than the single peak form, bimodal distributions have been observed in a wide range of experimental situations [15-20]. Also systems exhibiting this type of distribution are less well understood than those of the other type, and it does not appear that a theory based of Eq. (6) is the best way to proceed. In the next section we will describe a model system that shows bimodal behaviour, and explore its properties in some detail.

2. Bimodal systems

2.1. Model

The model that we will consider is one that was studied in some detail by computer simulations about a decade ago [3, 21, 22]. The model consists of hyperspherical droplets with dimensionality D. When a droplet of radius r_1 overlaps with another of radius r_2, a new droplet is formed with radius r given by

$$r^D = r_1^D + r_2^D. \tag{7}$$

New elementary droplets of radius r_0 (size $s_0 \sim r_0^D$) are introduced into the system at constant rate and at random positions. Droplets are immobile, and coalescence takes place when a new droplet impinges on one already in place. The system (substrate) has dimensionality d. Obviously $d = 2$ is the case of most interest, but there is physical interest also in 1 and 3 dimensions, and mathematical interest in arbitrary values of d.

The droplet evolution, from a computer simulation for $d = 2$, $D = 3$, is shown in Figure 2. The small number of large droplets embedded in a background of a large number of smaller droplets as described by the distribution of Figure 1b is apparent.

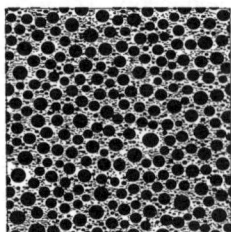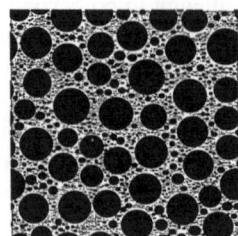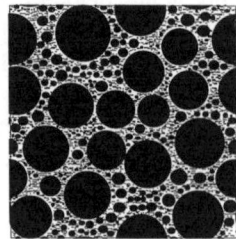

Figure 2. Typical droplet configurations for $d = 2$, $D = 3$ at fractional coverages of 69%, 77%, and 81% (left to right).

2.2. *Further scaling*

The fractional coverage, C, of the surface increases and approaches 1 asymptotically, as long as $D > d$. If $D = d$, complete coverage will occur after a finite time, but this case will not concern us here. The coverage, C, is proportional to one of the moments, $M_{d/D}$, of the distribution function. Assuming Eq. (4) is applicable, the fact that C approaches a limit gives another condition for determining the exponents which, with Eq. (5), yields the results

$$\theta = 1 + d/D \qquad (8)$$

$$z = D/(D - d) \qquad (9)$$

We now address the issue of the predominant small droplets and the lower cut-off to the integral in Eq. (2), to which we referred earlier. The small s behaviour is usually expressed through a new exponent, τ, in the universal function of Eq. (1)

$$f(x) \sim x^{\theta - \tau} \qquad (10)$$

for small x, so that $N_s \sim s^{-\tau}$ in this limit. From Eq. (2) it can be seen that for small n, the lower limit ($x_o = s_0/S$) of the integral dominates in the evaluation of M_n, and we need to replace Eq. (4) by

$$M_n \sim N \sim t^{-z'}, \qquad (11)$$

where

$$z' = z(\theta - \tau) \qquad (12)$$

To be precise, we should use Eq. (11) for the moments if $n < \tau - 1$, and Eq. (4) otherwise. The number density of the droplets is denoted by N (equal to the zeroth moment), and so all low moments scale with the droplet density.

The analytic expressions, Eqs. (8) and (9), for z and θ have been known for a number of years [21, 22]. Estimates of z' and τ have been obtained by computer simulations [21, 22]. Our interest has been to attempt to obtain values of these exponents analytically to complete the understanding of this type of growth process.

2.3. Dense packing

Figure 2 shows images that are reminiscent of the dense packing of hard discs. Clearly the growth process is one in which the droplets have to self-organise to efficiently fill the amount of space available. The most famous example of dense packing is the regular arrangement of discs of a precise sequence of sizes first studied by Appolonius of Perga around 200 BC. Random configurations have also been studied [19, 23, 24] and dense packing occurs with a continuous distribution of disc radii described by a power law, which is analogous to the one, $N_s \sim s^{-\tau}$, that we are observing. The radii distribution occurring in the dense packing is often characterised by a fractal dimensionality, d_f, and, in our notation, the relation between τ and the fractal dimensionality is

$$d_f = D(\tau - 1), \tag{13}$$

or, using Eqs. (8), (9) and (12),

$$z' = (d - d_f)/(D - d). \tag{14}$$

In practice, there will be a lower and upper cut-off in the sizes of the hard discs, and the porosity (uncovered surface) is determined by the relation

$$P = 1 - C = \left(r_{small}/r_{large} \right)^{d-d_f}. \tag{15}$$

Applied to the system of interest here, the smallest radius is that of the incident droplets, r_0. The largest radius is essentially just $S^{1/D}$, because as stated earlier most of the material deposited is in the largest droplets. Then using Eqs. (3), (9), (11) and (14), we can predict that the porosity will scale as the droplet density

$$P = 1 - C \sim N. \tag{16}$$

There is an alternative way of arriving at Eq. (16) by using the moments defined earlier. If we take the derivative of Eq. (2) with respect to S, bearing in mind that the lower limit of the integral is x_o $(= s_0/S)$, we obtain the relation

$$\left(\frac{d}{D} - n \right) \frac{M_n}{S} + \frac{dM_n}{dS} = S^{-1} s_0^{n-\theta+1} f(x_0). \tag{17}$$

Setting $n = d/D$ to focus on the coverage, we obtain

$$\frac{dC}{dS} \sim S^{-1} f(x_0), \tag{18}$$

which shows that it is the small clusters that determine the time dependence of C in its approach to its asymptotic value. We also have consistency with Eq. (16) by a trivial manipulation of Eq. (18) using Eqs. (3), (10) and (11). We will now relate this basic structure to computer simulations on this model.

3. Computer simulations

We describe the results of computer simulations and show how they can provide clues to a more complete analytic theory. The procedure is to introduce new droplets, one by one, at random positions on a substrate and, if any overlap occurs, coalescence according to Eq. (7) takes place. Periodic boundary condition are used. The time evolution of the number density and the porosity for a $d = 2$, $D = 3$ system is shown in Figure 3. Referring to the plot of N, the onset of the scaling regime occurs just after the maximum when we get into the straight line region of the curve plotted on a logarithmic scale. The second plot confirms the scaling of the porosity with N predicted by Eq. (16).

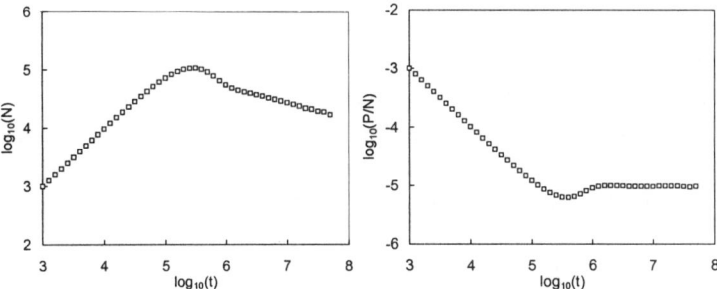

Figure 3. Simulations of $d = 2$, $D = 3$ droplet growth. A square substrate of side $1400r_0$ with periodic boundary conditions was used in the simulation. Droplet density, and ratio of porosity to density are plotted against time.

In the previous section we asserted that this system has something in common with the dense packing of hyperspherical objects, and introduced the concept of fractal dimensionality by analogy. Let us be a little more precise about the meaning of dense packing in this context. For simplicity, consider first $d = 1$, and the deposition of objects along a line. There is a distribution in the sizes of the gaps between neighbouring droplets, and this is shown in Figure 4 for the $D = 2$ and $D = 3$ case. We can see, first of all, that within statistical noise the distributions remain unchanged throughout the scaling regime. The average gap sizes for $D = 2$ and $D = 3$ are about 1.17 and 1.03 respectively in units of the diameter, $2r_0$, of the elementary droplet. Dense packing in this context means therefore droplets packed to within a separation of each other that scales with the diameter of the elementary droplet. This diameter and that of the mean cluster are the *two* characteristic length scales for this system.

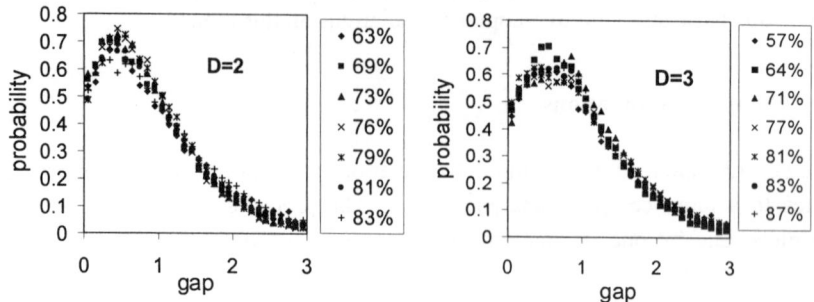

Figure 4. Distribution of gap sizes between neighbouring droplets for $d = 1$, and $D = 2$ and $D = 3$. The units on the horizontal axis are the diameter, $2r_0$ of the elementary droplet. The plots show the distributions at different coverages.

For higher dimensions, it is more convenient to express the packing in a slightly different way. We have used, as a measure, the number of droplets (neighbours) separated from a particular one by a distance (between edges) of less than a certain amount, and then taken the average over all droplets in the system. The results from computer simulation for $d = 2$, $D = 3$ are shown in Figure 5. In a random packing of hard disks, the average number of neighbours (discs in contact) is 6 [23]. This value in the present case is reached for a separation of about twice the elementary diameter, $2r_0$.

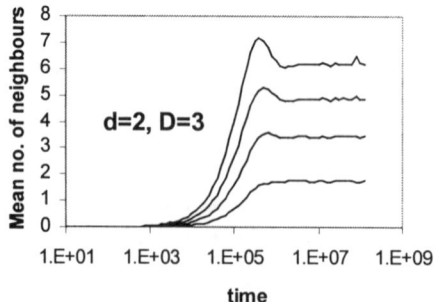

Figure 5. Time evolution of mean number of neighbours for the $d = 2$, $D = 3$ case. The plots are for neighbour distances 0.5, 1.0, 1.5, 2.0 (in units of $2r_0$).

To make progress in developing a theory we need to consider which are the dominant processes in the scaling regime and how they change the fractional coverage. When a new droplet is added, the following possibilities occur (as illustrated in Figure 6):

1. The droplet falls on the free surface without touching an existing droplet.

2. The new droplet falls on an existing one, whose radius increases as a result of the coalescence. There are then two possibilities: *either* (a) the expanding droplet does not touch another one, *or* (b) it does touch another resulting in a further coalescence.

3. The droplet falls so as to bridge two clusters that are close together causing them to coalesce. This process is rare during the early stages of growth, but it plays a major role in the scaling regime.

4. There are other possible scenarios. After coalescence has taken place by either (2b) or (3), the new droplet so formed could overlap with other droplets in the vicinity, resulting in further coalescence. It is found from simulations that such cascade processes almost never occur for $d = 1$. They do take place for $d \geq 2$, however, and often quite dramatically. We include all such cascade processes in category 4. An example of a cascade process is illustrated in Figure 7.

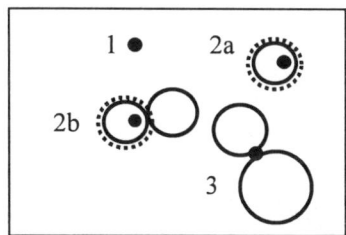

Figure 6. Processes (see text) resulting from the addition of a new droplet (filled circle). Open circles are droplets already present on the surface. Dotted circle indicates expansion of droplet after impact.

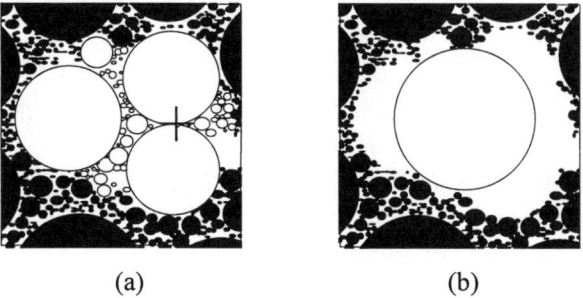

(a)   (b)

Figure 7. Example of a cascade process. In (a), the position of the impacting droplet is indicated by a cross. Open circles represent droplets involved in the cascade process. The configuration after the coalescence is shown in (b), where the open circle indicates the newly coalesced droplet.

Each process leads to a change in the surface coverage, C, with 1 and 2a causing an increase and 2b, 3, and 4 resulting in a decrease. If we denote the change in surface coverage due to each process since the start of the simulation with appropriately labelled lower case symbols then

$$C = c_1 + c_{2a} + c_{2b} + c_3 + c_4. \tag{19}$$

The time evolution of the individual contributions to the surface coverage was recorded during the simulations, and is shown in Figure 8a, along with the coverage itself. The net coverage, shown by the full line in the Figure 8a, had reached 84% at the end of the simulation. Process 2b gives the smallest contribution. The derivative of each c_α (where α=1, 2a, 2b, 3, 4) was obtained from the data in Figure 8a, and it was found that each scales with N. This is illustrated in Figure 8b. Therefore, to order $t^{z'}$, the growth behaviour must be determined by a dynamic balance between the various processes (note, $\dot{C} \sim t^{-z'-1}$):

$$\dot{c}_1 + \dot{c}_{2a} + \dot{c}_{2b} + \dot{c}_3 + \dot{c}_4 \sim 0 \tag{20}$$

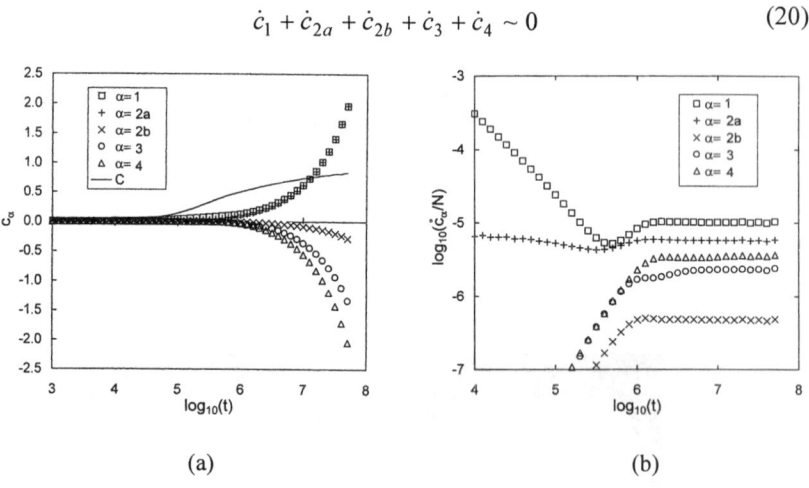

(a)   (b)

Figure 8. Simulations for $d = 2$, $D = 3$. (a) Time evolution of the coverage C, and the individual contributions, c_α, where $\alpha = 1, 2a, 2b, 3, 4$. (b) Evolution of derivatives of c_α divided by N. Magnitudes are shown.

This rather ubiquitous scaling with N provides us with a handle to develop the theory. It emerges that processes 2b and 3 are the ones that are, in the first instance, rate determining. We outline the argument for the most important case physically ($d = 2$, $D = 3$), but it can be generalised to arbitrary d and D.

Consider first process 2b. The probability that a new droplet will land on an island of radius r_i is proportional to $r_i^2 N_i$. If the initial coalescence causes the island to expand its radius by dr_i then the probability of a further coalescence with an island of radius r_j will scale like $N_j N^{-1}(r_i + r_j)dr_i$. A further factor describes the change of coverage. Summed over all pairs i and j, will give the rate of change of c_{2b}, which has been demonstrated to scale like N. We noted earlier (below Eq. (12)) that for M_n to scale like N the condition $\tau \geq n + 1$ had to apply. For dc_{2b}/dt to scale like N, again small islands have to dominate, and an argument that is somewhat more complex than the one applying to the moments because of the double summation over sizes leads to the condition $\tau \geq 3/2$.

The various factors occurring in the considerations above for 2b also appear in the discussion of process 3 – except for the r_i^2 factor. We now have to consider the probability that an impinging droplet bridges a pair of islands of radii r_i and r_j. It can be shown that this is proportional to $[r_i r_j/(r_i + r_j)]^{1/2}$. Going through an argument for dc_3/dt to scale like N, leads to the criterion $\tau \geq 19/12$.

We take process 3 as the one that is rate determining and take the equality as our prediction for τ. Using Eqs. (8), (9), (12), the analytic estimates of the exponents for $d = 2$, $D = 3$ are summarised as: $z = 3$, $\theta = 5/3$, $\tau = 19/12$, $z' = 1/4$. This agrees very well with the estimates from computer simulation: $z = 2.92$, $\tau = 1.54$, $z' = 0.26$. Note the computer estimates do not satisfy exactly Eqs. (5) and (12) because the scaling theory is exact only in the asymptotic limit.

The theory we have outlined can be generalised to arbitrary d and D

$$\tau = \frac{3d}{2D} + \frac{1}{2} \quad \text{if} \quad D \leq D_c$$
$$= \frac{5d-3}{4D} + 1 \quad \text{if} \quad D \geq D_c \tag{21}$$

and, from Eq. (12)

$$z' = \frac{1}{2} \quad \text{if} \quad D \leq D_c$$
$$= \frac{3-d}{4(D-d)} \quad \text{if} \quad D \geq D_c \tag{22}$$

D_c is a critical dimensionality given by

$$D_c = (d+3)/2. \tag{23}$$

For $D > D_c$, type 3 processes are rate determining; for $D < D_c$, type 2b processes dominate in determining the exponents.

These results were obtained about two years ago [25] and, as far as we are aware, are the first analytic estimates of τ and z' in the literature. Let us now look at them rather critically, starting with the deposition of droplets along a line ($d = 1$). It is instructive to express the results in terms of the fractal dimensionality as defined in Eqs. (13) or (14). The theoretical results are compared with those from simulations in Figure 10. The critical dimensionality, from Eq. (23) is 2, and it can be seen that the theoretical value of d_f is equal to the physical dimensionality of the substrate when $D = d$, falls to 0.5 at $D = D_c$, and remains constant for $D > D_c$. The values from the simulations agree very well with the theoretical values.

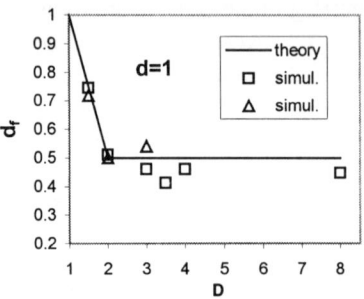

Figure 9. Fractal dimensionality for $d = 1$ and a range of values of D. Full line is from the theory. Data points are obtained from computer simulations using Eq. (14) (squares) and Eq. (13) (triangles).

The $d = 1$ case is of physical interest; droplets can gather along a step edge in a surface. Droplet formation on a surface ($d = 2$, $D = 3$) is probably the situation of most interest. We have demonstrated that the theory [25] provides a very good account of these 'physical' situations. However, there is no reason why the theory should not apply to the higher dimensions as well. Considering these cases will provide us with a deeper insight into what is happening.

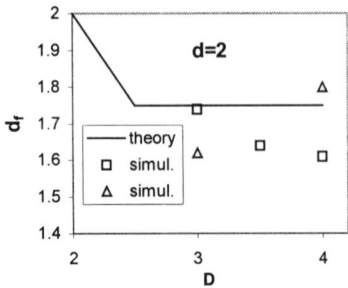

Figure 10. Fractal dimensionality for $d = 2$. Plot details are as in Figure 9.

The fractal dimensionality for the $d = 2$ case is shown in Figure 10. The theoretical value of d_f is equal to the physical dimensionality of the substrate when $D = d$, falls to 1.75 at $D = D_c$ (=2.5), and remains constant for $D > D_c$, as in the previous case. There is some uncertainty in the values obtained from computer simulations, and this manifests itself in a significant difference in the values of d_f deduced from Eqs. (13) and (14). This is because of the slow approach to the asymptotic regime, where the scaling relations of Eqs. (5) and (12) will hold exactly. The fractal dimensionality for $D = 3$, deduced from numerical evaluations of z', agrees well with theory, and this is more reliable than numerical estimates of τ. However there does appear some indication that the theoretical values may be a little too high, particularly at higher values of D.

Now let us consider $d = 3$. Eqs. (13) and (21) yield a fractal dimensionality equal to 3, the physical dimensionality and, for $d > 3$, the predicted d_f is larger than d. We need to consider whether $d = 3$ marks a critical dimensionality, above which different physics is occurring. Simulations were done for $d = 3$, $D = 4$ and the droplet density and porosity are shown as a function of time in Figure 11. Clearly, although z' is extremely small, the behaviour is not fundamentally different from that observed at lower dimensionality, and this indicates that the theory is not quite complete although, as already noted, any discrepancies are small for physically interesting dimensionalities.

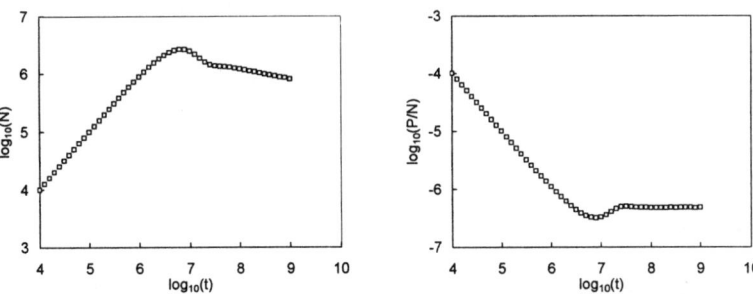

Figure 11. Droplet density and ratio of porosity to density plotted against time for $d = 3$, $D = 4$.

It appears that the analytic theory gives a fractal dimensionality that is somewhat larger than that obtained in the computer simulations. For $d = 1$, any discrepancy is comparable with the uncertainties in the values from the simulations. For $d = 2$, differences are small for $D = 3$, but increase with higher values of D. For $d \geq 3$ cases, however, the analytic theory is clearly inadequate. From Eq.(13), a high value of the fractal dimensionality implies a theoretical value of τ that is too large, which means that the theoretical size distribution is predicted to fall off slightly faster than in fact it does.

It should be borne in mind that the theory [25] is basically a mean field theory in the sense that no spatial fluctuations or correlations in the droplet distribution are included. We have noted in Section 1.3, that mean field theory, applied to droplet distributions of the type displayed in Figure 1(a), gives a partial description of the behaviour, but one has to go beyond this for a complete story. It is likely that we are pushing the limits of mean field theory in the current problem. Figure 7 is the clue to where we have to look to assess the limitations. After a coalescence there remains for some time a depletion of droplets around the droplet that has resulted from that coalescence. This effect is particularly large if the coalescence has involved a cascade. Such depletions will be much larger around large islands than around small ones, and this will effectively reduce the participation of large islands in the coalescence processes. After a cascade, such islands will be out of action, so to speak, until the depleted region has been restored to an average environment. As a consequence, we do not need quite such a large τ to ensure the dominance of small islands and the scaling like N that was observed in Figure 8b, and occurs for arbitrary dimensions.

Cascade processes are virtually non-existence for $d = 1$, which accounts for the excellent description of mean field theory in that case. For $d \geq 2$, cascade processes are certainly important and, furthermore, the size of the depletion region that appears after a coalescence increases with increasing droplet dimensionality, D. These points are entirely consistent with the observations described in this Section, and leads us to identify cascade processes as the feature that is responsible for the limitations of mean field theory.

4. Conclusions

This paper has attempted to give a broad overview of the essentials of scaling theory applied to aggregation processes. Two very different distributions were illustrated in Figure 1, and the points that they have in common were discussed. Understanding the behaviour of the bimodal distribution is less well developed. We outlined the progress that has been made in applying a mean field type of description to such systems, and showed that for physically relevant dimensions it does rather well. This is particularly true for a one dimensional substrate.

However, as we increase the dimensionality (both d and D) into more abstract realms, discrepancies cannot be ignored, and we have to address the limitations of mean field theory. Cascade processes were identified as the most likely suspect in causing the environment of the larger droplets to deviate from the mean. In the Introduction, we noted that systems exhibiting a single peak distribution required a treatment that went beyond mean field theory for a complete description. It appears that we are encountering a similar situation with the bimodal distribution. Work is underway on this problem, and we will report on further developments in due course.

References

1. S. K. Friedlander, *Smoke, Dust and Haze*, Wiley Interscience, New York, 1997.
2. H. E. Stanley and N. Ostrowsky (Eds), *On Growth and Form*, Nijhoff, Amsterdam, 1986.
3. P. Meakin, *Rep.Prog. Phys.* **35**, 157 (1992).
4. A.–L. Barabasi and H. E. Stanley, in: *Fractal Concepts in Surface Growth*, Cambridge University Press, 1995.
5. M Zinke-Allmang, *Thin Solid Films* **346**, 1 (1999).
6. H. E. Stanley, *Phase Transitions and Critical Phenomena*, Oxford University Press, 1971.
7. J. A. Stroscio and D. T. Pierce, *Phys. Rev. B* **49**, 12675 (1992).
8. J. A. Blackman and A. Wilding, *Europhys Lett.* **16**, 115 (1991).
9. P. A. Mulheran and J. A. Blackman, *Phys. Rev. B* **53**, 10261 (1996).
10. J. A. Blackman and P. A. Mulheran, *Phys. Rev. B* **54**, 11681 (1996).
11. P. A. Mulheran and J. A. Blackman, *Surf. Sci.* **376**, 403 (1997).
12. P. A. Mulheran and D. A. Robbie, *Europhys Lett.* **49**, 617 (2000).
13. J. A. Blackman, *Physica A* **220**, 85 (1995).
14. S. Cueille and C. Sire, *Phys. Rev. E* **57**, 881 (1998).
15. D. Beysens, *Atmos. Res.* **39**, 215 (1995).
16. E. Grantscharova and D. Dobrev, *Thin Solid Films* **161**, 213 (1988).
17. G. R. Carlow, R. J. Barel and M. Zinke-Allmang, *Phys. Rev. B* **56**, 12519 (1997).
18. L. Haderbache, R. Garrigos, R. Kofman, E. Søndergård, and P. Cheyssac, *Surf. Science* **410**, L748 (1998).
19. T. Aste, R. Botter, and D. Beruto, *Sensors and Actuators B* **24-25**, 826 (1995).
20. H. Lannibois, A. Hasmy, R. Botet, O.A. Chariol, and B. Cabane, *J. de Phys. II* **7**, 319 (1997).
21. F. Family and P. Meakin, *Phys. Rev. Lett.* **61**, 428 (1988).
22. F. Family and P. Meakin, *Phys. Rev. A* **40**, 3836 (1989).
23. T. Aste and D. Weaire, T*he Pursuit of Perfect Packing*, Institute of Physics Publishing, Bristol, 2000.
24. H. J. Herrmann, G. Mantica and D. Bessis, *Phys. Rev. Lett.* **65**, 3223 (1990).
25. J. A. Blackman and S. Brochard, *Phys. Rev. Lett.* **84**, 4409 (2000).

SYMMETRY NON-BREAKING PHASE TRANSITION AND CRITICAL POINT IN HEXAGONAL MIXED CRYSTAL $C_{70(1-x)}C_{60X}$.

P. ZIELIŃSKI,

*The H. Niewodniczański Institute of Nuclear Physics,
ul. Radzikowskiego 152, 31-342 Kraków, Poland*

W. SCHRANZ

*Institute for Experimental Physics,
University of Vienna,
Strudlhofgasse 4, A-1090 Wien, Austria*

A four-state pseudospin model for the isomorphous phase transition in mixed crystal C70(1-x)C60x is shown to be equivalent to a two-state Ising model with one state degenerate. The model predicts a metastable disordered phase corresponding to a glassy state. The effect of random fields and random bonds is to drive the system towards the critical point.

1. Introduction

Spin or pseudospin models are often used to describe phase transitions in solids. An individual pseudospin has a number of equivalent accessible orientations, which are occupied with equal probability at high enough temperatures. Below certain temperature a configuration is established so as the energy of the interactions be minimum. The new ordered structure has usually a lower symmetry. Thus, the ordering of spins falls into the category of spontaneous symmetry breaking phenomena. Two-state Ising pseudospins are used to model the simplest second order paramagnet-ferromagnet phase transition. The temperature of the phase transition lowers when the spins are diluted with non-magnetic atoms. At high enough concentration of such admixture the ordered phase gives room to a statically disordered structure called spin glass. This is a result of random fields and random bonds introduced by the alien atoms. The theoretical methods used in spin glasses are reviewed in refs. [1,2].

Similar methods are used in the description of quadrupolar glasses, where the disordered element: molecule or polyatomic ion has an ellipsoidal shape [3,4]. The phase transitions considered in refs. [3,4] were of first order but, in similarity with second order phase transitions, showed a symmetry breaking.

In contrast to that, the pure hexagonal crystal C_{70} and the mixed crystal $C_{70(1-x)}C_{60x}$, where the quadrupolar C_{70} molecules are diluted with globular C_{60}, undergo a phase transition, in which the symmetry does not change. Thus, this is an isomorphous phase transitions [5]. The space group of both phases rests the same: close packed hexagonal P6₃/mmc [6]. The orientational distribution of the C_{70} molecules is practically spherical in the high-temperature phase hcp-2. At about 337 K the long molecular axes become ordered along the crystallographic direction (0001). The ratio of the lattice constants c/a jumps abruptly from the value characteristic to spherical effective molecules: c/a = 1.63... (hcp-2 phase) to the value corresponding to the shape of a single C_{70} molecule c/a = 1.84 (hcp-1 phase) [7].

Theoretical treatments of glassy systems originating from first order phase transitions are scarce in literature. Ref. [8] presents general considerations on the impurity-induced rounding of the symmetry-breaking phase transition of first order without, however, studying particular models of interactions. Potts models, which exhibit discontinuous phase transitions for sufficiently high number of states, have been shown to turn to continuous phase transition at strong enough random fields [9]. This behaviour resembles the tricritical point [5] occurring in symmetry-breaking systems. Ref. [10] is the first attempt to study the interplay between the freezing induced by randomness and the critical point always allowed in isomorphous phase transitions. In the present note we show theoretical bases for a strongly anisotropic pseudospin model capable of reproducing the phenomena observed in $C_{70(1-x)}C_{60x}$ the glassy phase included.

It has been argued [10] that the one particle orientational potential of a C_{70} molecule in the hcp structure has a minimum for the orientation (0001) (state ①) and three equivalent minima ②, ③ and ④ 120° away from one another in the plane $\langle 0001 \rangle$.

One should notice that the ordering of all the molecules in the state ① produces close packed planes with the molecules perpendicular to the planes. This is not the case for all the remaining states. The actual configuration of the system is defined by the occupation numbers $N_i = 1, 0$, where $i = 1,...4$, for each state ①, ②, ③ and ④.

Following the idea of Pirc et al. [4,11,12] we introduce four symmetry-adapted combinations of the occupation numbers and label them by the irreducible representations of the site point group $\overline{6}m2$ (D_{3h}).

$$Z_1 = N_1 - \tfrac{1}{3}\left(N_2 + N_3 + N_4\right) \qquad (A_1')$$

$$Z_2 = N_1 + N_2 + N_3 + N_4 \qquad (A_1')$$

$$Z_3 = \tfrac{1}{\sqrt{2}}\left(N_3 - N_4\right) \qquad (E')$$

$$Z_4 = \tfrac{1}{\sqrt{6}}\left(-2N_2 + N_3 + N_4\right) \qquad (E') \tag{1}$$

The order parameter in the isomorphous phase transition hcp-2 $\rightarrow$ hcp-1 involves the mean value of the combination Z_1 only.

The general hamiltonian for the pseudospin-pseudospin interactions has the following form

$$H_{pseudo} = -\frac{1}{2}\sum_{i,j}\sum_{r,s} Z_{ri} J_{ij}^{rs} Z_{sj} \tag{2}$$

with a 4×4 matrix J_{ij}^{rs}, $r,s = 1,...4$ for each pair of sites ij. The matrices J_{ij}^{rs} are symmetrical

$$J_{ij}^{rs} = J_{ij}^{sr} = J_{ji}^{rs}, \tag{3}$$

as long as the molecules can be approximated by quadrupoles invariant by with respect to inversion.

In the mean field approximation the hamiltonian (2) bocomes

$$H_{MF} = -\sum_{i,j}\sum_{r,s}\langle Z_{ri}\rangle J_{ij}^{rs} Z_{sj} + \frac{1}{2}\sum_{i,j}\sum_{r,s}\langle Z_{ri}\rangle J_{ij}^{rs}\langle Z_{sj}\rangle. \tag{4}$$

When applying this kind of model to the mixed system KCN$_{(1-x)}$Br$_x$ the authors of refs. [4,11,12] assumed totally isotropic pseudospin interactions, i.e. all the matrices J_{ij}^{rs} diagonal with equal diagonal terms: $J_{ij}^{rs} = J_{ij}\delta_{rs}$. This assumption is incorrect for C$_{70}$, where the molecules stick together. The simplification we adopt here stems from the observation that all the detected ordered phases of C$_{70}$ consist of close packed hexagonal planes with the long molecular axes perpendicular to the planes. This is true for the ordered R$\overline{3}$m phase as well as for the hcp-1 phase. We then assume that the mean field at the site j has the general form $\sum_{i,r}\langle Z_{ri}\rangle J_{ij}^{rs} \sim \delta_{s1}$ and arises only when the neighbouring molecules are in the state ①, i.e. when $\langle Z_{ri} \rangle = \langle Z_1 \rangle \delta_{r1}$ for each site i. With eq. (3) in mind this is equivalent to the condition

$$\sum_i J_{ij}^{rs} = \delta_{r1}\delta_{s1}\sum_i J_{ij}^{11}, \tag{5}$$

which allows one to simplify the mean field Hamiltonian

$$H_{MF} = -\sum_{i,j}\langle Z_1\rangle J_{ij}^{11} Z_{1j} + \frac{1}{2}\sum_{i,j}\langle Z_1\rangle J_{ij}^{11}\langle Z_1\rangle. \tag{6}$$

The only relevant interaction parameters are now J_{ij}^{11}. Any long-range order with the molecules oriented in the states ②, ③ and ④ is precluded . The condition (5) is a drastic assumption similar to that of isotropic interactions but is better adapted to C_{70}.

Taking into account the coupling with strain and the full symmetry of the order parameters one arrives at the following mean-field hamiltonian

$$H = -\sum_{i,j}\langle Z_1\rangle J_{ij}^{11} Z_{1j} - h\sum_j Z_{1j} - r\varepsilon\sum_j Z_{1j} + \frac{1}{2}\sum_j c_0\varepsilon^2 + \frac{1}{2}\sum_{i,j}\langle Z_1\rangle J_{ij}^{11}\langle Z_1\rangle. \tag{7}$$

There is a straightforward relation between the system defined by the hamiltonian (7) and the usual two-state Ising model. This can be seen by adopting the transformation

$$Z_1 = (2s+1)/3$$
$$s = (3Z_1 - 1)/2. \tag{8}$$

When written in the variable s the hamiltonian (7) becomes

$$H = \sum_j \left(-\mathcal{H}s_j - \frac{1}{2}\bar{r}\varepsilon + \frac{1}{2}c_0\varepsilon^2 + \frac{1}{2}\bar{J}\langle s\rangle^2 \right), \tag{9}$$

where the molecular field is

$$\mathcal{H} = \langle s\rangle\bar{J} + \bar{h} + \bar{r}\varepsilon, \tag{10}$$

while the renormalized parameters are

$$\bar{J} = \frac{4}{9}\sum_i J_{i1}^{11}, \qquad \bar{h} = \frac{2}{3}h + \frac{2}{9}\sum_i J_{i1}^{11}, \qquad \bar{r} = \frac{2}{3}r \qquad (11)$$

The constant terms have been dropped in eq. (9) as only defining the absolute scale of energy. A further transformation

$$\varepsilon = \hat{\varepsilon} + \frac{\bar{r}}{2c_0} \qquad (12)$$

and

$$\hat{h} = \bar{h} + \frac{\bar{r}^2}{2c_0} \qquad (13)$$

allows one to write the hamiltonian (eq. (9)) in a simpler form

$$H = \sum_j \left(-\mathcal{H}s_j + \frac{1}{2}c_0\hat{\varepsilon}^2 + \frac{1}{2}\bar{J}\langle s \rangle^2 \right) \qquad (14)$$

with

$$\mathcal{H} = \langle s \rangle \bar{J} + \hat{h} + \bar{r}\hat{\varepsilon} \qquad (15)$$

The equilibrium values of the average $\langle s \rangle$ and of the strain $\hat{\varepsilon}$ are obtained by minimization of the free energy

$$F = -k_B T \ln \Omega, \qquad (16)$$

where the partition function Ω is

$$\Omega = \exp\left[\left(-\frac{1}{2}c_0\hat{\varepsilon}^2 - \frac{1}{2}\bar{J}\langle s \rangle^2 \right) / k_B T \right] \left[\exp(\mathcal{H}/k_B T) + 3\exp(-\mathcal{H}/k_B T) \right]. \qquad (17)$$

The resulting self-consistent equations of state can be written in the following form

$$\langle s \rangle = \frac{\left[\exp(\mathcal{H}/k_B T) - 3\exp(-\mathcal{H}/k_B T) \right]}{\left[\exp(\mathcal{H}/k_B T) + 3\exp(-\mathcal{H}/k_B T) \right]} = \tanh\left(\overline{\mathcal{H}}/k_B T \right) \qquad (18a)$$

$$\hat{\varepsilon} = \frac{\bar{r}}{c_0}\langle s \rangle, \qquad (18b)$$

where

$$\overline{\mathcal{H}} = \mathcal{H} - k_B T \ln(3)/2 \equiv \mathcal{H} + h_s, \qquad (19)$$

while

$$h_s = -k_B T \ln(3)/2. \qquad (20)$$

Equation (18a) with eqs. (15), (19) and (18b) is formally equivalent to that for the usual two-state Ising model in an external field $\hat{h} + h_s$. The field $\hat{h} + h_s$ is here a result of intermolecular interactions and is allowed by symmetry; no external field is needed. This term would necessarily vanish for a symmetry-breaking phase transition. A particularity of the present model is that the field $\hat{h} + h_s$ depends on (eq. (20)). The dependence is a result of the triple degeneracy of the state $s = -1$.

Because of the presence of the field-like term $\hat{h} + h_s$ the phase transition is generally of first order. The phase equilibrium occurs when the free energy (eq. (16)) with eq. (18b) inserted into eq. (18a) is an even function of $\langle s \rangle$ and has two minima of equal depths. The condition for the free energy being an even function is

$$\hat{h} + h_s = 0, \tag{21}$$

which is equivalent to

$$T = T_{eq} = \frac{2\hat{h}}{k_B \ln(3)}, \tag{22}$$

whereas the condition for the existence of two minima reads

$$T_{eq} < T_{ch} = (\overline{J} + \overline{r}^2 / \overline{c}_0) / k_B. \tag{23}$$

The particular case of $T = T_{eq} = T_{ch}$ describes the critical point. Whenever the condition (23) is fulfilled, the temperature T_{eq} (eq. (22)) describes the equilibrium coexistence of phases.

Eqs. (15), (19) and (20) define the molecular field acting on a given C_{70} molecule as a function of the quantity $\langle s \rangle$, which is the expectation value of s at a site occupied by a molecule C_{70}. In the diluted system the molecular field acting on a molecule depends on the volume average $\overline{s} = (1-x)\langle s \rangle$ rather than on the site expectation value $\langle s \rangle$:

$$\mathcal{H} = \overline{s}\overline{J} + \hat{h} + \overline{r}\,\hat{\varepsilon}. \tag{24}$$

Inserting eq. (24) into eq. (18a) and multiplying both sides of the resulting equation by *(1-x)* one obtains a self-consistent equation dependent on $\overline{s}$ only

$$\bar{s} = (1-x)\tanh\left(\overline{\mathcal{H}}/k_B T\right), \tag{25}$$

where the relations (19) and (20) have been applied. The deformation of the sample is also proportional to the volume average $\bar{s}$, so that the analogue of eq. (18b) now is

$$\hat{\varepsilon} = \frac{\hat{r}}{c_0}\langle s\rangle. \tag{26}$$

Eqs. (25) and (26) are consistent with the minimum condition of the free energy (eq. (16)) with the following one-particle partition function

$$\Omega = \exp\left[\left(-\frac{1}{2}c_0\hat{\varepsilon}^2 - \frac{1}{2}\overline{Js}^2\right)/k_B T\right]\left[\exp(\mathcal{H}/k_B T) + 3\exp(-\mathcal{H}/k_B T)\right]^{(1-x)}. \tag{27}$$

The critical temperature T_{ch} now is

$$T_{ch} = (1-x)\left(\bar{J} + \frac{\bar{r}^2}{c_0}\right)\bigg/k_B. \tag{28}$$

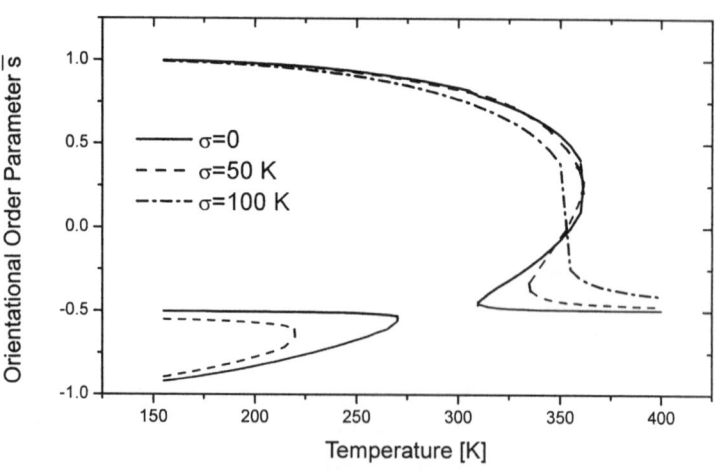

Figure 1. Temperature dependence of orientational order parameter (eq (29)) at presence of random bonds and random fields with distribution width σ.

In addition to the numerical proportionality of the volume average $\bar{s} = (1-x)\langle s\rangle$ discussed above, the presence of admixtures in a system with orientational degrees of freedom introduces random fields and random bonds [3,4,11,12]. If the phase transition in the non-diluted system is of second order the critical temperature lowers with increasing amplitude of random fields and with increasing distribution width of random bonds [1]. The resulting phase is replica-symmetric [1] provided that the random amplitudes are not too strong. The replica-symmetric state in the present case is obtained by replacing the self-consistent equation (25) by the following equation

$$\overline{s} = \frac{(1-x)}{\sqrt{2\pi}} \int\limits_{-\infty}^{+\infty} dz e^{-z^2/2} \tanh\left(\frac{\sigma z + \overline{\mathcal{H}}}{k_B T}\right),$$ (29)

where $\sigma = \sqrt{q_{EA} T_q^2 + T_\Delta^2}$, while q_{EA} is the Edwards-Anderson order parameter [11], T_q is the width of the random bonds distribution and T_Δ is the width of the random fields distribution (see ref. [13] for analogous formula in the case of a first order phase transition governed by a non-linear coupling with the strain). Fig. 1 shows the order parameter $\overline{s}$ calculated with eq. (29) for the parameters of the model: $c_0 = 7875000, \overline{J} = 39.0, \overline{r} = 52500, \hat{h} = 195$ corresponding to the pure C_{70} and for three values of the total width σ. Although a precise x-dependence of the parameter σ is not straightforward [2,8] one should expect it to be an increasing function. Fig. 1 shows that the main effect of the random fields and of random bonds is a narrowing of the hysteresis with practically no variation of the coexistence temperature. Now that the experiment shows a strong decrease in the transition temperature with increasing x the principal effect of the dilution lies in an x-dependence of the parameters of the model [14].

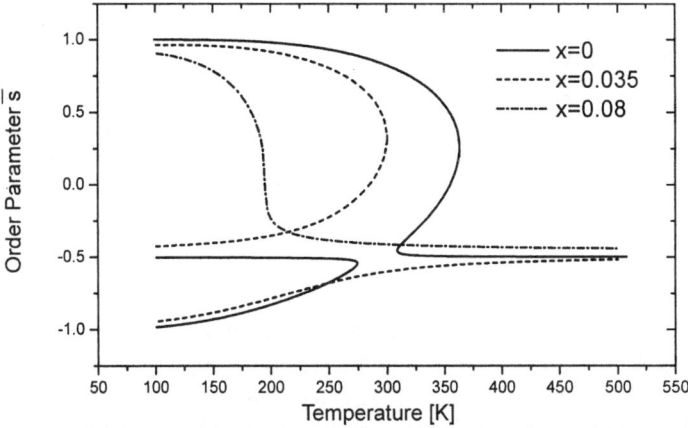

Figure 2. Temperature dependence of orientational order parameter $\overline{s}$ evaluated from eqs. (25) and (26) with parameters of eqs. (30).

The experimental data of ref. [14] allow one to establish the following x-dependence of the parameters of the model (in Kelvins):

$$c_0 = 7875000$$
$$\bar{J} = 39.0$$
$$\bar{r} = 52500 - 2460937.5x^2 \tag{30}$$
$$\hat{h} = 195 - 1098.56x$$

Fig. 2 shows the temperature dependence of the orientational order parameter $\bar{s}$ obtained numerically from eqs. (25), (26) with the values from eq. (30) for the concentrations x which were used in the dilatometric study [14]. An interesting and unexpected result is that the metastable phase with negative $\bar{s}$ reappears below 275K in addition to the stable ordered phase with $\bar{s} \approx 1$ for $x = 0$. The metastable phase is represented in Fig. 2 by the lowest branch of the continuous curve. For $x = 0.035$ and $x = 0.06$ the metastable phase subsists in the whole temperature range as a smooth continuation of the high-temperature disordered phase.

The above analysis shows that the role of random fields and random bonds in the behaviour of the glass-forming mixed crystals undergoing isomorphous phase transition is analogous to a parameter driving the system towards the critical point. Contrary to the second order phase transitions the temperature of the phase transition (coexistence temperature) does not drop with the parameter σ. The model predicts existence of a disordered metastable phase which should be identified with the glassy state of the system.

Acknowledgments

This work has been done within Austrian–Polish ÖAD-WTZ project Nr. 14/2002. Support from the FWF project P12226-PHY is gratefully acknowledged.

References

1. K. Binder and A.P. Young, *Rev. Mod. Phys.* **58**, 801 (1986).
2. K. Binder and J.D. Reger, *Adv. Phys.* **41**,547 (1992).
3. K.H. Michel, *Phys. Rev. Lett.* **57**, 2188 (1986).
4. B. Pirc. B. Tadić and R. Blinc, *Ferroelectrics* **183**, 235 (1996).
5. see e.g. P. Carpentier, R. Jakubas, J, Lefebvre, W. Zając and P. Zieliński, *Phase Transitions* **67**, 571 (1999).

6. C. Christides, I.M. Thomas, T.J.S. Dennis and K. Prassides, *Europhys. Lett.* **22**, 611 (1993).
7. G. van Tendeloo, S. Amelinckx, J.I. de Boer, S. van Smaalen, M.A. Varheijen, H. Meekes, and G. Meijer, *Europhys. Lett.* **21**, 329 (1993).
8. Y. Imry and M. Wortis, *Phys. Rev.* B **19**, 3580 (1979).
9. D. Blankschtein, Y. Shapir and A. Aharony, *Phys. Rev.* B **29**, 1263 (1984).
10. P. Zieliński, W.Schranz, D. Havlik and A. Kityk, *Eur. Phys. J.* B **24**, 155 (2001).
11. B. Tadić, B. Pirc. and R. Blinc, *Phys. Rev.* B **50**, 9824 (1994).
12. R. Pirc and B. Tadić, *Phys. Rev.* B **54**, 7121 (1998).
13. G. Papantopoulos, G. Papavassilou, F. Milia, V.H. Schmidt, J.E. Drumheller, N.J. Pinto, R. Blinc and B. Zalar, *Phys. Rev. Lett.* **73**, 276 (1994).
14. D. Havlik, W. Schranz, M. Haluška, H. Kuzmany and P. Rogl, *Solid State Comm.* **104**, 775 (1997).

STATIONARY FLUX IN MESOSCOPIC NOISY CYLINDERS

J. DAJKA, J. ŁUCZKA, M. SZOPA

Institute of Physics, University of Silesia,
40-007 Katowice,Poland

The aim of this paper is to investigate the existence of the stationary states of current in the mesoscopic cylinder. The dynamics of the flux is governed by a stochastic differential equation. We discuss both the influence of equilibrium (thermal) and non-equilibrium noise sources.

Small metallic cylinders threaded by a magnetic flux display a persistent, non-dissipative, currents run by coherent electrons. At the finite temperature T some of the electrons become "normal" i.e. non-coherent and the amplitude of the persistent current decreases. Our approach is based on the two fluid model where normal and coherent electrons coexist. The dissipative motion of normal electrons is affected by thermal equilibrium fluctuations of current - Nyquist noise. We assume that in the system under consideration only Nyquist noise plays a significant role.[1]

We consider the cylinder as a set of N one dimensional rings (current channels) stacked along certain axis. The coherent current as a function of magnetic flux depends on the parity of the number of coherent electrons in a channel. Let p denotes the probability of an even number of coherent electrons in a single current channel. The formula for coherent current reads

$$I_{coh}(\phi, T) = pI_{even}(\phi, T) + (1 - p)I_{odd}(\phi, T)$$

with

$$I_{even}(\phi, T) = I_{odd}(\phi - \phi_0/2, T) = NI_0 \sum_{n=1}^{\infty} A_n(T) \sin(2n\pi\phi/\phi_0),$$

flux quantum $\phi_0 = h/e$ and $I_0 = heN_e/(2l^2 m_e)$ where N_e is the number of coherent electrons in a single channel of a circumference l and m_e is the electron mass. The temperature dependent amplitude reads

$$A_n(T) = \frac{4T}{\pi T^*} \frac{\exp(-nT/T^*)}{1 - \exp(-2nT/T^*)} \cos(nk_F l),$$

where the characteristic temperature T^* is proportional to the energy gap at the Fermi surface [2] and k_F is the Fermi momentum. The current of normal electrons according to the Lenz's rule and the Ohm's law

$$RI_{nor}(\phi) = -\dot{\phi}$$

where R is effective resistance of the system.

The magnetic flux ϕ and the current in the cylinder are related via the expression

$$\phi = \phi_{ext} + L(I_{coh}(\phi, T) + I_{nor}(\phi)), \tag{1}$$

where L is a selfinductance of the system. eq. (1) equipped with Nyquist noise yields our basic evolution equation [1]

$$\frac{1}{R}\frac{d\phi}{dt} = -\frac{1}{L}(\phi - \phi_{ext}) + I_{coh}(\phi, T) + \sqrt{\frac{2k_B T}{R}}\,\Gamma(t), \tag{2}$$

with Gaussian white noise $\Gamma(t)$. Defining $\tau_0 = L/R$, $\varepsilon_0 = \phi_0^2/2L$, $D = k_B T/2\varepsilon_0$, and $\lambda = \phi_{ext}/\phi_0$, Eq. (2) can be rewritten in the dimensionless variables $x = \phi/\phi_0$ and $\tilde{t} = t/\tau_0$

$$\dot{x} = -V'(x) + \sqrt{2D}\Gamma(\tilde{t}) \tag{3}$$

with the generalized potential $V(x) := \frac{1}{2}x^2 - \lambda x - i_0 F(x)$ and $F(x) := \int \sum_{n=1}^{\infty} A_n(T)(p\sin(2\pi nx) + (1-p)\sin(2\pi n(x+\frac{1}{2})))dx$ with Gaussian white (with respect to $\tilde{t}$) noise of the unit strength.

In the following we focus on the stationary state of the system (3). The Langevin equation (3) defines a Markov process which probability density $p(x, \tilde{t})$ obeys the Fokker-Planck equation [4] with the natural boundary condition $\lim_{|x| \to \infty} p(x, \tilde{t}) = 0$. The stationary solution $p_s(x)$ takes the form

$$p_s(x) = N_0 e^{-V(x)/D} \tag{4}$$

with a normalization constant N_0. The stationary solution (4) is asymptotically stable.[4] Let us note that both the potential $V(x)$ and the noise intensity D depend on temperature T.

In the high temperature limit (where only normal electrons are present), the potential $V(x)$ is monostable with a minimum at $x_s = 0$ Lowering the temperature below some critical value T_c the potential $V(x)$ becomes symmetrically bistable.[1] The stable minima [3] correspond to the self-sustaining currents in the cylinder. Bifurcation in T_c is continuous i.e. it forms a pitchfork. The bistable system is hysteretic with respect to the external field λ.[1] This behaviour disappears for $T > T_c$. It is a hallmark of the first order phase transition.

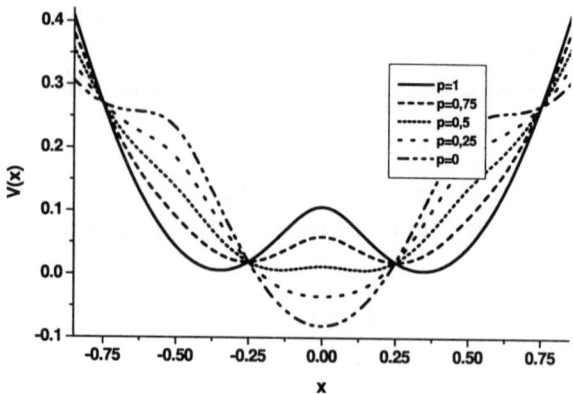

Figure 1. The generalized potential for several values of the probability p. The amplitude $i_0 = 1$ and the temperature T is fixed below T_c for $p = 1/2$.

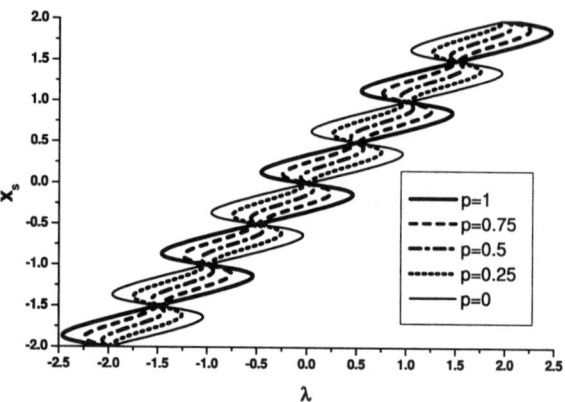

Figure 2. The stationary flux (corresponding to the selfsustaining current) determined by a minima x_s of the generalized potential $V(x)$. The part of the graph with negative slope corresponds to unstable x_s. The amplitude $i_0 = 1$.

The value of p describes (in the statistical sense) the electron structure of the cylinder i.e. the contribution of channels with an even number of coherent electrons in the system. The dependence of the generalized potential on the value of p is given in Fig. 1. Measurement of e.g. the hysteresis loop may allow to estimate p (Fig. 2.).

The maxima of probability density coincide with the minima of $V(x)$ i.e. the most probable values of flux are given by the minima of $V(x)$. In that sense critical temperature in presence of the thermal fluctuations still equals to T_c. However, as it was claimed in ref.[1] the most probable values of

x characterize system only in the regime where characteristic decay time to the minimum's basin of attraction is much greater than the mean time of a fluctuations-induced escape from the minimum of potential. Otherwise the system is characterized by a mean value of x which always vanishes since the potential, as well as the corresponding probability density, is always reflection symmetric.

In the following we discuss properties of the stationary and asymptotically stable probability distributions of magnetic flux in the cylinder with externally induced fluctuations of the amplitude i_0. Let us assume that the fluctuations are given by $i_0 \to (1+y(\tilde{t}))i_0$ with some stochastic process $y(\tilde{t})$. Experimentally such a situation can be achieved by imposing fluctuations of the Fermi level [5] or by indirect influence of Nyquist noise in the metallic cylinders [6] (e.g. carbon nanotubes [7]).

Let us consider the case when fluctuations are described by a Gaussian white noise $\xi(\tilde{t})$ with an amplitude σ, i.e. $y(\tilde{t}) = \sqrt{2\sigma}\xi(\tilde{t})$. The dynamics for the system with fixed $p = 1/2$ is governed by a stochastic equation

$$\dot{x} = -V'(x) + \sqrt{2\sigma}i_0 F'(x)\xi(\tilde{t}) + \sqrt{2D}\Gamma(\tilde{t}) \qquad (5)$$

with two independent white noise sources. This equation is interpreted in the Stratonovich sense.[8] Two independent noise sources can be replaced by a single one with an intensity $\sqrt{2\gamma(x)}$ for $\gamma(x) := \sigma(i_0 F'(x))^2 + D$.

Stationary probability density of the resulting process calculated via the Fokker-Planck equation reads

$$p_s(x) = \frac{N_0}{\sqrt{\gamma(x)}} \exp\left(\int \frac{-V'(x)}{\gamma(x)} dx\right). \qquad (6)$$

Again, according to ref.[4], it is asymptotically stable. Now, this probability distribution describes non-equilibrium state.

The presence of a multiplicative noise in a stochastic equation allows to expect the behaviour which is very different in comparison with the corresponding deterministic system. Indeed, for bistable generalized potential $V(x)$ (i.e. below T_c) neither the position nor the number of critical points coincide with extrema of corresponding stationary density (6) (Fig. 3).

We classify the possible stationary states (phases) with respect to the number of peaks of the corresponding stationary probability density (6). We denote the the state with i-peaked stationary density by P_i. For sufficiently large σ the stationary probability density can posses, except two symmetric peaks corresponding to non-zero fluxes, also a maximum at zero and is of type P_3 (Fig. 4). With further increasing of the amplitude σ one can also obtain the P_5-type stationary state. The reflection symmetry of $p_s(x)$ remains preserved i.e. the mean flux always vanishes.

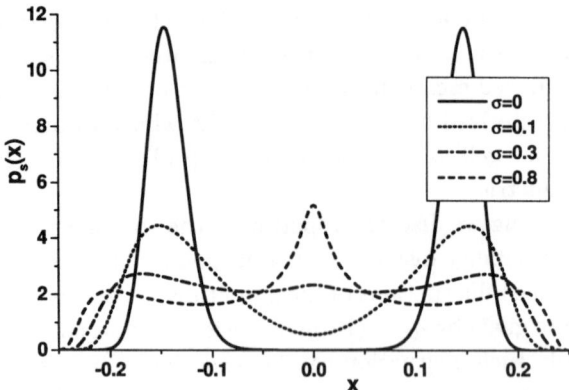

Figure 3.   The stationary probability density for different values of σ, $i_0 = 1$ and $T = 0.5T_c$

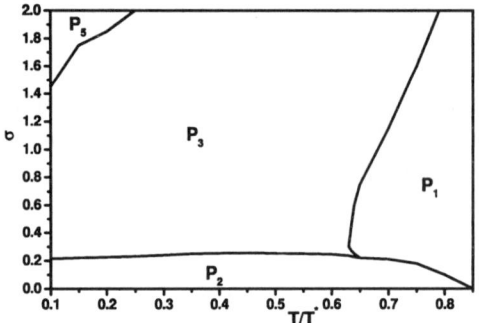

Figure 4.   The phase diagram for the system with multiplicative white noise source. P_i denotes the phase characterized by a probability density with i peaks. The amplitude $i_0 = 1$.

The possible transitions are depicted on the "phase diagram" in Fig. 4. The appearance of the additional (with respect to the equilibrium noise source case) peaks in $p_s(x)$ is a noise-induced effect, since it disappears if and only if $\sigma = 0$. For certain values of the temperature ($T \approx 0.64T^*$) there is a possible re-entrance behaviour in the system i.e. the transition $P_2 \rightarrow P_3 \rightarrow P_1 \rightarrow P_3$.

We conclude therefore that both equilibrium (thermal) and non-equilibrium noise are important in studying mesoscopic systems. The thermal additive fluctuations do not alter significantly the dynamical properties of the system provided the parameters are far from critical. The case of non-equilibrium multiplicative fluctuations is very different. We expect that the noise-induced effects are not only observable but also can serve as a method for the verification of the model. The proposed model can be applied not only to the metallic mesoscopic cylinders but also to matallic carbon nanotubes. The dependence of the coherent current on the magnetic flux is for such a nanotubes similar to the one of metallic cylinders.[7] Another system which can be described by means of our method is the system of carbon tori stacked along certain axis.[9] After certain modification of the generalized potential $V(x)$ (i.e. $F(x)$) the model can be applied to study currents in arbitrary carbon nanotubes or even superconducting rings.

The work supported by KBN Grant 5PO3B0320.

References

1. J. Dajka, J. Łuczka, M. Szopa, E. Zipper, cond-mat/0206478v1.
2. H.F. Cheung et al. Phys. Rev. B 37,6050 (1988),
   M. Szopa, E. Zipper, Int. J. Mod. Phys. B 9,161 (1995).
3. J. Hale, H. Koçak, *"Dynamics and Bifurcations"* (Springer, New York, 1991).
4. A. Lasota, M.C. Mackey *"Probabilisic Properties of Deterministic Systems"* (Cambridge University Press, Cambridge, 1985), Theorem **11.9.1**.
5. M. Krüger et al., Appl. Phys. Lett. 78,1291 (2001).
6. M. Stebelski, M. Szopa, E. Zipper, Z. Phys. B 103,79 (1997)
7. M. Szopa, M. Margańska, E. Zipper, Phys. Lett. A 299,593 (2002).
8. C.W. Gardiner *"Handbook of Stochastic Methods for Physics, Chemistry and Naturel Sciences"* (Springer, Berlin, 1998).
9. M. Margańska, M. Szopa, Acta Phys. Pol. B 32,427 (2001).

ELECTRON DIFFRACTION OF 3-D DEFECTS IN EPITAXIAL FILMS OF $A^{II}B^{VI}$ SEMICONDUCTORS

M. KUŹMA

Institute of Physics, Rzeszow University
Rejtana 16A, 35-959 Rzeszow, Poland

Thin epitaxial $A^{II}B^{VI}$ semiconducting films contain a lot of structural defects: zero-dimensional (point defects), one-dimensional (dislocations), two-dimensional (planar defects) as well as tree-dimensional (twins and included grains of second phase of material). Three-dimensional defects are detected in the layers by carefully inspection of irrational or satellite diffraction spots in reflection or transmission electron diffraction patterns. These additional spots can be caused by various mechanisms: 1) a different orientation of twins in respect to electron beam (primary diffraction); 2) double diffraction e.g. diffraction caused by system matrix/twins; 3) double positioning diffraction e.g. the diffraction caused simultaneously by two gains differently oriented on the substrate. We discuss the possible orientations of twins of the type {111} in f.c.c. (sphalerite) crystal structures (this is the usual structure of $A^{II}B^{VI}$ chalcogenides such as ZnTe, CdTe). Twinings in such crystals modify seriously their reciprocal lattice, what is the crucial argument in the interpretation of diffraction patterns of epitaxial layers. Reciprocal lattice for f.c.c. {111} twining crystals is addressed in details. Furthermore, the primary, double diffraction and double positioning diffraction is considered in details for [111] oriented sphalerite films formed on (001) NaCl substrates.

1. Introduction

The growth and structure of epitaxial films of $A^{II}B^{VI}$ compounds is the preferable task in modern electronic technologies like optoelectronic and nanotechnology. Over the last three decades a substantial body of work has been published on epitaxial films of these compounds. Experimental studies established technological parameters (like substrate temperature) of perfect layers growth by different methods like vacuum evaporation, [1,2,3] chemical vapour transport techniques, sputtering [4,5,6] or electron beam evaporation,[8] on several types of substrates made from mica, ionic crystals (e.g. NaCl, KCl) or Si, Ge and $A^{II}B^{VI}$ bulk crystals (see review papers [9,10] and references there). The structure of layers is investigated mainly by electron microscopy and electron diffraction. This methods allowed to determine not only the index of monocrystality of layers,[10] their orientation

178

but also to detect the structure defects and establish the kind of these defects.[9,10,11] The $A^{II}B^{VI}$ semiconductor compounds have mainly sphalerite structure; diamond-type crystals. Very popular defects in such crystals are twins. These defects influence very strongly on the quality of layer and their orientation. In present papers we will present crystallographic background of twining defects in diamond-type crystals and their relation to an electron diffraction experiments of layers. Considerations will be illustrated by transmission electron diffraction of HgCdTe layers obtained by PLD on KCl substrate.[12]

2. Twins in diamond-type crystals

Twining in diamond-type crystals invariably occurs on {111} planes [13,14]. Crystal structure of such crystals in the direction [111] is layered (Fig. 1a). The structure of chemical bondings between two layers of the type X' and Y

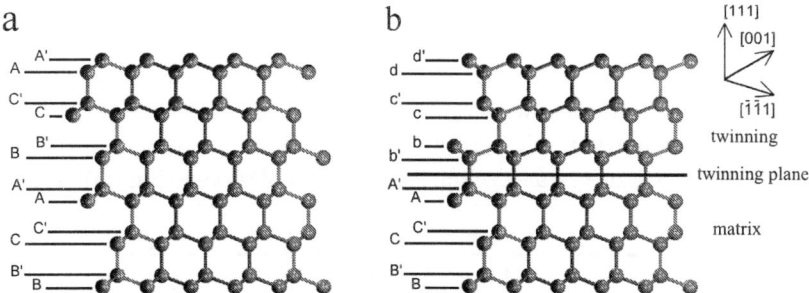

Figure 1.   Layered crystal structure of diamond-type crystals in [111] direction: perfect structure (a), twining structure (b).

(where X'=A',B',C' and Y=B,C,A repectively) does not need additional energy for 180° rotation of the layer Y about an axis containing the chemical bond of corresponding atoms of planes X' and Y. In this way the periodicity of structure in the [111] direction is disturbed in the growth process by planar defect: twinning.

In fact, there are four equivalent directions [111] in diamond-type crystal structure (Fig. 2). Therefore, one can observe four orientations of twinings formed in the way described above. Such twinings we call of the first order. Moreover, when twining occurs again in crystals already twinned, twins of second and successively higher order are formed.

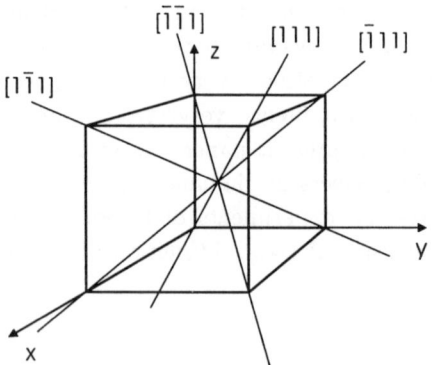

Figure 2.   Four equivalent axis [111] in face centered cubic lattice.

3. Quadruple positioning nucleation and texture structure of $A^{II}B^{VI}$ layers

Yasuda [15] studied by TEM the nucleation of CdSe evaporated onto surfaces of NaCl. Nuclei of seven preferred orientations were formed. Sphalerite structure nuclei were formed in (100) parallel orientations as well as in four azimuthal orientations (Fig. 3a) so called quadruple positioning of the (111) orientation. Two azimutal orientations, i.e. double positioning, of (0001) wurtzite-structure nuclei were also observed (Fig 3b). Other

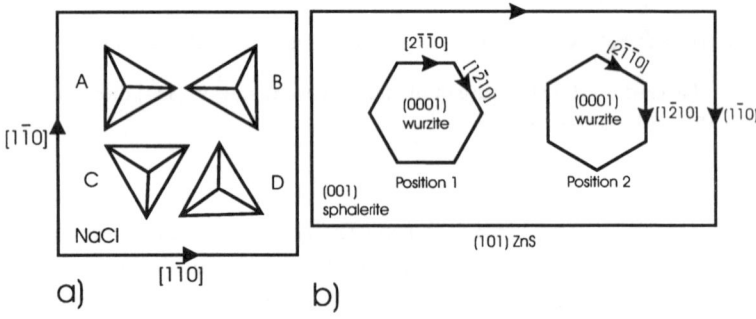

Figure 3.   Orientations of nucleous on (100) NaCl or KCl substrates a) quadruple positioning of sphalerite-structure, b) double positioning of wurtzite-structure.

studies [1] of ZnSe films crystallization on Ge substrates shoved nuclei to be formed with square, triangular and hexagonal shapes what provided not only (100) and (111) sphalerite orientation of films but also (0001) oriented wurtzite-structure material. These observations were discussed by Raven and Kirk [1] in term of the atomistic model of heterogeneous nucleation. Other published studies of the nucleation, orientation, and structure of $A^{II}B^{VI}$ compounds films are close to those presented above. In papers [16,17] Palatnik at al. investigated layers CdTe deposited thermally on NaCl and KCl (100) substrates. By low energy diffraction (LEED) they found that sphalerite-structure was predominant structure of layers. The layers were growing mainly in a [111] direction. Double as well as quadruple positioning provided a texture of films, which determines strongly electron diffraction patterns. Single wurtzite-structure oriented in the [111] direction, i.e. the plane (111) is perpendicular to an electron beam (B), should give the six spot array having six-fold symmetry (Fig. 4).[18]

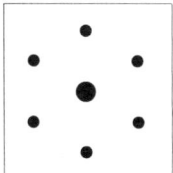

Figure 4.   An electron diffraction pattern from a single (111) oriented sphalerite film.

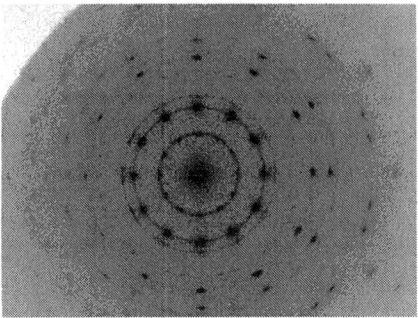

Figure 5.   Electron diffraction pattern from a textured film CdHgTe.

However, textured films show mainly twelve-folded symmetry of theirs electron-diffraction pattern in (111) orientation [12,16,17] (Fig. 5). This is caused mainly by quadruple positioning (Fig. 3) which is a reason of textured structure. In fact an electron diffraction pattern presented in Fig. 5 can be addressed as a superposition of two diffraction patterns presented in Fig. 4 one which is 180° rotated in respect to other one around axis perpendicular to the plane of array passing it in the point (000) (Fig. 6). Such rotation occurs as a result of the 30° different orientation of nucleus

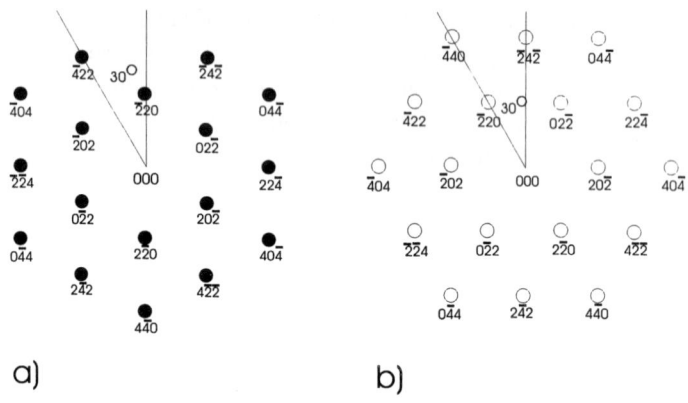

a)   b)

Figure 6. An electron diffraction patterns of grains of the type: a) A or B (Fig. 3a); b) C or D (Fig. 3a).

A and C (or B and D) (Fig. 3a) and finally the same different orientations posses grains in films as well as diffraction patterns of these two kinds of grains.

4. The orientation of twins in diamond-type crystals

Twining on the (111) plane may by considered either as reflection on that plane, or as a 180° rotation around an axis [111]. Both treatments assign the same index numbers of the twins, but the indices of all add orders twins carry opposite signs when twinning is interpreted as reflexion symmetry (Fig. 7). Wilhelm [19] derived four matrices which permit the calculation of the orientation indices of first and high-order twins in diamond–type crystals. The matrices were derived from three equations given by Slawson [14] for calculations of the twin orientation (h'k'l') from the orientation of the mother crystal (hkl) and the plane of reflection (uvw):

$$uh' + vk' + wl' = -uh - vk - wl \tag{1a}$$

$$vh' - uk' = vh - uk \tag{1b}$$

$$wh' - ul' = wh - ul \tag{1c}$$

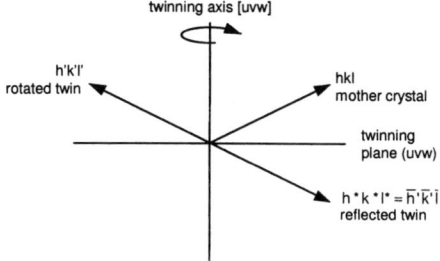

Figure 7. Relationship between the indices of reflected and rotated twins.

The π rotation around an axis $\langle uvw \rangle \equiv \langle 111 \rangle$ may be represented by rotation matrices D_i^*, where $i = 1, 2, 3, 4$ for four possible rotation axes (Table 1).

Table 1. Rotation matrices D_i^*.

i	Axis of rotation [uvw]	D_i^*	Matrix D_i
1	[111]	$\begin{bmatrix} -1/3 & 2/3 & 2/3 \\ 2/3 & -1/3 & 2/3 \\ 2/3 & 2/3 & -1/3 \end{bmatrix}$	$\begin{pmatrix} -1 & 2 & 2 \\ 2 & -1 & 2 \\ 2 & 2 & -1 \end{pmatrix}$
2	[$\bar{1}$11]	$\begin{pmatrix} -1/3 & -2/3 & -2/3 \\ -2/3 & -1/3 & 2/3 \\ -2/3 & 2/3 & 1/3 \end{pmatrix}$	$\begin{pmatrix} -1 & -2 & -2 \\ -2 & -1 & 2 \\ -2 & 2 & -1 \end{pmatrix}$
3	[$\bar{1}\bar{1}$1]	$\begin{pmatrix} -1/3 & 2/3 & -2/3 \\ 2/3 & -1/3 & -2/3 \\ -2/3 & -2/3 & -1/3 \end{pmatrix}$	$\begin{pmatrix} -1 & 2 & -2 \\ 2 & -1 & -2 \\ -2 & -2 & -1 \end{pmatrix}$
4	[1$\bar{1}$1]	$\begin{pmatrix} -1/3 & -2/3 & 2/3 \\ -2/3 & -1/3 & -2/3 \\ 2/3 & -2/3 & -1/3 \end{pmatrix}$	$\begin{pmatrix} -1 & -2 & 2 \\ -2 & -1 & -2 \\ 2 & -2 & -1 \end{pmatrix}$

If D_i^* acts on an orientation vector [h,k,l], it changes its direction to [h'k'l']

$$D_i^* \begin{bmatrix} h \\ k \\ l \end{bmatrix} = \begin{bmatrix} h' \\ k' \\ l' \end{bmatrix} \tag{2}$$

However, h'k'l' are often non-integer. Therefore, the general rule is applied that the lowest integer numbers are to be taken as Miller indices. Thus in the aim of simplicity, matrices $D_i = 3D_i^*$ are permissible for calculations (see Table 1):

$$D_i \begin{bmatrix} h \\ k \\ l \end{bmatrix} = 3 \begin{bmatrix} h' \\ k' \\ l' \end{bmatrix} = \begin{bmatrix} h^* \\ k^* \\ l^* \end{bmatrix} \qquad (3)$$

In Table 2 orientations of first and second-order twins originated from matrix layer oriented to [111], i.e. the plane (111) lies in the plane of a substrate are collected.

Let's consider transmission electron diffraction caused by twined $A^{II}B^{VI}$ layer in which matrix is (111) oriented layer parallel to the substrate. The basic transmission diffraction patterns of such single films in the case when electron beam B is directed towards to [111] direction of layer is presented in Fig 6a.[18] Basing spots of diffraction patterns are spots {220}. The complete pattern may be generated by repeating these spots according to the addition of vectors. Basing spots {220} inform that electron diffraction (ED) in the case B=z=[111] is caused mainly by crystal lattice planes {220} which are arranged towards to the six directions of the type < 220 > i.e. [2$\bar{2}$0], [$\bar{2}$20], [0$\bar{2}$2], [02$\bar{2}$], [$\bar{2}$02], [20$\bar{2}$]. In the twins these directions are changed by rotation formed twins (Table 3). However, one can notice in Table 3 that twining of the type D_1 (111) does not generate new directions from any direction < 220 >. Thus these directions are invariant in respect to twinnings of this type. Therefore we do not expect new spots on the Debay ring {220} which could be generated by such kind of multitwinings. These twinnings rise the diffraction spots {h*k*l*} in an equivalent positions of matrix spots {220} (Fig. 8). But other types of twinnings change the

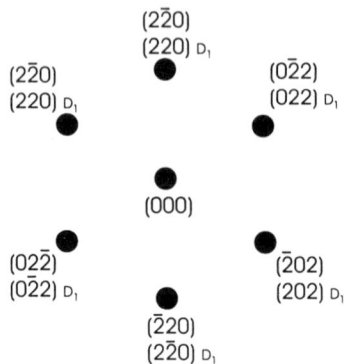

Figure 8. The indexing of the diffraction spots of the sector (220) from single layer oriented in [111] direction and from twings formed by twining axis D_1.

Table 2. Orientation of first- and second-order twins originated from matrix layer oriented to [111] i.e. plane (111) lies in the plane of a substrate.

First-order twins			Rotation axis	Second-order twins	
Rotation axis	Orientation	Denotation		Orientation	Denotation
$D_1 = [111]$	$\begin{pmatrix}1\\1\\1\end{pmatrix}$	$D_1\begin{pmatrix}1\\1\\1\end{pmatrix}$	D_1	$\begin{pmatrix}1\\1\\1\end{pmatrix}$	$D_1D_1\begin{pmatrix}1\\1\\1\end{pmatrix}$
			D_2	$\begin{pmatrix}-5\\-1\\-1\end{pmatrix}$	$D_2D_1\begin{pmatrix}1\\1\\1\end{pmatrix}$
			D_3	$\begin{pmatrix}-1\\-1\\-5\end{pmatrix}$	$D_3D_1\begin{pmatrix}1\\1\\1\end{pmatrix}$
			D_4	$\begin{pmatrix}-1\\-5\\-1\end{pmatrix}$	$D_4D_1\begin{pmatrix}1\\1\\1\end{pmatrix}$
$D_2 = [\bar{1}11]$	$\begin{pmatrix}-5\\-1\\-1\end{pmatrix}$	$D_2\begin{pmatrix}1\\1\\1\end{pmatrix}$	D_1	$\begin{pmatrix}1\\-11\\-11\end{pmatrix}$	$D_1D_2\begin{pmatrix}1\\1\\1\end{pmatrix}$
			D_2	$\begin{pmatrix}9\\9\\9\end{pmatrix}\equiv\begin{pmatrix}1\\1\\1\end{pmatrix}$	$D_2D_2\begin{pmatrix}1\\1\\1\end{pmatrix}$
			D_3	$\begin{pmatrix}5\\-7\\13\end{pmatrix}$	$D_3D_2\begin{pmatrix}1\\1\\1\end{pmatrix}$
			D_4	$\begin{pmatrix}5\\13\\-7\end{pmatrix}$	$D_4D_2\begin{pmatrix}1\\1\\1\end{pmatrix}$
$D_3 = [\bar{1}\bar{1}1]$	$\begin{pmatrix}-1\\-1\\-5\end{pmatrix}$	$D_3\begin{pmatrix}1\\1\\1\end{pmatrix}$	D_1	$\begin{pmatrix}-11\\-11\\1\end{pmatrix}$	$D_1D_3\begin{pmatrix}1\\1\\1\end{pmatrix}$
			D_2	$\begin{pmatrix}13\\-7\\5\end{pmatrix}$	$D_2D_3\begin{pmatrix}1\\1\\1\end{pmatrix}$
			D_3	$\begin{pmatrix}9\\9\\9\end{pmatrix}\equiv\begin{pmatrix}1\\1\\1\end{pmatrix}$	$D_3D_3\begin{pmatrix}1\\1\\1\end{pmatrix}$
			D_4	$\begin{pmatrix}-7\\13\\5\end{pmatrix}$	$D_4D_3\begin{pmatrix}1\\1\\1\end{pmatrix}$
$D_4 = [1\bar{1}1]$	$\begin{pmatrix}-1\\-5\\-1\end{pmatrix}$	$D_4\begin{pmatrix}1\\1\\1\end{pmatrix}$	D_1	$\begin{pmatrix}-11\\1\\-11\end{pmatrix}$	$D_1D_4\begin{pmatrix}1\\1\\1\end{pmatrix}$
			D_2	$\begin{pmatrix}13\\5\\-7\end{pmatrix}$	$D_2D_4\begin{pmatrix}1\\1\\1\end{pmatrix}$
			D_3	$\begin{pmatrix}-7\\5\\13\end{pmatrix}$	$D_3D_4\begin{pmatrix}1\\1\\1\end{pmatrix}$
			D_4	$\begin{pmatrix}9\\9\\9\end{pmatrix}\equiv\begin{pmatrix}1\\1\\1\end{pmatrix}$	$D_4D_4\begin{pmatrix}1\\1\\1\end{pmatrix}$

directions $\{220\}$ of matrix to other sector and therefore they originate to creation new spots in diffraction pattern.

Table 3. Orientation of matrix spots {220} and related twinnings spots $\{h^*k^*l^*\}$.

Twinning axis	Matrix spot {220}	Twinning spots $\{2^*2^*0^*\}$
$D_1=[111]$	$(2\bar{2}0)$	$(\bar{2}20)$
	$(\bar{2}20)$	$(2\bar{2}0)$
	$(\bar{2}02)$	$(20\bar{2})$
	$(20\bar{2})$	$(\bar{2}02)$
	$(02\bar{2})$	$(0\bar{2}2)$
	$(0\bar{2}2)$	$(02\bar{2})$
$D_2 = [\bar{1}11]$	$(2\bar{2}0)$	$(2\bar{2}\bar{8}) \equiv (1\bar{1}\bar{4})$
	$(\bar{2}20)$	$(\bar{2}28) \equiv (\bar{1}14)$
	$(\bar{2}02)$	$(\bar{2}82) \equiv (\bar{1}41)$
	$(20\bar{2})$	$(28\bar{2}) \equiv (14\bar{1})$
	$(02\bar{2})$	$(0\bar{6}6) \equiv (0\bar{2}2)$
	$(0\bar{2}2)$	$(06\bar{6}) \equiv (02\bar{2})$
$D_3=[\bar{1}\bar{1}1]$	$(2\bar{2}0)$	$(\bar{6}60) \equiv (\bar{2}20)$
	$(\bar{2}20)$	$(6\bar{6}0) \equiv (2\bar{2}0)$
	$(\bar{2}02)$	$(\bar{2}82) \equiv (\bar{1}41)$
	$(20\bar{2})$	$(28\bar{2}) \equiv (14\bar{1})$
	$(02\bar{2})$	$(82\bar{2}) \equiv (41\bar{1})$
	$(0\bar{2}2)$	$(\bar{8}22) \equiv (\bar{4}11)$
$D_4 = [1\bar{1}1]$	$(2\bar{2}0)$	$(2\bar{2}\bar{8}) \equiv (1\bar{1}\bar{4})$
	$(\bar{2}20)$	$(\bar{2}2\bar{8}) \equiv (\bar{1}1\bar{4})$
	$(\bar{2}02)$	$(60\bar{6}) \equiv (20\bar{2})$
	$(20\bar{2})$	$(\bar{6}06) \equiv (\bar{2}02)$
	$(02\bar{2})$	$(8\bar{2}\bar{2}) \equiv (4\bar{1}\bar{1})$
	$(0\bar{2}2)$	$(8\bar{2}2) \equiv (4\bar{1}1)$

5. Primary {111} twinning Electron Diffraction

Electron diffraction on twinning crystals can be interpreted basing on the reciprocal lattice of such crystals.[18] The reciprocal lattice for a face centered cubic crystals twinned on all four {111} planes consist of the superposition of five differently oriented body–centered cubic reciprocal lattices. Each of the four twin components is obtained from the matrix reciprocal lattice point (hkl) by 180° rotation about a twin axis $< 111 >$ perpendicular to the {111} twin plane. Thus point (hkl) is transformed to the point (h'k'l') (see eqs 1 and 2) where

$$h' = \frac{1}{3u}(-uh + 2vk + 2wl) \tag{4a}$$

$$k' = \frac{1}{3v}(2uh - vk + 2wl) \tag{4b}$$

$$l' = \frac{1}{3w}(2uh + 2vk - wl) \tag{4c}$$

From equations (4) it follows that the twin reciprocal lattice points either coincide with matrix points or are displaced from a matrix points (HKL) by vectors $\pm\frac{1}{3} < 111 >$:

$$(h'k'l') = (HKL) \pm \frac{1}{3} < 111 > \qquad (5)$$

Therefore, additional reciprocal lattice points originated from twins are placed on the $< 111 >$ lines of the reciprocal lattice in the distance $\pm$ one-third of $< 111 >$ from the matrix point. Moreover not all of these one third points are allowed.

Table 4. Twin points (h'k'l') in reciprocal lattice of f.c.c. twined crystals. (hkl) and (HKL) are matrix reciprocal lattice points; (uvw) is the plane of twininng). * Y, N – allowed or not allowed spots respectively, M – twin spot coinciding with matrix spots (always allowed).

(uvw)	(hkl)	$3(h'k'l')$	$3(h'k'l')$ $= (HKL)$ $\pm\begin{Bmatrix}0\\3(uvw)\end{Bmatrix}$	uH $+vK$ $+wL$	$\pm(3N+1)$	N	$\left.\begin{matrix}Y\\N\\M\end{matrix}\right\}*$
111	111	333	333=111	3	–	–	M
	$\bar{1}11$	$5\bar{1}\bar{1}$	$5\bar{1}\bar{1} = 200 - 111$	2	-(-2)	-1	Y
	$\bar{1}\bar{1}1$	$11\bar{5}$	$11\bar{5} = 00\bar{2} + 111$	-2	+(-2)	-1	Y
	$1\bar{1}1$	$\bar{1}5\bar{1}$	$\bar{1}5\bar{1} = 020 - 111$	2	-(-2)	-1	Y
	$2\bar{2}0$	$\bar{6}60$	$\bar{6}60 = \bar{2}20$	0	–	–	M
	$\bar{2}20$	$6\bar{6}0$	$6\bar{6}0 = 2\bar{2}0$	0	–	–	M
	$\bar{2}02$	$60\bar{6}$	$60\bar{6} = 20\bar{2}$	0	–	–	M
	$20\bar{2}$	$\bar{6}06$	$\bar{6}06 = \bar{2}02$	0	–	–	M
	$02\bar{2}$	$0\bar{6}6$	$0\bar{6}6 = 0\bar{2}2$	0	–	–	M
	$0\bar{2}2$	$06\bar{6}$	$06\bar{6} = 02\bar{2}$	0	–	–	M
$\bar{1}11$	111	$5\bar{1}\bar{1}$	$5\bar{1}\bar{1} = \bar{2}00 - \bar{1}11$	2	-(-2)	-1	Y
	$\bar{1}11$	333	$333 = \bar{1}11$	3	–	–	M
	$\bar{1}\bar{1}1$	$15\bar{1}$	$15\bar{1} = 020 - \bar{1}11$	2	-(-2)	-1	Y
	$1\bar{1}1$	$\bar{1}1\bar{5}$	$\bar{1}1\bar{5} = 00\bar{2} + \bar{1}11$	-2	+(-2)	-1	Y
	$2\bar{2}0$	$2\bar{2}\bar{8}$	$2\bar{2}\bar{8} = 1\bar{1}\bar{3} + \bar{1}11$	-5	+(-5)	-2	Y
	$\bar{2}20$	$\bar{2}2\bar{8}$	$\bar{2}2\bar{8} = \bar{1}1\bar{3} - \bar{1}11$	5	-(-5)	-2	Y
	$\bar{2}02$	$\bar{2}8\bar{2}$	$\bar{2}8\bar{2} = \bar{1}3\bar{1} - \bar{1}11$	5	-(-5)	-2	Y
	$20\bar{2}$	$28\bar{2}$	$28\bar{2} = 13\bar{1} + \bar{1}11$	-5	+(-5)	-2	Y
	$02\bar{2}$	$0\bar{6}6$	$0\bar{6}6 = 0\bar{2}2$	0	–	–	M
	$0\bar{2}2$	$06\bar{6}$	$06\bar{6} = 02\bar{2}$	0	–	–	M
$1\bar{1}1$	111	$11\bar{5}$	$11\bar{5} = 00\bar{2} + 1\bar{1}1$	-2	+(-2)	-1	Y
	$\bar{1}11$	$15\bar{1}$	$15\bar{1} = 0\bar{2}0 - 1\bar{1}1$	2	-(-2)	-1	Y
	$\bar{1}\bar{1}1$	$\bar{3}33$	$\bar{3}33 = \bar{1}\bar{1}1$	3	–	–	M
	$1\bar{1}1$	$5\bar{1}\bar{1}$	$5\bar{1}\bar{1} = \bar{2}00 - 1\bar{1}1$	2	-(-2)	-1	Y
	$2\bar{2}0$	$\bar{6}60$	$\bar{6}60 = \bar{2}20$	0	–	–	M
	$\bar{2}20$	$6\bar{6}0$	$6\bar{6}0 = 2\bar{2}0$	0	–	–	M
	$\bar{2}02$	$\bar{2}8\bar{2}$	$\bar{2}8\bar{2} = \bar{1}3\bar{1} - 1\bar{1}1$	5	-(-5)	-2	Y
	$20\bar{2}$	$28\bar{2}$	$28\bar{2} = 13\bar{1} + 1\bar{1}1$	-5	+(-5)	-2	Y
	$02\bar{2}$	$82\bar{2}$	$82\bar{2} = 31\bar{1} + 1\bar{1}1$	-5	+(-2)	-2	Y
	$0\bar{2}2$	$\bar{8}2\bar{2}$	$\bar{8}2\bar{2} = \bar{3}1\bar{1} - 1\bar{1}1$	5	-(-5)	-2	Y
$1\bar{1}1$	111	$15\bar{1}$	$15\bar{1} = 020 - 1\bar{1}1$	-2	+(-2)	-1	Y
	$\bar{1}11$	$11\bar{5}$	$11\bar{5} = 00\bar{2} + 1\bar{1}1$	-2	+(-2)	-1	Y
	$\bar{1}\bar{1}1$	$5\bar{1}\bar{1}$	$5\bar{1}\bar{1} = 200 - 1\bar{1}1$	2	-(-2)	-1	Y
	$1\bar{1}1$	$\bar{3}33$	$\bar{3}33 = 1\bar{1}1$	3	–	–	M
	$2\bar{2}0$	$2\bar{2}\bar{8}$	$2\bar{2}\bar{8} = 1\bar{1}3 - 1\bar{1}1$	5	-(-5)	-2	Y
	$\bar{2}20$	$\bar{2}2\bar{8}$	$\bar{2}2\bar{8} = \bar{1}1\bar{3} + 1\bar{1}1$	-5	+(-5)	-2	Y
	$\bar{2}02$	$60\bar{6}$	$60\bar{6} = 20\bar{2}$	0	–	–	M
	$20\bar{2}$	$\bar{6}06$	$\bar{6}06 = \bar{2}02$	0	–	–	M
	$02\bar{2}$	$\bar{8}2\bar{2}$	$\bar{8}2\bar{2} = \bar{3}1\bar{1} + 1\bar{1}1$	-5	+(-5)	-2	Y
	$0\bar{2}2$	$82\bar{2}$	$82\bar{2} = 3\bar{1}1 - 1\bar{1}1$	5	-(-5)	-2	Y

The true additional twining reciprocal lattice points should satisfy selection rule:[9]

$$uH + vK + wL = \pm(3N + 1) \tag{6a}$$

where N is integer, apart from twinning points (h'k'l') which coincides exactly with matrix points (HKL)

$$(h'k'l') \equiv (HKL); \tag{6b}$$

those are always allowed. The points satisfied rule (6) are exampled in Table 4. Let's consider twinning points originated from matrix points (111), $(\bar{1}11)$ and $(2\bar{2}0)$ generated by the twinning plane $(\bar{1}11) \equiv D_2^*$. According to (4) we get:

$$D_2^*(111) = \frac{1}{3}(\bar{5}1\bar{1}) = (\bar{2}00) - \frac{1}{3}(\bar{1}11) \tag{7a}$$

$$D_2^*(\bar{1}11) = \frac{1}{3}(\bar{3}33) = (\bar{1}11) \tag{7b}$$

$$D_2^*(2\bar{2}0) = \frac{1}{3}(2\bar{2}\bar{8}) = (11\bar{3}) + \frac{1}{3}(\bar{1}11) \tag{7c}$$

The selection rule (6) is satisfied for (7a) and (7c) because in the case (7a) the equation (6) is:

$$\bar{1} \cdot \bar{2} + 1 \cdot 0 + 1 \cdot 0 = 2 = -(3N + 1) \quad \text{for} \quad N = -1$$

and in the case (7c) the equation (6) is:

$$\bar{1} \cdot 1 + 1 \cdot \bar{1} + 1 \cdot \bar{3} = -1 - 1 - 3 = -5 = +(3N + 1) \quad \text{for} \quad N = -2$$

The arrangement of the twin reciprocal lattice points is exemplary plotted in Fig. 9.

6. Double-Twinning Electron Diffraction

Double diffraction occurs if a diffracted beam from the matrix passes into a twin and suffers a second diffraction, or vice versa. The direction of the double diffracted beam is determined by adding together the reciprocal lattice vectors corresponding to the two component diffractions. (hkl) and $(h'k'l')$ where (hkl) means matrix point and $(h'k'l')$ is primary-twinning reciprocal lattice point

$$
(hkl) + \left(h'k'l'\right) = (hkl) + \left((HKL) \pm \frac{1}{3}\langle uvw \rangle\right) =
$$
$$
= \left(\left(h + H \pm \frac{1}{3}u\right)\left(k + K \pm \frac{1}{3}v\right)\left(l + L \pm \frac{1}{3}w\right)\right) = \tag{8}
$$
$$
= \left(\left(H'' \pm \frac{1}{3}u\right)\left(K'' \pm \frac{1}{3}v\right)\left(L'' \pm \frac{1}{3}w\right)\right)
$$

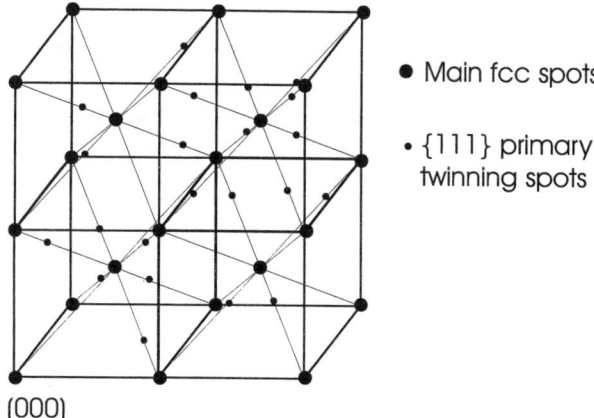

● Main fcc spots

• {111} primary
  twinning spots

(000)

Figure 9.   The reciprocal lattice for a f.c.c. crystal together with the some extra points from primary-twinnings diffraction.

The double diffraction does not produce extra diffraction spots in the case, when the diffraction in the twin coincides with a matrix reciprocal lattice point. Let's denote double diffracted reciprocal lattice point by $(h''k''l'')$ then

$$h'' = H'' \pm \frac{1}{3}u = h + H \pm \frac{1}{3}u \tag{9a}$$

$$k'' = K'' \pm \frac{1}{3}v = k + K \pm \frac{1}{3}v \tag{9b}$$

$$l'' = L'' \pm \frac{1}{3}w = l + L \pm \frac{1}{3}w \tag{9c}$$

These double diffraction reciprocal lattice points are plotted in Fig. 10 and they are also displaced from matrix points $(H''K''L'')$ by vectors of $\pm\frac{1}{3} < 111 >$. All points of this form are permissible.

7. Conclusions

Study of reciprocal lattice of twined f.c.c. structure is crucial in the interpretation of 3-D defected thin layers of f.c.c. crystals. The structure factors of the face-centered cubic structure and of the sphalerite structure give the same set of permitted reflections in ED. Therefore the sphalerite and f.c.c. diffraction patterns may be interpreted in the same way. The same equivalency exists between wurtzite an h.c.p. structures. Twining defects in diffraction pattern does occur mainly as satellite spots surrounding the original matrix spots. Often in the aim to observe this spots the sample

190

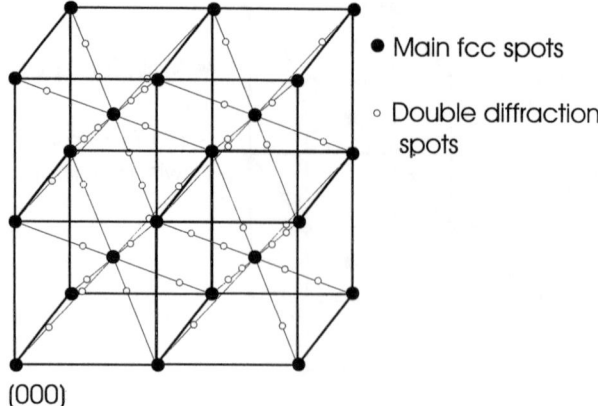

Figure 10. Double twinning-diffraction reciprocal lattice points of f.c.c. lattice.

have to be tilted, because additional reciprocal lattice points lies a little bit away from the main Ewald sphere on which the main point lies.

References

1. M.S. Raven, D.L. Kirk, *Thin Solid Films* **21**, 205-224 (1974).
2. D.M. Newbury, D.L. Kirk, *Thin Solid Films* **22**, 322-330 (1974).
3. P. Wilkes, J. Mater. Sci. 4, 91-93 (1969).
4. S.G. Parker, J.E. Pinnell, L.N. Swink, *J. Phys. Chem. Solids* **5**, 139-147 (1971)
5. P. Lilley, P.L. Jones, C.N.W. Litting, *J. Mater. Sci.* **5**,891-897 (1970).
6. P. Lilley, P.L. Jones, C.N.W. Litting, *J. Cryst. Growth* £13/14, 371 (1972).
7. T.G.R. Rawlins, R.J. Woodward, *J. Mater. Sci.* **7**, 257-264 (1972).
8. B.A. Unvala, J.M. Woodcock, D.B. Holt, *J.phys. D. Appl. Phys.* I 11-14 (1968).
9. D.W. Pashley, M.J. Stowel, *Phil. Mag.* **8**, 1605-1632 (1963).
10. D.B. Holt, *Thin Solid Films* **24**, 1-53 (1974).
11. D.B. Holt, *J. Mater. Sci.* **1**, 280-295 (1966).
12. I. Virt, M. Kuzma, G. Wisz, I. Rudyj, M. Fruginskii, I. Kurilo, I. Lopatynskii, *Mol. Phys. Report* **23**, 206-209 (1999).
13. J.A. Kohn, *Amer. Min.* **43**, 263 (1950).
14. C.B. Slawson, *Amer. Min.* **35**, 193 (1950).
15. Y. Yasuda, *Japan. J. Appl. Phys.* **7**, 1171-1180 (1968).
16. S. Palatnik, B.T. Boiko, W.K. Sorkin, W.E. Winogradov, *Nonorganic Materials VII*, **7**, 1132-1135 (1971).
17. S. Palatnik, B.T. Boiko, W.K. Sorkin, W.E. Winogradov, *Nonorganic Materials VII*, **7**, 1960-1965 (1971).

18. K.W. Andrews, D.j. Dyson, S.R. Keown, *Interpretation of electron diffraction patterns, Practical Electron Microscopy* (London 1968).
19. B.F. Wilhelm, *J. Appl. Cryst.* **4**, 521 (1971).
20. D.B. Holt, *J. Mater. Sci.* **4**, 935-943 (1969).

FRACTAL DESCRIPTION OF CLUSTERS FORMED IN THIN FILMS GROWTH SIMULATION

L. PYZIAK, K. ZEMBROWSKA, M. KUZMA, I. STEFANIUK

Institute of Physics, Rzeszow University
Rejtana 16 A, 35 - 310 Rzeszow, Poland

The first Si monolayer grown on Si (001) substrate by the pulsed laser deposition method (PLD) was simulated using the Monte Carlo procedure. Various growing conditions, such as temperature of substrate and time of annealing, were introduced into the simulation. Clearly, the shape of the clusters formed depends on these conditions. The best method for description of a cluster shape is their fractal characterisation. Box fractal dimension of these planar objects was studied. This parameter selects very well the shape of such clusters, which are most favourable for the epitaxial layer-growing mode. The relation between the fractal dimension of clusters and PLD parameters was established.

1. Introduction

The most important kinetic process in film growth is the diffusion of an adatom on a flat substrate surface. Smooth uniform films could not be formed without sufficient surface mobility. In the case of small or very high mobility, the growth front is always very rough.

The surface diffusion depends on a number of parameters, such as substrate temperature, crystal structure of substrate surface etc. A proper determination of these parameters in the film fabrication process is a crucial and difficult task. A very useful method to solve this problem is a computer simulation of layer growing based on a model of film growth. Macroscopic models of layer crystallisation, such as SOS [1] (solid on solid) determine parameters such as the layer growth rate. Nevertheless, they have several restrictions: they can be applied to high deposition rate only, numerical solving of complicated differential equations is very tedious or even impossible, etc. On the other hand, in the description of nanotechnological layer processes, another i.e. microscopic approach is more useful because of a microscopic scale of the object studied. Therefore, much recent attention has been focused on molecular dynamic [2] or the Monte Carlo [3,4,5] method of crystal layer growth simulation.

In this paper, we apply the Monte Carlo method for modelling of the first monolayer growth in the Si/(001) Si homoepitaxy provided by low energy pulse laser deposition [6,7]. The roughness of islands formed on the substrate in first stage of layer deposition was measured by their fractal dimension measurements [8]. The high values of these dimensions point to a smooth crystallisation front, which indicates a proper technological parameter assumption.

2. Surface diffusion in Monte Carlo simulation

The simulation was carried out on the $L \times L$ square flat lattice related to the (001) Si substrate surface. The Born-Karmann boundary conditions were imposed on the lattice. The first step in simulation was random deposition of a given portion of gas silicon flux consisting of N atoms on the substrate. This stage was performed by a series of random determinations of x, y coordinates for all N particles landing on the surface. The corresponding sites were denoted by number 1 (an empty site is denoted by zero). The atoms landed have kinetic energy determined by their source.

The second stage was based on surface diffusion of adatoms. In the Monte Carlo method, this process is based on the probability of an adatom jumping from one site to another (p_h). This probability is determined by potential barrier E_a between these sites and depends on the number of first (n_1) and second (n_2) near-neighbours of this adatom in the initial site

$$E_a = n_1 E_1 + n_2 E_2 \tag{1}$$

where E_1 and E_2 is an energy of interaction of adatom with the first and second near-neighbour, respectively.

The probability is given by the formula:

$$p_h = A \exp\left\{-\frac{E_a}{k_B T}\right\} \tag{2}$$

where A is a coefficient which value is in a relation to a time of hopping, and T is the substrate temperature.

We have assumed that an atom can jump to the nearest empty site only. For each direction of hopping, the probability of jumping in direction (p_i) depends on the number of the neighbours s_i in their prospective new positions:

$$p_i = \frac{(s_i + 1)}{\sum (s_i + 1)} \tag{3}$$

An adatom is moved to a new site if a randomly selected number belongs to the part "p_i" of the range $[0, 1]$ obtained by dividing it into parts proportional to p_i. Then, the whole procedure is repeated N times.

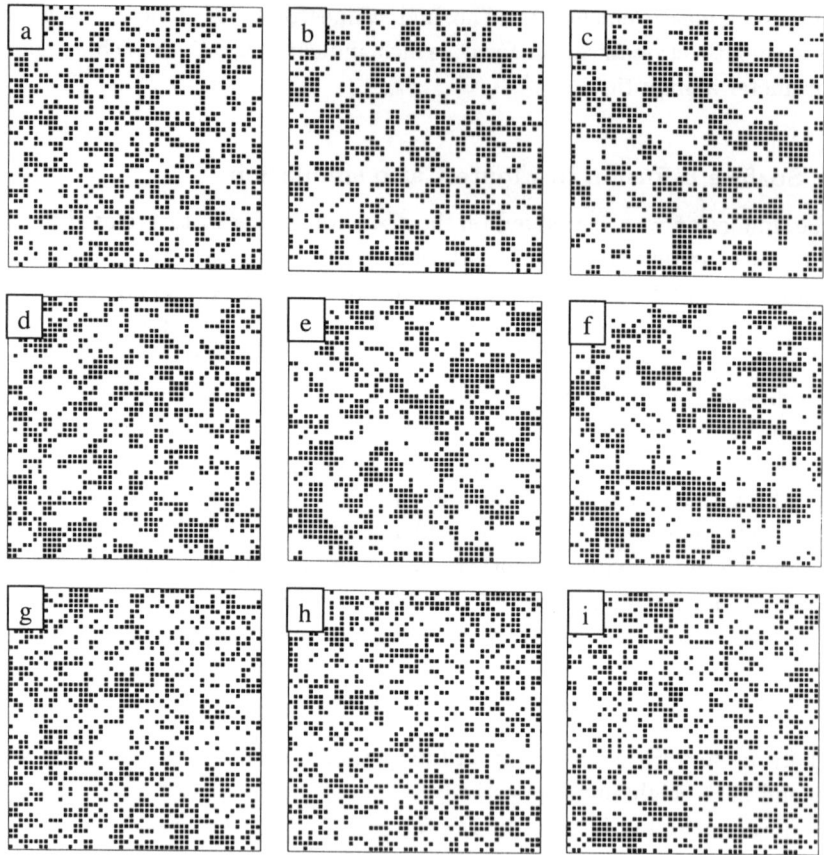

Figure 1. Si layer on (001) Si substrate growth at temperatures: 700 K (a,b,c); 760 K (d,e,f); 850 K (g,h,i) and annealed in times: 0,1 s (a,d,g); 0,5 s (b,e,h); 1 s (c,f,i).

3. Fractal characterisation of nonregular flat structures

The mathematical characterisation of the shape of irregular objects is done by their fractal dimension. In general, it is the Hausdorff dimension. In the particular case of quasi-homogeneous flat structures, it is a box-like dimension D_b, which is one of the Mandelbrot fractal dimensions [9].

On the surface, the box-like dimension D_b does never excess the value 2, which is the Euclidean dimension for two-dimensionality. It is also worth mentioning that for tightly packed structures (smooth shoreline, lack of holes) the box-like dimension is close to this value 2. On the contrary, for non-compact morphologies of flat clusters (rough edges, highly anisotropic shapes, a great number of lakes), this dimension is less then value 2 and not much greater then value 1.

Calculations were performed by a computer program. The program allows to determine the size of the clusters (the number of atoms in cluster), to calculate fractal dimension of cluster, and the average fractal dimension of clusters.

4. Results

Results of simulations were obtained for the following parameters: $L = 50$, $N = 1000$, $E_1 = 0, 6$ eV; $E_2 = 0, 05$ eV; $T = 650 - 850$ K. Fig. 1 presents examples of various layers composed of clusters growing at various temperatures and various times. The fractal dimension D_b, as well as the average fractal dimension $\langle D_b \rangle$, were calculated for clusters formed in different conditions (Fig. 2).

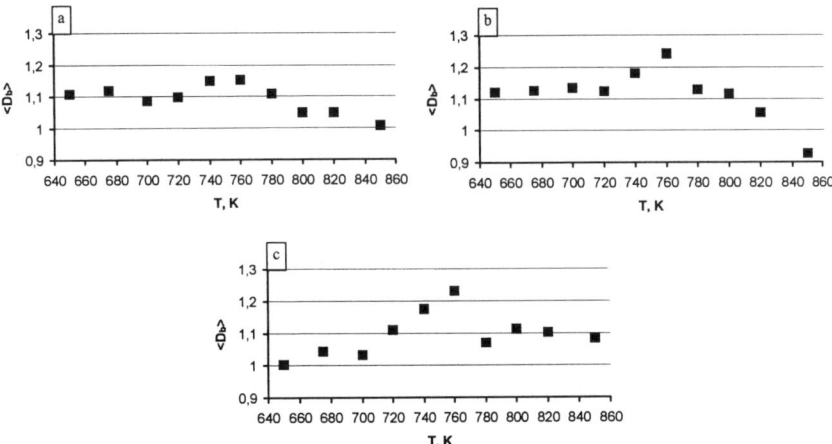

Figure 2. Averaged fractal dimension of Si clusters forming a layer in various temperatures annealed in time: 0,1 s (a); 0,5 s (b); 1 s (c).

5. Discussion and conclusions

One can see that in the range of temperature 720–780 K (temperature window), the fractal dimension increases clearly in comparison to the lower and higher temperatures. Furthermore, this range of temperature is the best for layer growth. Moreover, one can see that the crystallisation front of clusters clearly depends on the surface diffusion parameters determined by interaction of an adatom with its neighbours. For example, if in the simulation procedure we assume a very low probability for hopping from a site having two neighbours, than the shape of a cluster becomes more regular (Fig. 3). The crystallisation front is in this case smoother and evolves not only in {10} direction, but also in the directions {11} and {12}. In fact, this last direction becomes the most favourable one. Of course, the fractal dimension of such clusters approach value 2.

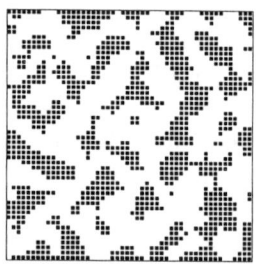

Figure 3. Clusters forming Si layer on (001) Si substrate obtained in a restricted surface diffusion conditions.

Simulations of the time relaxation of the layer formation point out to the laser pulse repetition time as a parameter important to the PLD method (this fact is frequently neglected in experiments). We have observed that at the temperature of the best growth, this time acts favourably for a good quality of layer. At other temperatures, the increase of the relaxation time does not change the morphology of a layer (the fractal dimension was not changed), or even shows a destructive action. The atomic mechanism of the early stage of thin-film growth can be easy applied to other method of layer crystallisation from vapour phase.

References

1. J.D. Weeks, G.H. Gilmer, K.A. Jackson, *J. Chemical Physics* **65**, 2 (1976) 712.
2. D.W. Hermann, Computer Simulation Methods in Theoretical Physics, *Springer-Verlag*, Berlin Heidelberg, 1986.
3. Z. Zhang, M.G. Lagally, *Science* **276** (1997) 377.
4. G.H. Gilmer, H. Huang, C. Roland, *Comp. Mater. Sci.* **12** (1998) 354.

5. M. Kuzma, M. Bester, L. Pyziak, I. Stefaniuk, I. Virt, *Applied Surface Science* **168** (2000) 132.
6. D.B. Chrisey, G.K. Hubler (Eds.), Pulsed Laser Deposition of Thin Films, *A Wiley-Interscience Publication*, New York, 1994.
7. I. Virt, M. Bester, L. Dumanski, M. Kuzma, I.O. Rudyj, M.S. Fruginskyi, I.V. Kurilo, *Applied Surface Science* **177** (2001) 201.
8. J. Feder, Fractals, *Plenum Press*, New York, 1988.
9. B.B. Mandelbrot, The fractal geometry of nature, *Freeman*, New York, 1982.

Part B

RIGGED STRING CONFIGURATIONS
YANG-BAXTER EQUATIONS
AND THEIR APPLICATIONS
IN SOLID STATE PHYSICS

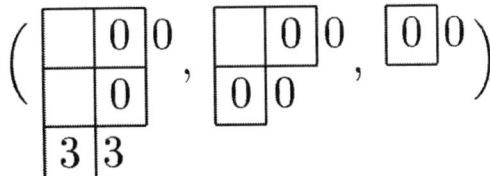

RIGGED CONFIGURATIONS AND THE BETHE ANSATZ

ANNE SCHILLING

Department of Mathematics,
University of California,
One Shields Avenue,
Davis, CA 95616-8633, U.S.A.
E-mail: anne@math.ucdavis.edu

This note is a review of rigged configurations and the Bethe Ansatz. In the first part, we focus on the algebraic Bethe Ansatz for the spin 1/2 XXX model and explain how rigged configurations label the solutions of the Bethe equations. This yields the bijection between rigged configurations and crystal paths/Young tableaux of Kerov, Kirillov and Reshetikhin. In the second part, we discuss a generalization of this bijection for the symmetry algebra $D_n^{(1)}$, based on work in collaboration with Okado and Shimozono.

1. Introduction

These notes arose from three lectures presented at the Summer School on Theoretical Physics "Symmetry and Structural Properties of Condensed Matter" held in Myczkowce, Poland, on September 11-18, 2002. We review the algebraic Bethe Ansatz in the simple setting of the spin 1/2 XXX model, explain the physical meaning of rigged configurations and give a bijection between rigged configurations and crystal bases for type $D_n^{(1)}$ [20] generalizing the bijection of Kerov, Kirillov and Reshetikhin [15,16] for type $A_n^{(1)}$.

The Bethe Ansatz originated in a paper by Bethe [2] in 1931 in which he studied the eigenvectors and eigenfunctions of the Hamiltonian of the Heisenberg antiferromagnet. The method he used is today often called the coordinate Bethe Ansatz, distinguishing it from the algebraic Bethe Ansatz that will be presented here. The algebraic Bethe Ansatz is a generalization of the coordinate Bethe Ansatz and is one of the most important outcomes of the quantum inverse scattering method introduced in [23,24,25]. The quantum inverse scattering method has unified the treatment of quantum integrable systems considering each model as a representation of the quantum monodromy matrix which satisfies certain commutation relations. The algebraic Bethe Ansatz is based on the idea of constructing eigenvec-

tors of the Hamiltonian (resp. trace of the monodromy matrix) by creation
and annihilation operators on a vacuum; the elements of the monodromy
matrix play the role of these operators. The eigenvectors are parametrized
by solutions of a system of algebraic equations, called the Bethe equations.
The solutions in turn are labeled by combinatorial objects called rigged
configurations.

We consider $\mathfrak{g}$-invariant models where $\mathfrak{g}$ is the symmetry algebra. The
Hilbert space is the tensor product of irreducible representations of $\mathfrak{g}$ denoted $\mathcal{H} = h_1 \otimes \cdots \otimes h_N$. The Bethe vectors are the highest weight vectors
in the decomposition into irreducible components of $\mathcal{H}$. It is known from
representation theory that the highest weight vectors are also labeled by
Young tableaux (see for example [5]) or certain paths in crystal theory (see for
example [7,8,18]). Assuming the completeness of the Bethe vectors, this suggests a bijection between rigged configurations and Young tableaux/crystal
paths. For $\mathfrak{g} = \mathfrak{gl}_n$ such a bijection was given by Kirillov and Reshetikhin
[16] and generalized in [17].

Analogous bijections for all $\mathfrak{g}$ of nonexceptional affine type were recently
proven in [20] for tensor products of the fundamental representation. An important property of all these bijections is that they preserve statistics that
can be defined on the set of rigged configurations and paths, respectively.
As a corollary it follows that one-dimensional configuration sums defined
in terms of crystal paths have fermionic formulas. Fermionic formulas reflect the quasiparticle structure of the underlying model and also reveal
the statistics of the quasiparticles. For general affine Kac-Moody algebras
fermionic formulas were conjectured by Hatayama, Kuniba, Okado, Takagi, Tsuboi and Yamada [7,8]. For type $A_n^{(1)}$ they were proven in [17] and for
nonexceptional types in special cases in [20].

The paper is organized as follows. In section 2 we review the algebraic
Bethe Ansatz for the spin $1/2$ XXX model and derive the Bethe equations.
In section 3 we present the solutions of the Bethe equations parametrized
by rigged configurations and discuss the bijection between rigged configurations and paths in section 4. Sections 2 and 3 follow the presentation of
Faddeev [4]. In sections 5-8 the bijection between paths and rigged configurations is generalized to types $A_n^{(1)}$ and $D_n^{(1)}$ based on work in collaboration
with Okado and Shimozono [20]. Crystal bases are introduced in section 6
and section 7 states the fermionic formula and rigged configurations in the
generalized set-up. The bijection is given explicitly in section 8.

2. Bethe Ansatz for the XXX model

In this section we discuss the algebraic Bethe Ansatz for the example of the spin 1/2 XXX Heisenberg chain. This is a one-dimensional quantum spin chain on N sites with periodic boundary conditions. It is defined on the Hilbert space $\mathcal{H}_N = \bigotimes_{n=1}^{N} h_n$ where in this case $h_n = \mathbb{C}^2$ for all n. Associated to each site is a local spin variable $\vec{s} = \frac{1}{2}\vec{\sigma}$ where

$$\vec{\sigma} = (\sigma^1, \sigma^2, \sigma^3) = \left(\begin{pmatrix} 0 & 1 \\ 1 & 0 \end{pmatrix}, \begin{pmatrix} 0 & -i \\ i & 0 \end{pmatrix}, \begin{pmatrix} 1 & 0 \\ 0 & -1 \end{pmatrix} \right)$$

are the Pauli matrices. The spin variable acting on the n-th site is given by

$$\vec{s}_n = I \otimes \cdots \otimes I \otimes \vec{s} \otimes I \otimes \cdots \otimes I$$

where I is the identity operator and $\vec{s}$ is in the n-th tensor factor. We impose periodic boundary conditions $\vec{s}_n = \vec{s}_{n+N}$.

The Hamiltonian of the spin 1/2 XXX model is

$$H_N = J \sum_{n=1}^{N} \left(\vec{s}_n \cdot \vec{s}_{n+1} - \frac{1}{4} \right).$$

Our goal is to determine the eigenvectors and eigenvalues of H_N in the antiferromagnetic regime $J > 0$ in the limit when $N \to \infty$.

The main tool will be the Lax operator $L_{n,a}(\lambda)$, also called the local transition matrix. It acts on $h_n \otimes \mathbb{C}^2$ where $\mathbb{C}^2$ is an auxiliary space and is defined as

$$L_{n,a}(\lambda) = \lambda I_n \otimes I_a + i\vec{s}_n \otimes \vec{\sigma}_a.$$

Here I_n and I_a are unit operators acting on h_n and the auxiliary space $\mathbb{C}^2$, respectively; λ is a complex parameter, called the spectral parameter. Writing the action on the auxiliary space as a 2×2 matrix, we have

$$L_n(\lambda) = \begin{pmatrix} \lambda + is_n^3 & is_n^- \\ is_n^+ & \lambda - is_n^3 \end{pmatrix} \tag{5}$$

where $s_n^{\pm} = s_n^1 \pm is_n^2$.

The crucial fact is that the Lax operator satisfies commutation relations in the auxiliary space $V = \mathbb{C}^2$. Altogether there are 16 relations which can be written compactly in tensor notation. Given two Lax operators $L_{n,a_1}(\lambda)$ and $L_{n,a_2}(\mu)$ defined in the same quantum space h_n, but different auxiliary spaces V_1 and V_2, the products $L_{n,a_1}(\lambda)L_{n,a_2}(\mu)$ and $L_{n,a_2}(\mu)L_{n,a_1}(\lambda)$ are defined on the triple tensor product $h_n \otimes V_1 \otimes V_2$. There exists an operator $R_{a_1,a_2}(\lambda - \mu)$ defined on $V_1 \otimes V_2$ such that

$$R_{a_1,a_2}(\lambda - \mu)L_{n,a_1}(\lambda)L_{n,a_2}(\mu) = L_{n,a_2}(\mu)L_{n,a_1}(\lambda)R_{a_1,a_2}(\lambda - \mu). \tag{6}$$

Explicitly, the R-matrix $R_{a_1,a_2}(\lambda)$ is given by

$$R_{a_1,a_2}(\lambda) = \left(\lambda + \frac{i}{2}\right)I_{a_1} \otimes I_{a_2} + \frac{i}{2}\vec{\sigma}_{a_1} \otimes \vec{\sigma}_{a_2}.$$

To deduce the 16 relations explicitly, one may write (6) as matrices in the auxiliary space $V_1 \otimes V_2$ using the convention $(A \otimes B)_{k\ell}^{ij} = A_{ij}B_{k\ell}$ where

$$M_{k\ell}^{ij} = \begin{pmatrix} M_{11}^{11} & M_{12}^{11} & M_{11}^{12} & M_{12}^{12} \\ M_{21}^{11} & M_{22}^{11} & M_{21}^{12} & M_{22}^{12} \\ M_{11}^{21} & M_{12}^{21} & M_{11}^{22} & M_{12}^{22} \\ M_{21}^{21} & M_{22}^{21} & M_{21}^{22} & M_{22}^{22} \end{pmatrix}.$$

In this notation the R-matrix reads

$$R(\lambda) = \begin{pmatrix} a(\lambda) & 0 & 0 & 0 \\ 0 & b(\lambda) & c(\lambda) & 0 \\ 0 & c(\lambda) & b(\lambda) & 0 \\ 0 & 0 & 0 & a(\lambda) \end{pmatrix}$$

where $a(\lambda) = \lambda + i$, $b(\lambda) = \lambda$ and $c(\lambda) = i$.

Geometrically, the Lax operator $L_{n,a}(\lambda)$ can be interpreted as the transport between sites n and $n+1$ of the quantum spin chain. Hence

$$T_{N,a}(\lambda) = L_{N,a}(\lambda)\cdots L_{1,a}(\lambda)$$

is the monodromy around the circle (recall that we assume periodic boundary conditions). In the auxiliary space write

$$T_N(\lambda) = \begin{pmatrix} A(\lambda) & B(\lambda) \\ C(\lambda) & D(\lambda) \end{pmatrix}$$

with entries in the full Hilbert space $\mathcal{H}_N$. From (6) it is clear that the monodromy matrix satisfies the following commutation relation

$$R_{a_1,a_2}(\lambda - \mu)T_{N,a_1}(\lambda)T_{N,a_2}(\mu) = T_{N,a_2}(\mu)T_{N,a_1}(\lambda)R_{a_1,a_2}(\lambda - \mu). \tag{12}$$

Explicitly, some of the relations contained in (12) are

$$[B(\lambda), B(\mu)] = 0$$
$$A(\lambda)B(\mu) = f(\lambda - \mu)B(\mu)A(\lambda) + g(\lambda - \mu)B(\lambda)A(\mu) \tag{13}$$
$$D(\lambda)B(\mu) = h(\lambda - \mu)B(\mu)D(\lambda) + k(\lambda - \mu)B(\lambda)D(\mu)$$

where

$$f(\lambda) = \frac{\lambda - i}{\lambda} \qquad g(\lambda) = \frac{i}{\lambda}$$
$$h(\lambda) = \frac{\lambda + i}{\lambda} \qquad k(\lambda) = -\frac{i}{\lambda}.$$

It is well-known [4,6,13] that the Hamiltonian is given in terms of the monodromy matrix as

$$H_N = \frac{i}{2} \frac{d}{d\lambda} \ln t_N(\lambda)|_{\lambda = i/2} - \frac{N}{2}$$

where $t_N(\lambda) = \mathrm{tr} T_N(\lambda) = A(\lambda) + D(\lambda)$.

Let $\omega_n = \begin{pmatrix} 1 \\ 0 \end{pmatrix}$. In the auxiliary space the Lax operator is triangular on ω_n

$$L_n(\lambda)\omega_n = \begin{pmatrix} \lambda + \frac{i}{2} & * \\ 0 & \lambda - \frac{i}{2} \end{pmatrix} \omega_n$$

where $*$ stands for an for us irrelevant quantity. This follows directly from (5). On the Hilbert space $\mathcal{H}_N$ we define $\Omega = \bigotimes_n \omega_n$ so that

$$T_N(\lambda)\Omega = \begin{pmatrix} \alpha^N(\lambda) & * \\ 0 & \delta^N(\lambda) \end{pmatrix} \Omega$$

where $\alpha(\lambda) = \lambda + \frac{i}{2}$ and $\delta(\lambda) = \lambda - \frac{i}{2}$. Equivalently this means that

$$C(\lambda)\Omega = 0$$
$$A(\lambda)\Omega = \alpha^N(\lambda)\Omega$$
$$D(\lambda)\Omega = \delta^N(\lambda)\Omega$$

so that Ω is an eigenstate of $A(\lambda)$ and $D(\lambda)$ and hence also of $t_N(\lambda) = A(\lambda) + D(\lambda)$.

The claim is that the other eigenvectors of $t_N(\lambda)$ are of the form

$$\Phi(\lambda, \Lambda) = B(\lambda_1) \cdots B(\lambda_n)\Omega$$

where the lambdas $\Lambda = \{\lambda_1, \ldots, \lambda_n\}$ satisfy a set of algebraic relations, called the Bethe equations. We will derive these now.

From the commutation relations (13) we find that

$$A(\lambda)B(\lambda_1) \cdots B(\lambda_n)\Omega = \prod_{k=1}^{n} f(\lambda - \lambda_k)\alpha^N(\lambda)B(\lambda_1) \cdots B(\lambda_n)\Omega$$

$$+ \sum_{k=1}^{n} M_k(\lambda, \Lambda)B(\lambda_1) \cdots \hat{B}(\lambda_k) \cdots B(\lambda_n)B(\lambda)\Omega.$$

The first term on the right hand side is obtained by using only the first term on the right hand side of (13). The other terms come from a combination of the application of the first and second term when moving A past the B's. In general the coefficients $M_k(\lambda, \Lambda)$ are quite involved using the explicit formulas. However, $M_1(\lambda, \Lambda)$ is obtained by using the second term in (13) moving $A(\lambda)$ past $B(\lambda_1)$ followed by applications of the first term in (13) only. This yields

$$M_1(\lambda, \Lambda) = g(\lambda - \lambda_1) \prod_{k=2}^{n} f(\lambda_1 - \lambda_k) \alpha^N(\lambda_1).$$

Note that the B's commute with each other by (13). Hence $M_j(\lambda, \Lambda)$ can be obtained from $M_1(\lambda, \Lambda)$ by replacing λ_1 by λ_j so that

$$M_j(\lambda, \Lambda) = g(\lambda - \lambda_j) \prod_{\substack{k=1 \\ k \neq j}}^{n} f(\lambda_j - \lambda_k) \alpha^N(\lambda_j).$$

Similarly,

$$D(\lambda)B(\lambda_1) \cdots B(\lambda_n)\Omega = \prod_{k=1}^{n} h(\lambda - \lambda_k)\delta^N(\lambda)B(\lambda_1) \cdots B(\lambda_n)\Omega$$

$$+ \sum_{k=1}^{n} N_k(\lambda, \Lambda)B(\lambda_1) \cdots \hat{B}(\lambda_k) \cdots B(\lambda_n)B(\lambda)\Omega$$

where

$$N_j(\lambda, \Lambda) = k(\lambda - \lambda_j) \prod_{\substack{k=1 \\ k \neq j}}^{n} h(\lambda_j - \lambda_k)\delta^N(\lambda_j).$$

For $\Phi(\lambda, \Lambda)$ to be an eigenvector of $t_N(\lambda) = A(\lambda) + D(\lambda)$ the terms

$$\sum_{k=1}^{n} M_k(\lambda, \Lambda)B(\lambda_1) \cdots \hat{B}(\lambda_k) \cdots B(\lambda_n)B(\lambda)\Omega$$

$$+ \sum_{k=1}^{n} N_k(\lambda, \Lambda)B(\lambda_1) \cdots \hat{B}(\lambda_k) \cdots B(\lambda_n)B(\lambda)\Omega$$

need to cancel. Since $g(\lambda - \lambda_j) = -k(\lambda - \lambda_j)$ this happens if the set of lambda's Λ satisfy the following set of equations

$$\prod_{\substack{k=1 \\ k \neq j}}^{n} f(\lambda_j - \lambda_k)\alpha^N(\lambda_j) = \prod_{\substack{k=1 \\ k \neq j}}^{n} h(\lambda_j - \lambda_k)\delta^N(\lambda_j)$$

for all $j = 1, 2, \ldots, n$. Explicitly this reads

$$\left(\frac{\lambda_j + \frac{i}{2}}{\lambda_j - \frac{i}{2}} \right)^N = \prod_{\substack{k=1 \\ k \neq j}}^{n} \frac{\lambda_j - \lambda_k + i}{\lambda_j - \lambda_k - i} \tag{23}$$

called the Bethe equations. In this case the eigenvalues of $\Phi(\lambda, \Lambda)$ are

$$\alpha^N(\lambda) \prod_{k=1}^{n} f(\lambda - \lambda_k) + \delta^N(\lambda) \prod_{k=1}^{n} h(\lambda - \lambda_k).$$

In the next section we will study solutions to (23) in the limit $N \to \infty$.

3. Solutions to the Bethe equations

Let us rewrite (23) in the following way

$$\left(\frac{\lambda + \frac{i}{2}}{\lambda - \frac{i}{2}}\right)^N = \prod_{\substack{\lambda' \in \Lambda \\ \lambda' \neq \lambda}} \frac{\lambda - \lambda' + i}{\lambda - \lambda' - i} \tag{25}$$

where $\lambda \in \Lambda = \{\lambda_1, \ldots, \lambda_n\}$.

Suggested by numerical analysis, it is assumed that in the limit $N \to \infty$ the λ's form strings. This hypothesis is called the string hypothesis. A string of length $\ell = 2M + 1$, where M is an integer or half-integer depending on the parity of ℓ, is a set of λ's of the form

$$\lambda_{jm}^M = \lambda_j^M + im$$

where $\lambda_j^M \in \mathbb{R}$ and $-M \leq m \leq M$ is integer or half-integer depending on M. The index j satisfies $1 \leq j \leq m_\ell$ where m_ℓ is the number of strings of length ℓ. A decomposition of $\{\lambda_1, \ldots, \lambda_n\}$ into strings is called a configuration. Each configuration is parametrized by $\{m_\ell\}$. It follows that

$$\sum_\ell \ell m_\ell = n.$$

Now take (25) and multiply over a string

$$\prod_{m=-M}^{M} \left(\frac{\lambda_j^M + i(m + \frac{1}{2})}{\lambda_j^M + i(m - \frac{1}{2})}\right)^N$$

$$= \prod_{m=-M}^{M} \prod_{\substack{M',j',m' \\ (M',j',m') \neq (M,j,m)}} \frac{\lambda_j^M - \lambda_{j'}^{M'} + i(m - m' + 1)}{\lambda_j^M - \lambda_{j'}^{M'} + i(m - m' - 1)}. \tag{28}$$

Many of the terms on the left and right cancel so that this equation can be rewritten as

$$e^{iN p_M(\lambda_j^M)} = \prod_{\substack{M',j' \\ (M',j') \neq (M,j)}} e^{iS_{MM'}(\lambda_j^M - \lambda_{j'}^{M'})}, \tag{29}$$

in terms of the momentum and scattering matrix

$$e^{ip_M(\lambda)} = \frac{\lambda + i(M + \frac{1}{2})}{\lambda - i(M + \frac{1}{2})}$$

$$e^{iS_{MM'}(\lambda)} = \prod_{m=|M-M'|}^{M+M'} \frac{\lambda + im}{\lambda - im} \cdot \frac{\lambda + i(m+1)}{\lambda - i(m+1)}.$$

Taking the logarithm of (29) using the branch cut

$$\frac{1}{i} \ln \frac{\lambda + ia}{\lambda - ia} = \pi - 2\arctan \frac{\lambda}{a}$$

we obtain

$$2N \arctan \frac{\lambda_j^M}{M + \frac{1}{2}} = 2\pi Q_j^M + \sum_{\substack{M',j' \\ (M',j') \neq (M,j)}} \Phi_{MM'}(\lambda_j^M - \lambda_{j'}^{M'}), \qquad (31)$$

where

$$\Phi_{MM'}(\lambda) = 2 \sum_{m=|M-M'|}^{M+M'} \left(\arctan \frac{\lambda}{m} + \arctan \frac{\lambda}{m+1} \right).$$

The first term on the right is absent for $m = 0$. Here Q_j^M is an integer or half-integer depending on the configuration.

In addition to the string hypothesis, we assume that the Q_j^M classify the λ's uniquely: λ_j^M increases if Q_j^M increases and in a given string no Q_j^M coincide. As we will see shortly with this assumption one obtains the correct number of solutions to the Bethe equations (25).

Using $\arctan \pm\infty = \pm\frac{\pi}{2}$ we obtain from (31) putting $\lambda_j^M = \infty$

$$Q_\infty^M = \frac{N}{2} - \left(2M + \frac{1}{2}\right)(m_{2M+1} - 1) - \sum_{M' \neq M} (2\min(M, M') + 1)m_{2M'+1}.$$

Since there are $2M + 1$ strings in a given string of length $2M + 1$, the maximal admissible $Q_{\max}^M$ is

$$Q_{\max}^M = Q_\infty^M - (2M + 1)$$

where we assume that if Q_j^M is bigger than $Q_{\max}^M$ then at least one root in the string is infinite and hence all are infinite which would imply $Q_j^M = Q_\infty^M$.

With the already mentioned assumption that each admissible set of quantum number Q_j^M corresponds uniquely to a solution of the Bethe equations we may now count the number of Bethe vectors. Since arctan is an odd function and by the assumption about the monotonicity we have

$$-Q_{\max}^M \leq Q_1^M < \cdots < Q_{m_{2M+1}}^M \leq Q_{\max}^M.$$

Hence defining P_ℓ as

$$P_\ell = N - 2 \sum_{\ell'} \min(\ell, \ell') m_{\ell'}$$

so that

$$P_\ell + m_\ell = 2Q_{\max}^M + 1 \qquad \text{with } \ell = 2M + 1.$$

With this the number of Bethe vectors with configuration $\{m_\ell\}$ is given by

$$Z(N, n | \{m_\ell\}) = \prod_{\ell \geq 1} \binom{P_\ell + m_\ell}{m_\ell}$$

where $\binom{p+m}{m} = (p+m)!/p!m!$ is the binomial coefficient. The total number of Bethe vectors is

$$Z(N, n) = \sum_{\substack{\{m_\ell\} \\ \sum_\ell \ell m_\ell = n}} \prod_{\ell \geq 1} \binom{P_\ell + m_\ell}{m_\ell}. \tag{39}$$

It should be emphasized that the derivation of (39) given here is not mathematically rigorous. Besides the various assumptions that were made we also did not worry about possible singularities of (28). However, as we shall see in the next section, (39) indeed yields the correct number of Bethe vectors.

4. Rigged configurations

In the last section we parametrized the Bethe vectors by solutions to the Bethe equations. As we have seen in section 2 the state space is the tensor product of irreducible representations of the underlying algebra, in our case the tensor product of $\mathbb{C}^2$ with underlying algebra being $\mathfrak{su}(2)$. The Bethe vectors are the highest weight vectors in the irreducible components in this tensor product.

In this section we will interpret (39) combinatorially in terms of rigged configurations. Since the Bethe vectors are also the irreducible components of the underlying tensor product which can be labeled by Young tableaux or crystal elements, one may expect a bijection between the rigged configurations and crystal elements. For the case A_n such a bijection is indeed known to exist [15,16,17]. For other types it was recently given in special cases in [20].

To interpret (39) combinatorially let us view the set $\{m_\ell\}$ as a partition ν. A partition is a set of numbers $\nu = (\nu_1, \nu_2, \ldots)$ such that $\nu_i \geq \nu_{i+1}$ and only finitely many ν_i are nonzero. The partition has part i if $\nu_k = i$ for some k. The size of partition ν is $|\nu| := \nu_1 + \nu_2 + \cdots$. In the correspondence between $\{m_\ell\}$ and ν, m_ℓ specifies the number of parts of size ℓ in ν. For example, if $m_1 = 1$, $m_2 = 3$, $m_4 = 1$ and all other $m_\ell = 0$ then $\nu = (4, 2, 2, 2, 1)$.

It is well-known (see e.g. [1]) that $\binom{p+m}{m}$ is the number of partitions in a box of size $p \times m$, meaning, that the partition cannot have more than m parts and no part exceeds p. Let $RC(N, n)$ be the set of all rigged configurations (ν, J) defined as follows. ν is a partition of size $|\nu| = n$ and J is a set of partition where J_ℓ is a partition in a box of size $P_\ell \times m_\ell$. Then (39) can be rewritten as

$$Z(N, n) = \sum_{(\nu, J) \in RC(N, n)} 1.$$

Example 4.1. Let $N = 5$ and $n = 2$. Then the following is the set of rigged configuration $RC(5, 2)$

The underlying partition on the left is (2) and on the right (1,1). The partitions J_ℓ attached to part length ℓ is specified by the numbers in each part. For example, the partition J_1 for the top rigged configuration on the right is (1,1) whereas for the one in the middle and bottom is $J_1 = (1)$ and $J_1 = \emptyset$, respectively. The numbers to the right of part ℓ is P_ℓ.

There exists a statistics on $RC(N, n)$, called cocharge. It is given by

$$cc(\nu, J) = cc(\nu) + \sum_\ell |J_\ell|$$

where

$$cc(\nu) = \sum_{j,k} \min(j, k) m_j m_k.$$

For example, the cocharge for the rigged configurations in Example 4.1 from top to bottom, left to right is 3, 2, 6, 5, 4, respectively.

As mentioned before, rigged configurations are in bijection with crystal elements. For our $\mathfrak{su}(2)$ example these are all sequences of 1's and 2's of length N such that the number of 2's never exceeds the number of 1's reading the sequence from right to left. The last condition is that of Yamanouchi words. The number n fixes the number of 2's in the sequence. Denote the set of all such sequences by $\mathcal{P}(N, n)$. For a path $p = p_N \cdots p_1 \in \mathcal{P}(N, n)$ define the energy as

$$E(p) = \sum_{j=1}^{N-1} (N - j)\chi(p_{j+1} > p_j) \tag{43}$$

where $\chi(\text{True}) = 1$ and $\chi(\text{False}) = 0$. The generating function of paths is given by

$$X(N, n) = \sum_{p \in \mathcal{P}(N,n)} q^{E(p)}.$$

Example 4.2. The set $\mathcal{P}(5, 2)$ is given by

$$\mathcal{P}(5, 2) = \{22111, 21211, 12211, 21121, 12121\}.$$

The energies are 2, 4, 3, 5 and 6, respectively. Hence $X(5, 2) = q^2 + q^3 + q^4 + q^5 + q^6$.

The bijection between $\mathcal{P}(N, n)$ and $\mathrm{RC}(N, n)$ is defined recursively. A path $p = p_N \cdots p_1 \in \mathcal{P}(N, n)$ is built up successively from right to left. The empty path is mapped to the empty rigged configuration. Assume that $p_{i-1} \cdots p_1$ corresponds to (ν^{i-1}, J^{i-1}). If $p_i = 1$, $(\nu^i, J^i) = (\nu^{i-1}, J^{i-1})$. If $p_i = 2$, then add a box to the largest singular string in (ν^{i-1}, J^{i-1}) and make it singular again. A string is singular if its label is equal to the vacancy number, in other words, if J_ℓ has a part of size P_ℓ. In the final rigged configuration (ν^N, J^N) take the complement of the partitions J_ℓ^N in the box $P_\ell^N \times m_\ell^N$. Let us call this map $\Psi : \mathcal{P}(N, n) \rightarrow \mathrm{RC}(N, n)$. We have the following theorem [15,16,17].

Theorem 4.1. *The map* $\Psi : \mathcal{P}(N, n) \rightarrow \mathrm{RC}(N, n)$ *is a bijection and* $E(p) = \mathrm{cc}(\Psi(p))$ *for all* $p \in \mathcal{P}(N, n)$.

Example 4.3. Take $p = 21121$. We get successively

p	(ν, J)
$\emptyset$	$\emptyset$
1	$\emptyset$
21	$\boxed{0}\,0$
121	$\boxed{0}\,1$
1121	$\boxed{0}\,2$
21121	$\begin{array}{l}\boxed{1}\,1\\[-2pt]\boxed{0}\end{array}$

Hence $\Psi(21121) = \begin{array}{l}\boxed{1}\,1\\[-2pt]\boxed{0}\end{array}$. Similarly,

$$\Psi(22111) = \boxed{}\,\boxed{0}\,1$$

$$\Psi(21211) = \begin{array}{l}\boxed{0}\,1\\[-2pt]\boxed{0}\end{array}$$

$$\Psi(12211) = \boxed{}\,\boxed{1}\,1$$

$$\Psi(12121) = \begin{array}{l}\boxed{1}\,1\\[-2pt]\boxed{1}\end{array}$$

Comparing with examples 4.1 and 4.2, the statistics match.

It follows immediately from Theorem 4.1 that

$$X(N, n) = \sum_{(\nu, J) \in \mathrm{RC}(N, n)} q^{\mathrm{cc}(\nu, J)}.$$

The q-binomial coefficient

$$\begin{bmatrix} p + m \\ m \end{bmatrix} = \frac{(q)_{p+m}}{(q)_p (q)_m},$$

where $(q)_m = \prod_{i=1}^{m}(1 - q^i)$, is the generating function of partitions in a box of size $p \times m$ [1]. Hence, defining $\mathrm{C}(N, n)$ to be the set of all partitions ν of n such that $P_\ell \geq 0$ for all ℓ the following corollary holds. The right-hand side is called fermionic formula.

Corollary 4.1.

$$X(N, n) = \sum_{\nu \in \mathrm{C}(N, n)} q^{\mathrm{cc}(\nu)} \prod_\ell \begin{bmatrix} P_\ell + m_\ell \\ m_\ell \end{bmatrix}.$$

5. Generalizations

So far we have only considered the spin $1/2$ XXX model and its counting. This model is based on the fundamental representation of $\mathfrak{su}(2)$. It turns out that the q-counting of Corollary 4.1 is associated with the Kac–Moody Lie algebra $A_1^{(1)}$. In the remainder of this note we will indicate how to generalize the q-counting that arises from the Bethe Ansatz.

The set of paths $\mathcal{P}(N, n)$, which is the set of Yamanouchi words in the letters 1 and 2 of length N with n twos, will be generalized to the set of highest weight elements in a tensor product of crystals of a given weight; the Yamanouchi condition is replaced by the highest weight condition and the condition on the number of twos becomes the requirement on the weight. Crystal bases were first introduced by Kashiwara [11] in connection with quantized universal enveloping algebras. The quantized universal enveloping algebra $U_q(\mathfrak{g})$ associated with a symmetrizable Kac–Moody Lie algebra $\mathfrak{g}$ was discovered independently by Drinfeld [3] and Jimbo [9] in their study of two dimensional solvable lattice models in statistical mechanics. The parameter q corresponds to the temperature of the underlying model. Kashiwara [11] showed that at zero temperature or $q = 0$ the representations of $U_q(\mathfrak{g})$ have bases, which he coined crystal bases, with a beautiful combinatorial structure and favorable properties such as uniqueness and stability under tensor products.

In the generalization from $\mathfrak{su}(2)$ to other types, rigged configurations become sequences of partitions with riggings. The number of partitions depends on the rank of the underlying algebra.

The generalization of the bijection from paths to rigged configurations to type $A_n^{(1)}$ is given in [16,17] and to other nonexceptional types in [20] in special cases. It was shown in [19,21] that all crystals can be realized as crystals of simply-laced type A, D, E. Hence the bijections for these types can be viewed as fundamental.

In the next section we will introduce crystal bases. The bijection algorithm for type $A_n^{(1)}$ and $D_n^{(1)}$ is presented in section 8.

6. Crystals

6.1. *Axiomatic definition of crystals*

Let $\mathfrak{g}$ be an affine Lie algebra and I the index set of its Dynkin diagram. Let α_i, h_i, Λ_i ($i \in I$) be the simple roots, simple coroots, and fundamental weights for $\mathfrak{g}$. Let $\delta = \sum_{i \in I} a_i \alpha_i$ denote the standard null root and $c = \sum_{i \in I} a_i^\vee h_i$ the canonical central element, where $a_i, a_i^\vee$ are the positive integers given in [10]. Let $P = \bigoplus_{i \in I} \mathbb{Z}\Lambda_i \oplus \mathbb{Z}\delta$ be the weight lattice and

$P^+ = \sum_{i \in I} \mathbb{Z}_{\geq 0} \Lambda_i \oplus \mathbb{Z}\delta$ the dominant weights.

A crystal B is a set $B = \sqcup_{\lambda \in P} B_\lambda$ (wt $b = \lambda$ if $b \in B_\lambda$) with the maps

$$e_i : B_\lambda \longrightarrow B_{\lambda + \alpha_i} \sqcup \{0\}, \quad f_i : B_\lambda \longrightarrow B_{\lambda - \alpha_i} \sqcup \{0\},$$

$$\varepsilon_i : B \longrightarrow \mathbb{Z} \sqcup \{-\infty\}, \quad \varphi_i : B \longrightarrow \mathbb{Z} \sqcup \{-\infty\}$$

for all $i \in I$ such that

for $b \in B_\lambda$, $\varphi_i(b) = \langle h_i, \lambda \rangle + \varepsilon_i(b)$,

for $b \in B$, we have

$$\begin{aligned}
\varepsilon_i(b) &= \varepsilon_i(e_i b) + 1 \text{ if } e_i b \neq 0, \\
&= \varepsilon_i(f_i b) - 1 \text{ if } f_i b \neq 0, \\
\varphi_i(b) &= \varphi_i(e_i b) - 1 \text{ if } e_i b \neq 0, \\
&= \varphi_i(f_i b) + 1 \text{ if } f_i b \neq 0,
\end{aligned}$$

for $b, b' \in B$, $e_i b' = b$ if and only if $b' = f_i b$,

for $b \in B$, $\varepsilon_i(b) = \varphi_i(b) = -\infty$ implies $e_i b = f_i b = 0$.

A crystal B can be regarded as a colored oriented graph by defining

$$b \xrightarrow{i} b' \quad \Longleftrightarrow \quad f_i b = b'.$$

If we want to emphasize I, B is called an I-crystal.

If B_1 and B_2 are crystals, then for $b_1 \otimes b_2 \in B_1 \otimes B_2$ the action of e_i is defined as

$$e_i(b_1 \otimes b_2) = \begin{cases} e_i b_1 \otimes b_2 \text{ if } \varepsilon_i(b_1) > \varphi_i(b_2), \\ b_1 \otimes e_i b_2 \text{ else,} \end{cases}$$

where $\varepsilon_i(b) = \max\{k \mid e_i^k b \text{ is defined}\}$ and $\varphi_i(b) = \max\{k \mid f_i^k b \text{ is defined}\}$. This is the opposite of the notation used by Kashiwara [11].

An element $b \in B$ is classically highest weight if $e_i b = 0$ for all $i = 1, 2, \ldots, n$. For $B = B_L \otimes \cdots \otimes B_1$ and $\Lambda \in P^+$, the set of paths is defined as follows

$$\mathcal{P}(B, \Lambda) = \{b \in B \mid e_i b = 0 \text{ for all } i = 1, 2, \ldots, n, \text{ wt} b = \Lambda\}.$$

In the following we will discuss the crystals of type $A_n^{(1)}$ and $D_n^{(1)}$ more explicitly.

6.2. Dynkin data of type A_n and D_n

Let ϵ_i be the i-th standard unit vector in $\mathbb{Z}^n$. Then for type A_{n-1}, the simple roots are

$$\alpha_i = \epsilon_i - \epsilon_{i+1} \qquad \text{for } 1 \leq i < n$$

and the fundamental weights are

$$\Lambda_i = \epsilon_1 + \cdots + \epsilon_i \qquad \text{for } 1 \le i < n.$$

For type D_n, the simple roots are

$$\alpha_i = \epsilon_i - \epsilon_{i+1} \qquad \text{for } 1 \le i < n$$
$$\alpha_n = \epsilon_{n-1} + \epsilon_n$$

and the fundamental weights are

$$\Lambda_i = \epsilon_1 + \cdots + \epsilon_i \qquad \text{for } 1 \le i \le n - 2$$
$$\Lambda_{n-1} = (\epsilon_1 + \cdots + \epsilon_{n-1} - \epsilon_n)/2$$
$$\Lambda_n = (\epsilon_1 + \cdots + \epsilon_{n-1} + \epsilon_n)/2.$$

6.3. *Affine crystals of type $A_n^{(1)}$ and $D_n^{(1)}$*

In [8] it is conjectured that there is a family of finite-dimensional irreducible $U_q'(\mathfrak{g})$-modules $\{W_i^{(a)} \mid a \in J, i \in \mathbb{Z}_{\ge 0}\}$ which, unlike most finite-dimensional $U_q'(\mathfrak{g})$-modules, have crystal bases $B^{a,i}$. Here $U_q'(\mathfrak{g})$ is the quantum universal enveloping algebra of the derived subalgebra of $\mathfrak{g}$, obtained by omitting the degree operator, and $J = I \backslash \{0\}$.

Here we will restrict our attention to the simplest affine crystals $B^{1,1}$ of type $A_n^{(1)}$ and $D_n^{(1)}$. As a set $B^{1,1}$ is $\{1 < 2 < \cdots < n+1\}$ for type $A_n^{(1)}$ and $\{1 < 2 < \cdots < n-1 < \frac{n}{\overline{n}} < \overline{n-1} < \cdots < \overline{1}\}$ for type $D_n^{(1)}$.

The crystal graphs are given in Figure 1.

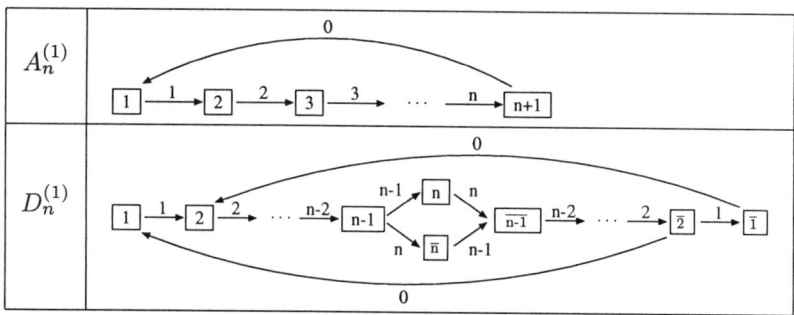

Figure 1. Crystals $B^{1,1}$

6.4. *One-dimensional sums*

The energy function (43) can be generalized to the crystal setting. In the case $B = (B^{1,1})^{\otimes L}$ it takes a simple form. There is [14] a unique (up to global additive constant) function $H : B^{1,1} \otimes B^{1,1} \to \mathbb{Z}$ called the local energy function, such that

$$H(e_i(b \otimes b')) = H(b \otimes b') + \begin{cases} -1 & \text{if } i = 0 \text{ and } e_0(b \otimes b') = b \otimes e_0 b' \\ 1 & \text{if } i = 0 \text{ and } e_0(b \otimes b') = e_0 b \otimes b' \\ 0 & \text{otherwise.} \end{cases}$$

We normalize H by the condition $H(1 \otimes 1) = 0$.

Example 6.1. Let $b \otimes b' \in B^{1,1} \otimes B^{1,1}$. Explicitly, the local energy function is given as follows. For type $A_n^{(1)}$, $H(b \otimes b') = -\chi(b > b')$. For type $D_n^{(1)}$, $H(b \otimes b') = 0$ if $b \le b'$, $H(b \otimes b') = -1$ if $b \otimes b' = n \otimes \bar{n}, \bar{n} \otimes n$ or $b > b'$ where $b \otimes b' \ne \bar{1} \otimes 1$, and $H(\bar{1} \otimes 1) = -2$.

For $b_L \otimes \cdots \otimes b_1 \in B = (B^{1,1})^{\otimes L}$

$$E(b_L \otimes \ldots \otimes b_1) = \sum_{j=1}^{L-1} (L - j)\, H(b_{j+1} \otimes b_j).$$

Define the one-dimensional sum $X(B, \lambda; q) \in \mathbb{Z}[q, q^{-1}]$ by

$$X(B, \lambda; q) = \sum_{b \in \mathcal{P}(B, \lambda)} q^{E(b)}.$$

Example 6.2. In the crystal language the set of paths of Example 4.2 corresponds to $B = (B^{1,1})^{\otimes 5}$ of type $A_1^{(1)}$ of weight $\lambda = \Lambda_1 + 2\Lambda_2$.

7. Fermionic formula and rigged configurations

Fermionic formulas associated to a Kac-Moody algebra $\mathfrak{g}$ were conjectured in [7,8]. We review the fermionic formulas for type $A_n^{(1)}$ and $D_n^{(1)}$.

Let $L_i^{(a)}$ with $a \in J$ and $i \in \mathbb{Z}_{\ge 0}$ denote the number of tensor factors $B^{a,i}$ in B and let λ be a dominant integral weight. Say that $\nu^{\bullet} = (m_i^{(a)})$ is a (B, λ)-configuration if

$$\sum_{\substack{a \in J \\ i \in \mathbb{Z}_{\ge 0}}} i\, m_i^{(a)} \alpha_a = \sum_{\substack{a \in J \\ i \in \mathbb{Z}_{\ge 0}}} i\, L_i^{(a)} \Lambda_a - \lambda. \tag{58}$$

The configuration $\nu^{\bullet}$ is admissible if all vacancy numbers are nonnegative

$$p_i^{(a)} \ge 0 \qquad \text{for all } a \in J \text{ and } i \in \mathbb{Z}_{\ge 0},$$

where

$$p_i^{(a)} = \sum_{k \in \mathbb{Z}_{\geq 0}} \left(L_k^{(a)} \min(i,k) - \sum_{b \in J} (\alpha_a | \alpha_b) \min(i,k) \, m_k^{(b)} \right). \tag{60}$$

Write $C(B, \lambda)$ for the set of admissible (B, λ)-configurations. Define

$$cc(\nu^\bullet) = \frac{1}{2} \sum_{a,b \in J} \sum_{j,k \in \mathbb{Z}_{\geq 0}} (\alpha_a | \alpha_b) \min(j,k) m_j^{(a)} m_k^{(b)}.$$

The fermionic formula is defined by

$$M(B, \lambda; q) = \sum_{\nu^\bullet \in C(B,\lambda)} q^{cc(\nu^\bullet)} \prod_{a \in J} \prod_{i \in \mathbb{Z}_{\geq 0}} \begin{bmatrix} p_i^{(a)} + m_i^{(a)} \\ m_i^{(a)} \end{bmatrix}. \tag{62}$$

The $X = M$ conjecture of [7,8] states that

$$X(B, \lambda; q^{-1}) = M(B, \lambda; q).$$

The fermionic formula $M(B, \lambda)$ can be interpreted using rigged configurations. Denote by $(\nu^\bullet, J^\bullet)$ a pair where $\nu^\bullet = (m_i^{(a)})$ is a matrix and $J^\bullet = (J^{(a,i)})$ is a matrix of partitions with $a \in J$ and $i \in \mathbb{Z}_{\geq 0}$. Then a rigged configuration is a pair $(\nu^\bullet, J^\bullet)$ such that $\nu^\bullet \in C(B, \lambda)$ and the partition $J^{(a,i)}$ is contained in a $m_i^{(a)} \times p_i^{(a)}$ rectangle for all a, i. The set of rigged (B, λ)-configurations for fixed λ and B is denoted by $RC(B, \lambda)$. Then (62) is equivalent to

$$M(B, \lambda) = \sum_{(\nu^\bullet, J^\bullet) \in RC(B,\lambda)} q^{cc(\nu^\bullet, J^\bullet)}$$

where $cc(\nu^\bullet, J^\bullet) = cc(\nu^\bullet) + |J^\bullet|$ and $|J^\bullet| = \sum_{(a,i)} |J^{(a,i)}|$. To emphasize the dependence on $\nu^\bullet$ we also write $m_i^{(a)}(\nu^\bullet)$ and $P_i^{(a)}(\nu^\bullet)$ for $m_i^{(a)}$ and $p_i^{(a)}$, respectively.

8. Bijection between rigged configurations and paths

In this section we give the description of the bijection $\Phi : RC(B, \lambda) \to \mathcal{P}(B, \lambda)$ for types $A_n^{(1)}$ and $D_n^{(1)}$ when $B = (B^{1,1})^{\otimes L}$.

Let $(\nu^\bullet, J^\bullet) \in RC(B, \lambda)$. We shall define a map $rk : RC(B, \lambda) \to B^{1,1}$ which associates to $(\nu^\bullet, J^\bullet)$ an element of $B^{1,1}$ called its rank. Denote by $RC_b(B, \lambda)$ the elements of $RC(B, \lambda)$ of rank b. We shall define a bijection $\delta : RC_b(B, \lambda) \to RC(\tilde{B}, \lambda - wt(b))$ where $\tilde{B} = (B^{1,1})^{\otimes(L-1)}$. The disjoint union of these bijections then defines a bijection $\delta : RC(B, \lambda) \to \bigcup_{b \in B^{1,1}} RC(\tilde{B}, \lambda - wt(b))$.

The bijection Φ is defined recursively as follows. For $b \in B^{1,1}$ let $\mathcal{P}_b(B, \lambda)$ be the set of paths in B that have b as leftmost tensor factor. For $L = 0$ the bijection Φ sends the empty rigged configuration (the only element of the set $\mathrm{RC}(B, \lambda)$) to the empty path (the only element of $\mathcal{P}(B, \lambda)$). Otherwise assume that Φ has been defined for $\tilde{B}$ and define it for B by the commutative diagram

$$
\begin{array}{ccc}
\mathrm{RC}_b(B, \lambda) & \xrightarrow{\ \Phi\ } & \mathcal{P}_b(B, \lambda) \\
\delta \downarrow & & \downarrow \\
\mathrm{RC}(\tilde{B}, \lambda - \mathrm{wt}(b)) & \xrightarrow{\ \Phi\ } & \mathcal{P}(\tilde{B}, \lambda - \mathrm{wt}(b))
\end{array}
$$

where the right hand vertical map removes the leftmost tensor factor b. In short,

$$
\Phi(\nu^\bullet, J^\bullet) = \mathrm{rk}(\nu^\bullet, J^\bullet) \otimes \Phi(\delta(\nu^\bullet, J^\bullet)).
$$

We also require the bijection $\tilde{\Phi} : \mathrm{RC}(B, \lambda) \to \mathcal{P}(B, \lambda)$ given by $\tilde{\Phi} = \Phi \circ \mathrm{comp}$ where $\mathrm{comp} : \mathrm{RC}(B, \lambda) \to \mathrm{RC}(B, \lambda)$ with $\mathrm{comp}(\nu^\bullet, J^\bullet) = (\nu^\bullet, \tilde{J}^\bullet)$ is the function which complements the riggings, meaning that $\tilde{J}^\bullet$ is obtained from $J^\bullet$ by complementing all partitions $J^{(a,i)}$ in the $m_i^{(a)}(\nu^\bullet) \times P_i^{(a)}(\nu^\bullet)$ rectangle.

Remark 8.1. The bijection Ψ of section 4 is the inverse of $\tilde{\Phi}$ for type $A_1^{(1)}$.

Theorem 8.1. $\Phi : \mathrm{RC}(B, \lambda) \to \mathcal{P}(B, \lambda)$ *is a bijection such that*

$$
cc(\nu^\bullet, J^\bullet) = -E(\tilde{\Phi}(\nu^\bullet, J^\bullet)) \qquad \text{for all } (\nu^\bullet, J^\bullet) \in \mathrm{RC}(B, \lambda).
$$

For type $A_n^{(1)}$ a generalization of this theorem for $B = B^{a_L, i_L} \otimes \cdots \otimes B^{a_1, i_1}$ was proven in [17]. For other types Theorem 8.1 is proved in. [20]

To describe the bijection explicitly for types $A_n^{(1)}$ and $D_n^{(1)}$, the following notation is needed. The matrix $\nu^\bullet = (m_i^{(a)})$ can be viewed as a sequence of partitions $\nu^\bullet = (\nu^{(1)}, \nu^{(2)}, \ldots, \nu^{(n)})$ where $m_i^{(a)}$ is the number of parts of size i in the partition $\nu^{(a)}$. Denote by $Q_i(\rho)$ the number of boxes in the first i columns of the partition ρ. Finally the partition $J^{(a,i)}$ is called singular if it has a part of size $p_i^{(a)}$.

8.1. *Bijection for type $A_n^{(1)}$*

Using the Dynkin data for type A_n the vacancy numbers (60) and the constraints (58) can be rewritten in the following explicit way

$$
P_i^{(a)}(\nu^\bullet) = Q_i(\nu^{(a-1)}) - 2Q_i(\nu^{(a)}) + Q_i(\nu^{(a+1)}) + L\delta_{a,1} \qquad \text{for } 1 \leq a \leq n
$$

and

$$|\nu^{(a)}| = L - \sum_{b=1}^{a} \lambda_b \qquad \text{for } 1 \le a \le n.$$

The algorithm δ is given as follows. Set $\ell^{(0)} = 0$ and repeat the following process for $a = 1, 2, \ldots, n$ or until stopped. Find the minimal index $i \ge \ell^{(a-1)}$ such that $J^{(a,i)}$ is singular. If no such i exists, set $b = a$ and stop. Otherwise set $\ell^{(a)} = i$ and continue with $a + 1$. If the process did not stop, set $b = n + 1$. Set all undefined $\ell^{(a)}$ to ∞.

The new rigged configuration is defined by

$$m_i^{(a)}(\tilde{\nu}^\bullet) = m_i^{(a)}(\nu^\bullet) + \begin{cases} 1 & \text{if } i = \ell^{(a)} - 1 \\ -1 & \text{if } i = \ell^{(a)} \\ 0 & \text{otherwise.} \end{cases}$$

The partition $\tilde{J}^{(a,i)}$ is obtained from $J^{(a,i)}$ by removing a part of size $P_i^{(a)}(\nu^\bullet)$ for $i = \ell^{(a)}$, adding a part of size $P_i^{(a)}(\tilde{\nu}^\bullet)$ for $i = \ell^{(a)} - 1$, and leaving it unchanged otherwise.

Example 8.1. Take $B = (B^{1,1})^{\otimes 7}$, $\lambda = \Lambda_3 + \Lambda_4$ and $(\nu^\bullet, J^\bullet) \in \mathrm{RC}(B, \lambda)$ as

$$\left(\begin{array}{|c|c} \hline 0 & 0 \\ \hline 0 \\ \hline 3 & 3 \\ \hline \end{array} \;, \begin{array}{|c|c} \hline 0 & 0 \\ \hline 0 & 0 \\ \hline \end{array} \;, \begin{array}{|c|c} \hline 0 & 0 \\ \hline \end{array} \right).$$

The algorithm for Φ on $\mathrm{comp}(\nu^\bullet, J^\bullet)$ yields

$(\nu^\bullet, J^\bullet)^{(1)}$	$(\nu^\bullet, J^\bullet)^{(2)}$	$(\nu^\bullet, J^\bullet)^{(3)}$	rk
0 0 / 0 / 0 3	0 0 / 0 0	0 0	3
0 0 / 2 2 / 0	0 0 / 0	0 0	4
0 0 / 0 2	0 0	$\emptyset$	2
1 1 / 0	0 0	$\emptyset$	3
0 1	$\emptyset$	$\emptyset$	1
0 0	$\emptyset$	$\emptyset$	2
$\emptyset$	$\emptyset$	$\emptyset$	1
$\emptyset$	$\emptyset$	$\emptyset$	

Hence $\tilde{\Phi}(\nu^\bullet, J^\bullet) = b = 3\otimes 4\otimes 2\otimes 3\otimes 1\otimes 2\otimes 1$ and $E(b) = \mathrm{cc}(\nu^\bullet, J^\bullet) = 12$.

8.2. Bijection for type $D_n^{(1)}$

Using the Dynkin data for type D_n the vacancy numbers (60) and the constraints (58) can be rewritten in the following explicit way

$$P_i^{(a)}(\nu^\bullet) = Q_i(\nu^{(a-1)}) - 2Q_i(\nu^{(a)}) + Q_i(\nu^{(a+1)}) + L\delta_{a,1} \text{ for } 1 \le a < n-2$$
$$P_i^{(n-2)}(\nu^\bullet) = Q_i(\nu^{(n-3)}) - 2Q_i(\nu^{(n-2)}) + Q_i(\nu^{(n-1)}) + Q_i(\nu^{(n)})$$
$$P_i^{(n-1)}(\nu^\bullet) = Q_i(\nu^{(n-2)}) - 2Q_i(\nu^{(n-1)})$$
$$P_i^{(n)}(\nu^\bullet) = Q_i(\nu^{(n-2)}) - 2Q_i(\nu^{(n)})$$

and

$$|\nu^{(a)}| = L - \sum_{b=1}^{a} \lambda_b \qquad \text{for } 1 \le a \le n-2$$

$$|\nu^{(n-1)}| = \frac{1}{2}(L - \sum_{b=1}^{n-1} \lambda_b + \lambda_n)$$

$$|\nu^{(n)}| = \frac{1}{2}(L - \sum_{b=1}^{n} \lambda_b).$$

The algorithm δ is given as follows. Set $\ell^{(0)} = 0$ and repeat the following process for $a = 1, 2, \ldots, n-2$ or until stopped. Find the minimal index $i \ge \ell^{(a-1)}$ such that $J^{(a,i)}$ is singular. If no such i exists, set $b = a$ and stop. Otherwise set $\ell^{(a)} = i$ and continue with $a+1$.

If the process has not stopped at $a = n-2$ continue as follows. Find the minimal indices $i, j \ge \ell^{(n-2)}$ such that $J^{(n-1,i)}$ and $J^{(n,j)}$ are singular. If neither i nor j exist, set $b = n-1$ and stop. If i exists, but not j, set $\ell^{(n-1)} = i$, $b = n$ and stop. If j exists, but not i, set $\ell^{(n)} = j$, $b = \bar{n}$ and stop. If both i and j exist, set $\ell^{(n-1)} = i$, $\ell^{(n)} = j$ and continue with $a = n-2$.

Now continue for $a = n-2, n-3, \ldots, 1$ or until stopped. Find the minimal index $i \ge \bar{\ell}^{(a+1)}$ where $\bar{\ell}^{(n-1)} = \max(\ell^{(n-1)}, \ell^{(n)})$ such that $J^{(a,i)}$ is singular (if $i = \ell^{(a)}$ then there need to be two parts of size $P_i^{(a)}(\nu^\bullet)$ in $J^{(a,i)}$). If no such i exists, set $b = \overline{a+1}$ and stop. If the process did not stop, set $b = \bar{1}$.

Set all yet undefined $\ell^{(a)}$ and $\bar{\ell}^{(a)}$ to ∞.

The new rigged configuration is defined by

$$m_i^{(a)}(\tilde{\nu}^\bullet) = m_i^{(a)}(\nu^\bullet) + \begin{cases} 1 & \text{if } i = \ell^{(a)} - 1 \\ -1 & \text{if } i = \ell^{(a)} \\ 1 & \text{if } i = \overline{\ell}^{(a)} - 1 \text{ and } 1 \le a \le n - 2 \\ -1 & \text{if } i = \overline{\ell}^{(a)} \text{ and } 1 \le a \le n - 2 \\ 0 & \text{otherwise} \end{cases}$$

The partition $\tilde{J}^{(a,i)}$ is obtained from $J^{(a,i)}$ by removing a part of size $P_i^{(a)}(\nu^\bullet)$ for $i = \ell^{(a)}$ and $i = \overline{\ell}^{(a)}$, adding a part of size $P_i^{(a)}(\tilde{\nu}^\bullet)$ for $i = \ell^{(a)} - 1$ and $i = \overline{\ell}^{(a)} - 1$, and leaving it unchanged otherwise.

Example 8.2. Take $B = (B^{1,1})^{\otimes 6}$, $\lambda = 2\Lambda_3$ and $(\nu^\bullet, J^\bullet) \in \mathrm{RC}(B, \lambda)$ as

Then the algorithm for Φ on $\mathrm{comp}(\nu^\bullet, J^\bullet)$ gives the following intermediate steps

so that $\tilde{\Phi}(\nu^\bullet, J^\bullet) = b = \overline{4} \otimes 3 \otimes \overline{1} \otimes 2 \otimes 1 \otimes 1$. The statistics in this case are $E(b) = \mathrm{cc}(\nu^\bullet, J^\bullet) = 8$.

Acknowledgements

Many thanks to Professor T. Lulek and the Organizing Committee for the invitation to the Summer School at Myczkowce and for providing excellent conditions for the meeting. I would also like to thank the Max-Planck-Institut für Mathematik in Bonn and the University of Wuppertal for hosting me while this work was completed. This work was partially supported by the Humboldt foundation and NSF grant DMS-0200774.

References

1. G. E. Andrews, *The Theory of Partitions*, Encyclopedia of Mathematics, vol. 2 (Addison-Wesley, Reading, Massachusetts, 1976).
2. H. Bethe, *Zur Theorie der Metalle, I. Eigenwerte und Eigenfunktionen der linearen Atomkette*, Z. Physik **71** (1931) 205–231.
3. V. G. Drinfeld, *Hopf algebra and the Yang–Baxter equation*, Soviet. Math. Dokl. **32** (1985) 254–258.
4. L. D. Faddeev, How the algebraic Bethe ansatz works for integrable models. Symétries quantiques (Les Houches, 1995), 149–219, North-Holland, Amsterdam, 1998 (hep-th/9605187).
5. W. Fulton, *Young tableaux*, London Mathematical Society Student texts 35, Cambridge University Press, 1997.
6. L. A. Takhtadzhyan and L. D. Faddeev, The spectrum and scattering of excitations in the one-dimensional isotropic Heisenberg model. (Russian) Differential geometry, Lie groups and mechanics, IV. Zap. Nauchn. Sem. Leningrad. Otdel. Mat. Inst. Steklov. (LOMI) 109 (1981), 134–178, 181–182, 184.
7. G. Hatayama, A. Kuniba, M. Okado, T. Takagi and Z. Tsuboi, *Paths, crystals, and fermionic formula*, MathPhys odyssey, 2001, 205–272, Prog. Math. Phys. **23**, Birkhuser Boston, Boston, MA, 2002.
8. G. Hatayama, A. Kuniba, M. Okado, T. Takagi, and Y. Yamada, *Remarks on fermionic formula*, Contemporary Math. **248** (1999) 243–291.
9. M. Jimbo, *A q-difference analogue of $U(\mathcal{G})$ and the Yang–Baxter equation*, Lett. Math. Phys. **10** (1985) 63–69.
10. V. G. Kac, *Infinite dimensional Lie algebras*, 3rd edition, Cambridge Univ. Press. Cambridge (1990).
11. M. Kashiwara, *Crystalizing the q-analogue of universal enveloping algebras*, Commun. Math. Phys. **133** (1990) 249–260.
12. M. Kashiwara and T. Nakashima, *Crystal graphs for representations of the q-analogue of classical Lie algebras*, J. Alg. **165** (1994) 295–345.
13. V. E. Korepin, N. M. Bogoliubov and A. G. Izergin, Quantum inverse scattering method and correlation functions, Cambridge Monographs on Mathematical Physics, Cambridge University Press 1993.
14. S-J. Kang, M. Kashiwara, K. C. Misra, T. Miwa, T. Nakashima and A. Nakayashiki, *Affine crystals and vertex models*, Int. J. Mod. Phys. **A7** (suppl. 1A) (1992) 449–484.
15. S. V. Kerov, A. N. Kirillov and N. Yu. Reshetikhin, *Combinatorics, the Bethe ansatz and representations of the symmetric group*, Zap.Nauchn. Sem.

(LOMI) **155** (1986) 50–64. (English translation: J. Sov. Math. **41** (1988) 916–924.)

16. A. N. Kirillov and N. Y. Reshetikhin, *The Bethe Ansatz and the combinatorics of Young tableaux*, J. Soviet Math. **41** (1988) 925–955.

17. A. N. Kirillov, A. Schilling and M. Shimozono, *A bijection between Littlewood-Richardson tableaux and rigged configurations*, Selecta Mathematica (N.S.) **8** (2002) 67–135.

18. T. Nakashima, *Crystal base and a generalization of the Littlewood-Richardson rule for the classical Lie algebras*, Comm. Math. Phys. **154** (1993) 215–243.

19. M. Okado, A. Schilling and M. Shimozono, *Virtual crystals and fermionic formulas of type* $D_{n+1}^{(2)}$, $A_{2n}^{(2)}$, *and* $C_n^{(1)}$, preprint math.QA/0105017.

20. M. Okado, A. Schilling and M. Shimozono, *A crystal to rigged configuration bijection for nonexceptional affine algebras*, preprint math.QA/0203163.

21. M. Okado, A. Schilling and M. Shimozono, *Virtual crystals and Kleber algorithm*, preprint math.QA/0209082.

22. M. Shimozono, *Affine type A crystal structure on tensor products of rectangles, Demazure characters, and nilpotent varieties*, J. Algebraic Combin. **15** (2002) 151–187.

23. E. K. Sklyanin, *Quantum version of the method of inverse scattering problem*, J. Sov. Math. **19** (1982) 1546–1596.

24. E. K.Sklyanin and L. D. Faddeev, *Quantum mechanical approach to completely integrable field theory models*, Sov. Phys. Dokl. **23** (1978) 902–904.

25. E. K. Sklyanin, L. A. Takhtajan and L. D. Faddeev, *Quantum inverse problem method I*, Theor. Math. Phys. **40** (1980) 688–706.

TOWARDS A GENERAL SOLUTION OF THE LINEAR HEISENBERG PROBLEM

W.J. CASPERS

Enschede, The Netherlands

T. LULEK, B. LULEK, M. KUZMA, A. WAL

Institute of Physics, University of Rzeszow, Poland

Our presentation of the general solution of the linear Heisenberg problem is based on two approaches: The representation of this solution in terms of the Bethe-Hulthén scheme and its asymptotic form in terms of strings. This asymptotic form is supposed to be complete. The relation between these approaches is explained.

1. Introduction

The construction of a *complete set* of stationary states of the linear Heisenberg system with periodic boundary conditions (rings) has been a topic of intensive research for many decades. Many eminent theoreticians have made their contribution to this joint effort, but we should mention in the first place the epoch-making work of Bethe [1] and Hulthén. [5] Their work is the basis of many papers on this topic that have appeared during the past 70 years.

We do no have the intention to give a complete survey of all this work but we will try to indicate that especially an *asymptotic approach* starting from the *Hypothesis of Strings* gives the prospect of a *general* and *complete* solution of the Heisenberg chain. [7]

In section 2 we first give a transformation of the general Bethe-Hulthén solution (BHS) into a form that is suitable for the formulation of the Hypothesis of Strings [7] and show that this transformation in first instance results in a simplification in the sense that the two-fold set of parameters $(k_j, \phi_{j,l})$ in BHS reduces to a single one in terms of the parameters Λ_j. For a presentation of the solution of the Heisenberg problem, however, BHS has the advantage of a picture of simple waves, which show the phenomenon of *reflection* without *distortion*.

The asymptotic form of the solution, the string solution (SS), has a simple presentation in terms of the parameter set $\{\Lambda_j\}$, and this is the contents of section 3, for the special case of one string. The generalization for more strings is explained in section 4. Having found these asymptotic solutions one may determine the corresponding BHS for a fixed but large number of spins N and find the corresponding set of integers $\{\lambda_j\}$ connected with the periodic boundary conditions. These integers classify a *family* of solutions for variable number of spins but a fixed number of inversions or deviations, i.e. *upturned* spins in a ferromagnetic reference state.

In section 5 we present a general method to derive the asymptotic solutions and demonstrate the relation with a corresponding BHS. The general idea is that all possible combinations of strings lead to an asymptotic solution and that the set of string solutions is complete.

We give in section 6 three simple examples of asymptotic solutions within the BHS and their relation with SS.

In several papers by other authors special cases are discussed in detail. [2,3,4,6,8]

Following a BHS in lowering N one observes a profusion of interesting algebraic phenomena, which will be discussed in some detail in the Appendix of this paper. Intensive numerical research during the past decade gave us the possibility to present the following general picture:

- for special values of N some sets $\{\lambda_n\}$ become redundant in the sense that they represent solutions that are also represented by other sets. This occurs if one of the quantities $|\mathrm{Re}\, k_j|$ or $|\mathrm{Re}\, \phi_{j,k}|$ reaches the value π. It turns out that these quantities may always be chosen within the interval $[0, \pi]$ for well-chosen sets $\{\lambda_n\}$. This will be clear after a careful look at the equations for k_j and $\phi_{j,k}$ given in the next section. The corresponding N values may be called *transition* points,

- for special values of N an originally *real* pair of wave numbers may change into a *complex conjugated* pair. These N values are called *critical* points,

- for special values of N the imaginary part of a wave number may become singular. This always occurs for pairs of complex conjugated wave numbers. For these N the corresponding BHS finds its *limit* point and for lower numbers of spins this solution need not be considered.

2. The Bethe-Hulthén solution in terms of the parameters Λ_j

The BHS for a ring of N spins with the standard Heisenberg Hamiltonian for nearest neighbours interaction (see for example Takahashi [7], formula

2.1 with substitutions $J = 4$, $H = 0$) has the general form:

$$\Psi = \sum_{1 \leq j_1 \leq \ldots \leq j_r \leq N} \sum_{P} \exp\left[i\left(\sum_{l=1}^{r} k_{P(l)} j_l + \frac{1}{2} \sum_{m < n} \phi_{P(m), P(n)}\right)\right] \Phi_{j_1 j_2 \ldots j_r}, \quad (1)$$

in which the symbol $\Phi_{j_1 j_2 \ldots j_r}$ denotes a state with r deviations from the ferromagnetic reference state, which deviations are at the positions $j_1, j_2, \ldots j_r$. The states with deviations at fixed positions are combined linearly to form a state Ψ for which the deviations have a wavelike character and are represented by the wave numbers or quasi momenta k_l. All permutations P of the wave numbers over the ordered deviations occur in (1), each permutation corresponding with a special phase. In the BHS the quasi momenta k_l and the phases $\phi_{l,m}$ obey a set of equations with r *fixed integers* $\{\lambda_j\}$, which identify a particular solution:

$$N k_l = 2\pi \lambda_l + \sum_{m(\neq l)} \phi_{l,m}, \quad (l, m = 1, 2 \ldots r), \quad (2)$$

$$-\frac{N}{2} \leq \lambda_1 \leq \lambda_2 \leq \ldots \leq \lambda_r \leq \frac{N}{2}, \quad (3)$$

and:

$$\cot\left(\frac{\phi_{l,m}}{2}\right) = \frac{\sin\left(\frac{k_l - k_m}{2}\right)}{\cos\left(\frac{k_l + k_m}{2}\right) - \cos\left(\frac{k_l - k_m}{2}\right)}. \quad (4)$$

It should be understood that the quasi momenta and the phases may have complex values and in special cases may be *singular*.

The BHS given in (1-4) may be transformed into a form suitable for the formulation of the SS, if we make use of the following substitution:

$$\Lambda_l = \cot\left(\frac{k_l}{2}\right), \quad (5)$$

which is equivalent to:

$$\exp(ik_l) = \frac{\Lambda_l + i}{\Lambda_l - i} = e(\Lambda_l) = [e(-\Lambda_l)]^{-1}, \quad (6)$$

and from (4) and (5) one may derive:

$$\exp(i\phi_{l,m}) = e\left(\frac{\Lambda_l - \Lambda_m}{2}\right). \quad (7)$$

In (6) the definition of the function $e(x)$ is given.

After multiplication by:

$$\prod_{l < m} \exp(i\phi_{l,m}) \quad (8)$$

the formula (1) may be transformed into:

$$\Psi = \sum_{1 \leq j_1 < j_2 < \ldots j_r \leq N} \sum_P \prod_{l=1}^{r} e(\Lambda_{P(l)})^{j_l} \prod_{\substack{1 \leq m < n \leq r \\ P(m) > P(n)}} e(\frac{\Lambda_{P(m)} - \Lambda_{P(n)}}{2}) \Phi_{j_1 j_2 \ldots j_r}, \quad (9)$$

with the condition:

$$e(\Lambda_l)^N = \prod_{\substack{m=1 \\ m(\neq l)}}^{r} e(\frac{\Lambda_l - \Lambda_m}{2}), \quad (10)$$

which follows from (2).

Because of the fact that the energy of all stationary states should be *real* it stands to reason that the quasi-momenta for any solution of the B.H. - problem should appear in *complex conjugated* pairs and the same may be said of the parameters Λ_i.

3. Solution with one single string

The *translational invariance* of an infinite chain suggests to explore the possible existence of a stationary state that corresponds with a *localized excitation* or a *wave packet* that moves along the chain without changing its form. In the asymptotic regime of large N this would result in a solution of (10) *independent* of the number of particles. This condition is the key to the two possibilities:

$$e(\Lambda_l)^N \Rightarrow 0 \text{ and } e(\Lambda_l)^N \Rightarrow \infty. \quad (11)$$

In both cases the imaginary part of Λ_l should be different from 0, according to (6). The consequence of the existence of a limit 0 for a given l, however, is that at least *one* of the factors in the right member of (10) is 0. But the condition:

$$e(\frac{\Lambda_l - \Lambda_m}{2}) = 0, \quad (12)$$

implies:

$$\frac{\Lambda_l - \Lambda_m}{2} = -i, \quad (13)$$

as a consequence of (6).

Now we take the following *ordered* set of Λ_l, i.e. the one given by Takahashi:[7]

$$\Lambda_l = \Lambda + i(r + 1 - 2l), \quad l = 1, 2, ..r, \quad (14)$$

in which r is the *length* of the string, the row of spin inversions in the wave packet. This implies that (13) is true for $m = l - 1$ and careful inspection of the right member of (9) shows that there is always such a factor in every term with the exception of the one with the natural order for which the permutation P is trivial, i.e.: $P(j) = j$. This leaves us with a simplified version of (9):

$$\Psi = \sum_{1 \leq j_1 < j_2 < \ldots j_r \leq N} \prod_{l=1}^{r} e(\Lambda_l)^{j_l} \Phi_{j_1 j_2 \ldots j_r},$$ (15)

with Λ_j given by (14). It should be kept in mind that this is an *asymptotic* solution, as indicated in Takahashi's paper.

So far the impression could exist that the parameter Λ could be given any value, but the *reality* of the total energy and the total quasi momentum makes it plausible that the individual quasi momenta k_j appear in *complex conjugated pairs*. This is not a strict proof but the experience teaches us that so far only B.H. states have been found that obey this condition. Then it follows that according to (5) that also the Λ_j form complex conjugated pairs and, because of the special way we have chosen to represent the Λ_j in (14), the parameter Λ should be *real* and:

$$\Lambda_j = \Lambda^*_{r-j+1}.$$ (16)

The actual value of Λ may be derived from (6):

$$\exp(ik) = \exp i(k_1 + k_2 + \ldots k_r) = \prod_{j=1}^{r} \frac{\Lambda_j + i}{\Lambda_j - i} = \frac{\Lambda + ir}{\Lambda - ir},$$ (17)

which turns out to be a complex number with modulus 1, as it should be, because the total quasi momentum k is real. From (17) we immediately derive a relation between Λ and k:

$$\cot(\frac{k}{2}) = \frac{\exp(ik) + 1}{\exp(ik) - 1} i = \frac{\Lambda}{r}.$$ (18)

The possible values of the total quasi momentum are determined by the well-known *periodic boundary condition:*

$$Nk = 2\pi\lambda, \quad \lambda = \sum_{j=1}^{r} \lambda_j,$$ (19)

or:

$$\left(\frac{\Lambda + ir}{\Lambda - ir}\right)^N = 1,$$ (20)

as follows from (2) and (17). The value of Λ as a function of k immediately follows from (18).

Our method is hybrid in the sense that the final result follows from two different arguments, one is asymptotic in character whereas the other is just the well-known periodic boundary condition for a finite system.

It may be concluded that, for the special set of Λ_j given in (14) with *real* Λ, all $\phi_{l,m}$ may be chosen to be *purely imaginary*, as follows from (7). The same may be said for the differences $k_l - k_m$. Then it follows from (2) that in the B.H.-representation all parameters λ_l *within one string* should be equal.

4. Generalization: States with 2 or more strings

A straightforward generalization of an asymptotic solution with *one* string can be made by just superposing *two or more strings* in the following way. We imagine a state with an asymptotic number of spins with two strings at a large distance:

$$\Psi =$$

$$\sum_{1 \leq j_1 < j_2 < \ldots j_r \leq M} \sum_{P \leq l_1 < l_2 < \ldots l_s \leq R} \prod_{n=1}^{r} \prod_{m=1}^{s} e(\Lambda_n)^{j_n} e(\Omega_m)^{l_s} \Phi_{j_1 j_2 \ldots j_r l_1 l_2 \ldots l_s},$$

$$M \gg 1,\ P - M \gg 1,\ R - P \gg 1,\ N - R \gg 1, \tag{21}$$

$$\Lambda_n = \Lambda + i(r + 1 - 2n),\ n = 1, 2, ..r \tag{22}$$

$$\Omega_m = \Omega + i(s + 1 - 2m), m = 1, 2, ...s, \tag{23}$$

in which Λ and Ω are *two different real numbers*. This last condition is the same as the condition that the two strings can be interchanged *without* resulting phase factors that are either 0 or ∞, as follows from (7). The resulting phase factor for the complete interchange of the two strings is now easily determined and an asymptotic form of a state with two strings may be represented by a superposition of two states of the form (21), with co-efficients:

$$1 \text{ and } \prod_{n=1}^{r} \prod_{m=1}^{s} e\left(\frac{\Omega_m - \Lambda_n}{2}\right), \tag{24}$$

and instead of (10) we now have the condition:

$$\prod_{m=1}^{s} e(\Omega_m)^N = \prod_{n=1}^{r} \prod_{m=1}^{s} e\left(\frac{\Omega_m - \Lambda_n}{2}\right). \tag{25}$$

In this case one should observe the analogy with the superposition of two single deviations with *real wave vectors* in the B.H.-scheme, by making the substitution:

$$r = s = 1, \quad \Lambda_n = k_1, \quad \Omega_m = k_2, \quad e(\frac{\Omega_m - \Lambda_n}{2}) = \exp(\phi_{2,1}). \qquad (26)$$

For the asymptotic states with 2 strings we also have a boundary condition of the type (20), which now takes the form:

$$\left(\frac{\Lambda + ir}{\Lambda - ir} \frac{\Omega + is}{\Omega - is}\right)^N = 1. \qquad (27)$$

Introducing the partial sums for the k_j of both strings:

$$k_\Lambda = \sum_{n=1}^{s} k_n, \quad k_\Omega = \sum_{m=1}^{s} k_m, \qquad (28)$$

the condition (27) may be written:

$$\exp[iN(k_\Lambda + k_\Omega)] = 1 \quad \text{or} \quad N(k_\Lambda + k_\Omega) = 2\pi\mu, \qquad (29)$$

in which expression μ again is an integer.

A generalization of the foregoing argument for states with more strings may be readily made by considering products of strings with all possible permutations of the group of indices corresponding to the different strings. Here again one observes an analogy with states with single deviations with *real k*.

For an asymptotic number of spins and a given number of deviations one generally has a complicated picture of the stationary states, but there only exist two possible configurations for neighbouring deviations:

-either neighbouring deviations stick together in a bound states or

-or they form a scattering state.

In the first case one has a string that moves along the chain without changing its form, but only showing a phase shift per lattice distance. Such a string may contain more than just the two deviations considered. It has a *real* total k and it may be considered as a generalization of one single deviation. In the second case one has individual deviations corresponding to different strings. So the asymptotic string solution (SS) will show a great analogy with BHS in that sense that its general form of the former is isomorphous with the latter with the restriction to *real k* for all strings.

5. A family of solutions starting from an asymptotic configuration

Starting from the Bethe-Hulthén equations (2-4) one may consider the following asymptotic regime:

$$\lambda = \lambda_0 N \qquad \lambda, N \Rightarrow \infty \qquad \lambda_0 \text{ fixed,} \tag{30}$$

which represents a situation for which the k_j and $\phi_{j,k}$ approach asymptotic values corresponding with SS. In this regime the different strings appear as wave packets of a form independent of N, and with an overall complex phase that depends on the position of the packets in the chain. This phase follows from the string solution given in the foregoing section.

Given the SS one has all the relevant information of an asymptotic solution within the BH-scheme, i.e. the wave numbers k_j, from which the $\phi_{j,k}$ may be derived. Choosing now a sufficiently large N one may determine the "exact" solution starting from these asymptotic values of the wave numbers and the phases.This last solution is characterized by the well-known set $\{\lambda_j\}$,which have a simple relation with the an analogous set of integers of the string solution, which will be given below. The wave numbers and the phases may be considered now as a "quasi-continuous" function of N in lowering this number of spins. This procedure will be illustrated in the next section for three simple examples.

The SS, being similar to BHS with real λ_j, may be characterized by the integers μ_l, as follows from the analysis in the foregoing section. So the relation (25) may be rewritten:

$$NK_2 = 2\pi\mu_2 + \Phi_{2,1} \tag{31}$$

in which the symbol K_2 denotes the total wave number of the string (23) and $\Phi_{2,1}$ the total phase shift accompanying the exchange of the two strings. As stated before there exists a simple relation between the string parameters μ_l and the parameters λ_j of the constituents waves in the asymptotic regime:

$$\mu_l = \sum_j \lambda_j, \tag{32}$$

the sum refers to the λ_j of the corresponding string.

6. Three simple examples of string solutions

6.1. $(r = 2, \lambda_1 = \lambda_2 = \lambda/2)$

Within the SS the relevant equation is given in (18), which gives the value of the quantity Λ in terms of the total wave number k. The example we consider is the series:

$$(N = 1000, \lambda_1 = \lambda_2 = -40), (N = 200, \lambda_1 = \lambda_2 = -8),$$
$$(N = 100, \lambda_1 = \lambda_2 = -4). \tag{33}$$

The B.H.-equations (2-4) may be solved with the Ansatz:

$$k_1 = A - ib, \quad k_2 = A + ib, \quad \phi_{1,2} = -iq, \quad A = \pi\lambda/N, \tag{34}$$

for which the equations may be transformed into:

$$\sinh(\frac{q}{2})\sinh(\frac{q}{N}) + \cosh(\frac{q}{2})[\cos(\frac{\pi\lambda}{N}) - \cosh(\frac{q}{N})] = 0, \quad Nb = q. \tag{35}$$

The equations (34) and (35) give us the BHS for given N and λ. The only parameter to be determined in the string solution is Λ in this case, which follows from (18). The value of b for the string solution may be found with (6), which results in:

$$\exp(ik_1) = \frac{\Lambda_1 + i}{\Lambda_1 - i} \quad \exp(ik_2) = \frac{\Lambda_2 + i}{\Lambda_2 - i}, \tag{36}$$

$$\exp(2b) = \exp(ik_1)\exp(-ik_2) = \frac{\Lambda_1 + i}{\Lambda_1 - i} \times \frac{\Lambda_2 - i}{\Lambda_2 + i} =$$

$$= \frac{\Lambda + 2i}{\Lambda} \frac{\Lambda - 2i}{\Lambda} = \frac{\Lambda^2 + 4}{\Lambda^2}, \tag{37}$$

$$E = -\frac{16}{\Lambda^2 + 4}. \tag{38}$$

In Table I and II we give the results for the exact as well as the string solution and show the convergence of the BHS to the string solution for $N \Rightarrow \infty$.

Table I. Solutions for $\lambda_1 = \lambda_2 = -N/25$. Bethe-Hulthén scheme.

λ	λ_1	λ_2	N	A	b	q	E
-8	-4	-4	100	$-2\pi/25$	0.03417	3.41660	-0.24681
-16	-8	-8	200	$-2\pi/25$	0.03203	6.40600	-0.24736
-80	-40	-40	1000	$-2\pi/25$	0.03192	31.9209	-0.24739

Table II. String solution for $A = -2\pi/25$.

Λ	A	b	E
-7.78949	$-2\pi/25$	0.03192	-0.24739

N.B.: In the string solution the value of $q = \infty$. The single parameter for the string solution is $\mu = \lambda$ for this case.

6.2. $(r = 2, s = 1, \ \lambda_1 = \lambda_2 = \lambda/2, \lambda_3 = \lambda/4)$

Within the SS the relevant equations can be found in section 4 and we consider the cases:

$$
\begin{aligned}
&(N = 1000, \lambda_1 = \lambda_2 = -40, \lambda_3 = -10), \\
&(N = 200, \lambda_1 = \lambda_2 = -8, \lambda_3 = -2), \\
&(N = 100, \lambda_1 = \lambda_2 = -4, \lambda_3 = -1).
\end{aligned} \tag{39}
$$

The B.H.-equations (2-4) may be solved with the Ansatz:

$$
k_1 = A - ib, k_2 = A + ib, k_3 = n, \phi_{1,2} = -iq, \phi_{1,3} = P - ip, \phi_{2,3} = P + ip, \tag{40}
$$

for which the equations may be transformed into:

$$
\begin{aligned}
NA &= 2\pi\lambda_1 + P, \\
Nb &= q + p, \\
Nn &= 2\pi\lambda_3 - 2P
\end{aligned} \tag{41}
$$

and:

$$
\begin{aligned}
&\sinh(q/2)\sinh(b) + \cosh(q/2)[\cos(A) - \cosh(b)] = 0, \\
&\sin(P)[\cos(A - n) - \cosh(b)] + \\
&+[\cos(P) + \cosh(p)][\sin(A) - \sin(n)\cosh(b) - \sin(A - n)] = 0, \\
&\sinh(q)[\cos(A - n) - \cosh(b)] + [\cos(P) + \cosh(p)] \times \\
&\times [-\cos(n)\sinh(b) + \sinh(b)] = 0.
\end{aligned} \tag{42}
$$

The solution of this set of equations may be found in Table III for the the three values of N given in (39).

For the two strings in the SS we have, according to (18):

$$
\cot(k/2) = \cot(A) = \Lambda/2, \ \cot(n/2) = \Omega, \tag{43}
$$

or:

$$
A = k/2 = \arctan(2/\Lambda), \ \ b = 1/2\ln(1 + 4/\Lambda^2), \ \ n = 2\arctan(1/\Omega), \tag{44}
$$

in which the expression for b is essentially the same as the one given in (37). For the string solution we now have to give the single boundary condition (25):

$$
\begin{aligned}
e(\Omega)^N &= \prod_{n=1}^{2} e(\frac{\Omega - \Lambda_n}{2}) = \frac{\Omega - \Lambda + i}{\Omega - \Lambda - i} \times \frac{\Omega - \Lambda + 3i}{\Omega - \Lambda - 3i} \\
&= \exp(iv)\exp(iw) = \exp(i\Phi_{2,1}),
\end{aligned} \tag{45}
$$

from which it follows:

$$
v = 2\arctan(\frac{1}{\Omega - \Lambda}), \ \ w = 2\arctan(\frac{3}{\Omega - \Lambda}). \tag{46}
$$

If we now treat the two strings as single waves in the BHS we have the boundary conditions:

$$Nk = -v - w + 2\pi\mu_1 = \Phi_{1,2} + 2\pi\mu_1, \quad \mu_1 \text{ integer,}$$

$$Nn = v + w + 2\pi\mu_2 = \Phi_{2,1} + 2\pi\mu_2, \quad \mu_2 \text{ integer,} \quad \Phi_{1,2} = -\Phi_{2,1}. \quad (47)$$

The equations (43), (46) and (47) now are a complete set of equations for the quantities k, n, Λ, Ω and w if we make a choice for the integers $\mu_1 = \lambda_1 + \lambda_2$ and $\mu_2 = \lambda_3$. The quantity b follows from (44). These string solution is given in Table IV

Table III. Solutions for $\lambda_1 = \lambda_2 = N/25, \lambda_3 = N/100$. Bethe-Hulthén scheme.

λ	λ_1	λ_2	λ_3	N	A	b	n	q
-9	-4	-4	-1	100	-0.24954	0.03379	-0.06641	3.36999
-18	-8	-8	-2	200	-0.25047	0.03181	-0.06455	6.35547
-90	-40	-40	-10	1000	-0.25116	0.03188	-0.06317	31.8711

λ	λ_1	λ_2	λ_3	N	P	p	E
-9	-4	-4	-1	100	0.17866	0.008545	-0.25218
-18	-8	-8	-2	200	0.17200	0.007417	-0.25403
-90	-40	-40	-10	1000	0.16698	0.006967	-0.25504

Table IV. String solution for $\lambda_1 = \lambda_2 = 4\lambda_3$.

λ	N	μ_1	μ_2	$A = k/2$	b
-9	100	-8	-1	-0.24954	0.03146
-18	200	-16	-2	-0.25047	0.03170
-90	1000	-80	-10	-0.25116	0.03188

λ	N	μ_1	μ_2	n	v	w
-9	100	-8	-1	-0.06641	-0.08979	-0.26793
-18	200	-16	-2	-0.06455	-0.08632	-0.25769
-90	1000	-80	-10	-0.06317	-0.08378	-0.25018

λ	N	μ_1	μ_2	Λ	Ω	E
-9	100	-8	-1	-7.84767	-30.10713	-0.25277
-18	200	-16	-2	-7.81739	-30.97210	-0.25406
-90	1000	-80	-10	-7.79489	-31.65217	-0.25504

6.3. $(r = 4, \lambda_1 = \lambda_2 = \lambda_3 = \lambda_4 = \lambda/4)$

In this example we make the supposition that the wave numbers constitute two complex conjugated pairs. Then the same can be said of the pairs $(\phi_{1,3}, \phi_{2,4})$ and $(\phi_{1,4}, \phi_{2,3})$:

$$k_1 = A - ib, \quad k_2 = A + ib, \quad k_3 = C - id, \quad k_4 = C + id,$$
$$\phi_{1,2} = -ip, \quad \phi_{1,3} = T - it, \quad \phi_{1,4} = V - iv, \quad (48)$$
$$\phi_{2,3} = V + iv, \quad \phi_{2,4} = T + it, \quad \phi_{3,4} = -ir.$$

The Bethe-Hulthén equations now take the form:

$$NA = 2\pi\lambda_1 + T + V, \quad Nb = p + t + v, \quad \lambda_1 = \lambda/4,$$
$$NC = 2\pi\lambda_3 - T - V, \quad Nd = -t + v + r, \quad \lambda_3 = \lambda/4, \quad (49)$$

$$\sinh(p/2)\sinh(b) + \cosh(p/2)[\cos(A) - \cosh(b)] = 0,$$
$$\sinh(r/2)\sinh(d) + \cosh(r/2)[\cos(C) - \cosh(d)] = 0,$$
(50)

$$\frac{\sin(T)}{\cos(T) + \cosh(t)} + \frac{\sin(A)\cosh(d) - \sin(C)\cosh(b) - \sin(A - C)}{\cos(A - C) - \cosh(b - d)} = 0,$$
$$\frac{\sinh(t)}{\cos(T) + \cosh(t)} + \frac{\cos(A)\sinh(d) - \cos(C)\sinh(b) + \sinh(b - d)}{\cos(A - C) - \cosh(b - d)} = 0,$$
(51)

$$\frac{\sin(V)}{\cos(V) + \cosh(v)} + \frac{\sin(A)\cosh(d) - \sin(C)\cosh(b) - \sin(A - C)}{\cos(A - C) - \cosh(b + d)} = 0,$$
$$\frac{\sinh(v)}{\cos(V) + \cosh(v)} + \frac{-\cos(A)\sinh(d) - \cos(C)\sinh(b) + \sinh(b + d)}{\cos(A - C) - \cosh(b + d)} = 0.$$
(52)

These equations have the solutions, presented in Table V, for a set of N values obeying $N = 100\lambda/4$.

Table V. Bethe-Hulthén solutions for $\lambda_1 = \lambda_2 = \lambda_3 = \lambda_4 = \lambda/4$, $N = 100\lambda/4$.

$\lambda/4$	N	A	b	C	d	p
20	2000	0.06258	0.005926	0.06308	0.001991	0.6931
10	1000	0.06258	0.006054	0.06308	0.002012	0.6776
4	400	0.06249	0.007937	0.06318	0.002559	0.5105
1	100	0.06171	0.015033	0.06395	0.004881	0.2699

$\lambda/4$	N	T	t	V	v	r	E
20	2000	−0.4979	10.0602	−0.00003	1.0986	12.9435	−0.03142
10	1000	−0.2431	4.3025	−0.004247	1.0743	5.2402	−0.03141
4	400	−0.1215	1.8680	−0.01745	0.7964	2.0952	−0.03130
1	100	−0.09036	0.8218	−0.02152	0.4116	0.8983	−0.03059

To illustrate the concepts of *transition point* and *limit point* we also give here the results for a fixed set $\lambda_1 = \lambda_2 = \lambda_3 = \lambda_4 = 1$ and variable N, i.e. from $N = 100$ downwards. For a certain value of the number of spins the quantity $\mathrm{Re}(\phi_{1,3}) = T$ reaches the value $-\pi$ and at that point the set of integers $\{\lambda_j\}$ can be replaced by an other one in order to keep $|T|$ within the boundaries $[0, \pi]$. The number of spins for which this "transition" occurs is not an integer, but it indicates a separation between two domains of integers N for which a different $\{\lambda_j\}$ set should be chosen. Finally one reaches a limit point for which the imiginary part of one wave number pair diverges: This is the natural ending of the solution with the given parameter set.

Table VI. Bethe-Hulthén solutions for the equivalent sets ($\lambda_1 = \lambda_2 = \lambda_3 = \lambda_4 = 1$) and ($\lambda_1 = \lambda_2 = 0$, $\lambda_3 = \lambda_4 = 2$); $N^* = 10.3180$ is the transition point; $N = 8$ is the limit point.

λ_1	λ_2	λ_3	λ_4	N	A	b	C	d
1	1	1	1	100	0.06171	0.01503	0.06395	0.004881
1	1	1	1	15	0.3269	0.2978	0.5109	0.1435
1	1	1	1	$N*$	0.3045	0.5382	0.9134	0.4934
0	0	2	2	$N*$	0.3045	0.53819	0.9134	0.4934
0	0	2	2	9	0.1948	0.6344	1.2014	1.01886
0	0	2	2	8	0	0.66622	$\pi/2$	∞

λ_1	λ_2	λ_3	λ_4	N	T	t	V	v
1	1	1	1	100	−0.09036	0.8218	−0.02152	0.4116
1	1	1	1	15	−1.3034	2.7496	−0.07690	1.0480
1	1	1	1	$N*$	$-\pi$	3.7041	0	1.1310
0	0	2	2	$N*$	π	3.7041	0	1.1310
0	0	2	2	9	1.7270	3.9105	0.02643	1.1026
0	0	2	2	8	0	3.6010	0	1.0625

λ_1	λ_2	λ_3	λ_4	N	p	r	E
1	1	1	1	100	0.2699	0.8983	−0.03059
1	1	1	1	15	0.6700	3.8540	−1.0346
1	1	1	1	N^*	0.7178	7.6639	−1.7403
0	0	2	2	N^*	0.7178	7.6639	−1.7403
0	0	2	2	9	0.6967	11.9776	−1.9967
0	0	2	2	8	0.6662	∞	−2.1580

The string solution, which will be compared with the results of Table V, is fully determined by:

$$r = 4, \quad k = \frac{2\pi * 4}{100} = \frac{2\pi}{25},$$
$$\Lambda = 4\cot(\frac{\pi}{25}), \tag{53}$$
$$\Lambda_1 = \Lambda + 3i, \quad \Lambda_2 = \Lambda + 1i, \quad \Lambda_3 = \Lambda - 1i, \quad \Lambda_4 = \Lambda - 3i.$$

The values for $\mathrm{Re}(k_j)$ ($j = 1,..4$) are all equal to $k/4 = \pi/50$. The imaginary parts follow from (6) and the the value of Λ given in (53):

$$\Lambda = 31.6633,$$
$$k_1 = A - ib, \quad k_2 = A + ib, \quad k_3 = C - id, \quad k_4 = C + id, \tag{54}$$
$$A = C = \pi/50, \quad b = 0.005926, \quad d = 0.001991.$$

This result shows a reasonable accordance with Table V for the values for $N = 2000$. Only the quantity C shows a considerable discrepancy.

7. Concluding remark

As is well-known from the literature the representation of the solutions of the linear Heisenberg problem in terms of the BHS is *not unique*. In contradistinction the representation with strings always is, i.e. a representation with a set Λ, Ω, etc. one real number for each string. In the BHS one always may choose a $\{\lambda_j\}$ set for which the sets $\{|\text{Re}\,k_j|\}, \{|\text{Re}\,\phi_{j,l}|\}$ are in the interval $[0, \pi]$. This implies the change of the $\{\lambda_j\}$ set for certain values of N, which are in general non-integer. An example of such a change may be found in Table VI: For $N = 10.3180$ the set has to be changed to keep the $\{|\text{Re}\,\phi_{j,l}|\}$ within the prescribed bounds. This special value of N we call a *transition point*. Another special value of N corresponds to a natural ending of a solution within the BHS, for which some quantities reach a *diverging imaginary part*. This occurs in the same example: For $N = 8$ the imaginary part of one pair of wave number diverges. Such a value of N we call a *limit point*. A third special point may occur for doublets with $\lambda_j = \lambda_{j+1} - 1$ for which an originally *real* pair (k_j, k_{j+1}) is transformated into a *complex conjugated* one. For this point the $\{\lambda_j\}$ set need not to be changed but the character of the BHS changes in a relevant way. The value of N represents a *critical point*.

These peculiarities of the BHS illustrate the drawbacks of this kind of representation, but it should be kept in mind that the changes are in effect not very fundamental because it only leads to a more convenient way of formulating a result, without changing its relevant properties, i.e. the values of the quantities Λ, Ω, etc.

Appendix: Special points in the BHS

a) Transition point

An example of such a point is found in our last example: $(r = 4, \lambda_1 = \lambda_2 = \lambda_3 = \lambda_4 = \lambda/4)$. In this example the 4 wave numbers are two by two complex conjugated. The Bethe-Hulthén equations (48-52) were solved for the range 8 to 100 and it was observed that for $N = 10.3180$ the set $\{\lambda_j\}$ has to be changed in order to keep $|T|$, the modulus of real part of $\phi_{1,3}$ and $\phi_{2,4}$, within the proper bounds $[0, \pi]$. That such a transformation of the set $\{\lambda_j\}$ has the desired effect follows immediately from (49). These are the only equations of the total set in which the λ_j appear, and it should be clear that these numbers are not unique for a given solution, but could be manipulated in order to get a special form of this solution. This is a general feature of the BHS and it constitutes a drawback as compared to the solution with strings. For more complicated examples one may follow an analogous procedure to keep all the $|\text{Re}\,k_j|$, $|\text{Re}\,\phi_{j,k}|$ within the interval $[0,\pi]$.

b) Critical point

This special value of N appears for the case that two λ_j have a difference 1. Now we have a pair of wave numbers that are real (and different) for N above this critical value N_{cr} and form a complex conjugated pair below this number of spins. We illustrate this with two simple and related examples, which are represented by the following sets:

$$(\lambda_1 = -3, \lambda_2 = -2) \quad (\lambda_1 = -3, \lambda_2 = -2, \lambda_3 = 1). \tag{55}$$

For both examples we make, for sufficiently large N, the following transformation:

$$k_1 = A - b, \quad k_2 = A + b, \quad \phi_{1,2} = \pi - p, \tag{56}$$

whereas for the second example we have to add:

$$k_3 = n, \quad \phi_{1,3} = R - r, \quad \phi_{2,3} = R + r. \tag{57}$$

All quantities (A, b, p, n, R, r) are *real*.

The transformed B.H.-equations now take the form:
case I: $r = 2$, $(\lambda_1 = -3, \lambda_2 = -2)$.

$$NA = -5\pi, \tag{58}$$

$$Nb = p, \tag{59}$$

$$\cos(\tfrac{p}{2})\sin(b) - \sin(\tfrac{p}{2})[\cos(b) - \cos(A)] = 0. \tag{60}$$

case II: $r = 3$, $(\lambda_1 = -3, \lambda_2 = -2, \lambda_3 = 1)$.

$$NA = -5\pi + R, \tag{61}$$

$$Nb = p + r, \tag{62}$$

$$Nk = 2\pi - 2R, \tag{63}$$

$$\cos(\tfrac{p}{2})\sin(b) - \sin(\tfrac{p}{2})[\cos(b) - \cos(A)] = 0, \tag{64}$$

$$\sin(R)[\cos(A - n) - \cos(b)] + $$
$$+[\cos(R) + \cos(r)][\sin(A) - \sin(n)\cos(b) - \sin(A - n)] = 0, \tag{65}$$

$$\sin(r)[\cos(A - n) - \cos(b)] + $$
$$+[\cos(R) + \cos(r)][-\cos(n)\sin(b) + \sin(b)] = 0. \tag{66}$$

Typical for a critical point is the *degeneracy* of the spectrum of wave numbers, i.e. in case I: $k_1 = k_2$ or $b = p = 0$ and in case II: $k_1 = k_2$ or $b = p = r = 0$. This occurs for a special value of N, which we call N_{cr}. This is not an integer and it does not correspond to a *real, physical system* but it separates 2 regions of N values with quantitatively different properties of the BHS. In our explanation here we approach N_{cr} from the upper side.

To find N_{cr} we linearize in case I the equation (60) in terms of b and p and combine it with (59), which results in the set:

$$Nb - p = 0, \tag{67}$$

$$b - \frac{p}{2}[1 - \cos(A)] = 0. \tag{68}$$

The determinant of this set of equations should be 0, for a solution of the B.H.-equations to exist in the neighbourhood of N_{cr}, which condition results in:

$$N[1 - \cos(A)] - 2 = 0. \tag{69}$$

Together with (58) this will give us the value of N_{cr} and the corresponding A.

In an analogous way we find for the case II the following set of equations as a result of linearizing (62),(64) and (66) in terms of b, p and r:

$$Nb - p - r = 0, \tag{70}$$

$$b - \frac{p}{2}[1 - \cos(A)] = 0, \tag{71}$$

$$b[1 - \cos(n)][1 + \cos(R)] - r[1 - \cos(A)] = 0. \tag{72}$$

Putting the value of the determinant of this set equal to 0 and combining this condition with (61), (63) and (65) we find N_{cr} and the corresponding values of A and n. In equation (65) we have to substitute the value 0 for b and r.

In this way we found the following critical values:

$$\text{case I: } N_{cr} = 61.3488 \qquad \text{case II: } N_{cr} = 62.3514. \tag{73}$$

Below the critical point the quantities b, p and r take purely imaginary values, and e.g. (60) should be replaced by:

$$\cosh(\frac{p}{2})\sinh(b) - \sinh(\frac{p}{2})[\cosh(b) - \cos(A)] = 0. \tag{74}$$

c) Limit point

The simplest example of such a point may be found in subsection 6.1. We now take a fixed value for the $\{\lambda_j\}$, e.g. $\lambda_1 = \lambda_2 = 4$ and follow the solution in lowering N. Now it turns out that b as well as q have the limit ∞ for N approaching the value 16, as a consequence of the fact that:

$$\cos(\frac{\pi\lambda}{N}) = 0 \text{ for } \lambda = 2\lambda_1 = 8 \text{ and } N_{\text{lim}} = 16. \tag{75}$$

For lower values of N the change of sign of $\cos(\pi\lambda/N)$ results in the vanishing of this solution. Analogous limit points may appear the cases in which there is a pair $\lambda_j = \lambda_{j+1} - 1$. We here consider the case I analyzed in Appendix b, which results in the equation (74), with $A = -5\pi/N$ and $b = p/N$, which gives $b = p = \infty$ for the limit value $N_{\text{lim}} = 10$.

References

1. H.A. Bethe, *Z. Phys.* **71**, 205 (1931).
2. W.J. Caspers, B. Lulek, T. Lulek, M. Kuzma and A. Wal, to be published.
3. A. Doikou, L. Mezincescu and R.I. Nepomechie, *Mod. Phys. Lett.* **12**, 2591 (1997).
4. F.H.L. Essler, V.E. Korepin and K. Schoutens, *J. Phys. A: Math. Gen.* **25**, 4115 (1992).
5. L. Hulthén, *Arkiv. Nat. Astron. Fys.* **26A**, 1 (1938).
6. S.N. Martynov, *Phys.Lett.A* **219**, 329 (1996).
7. M. Takahashi, *Prog. Theo. Phys.* **46**, 401 (1971).
8. A.A. Vladimirov, *Phys.Lett.* **105A,** 418 (1984).

SKEW GELFAND-TSETLIN PATTERNS, LATTICE PERMUTATIONS, AND SKEW PATTERN POLYNOMIALS

JAMES D. LOUCK

Los Alamos National Laboratory, Theoretical Division
Los Alamos, NM 87545 USA

A modification of the well-known Gelfand-Tsetlin patterns, which are one-to-one with Young-Weyl semistandard tableaux is introduced. These new patterns are in one-to-one correspondence with skew-tableaux, and with a slight modification can be used to enumerate lattice permutations. In particular, the coupling rule for angular momentum takes an elementary form in terms of these modified patterns. These interrelations will be presented, together with an outline of the construction of a class of polynomials that generalize the skew Schur functions.

1. Introduction and Review of Combinatorial Concepts

It is well known that semistandard Young-Weyl tableau of shape $\lambda \in \mathbb{P}ar_n$, where $\mathbb{P}ar_n$ denotes the set of all partitions $\lambda = (\lambda_1, \lambda_2, \ldots, \lambda_n), \lambda_1 \geq \lambda_2 \geq \ldots \geq \lambda_n \geq 0$, having n parts, counting 0 as a part, is bijective with the set of Gelfand-Tsetlin patterns of shape λ. The purpose of this article is to show how this result generalizes to a bijection between semi-standard skew tableau of shape $\lambda - \mu$ and skew Gelfand-Tsetlin patterns of shape $\lambda - \mu$, structures that are defined below. This result, in turn, allows the generalization of the well known D^λ−polynomials[1-4] to skew $D^{\lambda/\mu}$−polynomials. A consequence of this generalization is that the trace of the skew $D^{\lambda/\mu}$−polynomials yields the skew Schur functions $s_{\lambda/\mu}$, just as the trace of the ordinary D^λ−polynomials yields the ordinary Schur functions s_λ. We also formulate the concept of a modified Gelfand-Tsetlin pattern, and show their relation to lattice permutations and Littlewood-Richardson numbers. The properties of Littlewood-Richardson numbers are very important for the study of composite holistic multiparticle quantum systems to which the D^λ−polynomials and the $D^{\lambda/\mu}$−polynomials have applications through multiple Kronecker products of such polynomials and their reduction into irreducible forms. We do not address the latter in this article, but instead review and set forth the combinatorial approaches to the subject.

The interest in formulating tableau results in terms of Gelfand-Tsetlin patterns originates from the Weyl[5] group-subgroup significance of the conditions associated directly with Gelfand-Tsetlin[6] patterns, and the subsequent use of such patterns in numerous physical applications. We begin by a review of several well known results.[7]

1.1. *Semistandard Young-Weyl Tableaux and Gelfand-Tsetlin Patterns*

Semistandard Young-Weyl tableau ($SSYW$): A partition $\lambda \in \mathbb{P}ar_n$ is sometimes called a *shape* . A Young-Weyl tableau is a shape $\lambda \in \mathbb{P}ar_n$ in which the integers $1, 2, \ldots, n$ are distributed among the $|\lambda| = \lambda_1 + \cdots + \lambda_n$ boxes, one in each box, according to the rules:

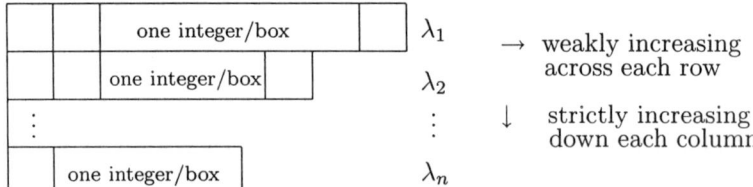

Notations:

$\mathbb{T}_\lambda$ = set of all semistandard tableaux of shape λ,

α = weight of a semistandard tableau = $(\alpha_1, \ldots, \alpha_n)$, α_i =number of i's,

$\mathbb{T}_\lambda(\alpha)$ = set of all semistandard tableaux of shape λ and weight α,

$\mathbb{W}_\lambda$ = set of all weights of semistandard tableaux of shape λ,

$K(\lambda, \alpha)$ = multiplicity of a weight $\alpha \in \mathbb{W}_\lambda$.

Gelfand-Tsetlin (GT) pattern: Let $\rho \in \mathbb{P}ar_r, \sigma \in \mathbb{P}ar_{r-1}, 1 < r \leq n$. The notation $\rho \succ \sigma$ means that the pair of partitions ρ and σ satisfy the betweenness conditions:

$$\rho_1 \geq \sigma_1 \geq \rho_2 \geq \sigma_2 \geq \cdots \geq \sigma_{r-1} \geq \rho_r \geq 0.$$

Two-rowed presentation:

$$\begin{pmatrix} \rho_1 & \rho_2 & \rho_3 & \cdots & \rho_{r-1} & \rho_r \\ & \sigma_1 & \sigma_2 & \cdots & & \sigma_{r-1} \end{pmatrix}.$$

The geometric placement of numbers is intended to suggest "betweenness." A *GT* pattern of shape λ is a sequence of n partitions $\lambda, \tau, \cdots \nu, \mu$ satisfying

$$\lambda \in \mathbb{P}ar_n, \ \tau \in \mathbb{P}ar_{n-1}, \cdots, \ \nu \in \mathbb{P}ar_2, \ \mu \in \mathbb{P}ar_1,$$

$$\lambda \succ \tau \succ \cdots \succ \nu \succ \mu.$$

A *GT* pattern of shape λ may be presented as triangular array with n rows:

$$\begin{pmatrix} \lambda_1 & \lambda_2 & \cdots & \lambda_{n-1} & \lambda_n \\ & \tau_1 & \tau_2 & \cdots & \tau_{n-1} \\ & & \vdots & & \\ & \nu_1 & \nu_2 & & \\ & & \mu_1 & & \end{pmatrix}, \ \lambda \succ \tau \succ \cdots \succ \nu \succ \mu.$$

The geometric placement of these partitions is intended to indicate "betweenness."

Notations:

$$\mathbb{G}_\lambda = \text{set of all } GT \text{ patterns of shape } \lambda \in \mathbb{P}ar_n,$$
$$\alpha = (|\mu|, |\nu| - |\mu|, \ldots, |\lambda| - |\tau|),$$
$$\mathbb{G}_\lambda(\alpha) = \text{ set of all } GT \text{ patterns of weight } \alpha,$$
$$\mathbb{W}_\lambda = \text{set of all weights of } GT \text{ patterns of shape } \lambda,$$
$$K(\lambda, \alpha) = \text{multiplicity of a weight } \alpha \in \mathbb{W}_\lambda.$$

The significant result is the bijection between the set of semistandard tableaux of shape λ and the set of *GT* patterns of shape λ of the same weight,

$$\mathbb{T}_\lambda(\alpha) \overset{BIJECTION}{\longleftrightarrow} \mathbb{G}_\lambda(\alpha),$$

as given by the following rules:

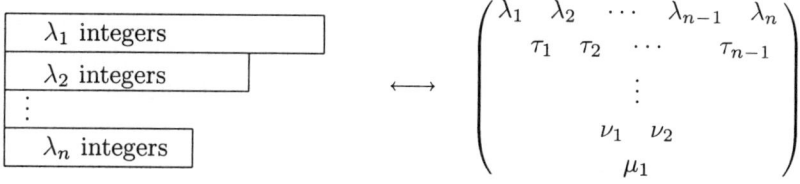

mapping rule: The shapes (partitions) in the *GT* pattern are obtained by sequential removal of integers from the semistandard tableau:

λ is the shape of the tableau;

τ is the shape of the tableau obtained by removing all n's;

$\vdots$

ν is the shape of the tableau obtained by removing all 3's;

μ is the shape of the tableau obtained by removing all 2's.

inverse rule: The semistandard tableau is obtained from the *GT* pattern by the rule

insert 1's in the shape μ;

followed by 2's in the new boxes contained in shape ν and not in shape μ;

$\vdots$

followed by n's in the new boxes contained in shape λ and not in shape τ.

Example:

The relation between *SSYW* tableaux and *GT* patterns of the same shape is illustrated for $\lambda = (6, 5, 3, 0)$ by

1	1	1	2	3	4
2	2	3	3	4
3	4	4

$\longleftrightarrow$

$$\begin{pmatrix} 6 & & 5 & & 3 & & 0 \\ & 5 & & 4 & & 1 & \\ & & 4 & & 2 & & \\ & & & 3 & & & \end{pmatrix}$$

The bijection described above extends, of course, to the full sets of semistandard Young-Weyl tableaux and Gelfand-Tsetlin patterns of shape λ:

$$\mathbb{T}_\lambda \overset{BIJECTION}{\longleftrightarrow} \mathbb{G}_\lambda.$$

2. The D^λ–Polynomials

One of the significant applications of SSYW tableaux, or, equivalently, of *GT* patterns is their role as indexing sets or labels of a class of polynomials that arise in many different contexts. These polynomials have been discussed in a number of articles.[1-4] Some of their principal properties are recalled here for the purpose of showing how they recur in the context of skew *GT* and modified *GT* patterns in Sections 3 and 6.2 .

2.1. *Basic Orthogonal Maclaurin Polynomials*

We adopt the following notations and definitions:

$$\frac{Z^A}{A!} = \prod_{i,j=1}^{n} \frac{z_{ij}^{a_{ij}}}{a_{ij}!},$$

$$\text{variables} = Z = \begin{pmatrix} z_{11} & z_{12} & \cdots & z_{1n} \\ z_{21} & z_{22} & \cdots & z_{2n} \\ & & \vdots & \\ z_{n1} & z_{n2} & \cdots & z_{nn} \end{pmatrix},$$

$$\text{exponents} = A = \begin{pmatrix} a_{11} & a_{12} & \cdots & a_{1n} \\ a_{21} & a_{22} & \cdots & a_{2n} \\ & & \vdots & \\ a_{n1} & a_{n2} & \cdots & a_{nn} \end{pmatrix} \begin{matrix} \alpha_1 \\ \alpha_2 \\ \vdots \\ \alpha_n \end{matrix} ,$$

$$\alpha_1' \quad \alpha_2' \quad \cdots \quad \alpha_n'$$

$$M_{n \times n}^p(\alpha, \alpha') = \left\{ n \times n \text{ arrays } A \ \middle| \right.$$

$$\left. \text{row sums} = \alpha, \text{ column sums} = \alpha', |\alpha| = |\alpha'| = p \right\}.$$

The Maclaurin polynomials are homogeneous as follows:

$$\alpha_i = \sum_{j=1}^{n} a_{ij} \text{ in } z_i = (z_{i1}, z_{i2}, \ldots, z_{in}) = \text{row } i \text{ of } Z,$$

$$\alpha_j' = \sum_{i=1}^{n} a_{ij} \text{ in } z^j = (z_{1j}, z_{2j}, \ldots, z_{nj}) = \text{column } j \text{ of } Z.$$

The Maclaurin polynomials are orthogonal in the inner product (,) described in detail by Louck.[4]

$$(Z^A, Z^B) = \delta(A, B) A!.$$

2.2. Basic Orthogonal D^λ−Polynomials

We now define a class of invertible real linear transformations of the Z^A polynomials that preserve their homogeneity properties in the rows and columns of the variable matrix Z :

$$D \begin{pmatrix} m' \\ \lambda \\ m \end{pmatrix} (Z) = \sum_{A \in M_{n \times n}^p(\alpha, \alpha')} C \begin{pmatrix} m' \\ \lambda \\ m \end{pmatrix} (A) \frac{Z^A}{A!},$$

$$\frac{Z^A}{A!} = \sum_{\substack{|\lambda|=p \\ m, m' \in \mathbb{G}_\lambda(\alpha, \alpha')}} \frac{1}{M(\lambda) A!} C \begin{pmatrix} m' \\ \lambda \\ m \end{pmatrix} (A) D \begin{pmatrix} m' \\ \lambda \\ m \end{pmatrix} (Z).$$

The notations in this definition are:

(1) $\binom{\lambda}{m}$ is a Gelfand-Tsetlin (GT) pattern, and $\binom{m'}{\lambda}$ is an inverted GT pattern. We now write a GT pattern in the more detailed form

$$
\binom{\lambda}{m} = \begin{pmatrix} \lambda_1 & \lambda_2 & \cdots & & \lambda_n \\ & m_{1,n-1} & m_{2,n-1} & \cdots & m_{n-1,n-1} \\ & & \vdots & & \\ & & m_{1,2} & m_{2,2} & \\ & & m_{1,1} & & \end{pmatrix},
$$

in which row j is given by

$$
m_j = (m_{1,j}, m_{2,j}, \cdots, m_{j,j}) \in \mathbb{P}ar_j, j = 1, 2, \ldots, n, \lambda = m_n.
$$

(2) α, α' are weights of the GT patterns:

$$
\alpha = W\binom{\lambda}{m} = (\alpha_1, \alpha_2, \ldots, \alpha_n), \ \alpha' = W\binom{\lambda}{m'} = (\alpha_1', \alpha_2', \ldots, \alpha_n').
$$

(3) $\mathbb{G}_\lambda(\alpha, \alpha')$ is the set of double GT patterns of weight (α, α') and shape λ with cardinality

$$
|\mathbb{G}_\lambda(\alpha, \alpha')| = K(\lambda, \alpha)K(\lambda, \alpha').
$$

(4) $M(\lambda)$ is the invariant normalizing factor defined by

$$
M(\lambda) = \left(\prod_{i=1}^n (\lambda_i + n - i)! \right) / 1!2! \cdots (n-1)! Dim\lambda,
$$

in which $Dim\lambda$ is the Weyl dimension formula:

$$
Dim\lambda = |\mathbb{G}_\lambda| = \left(\prod_{1 \le i < j \le n} (\lambda_i - \lambda_j + j - i) \right) / 1!2! \cdots (n-1)!.
$$

The orthogonality and normalization of the D^λ−polynomials are expressed by

$$
\left(D\begin{pmatrix} m''' \\ \lambda \\ m'' \end{pmatrix}(Z), \ D\begin{pmatrix} m' \\ \lambda' \\ m \end{pmatrix}(Z) \right) = \delta_{m,m''}\delta_{m',m'''}\delta_{\lambda,\lambda'} M(\lambda).
$$

2.3. Combinatorial Definition of the $C-$Coefficients

$$
\left(D\begin{pmatrix} m''' \\ \lambda + e_\tau \\ m'' \end{pmatrix}(Z), z_{ij} D\begin{pmatrix} m' \\ \lambda \\ m \end{pmatrix}(Z) \right)
$$

$$
= \left\langle \begin{matrix} \lambda + e_\tau \\ m'' \end{matrix} \middle| t_{i\tau} \middle| \begin{matrix} \lambda \\ m \end{matrix} \right\rangle \left\langle \begin{matrix} \lambda + e_\tau \\ m''' \end{matrix} \middle| t_{j\tau} \middle| \begin{matrix} \lambda \\ m' \end{matrix} \right\rangle,
$$

e_τ = unit row vector of length $n = (0, \ldots, 0, 1, 0, \ldots, 0)$, 1 in position τ,

The n^2 quantities $t_{i\tau}$ are called fundamental shift operators. The coefficients

$$\left\langle \begin{matrix} \lambda + e_\tau \\ m'' \end{matrix} \middle| t_{i\tau} \middle| \begin{matrix} \lambda \\ m \end{matrix} \right\rangle, i, j = 1, 2, \ldots n$$

are fully defined as functions over arc digraphs[8] (earlier called the pattern calculus[9]). The above matrix element relation and the arc digraphs determine uniquely the D^λ–polynomials. It is the algebra of these fundamental shift operators that places the construction of the C–coefficients, hence, the D^λ–polynomials on a fully combinatorial basis.[3-4]

For the purposes of this article, we henceforth take the D^λ–polynomials as fully known. It is useful to summarize some of their principal properties.

2.4. Summary of Structural Properties of the D^λ–Matrices

Matrix form of basic polynomials:

$$D^\lambda(Z) = \sum_{A \in M^p_{n \times n}} \frac{Z^A}{A!} C^\lambda(A), \text{each } \lambda \in \mathbb{P}ar_n$$

$$M^p_{n \times n} = \bigcup_{\alpha, \alpha' \in W_\lambda} M^p_{n \times n}(\alpha, \alpha').$$

In this relation, $D^\lambda(Z)$ and $C^\lambda(A)$ are matrices of dimension $Dim(\lambda)$ given by the Weyl dimension formula, and the coefficients in this matrix relation are the Maclaurin monomials $Z^A/A!$. The following properties offer but a glimpse at the important structural properties of these matrices, whose elements are the polynomials

$$D \begin{pmatrix} m' \\ \lambda \\ m \end{pmatrix} (Z),$$

whose rows and columns are enumerated by the Gelfand-Tsetlin patterns $\binom{\lambda}{m}$ and $\binom{\lambda}{m'}$.

(1) Multiplication of $D^\lambda(Z)$ matrices:

$$D^\lambda(X) D^\lambda(Y) = D^\lambda(XY),$$

for arbitrary matrices X and Y. A full combinatorial proof of this relation can be given, which starts with the proof given in Ref.3, and then uses[4] Pieri's rule for multiplying certain of these polynomials.

(2) Multiplication of $C^\lambda(A)$ matrices:

$$C^\lambda(A)C^\lambda(B) = \sum_{C \in M^p_{n \times n}(\alpha, \gamma)} \begin{Bmatrix} C \\ A\,B \end{Bmatrix} C^\lambda(C),$$

where

$$A \in M^p_{n \times n}(\alpha, \beta), B \in M^p_{n \times n}(\beta, \gamma),$$

and the structure constants are given by the double inner product of Maclaurin polynomials.[4]

$$\begin{Bmatrix} C \\ A\,B \end{Bmatrix} = \left(X^A, \left(Y^B, (XY)^C \right) \right) / C!.$$

(3) Matrix Schur function:

$$D^\mu(Z) \otimes D^\nu(Z) \sim \sum_\lambda c^\lambda_{\mu\nu} D^\lambda(Z),$$

in which the coefficients $c^\lambda_{\mu,\nu}$ are the Littlewood-Richardson numbers.

(4) Diagonal properties:

$$D^\lambda(diag(x_1, x_2, \ldots, x_n)) = \sum_{\alpha \in W_\lambda} \oplus x_1^{\alpha_1} x_2^{\alpha_2} \cdots x_n^{\alpha_n} I_{K(\lambda, \alpha)},$$

$$I_{K(\lambda, \alpha)} = \text{ identity matrix of dimension } K(\lambda, \alpha),$$

$$D^\lambda(I_n) = I_{Dim\lambda}.$$

(5) Relation to Schur functions:

$$s_\mu(x)s_\nu(x) = \sum_\lambda c^\lambda_{\mu\nu} s_\lambda(x),$$

$$Trace D^\lambda(diag(x_1, x_2, \ldots, x_n)) = s_\lambda(x).$$

(6) Importance for physics: The matrices $D^\lambda(Z)$ give all inequivalent integer representations of $GL(n, \mathbf{C})$ for $Z \in GL(n, \mathbf{C})$. Thus, the matrices $D^\lambda(Z)$ also give all inequivalent unitary representations of $U(n)$ for $Z \in U(n)$. Moreover, these matrices and their elements occur as state vectors for composite physical systems and have a wealth of properties for various choices of the variables Z. We will see some of this in the following sections.

3. Skew Tableaux and Skew Gelfand-Tsetlin Patterns

Semistandard skew Young tableaux: Let $\lambda, \mu \in \mathbb{P}ar_n$ with $\lambda_i \geq \mu_i$, $i = 1, 2, \ldots, n$. This condition is denoted $\lambda \supseteq \mu$, and means that the shape μ "fits inside" the shape λ. The *skew shape* $\lambda - \mu$ refers to the shape of the "staggered" rows of boxes that remain after deleting all the boxes of shape μ from the shape λ. As with a semistandard Young-Weyl tableau, the boxes in the shape $\lambda - \mu$ are filled in with the integers $1, 2, \ldots, n$ with one integer per box such that the sequence in each row is weakly increasing and the sequence in each column is strictly increasing:

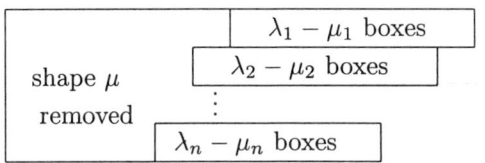

$\rightarrow$ weakly increasing across each row

$\downarrow$ stictly increasing down each column

Notations:

$\mathbb{T}_{\lambda/\mu}$ = set of all skew tableaux of shape $\lambda - \mu$,

α = weight of a skew tableau = $(\alpha_1, \ldots, \alpha_n), \alpha_i$ = number of i's ,

$\mathbb{T}_{\lambda/\mu}(\alpha)$ = set of all skew tableaux of shape $\lambda - \mu$ and weight α ,

$\mathbb{W}_{\lambda/\mu}(\alpha)$ = set of all weights in $\mathbb{T}_{\lambda/\mu}(\alpha)$,

$\mathbb{W}_{\lambda/\mu}$ = set of all weights in $\mathbb{T}_{\lambda/\mu}$,

$K_{\lambda/\mu}(\alpha)$ = multiplicity of weight $\alpha \in \mathbb{W}_{\lambda/\mu}$.

Skew Gelfand-Tsetlin patterns: The notation $\rho \sqsupseteq \sigma$ means that the pair of partitions $\rho, \sigma \in \mathbb{P}ar_n$ satisfy the betweenness conditions:

$$\rho_1 \geq \sigma_1 \geq \rho_2 \geq \sigma_2 \geq \cdots \geq \rho_n \geq \sigma_n \geq 0.$$

Two-rowed presentation:

$$\begin{pmatrix} \rho_1 & \rho_2 & \rho_3 & \cdots & \rho_{n-1} & \rho_n \\ \sigma_1 & \sigma_2 & & \cdots & \sigma_{n-1} & \sigma_n \end{pmatrix}.$$

A skew GT pattern of shape $\lambda - \mu$, with $\mu \subseteq \lambda$, is a sequence of $n + 1$ partitions $\lambda, \tau, \ldots, \nu, \mu$ beginning with λ and ending with μ, all of which belong to $\mathbb{P}ar_n$, and which satisfy the conditions $\lambda \sqsupseteq \tau \sqsupseteq \cdots \sqsupseteq \nu \sqsupseteq \mu$. A skew GT pattern of shape $\lambda - \mu$ may be presented as a parallelogram with

$n + 1$ rows:

$$\begin{pmatrix} \lambda_1 & \lambda_2 & \cdots & \lambda_n & \\ & \tau_1 & \tau_2 & \cdots & \tau_n \\ & & \vdots & & \\ & \nu_1 & \nu_2 & \cdots & \nu_n \\ & \mu_1 & \mu_2 & \cdots & \mu_n \end{pmatrix}, \quad \lambda \sqsupseteq \tau \sqsupseteq \cdots \sqsupseteq \nu \sqsupseteq \mu.$$

The placement of the entries in this configuration is intended to be suggestive of the betweenness relations.

Notations:

$\mathbb{G}_{\lambda/\mu}$ = set of all skew GT patterns of shape $\lambda - \mu$,

α = weight of a skew GT pattern = $(\alpha_1, \ldots, \alpha_n)$
$$= (|\nu| - |\mu|, \ldots, |\tau| - |\sigma|, |\lambda| - |\tau|),$$

$\mathbb{G}_{\lambda/\mu}(\alpha)$ = set of all skew GT patterns of shape $\lambda - \mu$ and weight α ,

$\mathbb{W}_{\lambda/\mu}(\alpha)$ = set of all weights in $\mathbb{G}_{\lambda/\mu}(\alpha)$,

$\mathbb{W}_{\lambda/\mu}$ = set of all weights in $\mathbb{G}_{\lambda/\mu}$,

$K_{\lambda/\mu}(\alpha)$ = multiplicity of weight $\alpha \in \mathbb{W}_{\lambda/\mu}$.

These notations for weights anticipate that these sets are identical to those defined for semistandard skew tableau.

Again, we have the bijection between the sets of semistandard skew tableaux and skew GT patterns of the same weight,

$$\mathbb{T}_{\lambda/\mu}(\alpha) \overset{BIJECTION}{\longleftrightarrow} \mathbb{G}_{\lambda/\mu}(\alpha),$$

as given by the following rules:

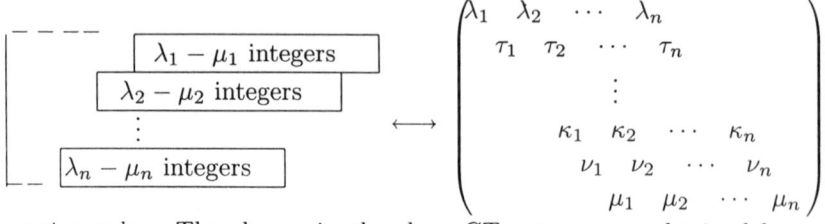

mapping rule: The shapes in the skew GT pattern are obtained by sequential removal of integers from the skew tableau:

λ is the shape of the tableau;

τ is the shape of the tableau obtained by removing all n's;

$\vdots$

ν is the shape of the tableau obtained by removing all 2's;

μ is the shape of the tableau obtained by removing all 1's.

inverse rule: The skew tableau is obtained from the skew GT pattern by the rule

insert 1's in the shape $\nu - \mu$; followed by 2's in the shape $\kappa - \nu$;

$\cdots$; followed by n's in the shape $\lambda - \tau$.

Example: $\lambda = (6530), \mu = (3210)$:

$$\begin{array}{cccc} 6 & 5 & 3 & 0 \\ 6 & 4 & 2 & 0 \\ 5 & 3 & 2 & 0 \\ 5 & 3 & 1 & 0 \\ 3 & 2 & 1 & 0 \end{array}$$

The bijection described above extends, of course, to the full sets of semistandard skew Young tableaux and skew Gelfand-Tsetlin patterns of shape $\lambda - \mu$:

$$\mathbb{T}_{\lambda/\mu} \overset{BIJECTION}{\longleftrightarrow} \mathbb{G}_{\lambda/\mu}.$$

4. The $D^{\lambda/\mu}$−Polynomials

It is somewhat surprising that the set of skew GT patterns $\mathbb{G}_{\lambda/\mu}$ can all be realized as ordinary triangular GT patterns[10] in $2n$ rows corresponding to partitions $(\lambda, \, 0^n), \lambda \in \mathbb{P}ar_n$, with further specialization of the patterns, as we describe in this section.

4.1. *Relation between Skew GT Patterns and GT Patterns*

We introduce the more detailed $m_{i,j}$ notation as follows for a skew GT pattern:

$$\begin{bmatrix} \lambda/\mu \\ m \end{bmatrix} = \begin{pmatrix} \lambda_1 & \lambda_2 & \cdots & \lambda_n & \\ & m_{1,n-1} \, m_{2,n-1} & \cdots & m_{n,n-1} & \\ & & \vdots & & \\ & m_{1,2} & m_{2,2} & \cdots & m_{n,2} & \\ & & m_{1,1} & m_{2,1} & \cdots & m_{n,1} \\ & & \mu_1 & \mu_2 & \cdots & \mu_n \end{pmatrix},$$

in which

$$\lambda \sqsupseteq m_{n-1} \sqsupseteq \cdots \sqsupseteq m_1 \sqsupseteq \mu,$$

$$m_j = (m_{1,j}, m_{2,j}, \ldots, m_{n,j}) \in \mathbb{P}ar_n, j = 0, 1, \ldots, n,$$

$$\lambda = (m_{1,n}, m_{2,n}, \ldots, m_{n,n}),$$
$$\mu = (m_{1,0}, m_{2,0}, \ldots, m_{n,0}).$$

Consider next the triangular GT pattern with $2n$ rows and partition $(\lambda, 0^n), \lambda \in \mathbb{P}ar_n$ as follows:

$$\begin{pmatrix} \lambda_1 & \lambda_2 & \cdots & \lambda_n & 0 & \cdots & & 0 \\ m_{1,n-1} & m_{2,n-1} & \cdots & m_{n,n-1} & 0 & \cdots & & 0 \\ & & \vdots & \vdots & & & \\ & m_{1,2} & m_{2,2} & \cdots & m_{n,2} & 0 & 0 \\ & & m_{1,1} & m_{2,1} & \cdots & m_{n,1} & 0 \\ & & & \mu_1 & \mu_2 & \cdots & \mu_n \\ & & & & & & \\ & & & & l & & \end{pmatrix} = \left(\begin{bmatrix} \lambda/\mu \\ m \end{bmatrix} \begin{pmatrix} \mu \\ l \end{pmatrix} \right),$$

where $\begin{pmatrix} \mu \\ l \end{pmatrix}$ is an ordinary triangular GT pattern with n rows. Notice that in this notation μ is included both as the bottom row in the skew pattern $\begin{bmatrix} \lambda/\mu \\ m \end{bmatrix}$ and as the top row in $\begin{pmatrix} \mu \\ l \end{pmatrix}$. We then have the one-to-one correspondence between patterns in which we choose the entries $l_{i,j}$ in $\begin{pmatrix} \mu \\ l \end{pmatrix}$ to be maximal, that is, $l_{i,j} = \mu_i, j = i, i+2, \ldots, n-1; i = 1, 2, \ldots, n-1$.

$$\begin{bmatrix} \lambda/\mu \\ m \end{bmatrix} \longleftrightarrow \left(\begin{bmatrix} \lambda/\mu \\ m \end{bmatrix} \begin{pmatrix} (0) \\ \mu \\ max \end{pmatrix} \right).$$

The following $D^{(\lambda\,0^n)}$−polynomials, which are labeled by a pair of Gelfand-Tsetlin pattern having partition $(\lambda, 0^n), \lambda \in \mathbb{P}ar_n$, are fully defined, as discussed in Section 2.2:

$$D \left(\begin{pmatrix} l' \\ \mu \end{pmatrix} \\ \begin{bmatrix} m' \\ \lambda/\mu \\ m \end{bmatrix} \langle 0 \rangle \\ \begin{pmatrix} \mu \\ l \end{pmatrix} \right) (Z_{2n}),$$

in which $\langle 0 \rangle$ denotes the double GT patterns for partition 0^n.

We now define the following polynomials, which are labeled by double skew GT patterns, as the special case of these polynomials corresponding to choosing the patterns $\binom{\mu}{l}$ and $\binom{l'}{\mu}$ to be maximal:

$$
D\begin{bmatrix} m' \\ \lambda/\mu \\ m \end{bmatrix}(Z) = D\left(\binom{max}{\mu} \middle| \begin{bmatrix} m' \\ \lambda/\mu \\ m \end{bmatrix} \middle| \langle 0 \rangle \quad \binom{\mu}{max} \right)\begin{pmatrix} I_n & 0 \\ 0 & Z \end{pmatrix},
$$

with weights of lower and upper patterns given by

$$
\gamma = (\mu, \alpha), \quad \alpha = \text{weight of } \begin{bmatrix} \lambda/\mu \\ m \end{bmatrix},
$$

$$
\gamma' = (\mu, \alpha'), \quad \alpha' = \text{weight of } \begin{bmatrix} \lambda/\mu \\ m' \end{bmatrix}.
$$

These polynomials then have the following properties, as expressed in terms of the matrices of dimension

$$
Dim D^{\lambda/\mu}(Z) = \sum_{\alpha \in W_{\lambda/\mu}} K(\lambda/\mu, \alpha),
$$

which are obtained by using the lower patterens $\begin{bmatrix} \lambda/\mu \\ m \end{bmatrix}$ to label rows and the upper patterns $\begin{bmatrix} m' \\ \lambda/\mu \end{bmatrix}$ to label columns:

(1) Multiplication of $D^{\lambda/\mu}(Z)$ matrices:

$$
D^{\lambda/\mu}(X)D^{\lambda/\mu}(Y) = D^{\lambda/\mu}(XY),
$$

for arbitrary matrices X and Y. This multiplication property is a direct consequence of the definition of these matrices, the multiplication property for ordinary $D^{\lambda}(Z)$ matrices, and the fact that the restriction to maximal labels propagates through products.

(2) Matrix skew Schur function:

$$
D^{\lambda/\mu}(Z) \sim \sum_{\nu} c^{\lambda}_{\mu\nu} D^{\nu}(Z).
$$

(3) Diagonal properties:

$$D^{\lambda/\mu}(diag(x_1, x_2, \ldots, x_n)) = \sum_{\alpha \in W_{\lambda/\mu}} \oplus x_1^{\alpha_1} x_2^{\alpha_2} \cdots x_n^{\alpha_n} I_{K(\lambda/\mu,\alpha)},$$

$I_{K(\lambda/\mu,\alpha)} = $ identity matrix of dimension $K(\lambda/\mu, \alpha)$,

$D^{\lambda/\mu}(I_n) = I_{Dim\lambda/\mu}.$

(4) Relation to skew Schur functions:

$$s_{\lambda/\mu}(x) = \sum_{\nu} c_{\mu\nu}^{\lambda} s_{\nu}(x),$$

$$Trace D^{\lambda/\mu}(diag(x_1, x_2, \ldots, x_n)) = s_{\lambda/\mu}(x).$$

(5) Importance for physics: The matrices $D^{\lambda/\mu}(Z)$ give a new class of reducible integer representations of $GL(n, \mathbf{C})$ for $Z \in GL(n, \mathbf{C})$, and, similarly, for $U(n)$ by choosing $Z \in U(n)$. We expect to uncover a wealth of other properties for various choices of the variables Z.

5. Words of Semistandard Skew Tableaux and Skew GT Patterns

The notion of a word of a semistandard skew tableau or a skew GT pattern gives an alternative enumeration of the latter objects that includes the concept of a lattice permutation. The enumeration of lattice permutations of a certain type gives the Littlewood-Richardson numbers $c_{\mu\nu}^{\lambda}$, which, in turn, are needed in the reduction of Kronecker products of irreducible representations of $U(n)$. In this section, we give some basic definitions and properties of words.

The standard form of a sequence of repeated $1's, \ldots, n's$: Let $\alpha = (\alpha_1, \alpha_2, \ldots, \alpha_n)$ be a sequence of non-negative integers with $|\alpha| = k$. The weakly increasing sequence defined by

$$(1, 2, \cdots, n)^{\alpha} = 1^{\alpha_1} 2^{\alpha_2} \cdots n^{\alpha_n}$$

is called the *standard form* or *type* of any sequence

$$A_k = a_1 a_2 \ldots a_k, \quad k = \alpha_1 + 2\alpha_2 + \cdots + n\alpha_n$$

containing α_1 $1's, \alpha_2$ $2's, \ldots, \alpha_n$ $n's$. We denote the set of all such sequences by $\mathbb{A}_k(\alpha)$.

5.1. *Word of a Semistandard Skew Tableau of Shape $\lambda - \mu$*

Row i of the semistandard skew tableau

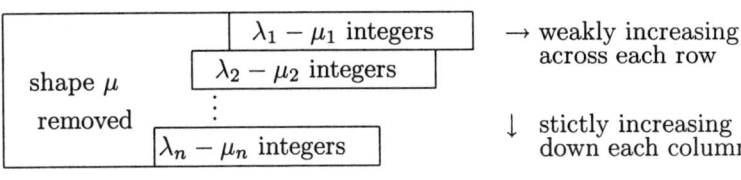

is filled out as follows:

$l_{i,1}$ 1's			$l_{i,2}$ 2's						$l_{i,n}$ n's		
1	$\cdots$	1	2	$\cdots$	2	$\cdots$	n	$\cdots$	n	$\lambda_i - \mu_i$

Then, by reverse (right-to-left) reading of row i we obtain the sequence

$$L^{(i)}_{\lambda/\mu} = n^{l_{i,n}} \cdots 2^{l_{i,2}} 1^{l_{i,1}}.$$

The *word* of this semistandard skew tableau of weight α is the sequence defined by

$$L_{\lambda/\mu}(\ell) = L^{(1)}_{\lambda/\mu} L^{(2)}_{\lambda/\mu} \cdots L^{(n)}_{\lambda/\mu},$$

where the weight α is given by

$$\alpha = (\alpha_1, \alpha_2, \ldots, \alpha_n), \ \alpha_j = l_{1,j} + l_{2,j} + \ldots + l_{n,j}, j = 1, 2, \ldots, n.$$

This word sequence is of type $(1, 2, \ldots, n)^\alpha$. The n^2 non-negative integers

$$\ell = (l_{ij})_{1 \leq i,j \leq n}$$

appearing in this word must, of course, fulfill the rules for a semistandard skew tableau.

5.2. *Word of a Skew Gelfand-Tsetlin Pattern of Shape* $\lambda - \mu$

Consider again the skew GT pattern

$$
\begin{bmatrix} \lambda/\mu \\ m \end{bmatrix} =
\begin{matrix}
\lambda_1 & \lambda_2 & \cdots & \lambda_n \\
& & \vdots & \\
m_{1,j-1} & m_{2,j-1} \cdots m_{n,j-1} & \\
& & \vdots & \\
m_{1,1} & m_{2,1} & \cdots & m_{n,1} \\
\mu_1 & \mu_2 & \cdots & \mu_n
\end{matrix} ,
$$

which has weight

$$\alpha = (|m_1| - |m_0|, |m_2| - |m_1|, \ldots, |m_n| - |m_{n-1}|).$$

The i—th left diagonal ($i = 1, 2, \ldots, n$) of this pattern is mapped to the sequence $L_{\lambda/\mu}^{(i)}$, that is,

$$
\begin{array}{l}
m_{i,n} \\
\quad \ddots \\
\qquad m_{i,1} \\
\qquad\quad m_{i,0}
\end{array}
\longrightarrow L_{\lambda/\mu}^{(i)} = n^{l_{i,n}} \cdots 2^{l_{i,2}} 1^{l_{i,1}},
$$

where the non-negative integers $l_{i,j} = m_{i,j} - m_{i,j-1}, j = 1, 2, \ldots, n$, are read as successive differences along this diagonal. This sequence of type $(1, 2, \ldots, n)^\alpha$ defines the *word* of this skew pattern:

$$
L \begin{bmatrix} \lambda/\mu \\ m \end{bmatrix} = L_{\lambda/\mu}^{(1)} L_{\lambda/\mu}^{(2)} \cdots L_{\lambda/\mu}^{(n)}.
$$

Since the bijection

$$
\mathbb{T}_{\lambda/\mu} \overset{BIJECTION}{\longleftrightarrow} \mathbb{G}_{\lambda/\mu}
$$

is exactly the one-to-one correspondence given by setting $l_{i,j} = m_{i,j} - m_{i,j-1}, j = 1, 2, \ldots, n$ in row i of the semistandard skew tableau, each $i = 1, 2, \ldots, n$, the set of words of $\mathbb{T}_{\lambda/\mu}$ and $\mathbb{G}_{\lambda/\mu}$ are all identical (by design):

$$
L_{\lambda/\mu}(\ell) = L \begin{bmatrix} \lambda/\mu \\ m \end{bmatrix}, \text{ for all patterns } m.
$$

Notations:

$\mathbb{L}_{\lambda/\mu}(\alpha) = $ set of all words of type $(1, 2, \ldots, n)^\alpha$ of semistandard skew tableaux or skew GT patterns of shape $\lambda - \mu$,

$\mathbb{L}_{\lambda/\mu} = $ set of all words of semistandard skew tableaux or skew GT patterns of shape $\lambda - \mu$.

6. Lattice Permutations and Littlewood-Richardson Numbers

The properties of the Littlewood-Richardson numbers $c_{\mu\nu}^{\lambda}$ are basic to the reduction of Kronecker products of D^λ–polynomials and to the reduction of skew $D^{\lambda/\mu}$–polynomials, which is one of the reasons we study them. Their relation to lattice permutations is one such fundamental property. First, we define a lattice permutation.

6.1. *Lattice Permutations*

A word

$$A_k = a_1 a_2 \ldots a_j \ldots a_k \in \mathbb{A}_k(\alpha)$$

of type $(1, 2, \ldots, n)^\alpha$ is a lattice permutation if and only if Rule L as follows is true:

Rule L : In each left factor $A_j = a_1 a_2 \ldots a_j, 1 \le j \le k$, of A_k the number of $i's$ is greater than or equal to the number of $i + 1's$

It follows from this rule that the sequence α is a **partition**, which we henceforth denote by ν.

From now on, we develop all results in the language of skew Gelfand-Tsetlin patterns , although all results could be rephrased in the language of semistandard skew tableau.

Notations:

Let $\lambda, \mu, \nu \in \mathbb{P}ar_n$ with $\mu, \nu \subseteq \lambda, |\lambda| = |\mu| + |\nu|$. Define the following sets:

$$\mathbb{G}_{\lambda/\mu}(\nu) = \left\{ \text{subset of } \mathbb{G}_{\lambda/\mu} \text{ of weight } \nu \right\},$$

$\mathbb{G}^\lambda_{\mu,\nu} = \{$subset of $\mathbb{G}_{\lambda/\mu}(\nu)$ having words that are lattice

permutations$\}$,

$\mathbb{L}_{\lambda/\mu}(\nu)=\{$set of words corresponding to all patterns in $\mathbb{G}_{\lambda/\mu}(\nu)\}$,

$\mathbb{L}^\lambda_{\mu,\nu} = \{$set of lattice permutations corresponding to all

patterns in $\mathbb{G}^\lambda_{\mu,\nu}\}$

6.2. *Littlewood-Richardson Numbers*

Let $\lambda, \mu, \nu \in \mathbb{P}ar_n$. The Littlewood-Richardson number $c^\lambda_{\mu\nu}$ are given by

$$c^\lambda_{\mu\nu} = \begin{cases} |\mathbb{L}^\lambda_{\mu,\nu}|, & \text{for } \mu, \nu \subseteq \lambda, \ |\lambda| = |\mu| + |\nu| \\ 0, \text{otherwise} \end{cases}$$

(See Littlewood and Richardson,[11] Macdonald,[12] Stanley.[7])

Thus, a basic problem is:

Find the set of words $\mathbb{L}^\lambda_{\mu,\nu}$.

We shall not solve this problem here, but we can find a subset of $\mathbb{G}_{\lambda/\mu}(\nu)$ that contains $\mathbb{L}^\lambda_{\mu,\nu}$, and this narrows considerably the problem. For this, we make the following two observations:

Observation 1: A skew GT pattern can be split by the minor diagonal of the parallelogram into two triangular patterns as follows:

$$\begin{bmatrix} \lambda/\mu \\ m \end{bmatrix} = \left[\binom{\lambda}{m'} \Big/ \binom{m''}{\mu} \right].$$

Examples for $n = 2, 3$:

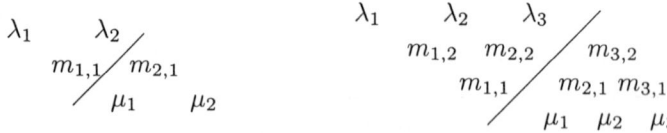

The pattern $\binom{\lambda}{m'}$ is a normal triangular GT pattern with n rows in which $m'_{i,j} = m_{i,j}, j = 1, 2, \ldots, n; i = 1, 2, \ldots, j$. The pattern $\binom{m''}{\lambda}$ is a normal inverted triangular GT pattern with n rows in which

$$m''_{i,j} = m_{n-j+i,n-j}, j = 1, 2, \ldots, n; i = 1, 2, \ldots, j.$$
$$\mu = (m''_{1,n}, m''_{2,n}, \ldots, m''_{n,n}).$$

Of course, the entries in these two triangular patterns are constrained by the betweenness relations for the full skew pattern.

Observation 2: For word of a skew GT pattern to be a lattice permutation, it is necessary that the pattern $\binom{\lambda}{m'}$ be maximal, that is,

$$\begin{matrix} \lambda_1 & \lambda_2 & \cdots & & \lambda_n \\ & \lambda_1 & \lambda_2 & \cdots & \lambda_{n-1} \\ & & \vdots & & \\ & & \lambda_1 & \lambda_2 & \\ & & & \lambda_1 & \end{matrix} \quad = \text{maximal pattern}$$

Proof. The word of a skew GT pattern is

$$L \begin{bmatrix} \lambda/\mu \\ m \end{bmatrix} = L_{\lambda/\mu} = L^{(1)}_{\lambda/\mu} L^{(2)}_{\lambda/\mu} \cdots L^{(n)}_{\lambda/\mu}.$$

In order that this word be a lattice permutation, the following conditions are necessary:

$L^{(1)}_{\lambda/\mu} = n^{l_{1,n}} \cdots 2^{l_{1,2}} 1^{l_{1,1}}$ can have no entries to the left of $1^{l_{1,1}}$; hence,

$$l_{1,n} = l_{1,n-1} = \cdots = l_{1,2} = 0;$$

$L^{(2)}_{\lambda/\mu} = n^{l_{2,n}} \cdots 2^{l_{2,2}} 1^{l_{2,1}}$ can have no entries to the left of $2^{l_{2,2}}$;
hence,

$$l_{2,n} = l_{2,n-1} = \cdots = l_{2,3} = 0;$$

$\vdots$

$L^{(i)}_{\lambda/\mu} = n^{l_{i,n}} \cdots 2^{l_{i,2}} 1^{l_{i,1}}$ can have no entries to the left of $i^{l_{i,i}}$;
hence,

$$l_{i,n} = l_{i,n-1} = \cdots = l_{i,i+1} = 0;$$

$\vdots$

$L^{(n)}_{\lambda/\mu} = n^{l_{n,n}} \cdots 2^{l_{n,2}} 1^{l_{n,1}}$ can have no entries to the left of $n^{l_{n,n}}$;
hence,

no condition.

This collection of conditions

$l_{i,j} = 0$, equivalently, $m'_{i,j-1} = m'_{i,j}, i = 1, 2, \cdots, n-1; j = i+1, i+2, \cdots, n,$

are just the conditions that the pattern $\binom{\lambda}{m'}$ be maximal, so that

$$\begin{bmatrix} \lambda/\mu \\ m \end{bmatrix} = \left[\binom{\lambda}{max} \Big/ \binom{m''}{\mu} \right].$$

The word of this special skew GT pattern is

$$L\left[\binom{\lambda}{max} \Big/ \binom{m''}{\mu} \right]$$

$$= 1^{l_{1,1}} \quad 2^{l_{2,2}} 1^{l_{2,1}} \quad 3^{l_{3,1}} 2^{l_{3,2}} 1^{l_{3,3}} \quad \cdots \quad n^{l_{n,n}} \cdots 2^{l_{n,2}} 1^{l_{n,1}}. \quad \square$$

This last result can be put in better form: The information is fully encoded in the inverted pattern $\binom{m''}{\mu}$ and the boundary right diagonal $(\lambda_1, \lambda_2, \cdots, \lambda_n)$ of the pattern $\binom{\lambda}{max}$. We rearrange as follows: The inverted pattern is put in normal non-inverted form, keeping the λ−diagonal, and removing the double prime from $m''_{i,j}$ to arrive at the *modified Gelfand-Tsetlin pattern* given by

$$\binom{\lambda/\mu}{m} = \begin{matrix} \emptyset & \mu_1 & \mu_2 & \mu_3 & \cdots & \mu_n \\ & \lambda_1 & m_{1,n-1} & m_{2,n-1} & \cdots & m_{n-1,n-1} \\ & & \lambda_2 & m_{1,n-2} & m_{2,n-2} \cdots m_{n-2,n-2} \\ & & & \vdots & \\ & & & \lambda_{n-1} & m_{1,1} \\ & & & & \lambda_n \end{matrix},$$

in which $\emptyset$ designates that no entry is placed in the indicated position. The word of this pattern is now given by

$$L\binom{\lambda/\mu}{m} = \underbrace{1^{r_{1,1}}}_{} \ \underbrace{2^{r_{2,2}} 1^{r_{2,1}}}_{} \ \cdots \ \underbrace{i^{r_{i,i}} \cdots 2^{r_{i,2}} 1^{r_{i,1}}}_{} \ \cdots \ \underbrace{n^{r_{n,n}} \cdots 2^{r_{n,2}} n^{r_{n,1}}}_{},$$

where the exponents $r_{i,j}, j = 1, 2, \ldots, i$, of the i-th partial word are now read off the i-th *right* diagonal:

$$
\begin{aligned}
&\mu_i = m_{i,n} \\
&\quad m_{i-1,n-1} \\
&\quad\ m_{i-2,n-2} \\
&\qquad \vdots \qquad\qquad \longrightarrow i^{r_{i,i}} \cdots 2^{r_{i,2}} 1^{r_{i,1}}, \\
&\quad\ m_{1,n-i+1} \\
&\lambda_i = m_{0,n-i}
\end{aligned}
$$

where the

$$r_{i,j} = m_{i-j,n-j} - m_{i-j+1,n-j+1}, \quad j = i, i-1, \ldots, 1,$$

are the successive difference between the entries along the diagonal.

The weight of a modified Gelfand-Tselin pattern is defined by

$$\alpha = W\binom{\lambda/\mu}{m} = (\alpha_1, \alpha_2, \ldots, \alpha_n), \alpha_i = r_{i,i} + r_{i+1,i} + \ldots + r_{n,i},$$

in which α_i denotes the number of $i's$ that appear in the word $L\binom{\lambda/\mu}{m}$. Thus, the word $L\binom{\lambda/\mu}{m}$ is of type $(1, 2, \ldots, n)^\alpha$.

Given $\mu, \lambda \in \mathbb{P}ar_n$ with $\mu \subseteq \lambda$, the entries in a modified Gelfand-Tsetlin patterns satisfy all the standard betweenness relations, and the presence of $\emptyset$ serves only to fill out the triangular array, and be a reminder that λ_1 may be as large as one pleases.

Notations:

$$M\mathbb{L}_{\lambda/\mu} = \left\{ \text{set of words } L\binom{\lambda/\mu}{m} \ \middle|\ m \text{ runs over all values} \right.$$

$$\left. \text{for which } \binom{\lambda/\mu}{m} \text{ is a modified } GT \text{ pattern} \right\},$$

$$M\mathbb{L}_{\lambda/\mu}(\nu) = \left\{ \text{subset of } M\mathbb{L}_{\lambda/\mu} \ \middle|\ W\binom{\lambda/\mu}{m} = \nu \supseteq \lambda \right\},$$

$$M\mathbb{L}^{\lambda}_{\mu,\nu} = \{ \text{subset of words in } M\mathbb{L}_{\lambda/\mu}(\nu) \text{ that are lattice}$$

$$\text{permutations.}\}$$

By design, a modified GT pattern captures all those words in $\mathbb{L}_{\lambda/\mu}(\nu)$ that can possibly be lattice permutations, that is, $M\mathbb{L}^{\lambda}_{\mu,\nu} = \mathbb{L}^{\lambda}_{\mu,\nu}$. Accordingly, we have that, for $\mu, \nu, \lambda \in \mathbb{P}ar_n$ with $\mu, \nu \subseteq \lambda$ and $|\lambda| = |\mu| + |\nu|$, then

$$c_{\mu\nu}^{\lambda} = |M\mathbb{L}_{\mu,\nu}^{\lambda}| = |\mathbb{L}_{\mu,\nu}^{\lambda}|.$$

Examples:

n=2:

$$\begin{array}{ccc} \emptyset & \mu_1 & \mu_2 \\ & \lambda_1 & m_{1,1} \\ & \lambda_2 & \end{array}$$

$$\text{word} = 1^{\lambda_1 - \mu_1} \, 2^{\lambda_2 - m_{1,1}} \, 1^{m_{1,1} - \mu_2};$$

word = lattice permutation if and only if $\lambda_1 - \mu_1 \geq \lambda_2 - m_{1,1}$.

Recall the familiar angular momentum addition rule as given by the Clebsch-Gordan series

$$(2j_1 \, 0) \otimes (2j_2 \, 0) = \sum_{j=j_1-j_2}^{j_1+j_2} \oplus (j_1 + j_2 + j, j_1 + j_2 - j), \quad \text{for } j_1 \geq j_2$$

for which $\mu = (2j_1 \, 0), \nu = (2j_2 \, 0), \lambda = (j_1 + j_2 + j, j_1 + j_2 - j)$. The unique modified GT pattern of weight $\nu = (2j_2 \, 0)$ is

$$\begin{array}{ccc} \emptyset & 2j_1 & 0 \\ j_1 + j_2 + j & j_1 + j_2 - j & \\ & j_1 + j_2 - j & \end{array} \mapsto \text{word} = 1^{j_2 - j_1 + j} \, 2^0 \, 1^{j_1 + j_2 - j} = 1^{2j_2}.$$

Thus, the values of the total angular momentum j are *exactly* those for which the GT pattern satisfies betweenness, each such j yielding the same lattice permutation of weight $(2j_2 \, 0)$. We conclude:

$$c_{(2j_1 \, 0),(2j_2 \, 0)}^{(j_1+j_2+j,j_1+j_2-j)} = \begin{cases} 1, \text{each } j = j_1 - j_2, \, j_1 - j_2 + 1, \ldots, j_1 + j_2; \\ 0, \text{otherwise} \end{cases}$$

n=3:

$$\begin{array}{cccc} \emptyset & \mu_1 & \mu_2 & \mu_3 \\ & \lambda_1 & m_{1,2} & m_{2,2} \\ & & \lambda_2 & m_{1,1} \\ & & \lambda_3 & \end{array}$$

$$\text{word} = 1^{\lambda_1 - \mu_1} 2^{\lambda_2 - m_{1,2}} 1^{m_{1,2} - \mu_2} 3^{\lambda_3 - m_{1,1}} 2^{m_{1,1} - m_{2,2}} 1^{m_{2,2} - \mu_3}.$$

$$\text{weight} = \nu = (\lambda_1 - \mu_1 + m_{1,2} - \mu_2 + m_{2,2} - \mu_3,$$

$$\lambda_2 - m_{1,2} + m_{1,1} - m_{2,2}, \lambda_3 - m_{1,1}).$$

Numerical example 1: $\mu = (4,2,0), \nu = (5,4,1), \lambda = (7,6,3)$. There are three modified GT patterns of type $(1,2,3)^{(5,4,1)} = 1^5 2^4 3^1$:

$$
\begin{array}{cccc}
\emptyset & 4 & 2 & 0 \\
& 7 & 2 & 2 \\
& & 6 & 2 \\
& & & 3
\end{array}
\qquad \mapsto \qquad 1^3 2^4 3^1 1^2 \quad \text{(non lattice)}
$$

$$
\begin{array}{cccc}
\emptyset & 4 & 2 & 0 \\
& 7 & 3 & 1 \\
& & 6 & 2 \\
& & & 3
\end{array}
\qquad \mapsto \qquad 1^3 2^3 1^1 3^1 2^1 1^1 \quad \text{(lattice)}
$$

$$
\begin{array}{cccc}
\emptyset & 4 & 2 & 0 \\
& 7 & 4 & 0 \\
& & 6 & 2 \\
& & & 3
\end{array}
\qquad \mapsto \qquad 1^3 2^2 1^2 3^1 2^2 \quad \text{(lattice)}
$$

Therefore: $c_{(420),(541)}^{(763)} = 2$.

Numerical example 2: $\mu = (2,2,2), \nu = (5,4,1), \lambda = (7,6,3)$. There is one modified GT patterns of type $(1,2,3)^{(5,4,1)} = 1^5 2^4 3^1$:

$$
\begin{array}{cccc}
\emptyset & 2 & 2 & 2 \\
& 7 & 2 & 2 \\
& & 6 & 2 \\
& & & 3
\end{array}
\qquad \mapsto \qquad 1^5 2^4 3^1 \quad \text{(lattice)}
$$

Therefore: $c_{(222),(541)}^{(763)} = 1$.

It is tractable to carry the above forward and derive necessary and sufficient conditions that a pattern in the set of modified GT patterns corresponds to a lattice permutation of given weight. This is important for the theory of tensor operators in $U(n)$.

The principal result is: Necessary and sufficient conditions that the word $L\binom{\lambda/\mu}{m}$ be a lattice permutation are that the exponents $r_{i,j}$ satisfy the following relations

$$r_{i,i} + r_{i+1,i} + \ldots + r_{k,i} \geq r_{i+1,i+1} + r_{i+2,i+1} + \ldots + r_{k+1,i+1},$$

where for each fixed $i = 1, 2, \ldots, n - 1$, the index k has values $k = i, i+1, \ldots, n-1$. In terms of the conditions on the entries $m_{i,j}$ specifying the word $L\binom{\lambda/\mu}{m}$, the necessary and sufficient conditions for a lattice permutation may be written as

$$\lambda_i - \lambda_{i+1} \geq \sum_{l=1}^{h+1}(m_{l,n-i+1} - m_{l,n-i}) - \sum_{l=1}^{h}(m_{l,n-i} - m_{l,n-i-1}), \ (m_{l,n} = \mu_l), *$$

where for each fixed $i = 1, 2, \ldots, n - 1$, the index h has values $h = 0, 1, \ldots, n - i - 1$, where the second summation term is 0 for $h = 0$. Thus, the set $\mathbb{L}_{\mu,\nu}^{\lambda}$ if given by

$$\mathbb{L}_{\mu,\nu}^{\lambda} = \left\{ L \binom{\lambda/\mu}{m} \in M\mathbb{L}_{\lambda/\mu} \middle| \text{ conditions } * \text{ are satisfied} \right\}.$$

The D−polynomials corresponding to modified GT patterns can also be given, since they occur as specializations of the general polynomials discussed in Section 2:

$$D \begin{pmatrix} l' \\ \lambda/\mu \\ l \end{pmatrix} (Z_{2n}) = D \begin{pmatrix} \binom{l'}{\mu} \\ \begin{bmatrix} max \\ \lambda/\mu \\ max \end{bmatrix} \langle 0 \rangle \\ \binom{\mu}{l} \end{pmatrix} (Z_{2n}) = D \begin{pmatrix} l' \\ \mu \\ \lambda \quad 0 \\ \mu \\ l \end{pmatrix} (Z_{n+1}).$$

The dependence of these polynomials on only those variables occurring in the $(n + 1) \times (n + 1)$ submatrix Z_{n+1} of the $2n \times 2n$ matrix Z_{2n} is a consequence of the choice of maximal labels im the skew pattern $\begin{bmatrix} max \\ \lambda/\mu \\ max \end{bmatrix}$,

which gives the weights of the lower and upper patterns as

$$\gamma = (\alpha, 0^n), \quad \gamma' = (\alpha', 0^n),$$

$$\alpha = W \binom{\mu}{l}, \quad \alpha' = W \binom{\mu}{l'}.$$

We have not yet worked out the properties of these polynomials labeled by pairs of modified Gelfand-Tsetlin patterns.

Acknowledgements

Work performed under the auspices of The U. S. Department of Energy, contract W-7405-ENG-36. We thank the organizers for the opportunity for discussions with conference members, for the presentation of the ideas in this article, and for their timely publication by World Scientific.

References

1. L.C. Biedenharn and J. D. Louck, *Angular Momentum in Quantum Physics; The Racah-Wigner Algebra in Quantum Theory*, in: *Encycl. of Mathematics and Its Applications*, ed., G.-C. Rota, Vols. 8 and 9 (Cambridge Univ. Press, Cambridge, 1981).
2. J.D. Louck and L.C. Biedenharn, *Adv. Quant. Chem.* **23**, 127 (1992).
3. W.Y.C. Chen and J. D. Louck, *Adv. Math.* **140**, 207 (1998).
4. J.D. Louck, *New Perspectives on the Unitary Group and its Tensor Operators*, in: *Symmetry and Structural Properties of Condensed Matter*, eds., T. Lulek, B. Lulek, and A. Wal; Proc. sixth SSCPM (World Scientific, Singapore, 2001) pp. 23-36.
5. H. Weyl, *The Theory of Groups and Quantum Mechanics*, Dover, New York, 1949.
6. I.M. Gelfand and M. L. Tsetlin, *Dokl. Akad. Nauk SSSR* **71**, 825 (1950); reproduced in I.M. Gelfand, R.A. Minlos, and Z. Ya. Shapiro, *Representations of the Rotation and Lorentz Groups and their Applications* (Pergamon, New York, 1963).
7. R.P. Stanley, *Enumerative Combinatorics*, Vol. II, Cambridge University Press, United Kingdom, 1999.
8. M.A. Mendez, *Ann. Comb.* **5**, 459 (2001).
9. J.D. Louck and L. C. Biedenharn, *Commun. Math. Phys.* **8**, 89 (1968).
10. J.D. Louck and L. C. Biedenharn, *Adv. Appl. Math. Suppl. Issue* **10** , 239 (1981).
11. D.E. Littlewood and A. R. Richardson, *Phil. Trans. A* **233**, 99 (1934).
12. I.G. Macdonald, *Symmetric Functions and Hall Polynomials*, Oxford University Press, London/New York, 1979.

TOWARDS A COMBINATORIAL DESCRIPTION OF THE MATRICES CORRESPONDING TO IRREDUCIBLE REPRESENTATIONS OF THE UNITARY AND GENERAL LINEAR GROUPS

MIGUEL MÉNDEZ

IVIC,
Lab. Anal. Mat.-Matemáticas, Carretera Panamericana, Km.11
Caracas, Venezuela
and
UCV,
Dpto. de Matemáticas, Facultad de Ciencias
Caracas, Venezuela
E-mail: mmendez@cauchy.ivic.ve

We present a combinatorial model for the matrices corresponding to polynomial representations of the unitary and general linear groups. Our model relies on the concept of eulerian bijections over directed graphs. Using this approach we obtain combinatorial proofs of the multiplication property for the totally symmetric and totally anti-symmetric representations.

1. Introduction

This article is motivated by many problems related to the combinatorics of the matrices $D^\lambda(\mathbf{X})$, λ a partition of k, and $\mathbf{X} = (x_{i,j})_{i,j=1}^n$, that give the irreducible representations of the unitary groups (see for example Chen and Louck[3], and Louck[8]). It represents the very first step in a program that aims to the understanding of the combinatorics behind those matrices, their recursive formulas, and the pattern calculus of Biedenharn and Louck.[2] Our paradigm is that the most natural combinatorial analog of a matrix is a directed graph. Using this paradigm we have found a very simple combinatorial rule to multiply the matrix coefficients of any matrix of polynomials $R(\mathbf{X})$ satisfying the multiplication property $R(\mathbf{XY}) = R(\mathbf{X}).R(\mathbf{Y})$. This combinatorial rule leads to a simple recipe to multiply basic elements of the Schur algebra.[4] We give combinatorial proofs for the multiplication property of the matrices $D^\lambda(\mathbf{X})$ corresponding to the partitions $\lambda = (k)$, and $\lambda = (1^k)$.

2. Directed Graphs

Definition 2.1. Let V be a finite set. A digraph(directed graph) G over the set of vertices V consists of a set of arrows A, and two functions $t, h :$ $A \rightarrow V$ that assign to each arrow $a \in A$ its initial(tail) and final(head) endpoints, $t(a)$ and $h(a)$ respectively. A weighted digraph G^w is a digraph where each arrow $a \in A$ has a weight $w(a) \in \mathbb{C}$. Observe that an ordinary digraph might be considered as weighted with the trivial weight $w(a) = 1$ for every arrow $a \in A$.

Let $A_{i,j}$ be the set of arrows going from the vertex i to vertex j, $i, j \in V$

$$A_{i,j} = t^{-1}(i) \cap h^{-1}(j)$$

in a weighted digraph G^w. Let $a_{i,j}$ be the sum of the weights of the arcs in $A_{i,j}$. The matrix $(a_{i,j})_{i,j \in V}$ is called the adjacency matrix of G^w and denoted by $|G^w|$.

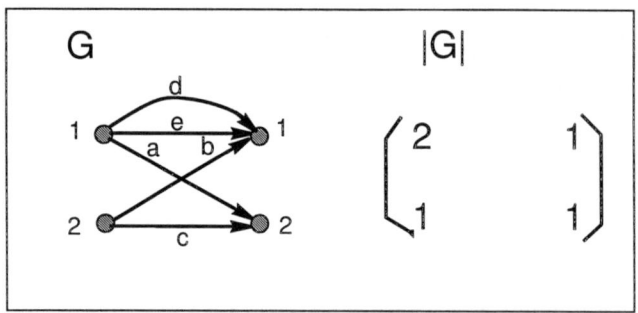

Figure 1.   Directed Graph with respective adjacency matrix $|G|$.

The digraphs are the combinatorial counterpart of matrices. If we have a matrix $(a_{i,j})_{i,j \in V}$ with nonnegative integer entries, we may identify it with the ordinary digraph with exactly $a_{i,j}$ arrows from vertex i to vertex j. In the more general case where the entries of $(a_{i,j})_{i,j \in V}$ are complex numbers, we can identify it with with the weighted digraph that has one arrow from i to j with weight $a_{i,j}$, for each entry $a_{i,j} \neq 0$. The operations of the algebra of matrices can also be represented as combinatorial operations with digraphs.

Definition 2.2. Let $G_1^{w_1}$ and $G_2^{w_2}$ be two weighted digraphs over the same set of vertices. The sum $G_1^{w_1} + G_2^{w_2}$ is defined as the weighted digraph whose set of arrows(with respective weights) is the disjoint union of the set of arrows of $G_1^{w_1}$ with that of $G_2^{w_2}$. The product $G_1^{w_1}.G_2^{w_2}$ is the digraph whose arcs are the paths of length two, first arrow in $G_1^{w_1}$ and second arrow in $G_2^{w_2}$. The weight of each 2-path of $G_1^{w_1}.G_2^{w_2}$ is the product of the respective weights of the two arrows forming the path.

It is easy to verify the following proposition.

Proposition 2.1. *The sum and product of weighted digraph are preserved to the corresponding matrix operations by taking adjacency matrices:*

$$|G_1^w + G_2^w| = |G_1^w| + |G_2^w| \tag{1}$$

$$|G_1^{w_1}.G_2^{w_2}| = |G_1^{w_1}|.|G_2^{w_2}| \tag{2}$$

Definition 2.3. Let G be a digraph over the set of vertices $[n] := \{1, 2, 3, \ldots, n\}$. The outdegree of a vertex $i \in [n]$ is the number of arrows of G leaving i. The indegree of i is the number of arrows in G pointing to i. Formally $\text{od}(i) = |t^{-1}(i)|$, and $\text{id}(i) = |h^{-1}(i)|$. The outdegree and indegree sequences of G are defined by

$$\text{od}(G) = (|t^{-1}(1)|, |t^{-1}(2)|, \ldots, |t^{-1}(n)|) \tag{3}$$

$$\text{id}(G) = (|h^{-1}(1)|, |h^{-1}(2)|, \ldots, |h^{-1}(n)|). \tag{4}$$

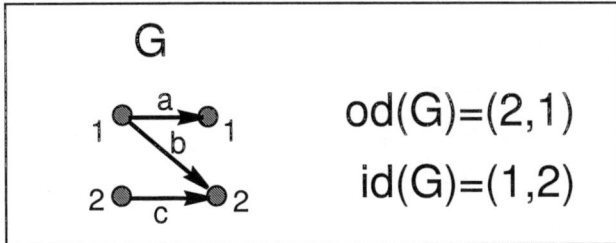

Figure 2. Outdegree and indegree sequences of a digraph G.

Let $\mathbf{X} = (x_{i,j})_{i,j=1}^n$ be a matrix of variables and $\mathbf{A} = (a_{i,j})_{i,j=1}^n$ a matrix of non-negative integer entries. We use the notation

$$\mathbf{X}^{\mathbf{A}} = \prod_{i,j=1}^n x_{i,j}^{a_{i,j}}$$

and

$$\mathbf{A}! = \prod_{i,j=1}^{n} a_{i,j}!$$

Let $R(\mathbf{X})$ be a $V \times V$ matrix of polynomials depending on the matrix of variables $\mathbf{X}$. We can expand it as a Taylor series,

$$R(\mathbf{X}) = \sum_{\mathbf{A}} R_{\mathbf{A}} \frac{\mathbf{X}^{\mathbf{A}}}{\mathbf{A}!}$$

where for each $\mathbf{A}$, $R_{\mathbf{A}}$ is a $V \times V$ complex matrix coefficient. $R(\mathbf{X})$ defines a correspondence

$$\mathbf{A} \mapsto R_{\mathbf{A}} \tag{5}$$

that assigns to the $n \times n$ integer matrix $\mathbf{A}$, the complex matrix $R_{\mathbf{A}}$. Recall that we can identify integer matrices with digraphs and complex matrices with weighted digraphs. Using this identification, the correspondence (5) can be thought of as a rule assigning to a directed graph G over the set of vertices $[n]$, a weighted digraph R_G over the set of vertices V. Formally, the rule

$$R : G \mapsto R_G$$

is a functor from the category of digraphs to the category of weighted digraphs. In other words, a special kind of a multisort species in the sense of Joyal.[5] See also Bergeron-Labelle-Leroux, [1] and Méndez.[7]

3. Product Rule

3.1. *Eulerian bijections*

Definition 3.1. Let $G_1 = (A_1, t_1, h_1)$ and $G_2 = (A_2, t_2, h_2)$ be two digraphs over the set of vertices $[n]$. An *eulerian bijection from G_1 to G_2 is a bijection $f : A_1 \to A_2$ such that the tail of $f(a)$ in G_2 is equal to the head of a in G_1, for every $a \in A_1$.*

For every $i \in [n]$, an eulerian bijection f sends the elements of $h_1^{-1}(i)$ into the elements of $t_2^{-1}(i)$. For this reason, the set of eulerian bijections $\text{Eul}[G_1, G_2]$ is empty if $\text{id}(G_1) \neq \text{od}(G_2)$. Assuming that $\text{id}(G_1) = \text{od}(G_2) = \beta = (\beta_1, \beta_2, \ldots, \beta_n)$, the number of elements of $\text{Eul}[G_1, G_2]$ is equal to $\beta! = \prod_{i=1}^{n} \beta_i!$.

Definition 3.2. Let G_1 and G_2 be as in the previous definition. Let $f \in \text{Eul}[G_1, G_2]$. The composition $G_2 \circ_f G_1$ of the two digraphs according to f is the digraph whose arrows are the paths of length two of the form $(a, f(a))$, $a \in A_1$. The tail of the arrow $(a, f(a))$ is the tail of a in G_1, the head of $(a, f(a))$ is the head of $f(a)$ in G_2.

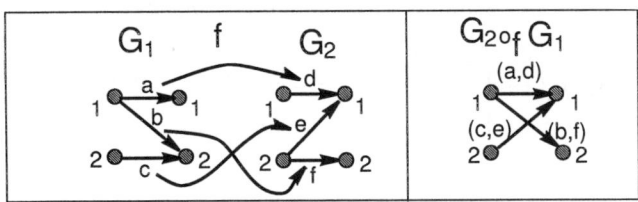

Figure 3.   Eulerian bijection f and respective composition digraph.

Remark 3.1. Observe that $\mathrm{od}(G_2 \circ_f G_1) = \mathrm{od}(G_1)$ and $\mathrm{id}(G_2 \circ_f G_1) = \mathrm{id}(G_2)$.

See Méndez[6] for a proof of the following theorem.

Theorem 3.1. *Let* $R(\mathbf{X})$ *be a matrix polynomial. The multiplication property*

$$R(\mathbf{X}.\mathbf{Y}) = R(\mathbf{X}).R(\mathbf{Y})$$

is equivalent to the following combinatorial rule to multiply the matrix coefficients of $R(X)$,

$$R_{G_1}.R_{G_2} = \sum_{f \in \mathrm{Eul}[G_1, G_2]} R_{G_2 \circ_f G_1} \tag{6}$$

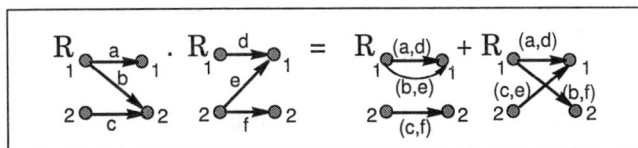

Figure 4.   Product of two coefficients of a multiplicative matrix polynomial $R(\mathbf{X})$.

From the previous theorem we immediately obtain the following,

Corollary 3.1. *Let* $\mathbf{A}$, $\mathbf{B}$ *and* $\mathbf{C}$ *be integer matrices. Introducing the co-efficient*

$$\begin{bmatrix} \mathbf{C} \\ \mathbf{A}, \mathbf{B} \end{bmatrix} = |\{f \in \mathrm{Eul}[G_1, G_2] : |G_2 \circ_f G_1| = \mathbf{C}\}|,$$

G_1 *and* G_2 *being the digraphs satisfying* $|G_1| = \mathbf{A}$, $|G_2| = \mathbf{B}$, *we have*

$$R_{\mathbf{A}}.R_{\mathbf{B}} = \begin{cases} \sum_{\mathbf{C}} \begin{bmatrix} \mathbf{C} \\ \mathbf{A}, \mathbf{B} \end{bmatrix} R_{\mathbf{C}} & \text{if } \mathrm{col}(\mathbf{A}) = \mathrm{row}(\mathbf{B}) \\ 0 & \text{otherwise} \end{cases} \tag{7}$$

4. $D^{(k)}(\mathbf{X})$, and $D^{(1^k)}(\mathbf{X})$

We now study the totally symmetric representation, corresponding to the partition with only one component $\lambda = (k)$. The rows and columns of $D^{(k)}(\mathbf{X})$ are indexed by n-sequences of non negative integers adding up to k. The entry α, β of $D^{(k)}(\mathbf{X})$ is the polynomial

$$D^{(k)}_{\alpha,\beta} = (\mathbf{X}) \sqrt{\alpha! \beta!} L_{\alpha,\beta}(\mathbf{X})$$

where

$$L_{\alpha,\beta}(\mathbf{X}) = \sum_{\mathrm{row}(\mathbf{A})=\alpha, \mathrm{col}(\mathbf{A})=\beta} \frac{\mathbf{X}^{\mathbf{A}}}{\mathbf{A}!}.$$

Expanding $D^{(k)}(\mathbf{X})$ as a Taylor series, we get

$$D^{(k)}(\mathbf{X}) = \sum_{|\alpha|=|\beta|=k} \mathbf{E}_{\alpha,\beta} \sum_{\mathrm{row}(\mathbf{A})=\alpha, col(\mathbf{A})=\beta} \frac{\mathbf{X}^{\mathbf{A}}}{\mathbf{A}!} \tag{8}$$

where $\mathbf{E}_{\alpha,\beta}$ is the matrix with $\sqrt{\alpha! \beta!}$ in the α, β-entry, and zero in the rest of the entries. It defines the correspondence

$$\mathbf{A} \mapsto \mathbf{E}_{\alpha,\beta},$$

where $\alpha = \mathrm{row}(A)$ and $\beta = \mathrm{col}(A)$. In terms of digraphs,

$$D^{(k)} : G \mapsto (\alpha \overset{\sqrt{\alpha! \beta!}}{\to} \beta),$$

where $\alpha = \mathrm{od}(G)$, and $\beta = \mathrm{id}(G)$. In words, $D^{(k)}$ maps a digraph G to the singleton weighted digraph with only one arrow from $\alpha = \mathrm{od}(G)$ to $\beta = \mathrm{id}(G)$, and weight $\sqrt{\alpha! \beta!}$.

Now we give a combinatorial proof of the multiplication property of $D^{(k)}(\mathbf{X})$ using theorem (3.1). The first combinatorial proof of this fact was given by Chen and Louck.[3]

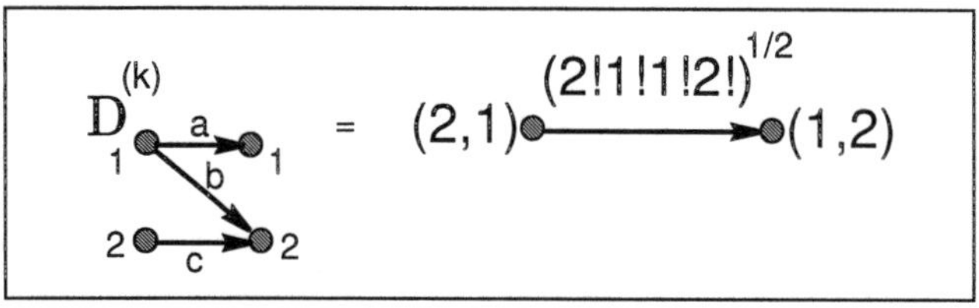

Figure 5. An example of the digraph correspondence given by the matrix polynomial $D^{(3)}(\mathbf{X})$.

Theorem 4.1. *The matrix polynomial $D^{(k)}(\mathbf{X})$ satisfies the multiplication property*

$$D^{(k)}(\mathbf{XY}) = D^{(k)}(\mathbf{X})D^{(k)}(\mathbf{Y}) \qquad (9)$$

Proof. From theorem(3.1) all we have to prove is that

$$D^{(k)}_{G_1} \cdot D^{(k)}_{G_2} = \sum_{f \in \mathrm{Eul}[G_1, G_2]} D^{(k)}_{G_2 \circ_f G_1}. \qquad (10)$$

Let $\alpha = \mathrm{od}(G_1)$, $\gamma = \mathrm{id}(G_1)$, $\delta = \mathrm{od}(G_2)$, and $\beta = \mathrm{id}(G_2)$. Assume that $\gamma \neq \delta$. In that case the left hand side is zero, by the definition of product of weighted digraphs. Since $|\mathrm{Eul}[G_1, G_2]| = 0$, the sum of the right hand side is also zero. In the case where $\gamma = \delta$, by remark (3.1) the right hand side of (10) is equal to

$$\sum_{f \in \mathrm{Eul}[G_1, G_2]} \alpha \xrightarrow{\sqrt{\alpha! \beta!}} \beta = \alpha \xrightarrow{\gamma! \sqrt{\alpha! \beta!}} \beta$$

The last identity because $|\mathrm{Eul}[G_1, G_2]| = \gamma!$. The left hand side is equal to the product of weighted digraphs

$$\alpha \xrightarrow{\sqrt{\alpha! \gamma!}} \gamma \cdot \gamma \xrightarrow{\sqrt{\gamma! \beta!}} \beta = \alpha \xrightarrow{\gamma! \sqrt{\alpha! \beta!}} \beta \qquad \square$$

We consider now the skew symmetric representation corresponding to the partition $\lambda = (1^k) = \overbrace{(1, 1, \ldots, 1)}^{k}$. The rows and columns of $D^{(1^k)}(\mathbf{X})$ are indexed by the k-subsets of $[n]$. Explicitly, for two k-subsets I, J, the I, J entry of $D^{(1^k)}(\mathbf{X})$ is the minor $\det_{I,J}(\mathbf{X})$ of the matrix $\mathbf{X}$. Expanding this minor we obtain

$$\det_{I,J}(\mathbf{X}) = \sum_{\sigma: I \to J} \mathrm{sig}(\sigma) \prod_{i=1}^{n} x_{i, \sigma i} = \sum_{\sigma: I \to J} \mathrm{sign}(\sigma) \frac{\mathbf{X}^{\mathbf{P}_\sigma}}{\mathbf{P}_\sigma!}$$

where σ ranges over all the bijections from I to J, and $\mathbf{P}_\sigma$ is the matrix of zeros and ones corresponding to σ. In this way we obtain the expansion of the matrix polynomial

$$D^{(1^k)}(\mathbf{X}) = \sum_{|I|=|J|=k} \sum_{\sigma: I \to J} \mathbf{F}_{I,J}^\sigma \frac{\mathbf{X}^{\mathbf{P}_\sigma}}{\mathbf{P}_\sigma!}$$

where $\mathbf{F}_{I,J}^\sigma$ is the matrix with coefficient $\mathrm{sign}(\sigma)$ in its I, J-entry, and zero in the rest. Identifying $\mathbf{F}_{I,J}^\sigma$ with the corresponding weighted digraph, we obtain the rule

$$D_G^{(1^k)} = \begin{cases} I \xrightarrow{\mathrm{sign}(G)} J & \text{if } G \text{ is a bijective digraph} \\ \emptyset & \text{otherwise} \end{cases} \tag{11}$$

where $I = domain(G)$, and $J = range(G)$. We are now ready to give a

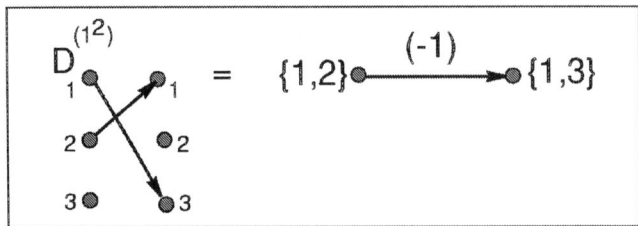

Figure 6. An example of the digraph correspondence defined by the matrix polynomial $D^{(1^2)}(\mathbf{X})$.

combinatorial proof of the following theorem.

Theorem 4.2. *(Binnet-Cauchy theorem) The matrix polynomial $D^{(k)}(\mathbf{X})$ satisfies the multiplication property*

$$D^{(1^k)}(\mathbf{XY}) = D^{(1^k)}(\mathbf{X}) D^{(1^k)}(\mathbf{Y}) \tag{12}$$

The proof of this theorem is analogous to that of (4.1) and is left to the reader.

References

1. F. Bergeron, G. Labelle, P. Leroux, Combinatorial Species and Tree-like Structures, Encyclopedia of Mathematics and its Applications,**67**, (Cambridge University Press, 1998).
2. L. C. Biedenharn and J. D. Louck, Angular Momentum in Quantum Physics, Encycl. of Mathematics and Its Applications, (G. C. Rota ed.), vol. 8. Addison-Wesley Publishing co., Reading, MA, 1981.

3. W. Y. C. Chen and J. D. Louck, The combinatorics of a class of representation functions, *Adv. Math.* **140** (1998), 207-236.
4. J. A. Green, Polynomial Representations of GL_n, Lecture Notes in Mathematics, **830** (Springer-Verlag, Berlin, 1980).
5. A. Joyal, Une théorie combinatoire des séries formelles, *Adv. Math.* **42** (1981), 1-82.
6. M. Méndez, Directed graphs and the combinatorics of the polynomial representations of $GL_n(\mathbb{C})$, *Annals of Combinatorics*, **5** (2001), 459-478.
7. M. Méndez, Species on digraphs, *Adv. in Math.* **123** (1996), 243-275.
8. J. D. Louck, MacMahon's master theorem, double tableau polynomials, and representations of groups, *Adv. in Appl. Math.* **17** (1996), 143-168.

NEW EXCITATION MODES IN A FINITE SIZE FERROMAGNET

S. COJOCARU* AND A. CEULEMANS

Department of Chemistry, University of Leuven, Celestijnenlaan 200F,
B-3001 Leuven, Belgium
* Institute of Applied Physics, Chişinău, Moldova.

We consider the bound magnon states in the isotropic Heisenberg $S = 1/2$ fer-romagnet by means of a refined continuum approach. The approach is based on an explicit incorporation of the symmetries of the discrete lattice into its contin-uum representation. New excitation modes are found for the finite square lattice with periodic boundary conditions. In contrast to the known ones these modes are strongly size dependent. The new modes are "genealogically" related to the antisymmetric two-magnon mode found by Bethe for the finite spin chain. In this way we establish a relation of the new modes in $2D$ with the non-string solutions in $1D$.

1. Introduction

We consider the spin excitations of the finite size two-dimensional spin-$1/2$ Heisenberg ferromagnet with nearest neighbour interaction $J_{ij} = J$ and periodic boundary conditions within a refined continuum approach. It is often claimed that for the thermodynamic limit the solutions of this model are known exactly for a lattice of any dimensionality. This opinion is based on extensive studies carried out within the usual continuum approach (see, e.g. ref. [1]). However some deficiencies of the latter have been pointed out in our earlier paper [2], e.g., the existence of certain discrepancies with the exact solution found by Bethe for a spin chain.[3] It was shown that our refined continuum treatment is able to remove these discrepancies and, in addition, allows to consider the finite size corrections to the thermody-namic limit. For instance, it captures the so called non-string behaviour of bound state (soliton) solutions characteristic of a finite chain at small values of the total momentum of the multimagnon excitation. Unlike the usual continuum treatment, it predicts the right number of the bound ex-citations and describes the correct behaviour of their wave functions. For a two dimensional finite system the exact solution is not available and the interpretation of numerical simulations is a non-trivial problem in itself.[4]

This motivates the extension of our treatment to the $2D$ case. We consider mainly the two-magnon excitation sector for a $N \times N$ square lattice because it contains the main features of the general interaction problem. Since the paper by Wortis [5] there have been known two bound state solutions of s- and d-wave symmetry. The latter solution is confined to the corners of the Brillouin zone, while the former is believed to exist for any value of the total momentum. We show below that the correct picture is much reacher and illustrate the behaviour of several new modes.

2. Bound states on a square lattice

The amplitude of the two spin deviations from the ferromagnetic ground state on the sites $\mathbf{n}_1$ and $\mathbf{n}_2$ of the lattice can be factorized as follows

$$\Psi(\mathbf{n}_1, \mathbf{n}_2) = \exp(i\mathbf{P} \cdot \mathbf{R})\, a(\mathbf{P}\,|\,\mathbf{r}),$$

where $\mathbf{P}$ is the total momentum and $\mathbf{R}$ and $\mathbf{r}$ the "center of mass" and the relative distance respectively. The amplitude of the relative motion is expanded into the finite Fourier series

$$a(\mathbf{P}\,|\,\mathbf{r}) = \frac{1}{N^2} \sum_{\mathbf{Q}} b(\mathbf{P}\,|\,\mathbf{Q}) \exp\left(i\mathbf{r} \cdot \mathbf{Q}\right). \tag{1}$$

From the boundary conditions we find the quantization of the total momentum ($P_{x,y} = 2\pi l_{x,y}/N$) and the constraints on the discrete quantities $\mathbf{Q}$. These constraints determine the allowed values of each component of the vector $\mathbf{Q}$ depending on the parity of the quantum numbers l_x and l_y in analogy with the $1D$ case [2]:

$$Q^s_{x,y} = \frac{2\pi m_{x,y}}{N}\,,\ Q^a_{x,y} = \frac{2\pi m_{x,y}}{N} + \frac{\pi}{N}\,;\ m_{x,y} = 0, 1, ..., N - 1. \tag{2}$$

The symmetry group of the lattice implies that the excitation modes have to be classified according to the irreducible representations of the group (C_{4v}). This can be achieved by using the standard projection operator technique. However after such a projection only those modes are considered which comply with the boundary conditions as described above. For instance, after the projection onto the A_2 irreducible representation the relative amplitude satisfies the following relations

$$a_{A_2}\left(P_x, P_y | X, Y\right) = -a_{A_2}\left(P_x, P_y | X, N - Y\right) = -a_{A_2}\left(P_x, P_y | N - X, Y\right)$$

$$= -a_{A_2}\left(P_y, P_x | N - X, N - Y\right) = a_{A_2}\left(P_y, P_x | Y, N - X\right). \tag{3}$$

The other irreducible representations are considered in a similar way. The above relations together with the boundary conditions imposed on the amplitudes (1) determine the quantum numbers and the symmetry properties of the respective Fourier amplitudes:

$$b_{A_1}\left(P_x, P_y | Q_x, Q_y\right) = b_{A_1}\left(P_y, P_x | Q_y, Q_x\right); \quad \{l_x, l_y\} = even;$$

$$b_{B_2}\left(P_x, P_y | Q_x, Q_y\right) = b_{B_2}\left(P_y, P_x | Q_y, Q_x\right); \quad \{l_x, l_y\} = odd;$$

$$b_{B_1}\left(P_x, P_y | Q_x, Q_y\right) = -b_{B_1}\left(P_y, P_x | Q_y, Q_x\right); \quad \{l_x, l_y\} = even;$$

$$b_{A_2}\left(P_x, P_y | Q_x, Q_y\right) = -b_{A_2}\left(P_y, P_x | Q_y, Q_x\right); \quad \{l_x, l_y\} = odd. \quad (4)$$

The remaining values of the total momentum correspond to the two components of the $E-$mode: $a_E^y\left(P_y, P_x | X, Y\right)$ has $P_x = even, P_y = odd$ and $a_E^x\left(P_y, P_x | X, Y\right)$ has $P_y = even, P_x = odd$. The continuum limit is then considered by taking the projected quantities which automatically satisfy the symmetry requirements similar to (3) and replacing the sums by integrals with the cut-off determined by the respective $N-$dependent shifts from (2). E.g., for the A_2 mode one takes

$$\frac{1}{N^2} \sum_{Q_{x, y}^a} \to \frac{1}{\pi^2} \int_{\pi/N}^{\pi+\pi/N} \int_{\pi/N}^{\pi+\pi/N} dQ_x dQ_y.$$

Then the energy of each mode is obtained by solving the Schrödinger equation (see, e.g., ref.[1]) for the irreducible amplitudes. It is easier to consider the solutions for the symmetric direction in the Brillouin zone where $P = P_x = P_y$. It then turns out that for a finite N one gets 6 different branches in agreement with the number of possible modes. The s- and d-wave solutions mentioned before correspond to the A_1 and B_1 modes and we recover the expressions obtained in ref.[5] Thus these modes can only be excited if both quantum numbers are even integers. All the other modes contain at least one "antisymmetric" component. For instance, the B_2 mode is determined by the expressions

$$a_{B_2}\left(r_x, r_y\right) =$$

$$\frac{C}{\pi^2} \int_{\pi/N}^{\pi+\pi/N} \int_{\pi/N}^{\pi+\pi/N} \frac{\left[\cos\left(Q_x\right) + \cos\left(Q_y\right)\right] \cos\left(r_x Q_x\right) \cos\left(r_y Q_y\right) dQ_x dQ_y}{2 - \varepsilon/2 - \cos\left(\frac{P_x}{2}\right) \cos\left(Q_x\right) - \cos\left(\frac{P_y}{2}\right) \cos\left(Q_y\right)},$$

where

$$C = \frac{1}{\pi^2} \int_{\pi/N}^{\pi+\pi/N} \int_{\pi/N}^{\pi+\pi/N} \left[\cos\left(\frac{P}{2}\right) - \cos\left(Q_x\right)\right] dQ_x dQ_y.$$

The eigenenergy ε is found as a solutions of the compatibility equation

$$1 + 2A \cos\left(\frac{P}{2}\right) - D = 0, \tag{5}$$

where

$$A = \frac{1}{\pi^2} \int_{\pi/N}^{\pi+\pi/N} \int_{\pi/N}^{\pi+\pi/N} \frac{\cos(Q_x)\, dQ_x dQ_y}{2 - \cos\left(\frac{P}{2}\right)(\cos(Q_x) + \cos(Q_y)) - \varepsilon/2},$$

$$D = \frac{1}{\pi^2} \int_{\pi/N}^{\pi+\pi/N} \int_{\pi/N}^{\pi+\pi/N} \frac{\cos(Q_x)[\cos Q_x + \cos Q_y]\, dQ_x dQ_y}{2 - \cos\left(\frac{P}{2}\right)(\cos Q_x + \cos Q_y) - \varepsilon/2}.$$

One can calculate the asymptotic expansion of (5) analytically. The most important difference of the B_2 mode from the "Wortis" modes consists in a very significant finite size dependence of its energy and of the critical line in the Brillouin zone where this bound state becomes unstable. We give below the first few terms of the asymptotic expansion defining the point on this line for the chosen direction in the Brillouin zone $P = P_c$:

$$2\left(1 - \cos\left(\frac{P_c}{2}\right)\right) \simeq \frac{1}{\frac{1}{\pi}\ln N - 0.22}. \tag{6}$$

One then can easily see from the above expression that the dispersion of the B_2 mode is well separated from the above two modes ($P_{B_1} \simeq 2\arccos\left(\frac{4}{\pi} - 1\right)$ and $P_{A_1} = 0$) even at macroscopic numbers of spins in the system. A more complicated behaviour is observed for the $E-$mode since the bound state spectrum for each component is split in two branches. One of the branches is close to the energy of the B_2-mode and the other to that of B_1 and A_2. This difference is due to a more rapid angular variation of the wave function for the latter branch (see Fig. 1). For the new modes at least one of the quantum numbers is required to be an odd integer. In this sense these modes can be considered as "descendents" of the antisymmetric Bethe mode. As we have shown earlier [2] for a finite N this mode becomes unstable at some critical value of the total momentum and this instability is directly related to the so called "non-string" solutions of the Bethe ansatz equations [6,7]. For the finite chain the value of the critical momentum is very small and so is the deviation from the string solution. For the two dimensional lattice the critical points are determined by the logarithmic dependence, as is shown in (6), and the "non-string" behaviour can not be ignored at any finite number of spins. It is also important to stress that the new modes occupy a larger space in the Brillouin zone than those known earlier. By generalizing the above arguments for the multimagnon excitations one can conclude that the antisymmetric modes are responsible

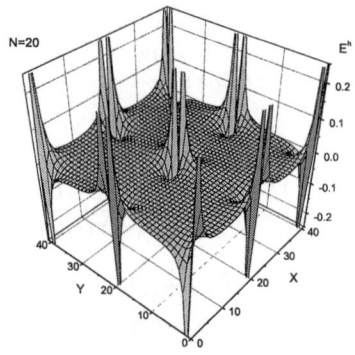

Figure 1. The relative distance amplitude of the two magnon bound state for the high energy branch of the E-mode at $N = 20$. The total momentum of the excitation is close to the corner of the Brillouin zone.

the logarithmic size dependences of the thermodynamic quantities of the $2D$ ferromagnet.

3. Acknowledgments

Financial support from the Concerted Action Scheme of the Flemish Government and from the National Science Foundation (FWO) is gratefully acknowledged. One of us, S.C., also acknowledges support from the Supreme Council for Scientific and Technological Development of Moldova.

References

1. D. C. Mattis, *The Theory of Magnetism. vol. I* (Springer Verlag, Berlin, Heidelberg, 1981), p. 300.
2. A. Ceulemans, S. Cojocaru and L.F. Chibotaru, Eur. Phys. J. B **21**, 511 (2001).
3. H. Bethe, Z. Phys. **71**, 205 (1931).
4. M. Hood and P.D. Loly, J. Phys. C **19**, 4729 (1986).
5. M. Wortis, Phys. Rev. **132**, 85 (1963).
6. F. Woynarovich, J. Phys. A **15**, 2985 (1982).
7. A.A. Vladimirov, Phys. Lett. A **105**, 418 (1984).

BETHE ANSATZ, YOUNG TABLEAUX, AND THE SPECTRUM OF JUCYS-MURPHY OPERATORS

TADEUSZ LULEK

Institute of Physics, University of Rzeszów,
ul. Rejtana 16A, 35-310 Rzeszów
Poland
E-mail: tadlulek@univ.rzeszow.pl

We emphase the role of a maximal commutative subalgebra of the group algebra of the symmetric group in a classification of quantum states of a finite one-dimensional magnetic ring. This subalgebra is spanned on the Jucys-Murphy operators and provides a complete classification of the basis of standard Young tableaux. The spectrum of Jucys-Murphy operators provides in this way the maximal available information on statistics of a system of N identical distinguishable quantum objects, like atomic spins in the Heisenberg model of magnetism.

1. Introduction

It is well known that Bethe Ansatz (BA) provides a nice mathematical tool for an exact solution of the Heisenberg Hamiltonian eigenproblem for a magnetic ring. We have listed both the advantages and some related open problems in our reports on two previous SSPCM meetings.[1,2] Here we aim to point out a particular aspect of BA, related to the symmetric group Σ_N of all permutations of nodes of the magnet, within the general scheme of the duality of Weyl. We are going to emphase the role of a maximal Abelian subalgebra within the group algebra $\mathbb{C}(\Sigma_N)$, related to Jucys-Murphy operators. The latter operators form a complete set of commuting quantum mechanical observables, and provide thus an exhausting interpretation of statistical properties of the Bethe wavefunction.

2. The Weyl duality and Robinson-Schensted-Knuth algorithm

We consider a magnetic ring consisting of N nodes, each node with the spin s. We put $n = 2s + 1$ so that the general scheme of Weyl duality is established by the two dual groups, the symmetric group Σ_N on the set $\tilde{N} = \{j = 1, 2, ..., N\}$ of nodes, and the unitary group $U(n)$ in the space

h spanned on the n eigenstates $|sm_s>$ of the single-node spin. In the following each such a state $|sm_s>\in h$ is treated as a letter in the alphabet A, $|A| = n$, with the linear ordering according to descending values of $m_s = s, s - 1, ..., -s$. In this way, a word w of the length N in the alphabet A corresponds to a magnetic configuration, and the set

$$A^{\tilde{N}} = \{w = a_1 a_2 ... a_N | a_j \in A, j \in \tilde{N}\} \tag{1}$$

of all such words spans the space $\mathcal{H}$ of all quantum states of the magnet and provides an orthonormal basis in this space.

The set $A^{\tilde{N}}$ of all magnetic configurations constitues the initial basis in the space $\mathcal{H}$. The most desired basis in condensed matter theory is that of all eigenstates of the Heisenberg Hamiltonian, but it cannot be given, at least explicitely, for an arbitrary N. In the termodynamic limit $N \to \infty$, one usually assumes the hypothesis of strings to be valid, which yields a complete combinatoric classification of states in terms of rigged configurations of Kerov, Kirillov and Reshetikhin. For small N, the basis of wavelets is well suited for an immediate diagonalization of the Hamiltonian within each subspace of $\mathcal{H}$, corresponding to a definite number of spin deviations from the saturation configuration $w_0 = ss...s$. But perhaps the best insight into the physical content of Bethe eigenfunctions for finite N can be reached by an analysis of details of the Weyl duality scheme. In particular, we consider here the statistical properties of these functions, i.e. properties with respect to the permutation of nodes by the symmetric group Σ_N.

Let $\mu = (\mu_1, \mu_2, ..., \mu_n)$ be the weight of a word $w \in A^{\tilde{N}}$, so that μ_i is the number of occurences of the letter $i \in A$ in w. Clearly, all words with the weight μ constitue an orbit $O_\mu \subset A^{\tilde{N}}$ of the natural action of the symmetric group Σ_N on the set $A^{\tilde{N}}$ of all magnetic configurations. The orbit O_μ spans therefore a subspace $\mathcal{H}_\mu$ in $\mathcal{H}$, which is invariant under Σ_N. In other words, the $\mathcal{H}_\mu$ is a carrier space of the transitive representation $R^{\Sigma_N:\Sigma^\mu}$, where

$$\Sigma^\mu = \Sigma_{\mu_1} \times \Sigma_{\mu_2} \times ... \times \Sigma_{\mu_n} \subset \Sigma_N \tag{2}$$

is the Young subgroup defined by the weight μ.

The formula

$$R^{\Sigma_N:\Sigma^\mu} = \sum_{\lambda \vdash N} K_{\lambda\mu} \Delta^\lambda \tag{3}$$

yields the decomposition of $R^{\Sigma_N:\Sigma_\mu}$ into irreps Δ^λ of Σ_N, with λ being a partition of N into (not more than) n parts, and the multiplicities $K_{\lambda\mu}$

being the Kostka numbers. In particular, for $s = 1/2$ and thus $n = 2$, we have $\mu = \{N - r, r\}$ with r being the number of spin deviations from the configuration $+ + ... +$ of ferromagnetic saturation and

$$R^{\Sigma_N : (\Sigma_N - r \times \Sigma_r)} = \sum_{r'=0}^{r} \Delta^{\{N-r', r'\}} = \Delta^{\{N\}} + \Delta^{\{N-1,1\}} + ... + \Delta^{\{N-r,r\}} \quad (4)$$

The last term in rhs of Eq. (4) corresponds in BA to the highest weight vectors, i.e. to those eigenstates with r spin deviations for which $S = M = N/2 - r$, where S and M is the total spin and its z-projection respectively. Just this last term corresponds to the main target of BA - the highest weight eigenstate of the Heisenberg Hamiltonian. All other eigenstates can be derived from them by a routine application of powers of the lowering operator of the total spin.

A basis adapted to the decomposition (3) or (4), that is, an irreducible basis of the symmetric group Σ_N, is provided by the Robinson-Schensted-Knuth algorithm (RSK). We do not describe this algorithm in detail here (cf.,e.g., ref.[10]) but only mention that it associates each word $w \in O_\mu$ with a pair $(P(w), Q(w))$, where $P(w)$ is a semistandard tableaux in alphabet A, whereas $Q(w)$ is a standard tableaux in the alphabet $\tilde{N}$ of nodes of the magnet. These tableaux are composed according to some rules concerning weakly increasing subwords of w, such that the pair $(P(w), Q(w))$ preserves bijectively all the combinatoric information of the magnetic configuration w, and at the same time, provides a consistent irreducible basis for the symmetric group Σ_N. Moreover, an appropriate collection of orbits $O_{SYT\mu'}$ yields an irreducible basis for the irrep λ of the unitary group $U(n)$, along the general duality of Weyl. We refer to $P(w)$ and $Q(w)$ as to the Weyl and Young tableaux, respectively. With respect to the symmetric group Σ_N, they play the role of the repetition label and the standard basis function, respectively, of the irrep Δ^λ in $\mathcal{H}_\mu$.

In a more detail, let $WT(\lambda, \mu)$ be the set of all semistandard Young tableaux of the shape λ and the weight μ (Weyl tableaux), and $STY(\lambda, 1^N)$ - the set of all standard Young tableaux of the shape λ and the weight 1^N (i.e. standard tableaux in the alphabet $\tilde{N}$ of nodes). Then the restriction of the RSK algorithm to the orbit O_μ yields the bijection

$$RSK \mid_\mu : C_\mu \longrightarrow \cup_{\{\lambda \vdash N \mid K_{\lambda\mu} \neq 0\}} WT(\lambda, \mu) \times SYT(\lambda, 1^N) \quad (5)$$

such that

$$w \longmapsto (P(w), Q(w)). \quad (6)$$

The pair $(P(w), Q(w))$, $P(w) \in WT(\lambda, \mu)$, $Q(w) \in SYT(\lambda, 1^N)$, encloses thus the whole combinatoric information of the magnetic configuration w.

However, when considering the linear structure of the space $\mathcal{H}$ of quantum states of the magnet, the RSK algorithm defines a unitary transformation in the space $\mathcal{H}_\mu$ which can be written explicitely as

$$| P(w), Q(w) >= \sum_{w' \in O_\mu} A_{w'w} \mid w' >, \qquad (7)$$

which defines a unitary matrix A in $\mathcal{H}_\mu$. The set $\{| P(w), Q(w) >| Q(w) \in SYT(\lambda, 1^N)\}$ constitues the set of all standard Young tableaux in the alphabet $\tilde{N}$, and thus the basis for the irreducible Σ_N - module specified by λ. Weyl tableau P(w) labels distinct copies of such a module in the space $\mathcal{H}_\mu$. Eq.(4) implies that for $s = 1/2$ (the orginal BA) the repetition label P(w) becomes redundant, and thus Young tableaux provide a complete classification of quantum states.

3. Jucys-Murphy operators

Jucys [3,4] and independently, Murphy [5] introduced the following operators

$$M_j = \sum_{j'=1}^{j-1} (j', j), \quad j = 2, 3, ..., N, \qquad (8)$$

within the group algebra $C(\Sigma_N)$ of the symmetric group (cf.also refs.[6,7]). An operator M_j constitues thus the sum of all transpositions (j', j) for $1 \leq j' < j$, i.e. for all positive integers which procede j. These operators have a remarkable property

$$[M_j, M_{j'}] = 0, \qquad (9)$$

so that they span a commutative subalgebra in $\mathbb{C}(\Sigma_N)$. Moreover, the Jucys-Murphy operators are hermitian and the algebra spanned on them is maximal. In the quantum-mechanical language, Jucys-Murphy operators constitue a complete set of commuting observables for a carrier space of an irrep Δ^λ of the symmetric group Σ_N. In other words, the spectrum of Jucys-Murphy operators provides a complete classification scheme for quantum-statistical properties of any physical system of N identical objects.

Jucys-Murphy operators satisfy also the recurrence relation

$$M_{j+1} = U_j M_j U_j + U_j, \quad j = 2, 3, ..., N - 1, \qquad (10)$$

where

$$U_j = (j, j + 1) \qquad (11)$$

are elementary transpositions - the Coxeter generators of Σ_N. The sum

$$\sum_{j=2}^{N} M_j = C^{(2)} \tag{12}$$

of all these operators for a given N is the class operator for all transpositions.

It is a remarkable fact that the eigenfunction of each Jucys-Murphy operator M_j are labelled uniquely by standard Young tableaux. Namely, we have

$$M_j \mid t >= m_j(t) \mid t >, \quad j = 2, 3, ..., N, \quad t \in SYT(\lambda, 1^N), \tag{13}$$

where

$$m_j(t) = c_j(t) - r_j(t), \tag{14}$$

and the pair of positive integers $(r_j(t), c_j(t))$ denotes the location of the item "j" in the standard Young tableaux $t \in SYT(\lambda, 1^N)$, i.e. the row and the column of "j" in t, respectively (cf.Fig.1.).

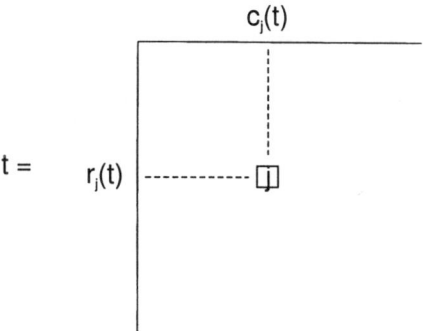

Figure 1. Location of the node $j \in \tilde{N} (j \neq 1)$ in a standard Young tableau $t \in SYT(\lambda, 1^N)$.

In this way, due to the commutativity (9) and completeness of Jucys-Murphy operators, each basis function $\mid t >$ of an irrep Δ^λ of Σ_N, corresponding to a particular Young tableau $t \in SYT(\lambda, 1^N)$ can be presented in the form

$$\mid t >=\mid m_2(t), m_3(t), ..., m_N(t) >=\mid \mathbf{m}(t) >, \tag{15}$$

as a common eigenfunction of all Jucys-Murphy operators.

Eq. (14) implies that the spectrum of the set of all Jucys-Murphy operators for a given irrep Δ^λ is fully determined by the shape of the Young diagram λ. For a given tableau $t \in SYT(\lambda)$, each box

$$(\alpha, \beta) = (r(j), c(j)) \tag{16}$$

yields the eigenvalue

$$m(j) = c(j) - r(j) = \beta - \alpha, \tag{17}$$

with the exception of the first box,

$$(1, 1) = (r(1), c(1)), \tag{18}$$

which is fixed by conditions of standardness. Each particular eigenvalue m of a Jucys-Murphy operator M_j, $j = 2, 3, ..., N$, is associated therefore with a box $(\alpha, \alpha + m)$ of the Young diagram (frame) λ. All boxes related to the eigenvalue m are located in the diagonal

$$D_\lambda(m) = \{(\alpha, \alpha + m) \mid P_l(m) \le \alpha \le P_u(m)\} \subset \lambda, \tag{19}$$

where $P_l(m)$ and $P_u(m)$ is the lower and upper limit, respectively, for the row label α of the diagonal. In particular, the eigenvalue $m = 0$ is associated with the main diagonal of the frame λ (cf. Fig. 2).

In this way, each Young diagram λ is uniquely associated with its table of contents, i. e. with the frame λ whose boxes are filled in by appropriate eigenvalues m of Jucys-Murphy operators, on accordance with Eqs. (13)-(17) and Figs. 1-2. For example, the frame $\lambda = \{764^3 2^2\}$ yields the table of contents.

0	1	2	3	4	5	6
-1	0	1	2	3	4	
-2	-1	0	1			
-3	-2	-1	0			
-4	-3	-2	-1			
-5	-4					
-6	-5					

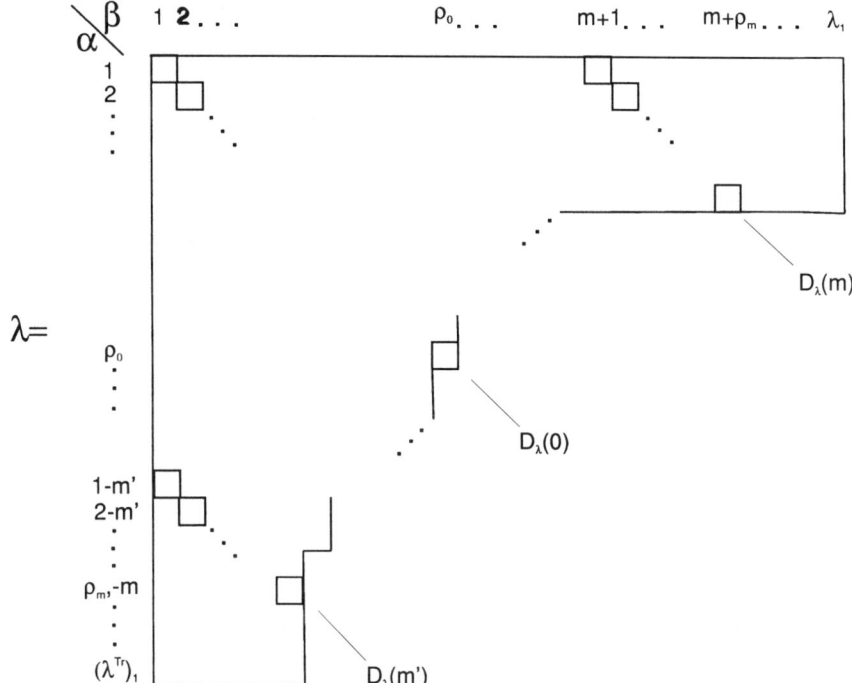

Figure 2. The diagonals $D_\lambda(m)$, $D_\lambda(0)$, and $D_\lambda(m')$, with the corresponding multi-plicities ρ_m, ρ_0, and $\rho_{m'}$, respectively. Each diagonal starts either at the top or the left border of the frame λ, and terminates either at a bottom or a right border (may be on both), depending on the eigenvalue m and the shape of the frame λ. Indices α and β label respectively the rows and columns of the frame λ.

Clearly, the whole spectrum of Jucys-Murphy operators remains constant, together with the multiciplities ρ_m of an eigenvalue m, within each Young tableau t for a fixed Young diagram λ. The action of the symmetric group Σ_N on a carrier space of the irrep Δ^λ involves therefore only some admissible permutations of components of vectors $\mathbf{m}(t)$, $t \in SYT(\lambda)$.

The range of admissible eigenvalues m for a given λ is

$$-(\lambda^{Tr})_1 \le m \le \lambda_1, \tag{20}$$

i. e. is constrained by the number

$$(\lambda^{Tr})_1 \le n = 2s + 1 \tag{21}$$

of rows of λ from the bottom, and by the number λ_1 of columns from the top. The lower limit

$$P_l(m) = \begin{cases} 1 & \text{for } m > 0, \\ 2 & \text{for } m = 0, \\ 1 - m & \text{for } m < 0, \end{cases} \tag{22}$$

depends upon m, and the upper limit $P_u(m)$ depends in a unique but nasty way upon the shape of λ, i. e. on the length λ_i of its appropriate parts. Namely, the multiplicity $\rho(m)$

$$\rho(m) = P_u(m) - P_l(m) + 1 \tag{23}$$

of the eigenvalue m in the whole spectrum is uniquely determined from the following conditions:

$$\left. \begin{array}{l} \lambda_{\rho(m)} \geq 1 + m + \rho(m) \\ \lambda_{\rho(m)+1} < 2 + m + \rho(m) \end{array} \right\} \text{ for } m > 0, \tag{24}$$

$$\left. \begin{array}{l} \lambda_{\rho(0)+1} \geq 1 + \rho(0) \\ \lambda_{\rho(0)+2} < 2 + \rho(0) \end{array} \right\} \text{ for } m = 0, \tag{25}$$

(observe that the eigenvalue $m = 0$ occurs in the whole spectrum of Jucys-Murphy operators with the multiplicity

$$\rho(0) = | D_\lambda(0) | - 1, \tag{26}$$

since the initial node $j = 0$, located by standardness in the box $(\alpha, \beta) = (1, 1)$, is not associated with any Jucys-Murphy operator), and

$$\left. \begin{array}{l} \lambda_{1-m+\rho(m)} \geq \rho(m) \\ \lambda_{2-m+\rho(m)} < \rho(m) + 1 \end{array} \right\} \text{ for } m < 0. \tag{27}$$

In particular, all values m are non-degenerative, iff λ has the shape of a hook, that is

$$\rho(m) = 1 \text{ iff } \lambda_\alpha = 1 \text{ for } \alpha > 1. \tag{28}$$

It is worthwhile to point out that the data involved in a Young tableau $t \in SYT(\lambda)$, and the corresponding λ - admissible Jucys-Murphy vector m, i.e. the vector given by Eq. (15) and satisfying conditions (16)-(27), are combinatorially equivalent.
Let

$$Y(\lambda) = \{(\alpha, \beta) | \alpha \in \{1, 2, ..., (\lambda^{Tr})_1, \ \beta \in \{1, 2, ..., \lambda_\alpha\}\} \tag{29}$$

be the set of all boxes of the Young diagram λ. Then any $t \in SYT(\lambda)$ can be looked at as a bijection $t : Y(t) \longrightarrow \tilde{N}$, i.e.

$$t(\alpha, \beta) = j \in \tilde{N}, \ (\alpha, \beta) \in Y(\lambda), \tag{30}$$

which satisfies the standardness

$$t(\alpha', \beta) > t(\alpha, \beta) \text{ if } \alpha' > \alpha,$$
$$t(\alpha, \beta') > t(\alpha, \beta) \text{ if } \beta' > \beta \tag{31}$$

and initial conditions

$$t(1, 1) = 1. \tag{32}$$

Clearly, Eqs. (16)-(17) determine uniquely $\mathbf{m}$ from t. Conversely, the knowledge of $\mathbf{m}$ implies a unique filling of the Young diagram $Y(\lambda)$ by elements of the set $\tilde{N}$ of nodes. To perform this filling, we decompose the diagram $Y(\lambda)$ into diagonals,

$$Y(\lambda) = \bigcup_{-(\lambda^{Tr})_1 \leq m \leq \lambda_1} D_\lambda(m), \tag{33}$$

and fill in each box of the diagonal $D_\lambda(m)$ by elements of the set

$$\tilde{N}_m = \{j \in \tilde{N} | m_j = m\} \subset \tilde{N}, \tag{34}$$

consecutively from the left top to the right bottom, in increasing order of the set $\tilde{N}$. This procedure, together with the initial condition

$$\xi_{\mathbf{m}}(1) = (1, 1), \tag{35}$$

establishes the bijection $\xi_{\mathbf{m}} : \tilde{N} \longrightarrow Y(\lambda)$ which produces a standard Young tableau $t \in SYT(\lambda)$ from given λ - admissible Jucys-Murphy vector $\mathbf{m}_g$ i.e. locates standardly each $j \in \tilde{N}$ in the Young frame $Y(\lambda)$. For example, the standard Young tableau

$$
t= \quad
\begin{array}{cccc}
1 & 2 & 4 & 7 \\
3 & 5 & 9 & \\
6 & 8 & &
\end{array}
$$

$$\tag{36}$$

for $\lambda = \{432\}$ yields the Jucys-Murphy vector

$$\mathbf{m} = (1, -1, 2, 0, -2, 3, -1, 1), \tag{37}$$

according to the table of contens

$$
\begin{array}{cccc}
0 & 1 & 2 & 3 \\
-1 & 0 & 1 & \\
-2 & -1 & &
\end{array}
$$

(38)

In this case, the values $m = -2, 0, 2$, and 3 are nondegenerate, whereas $m = \pm 1$ are doubly degenerated. The diagonals $D_\lambda(0)$ and the corresponding subsets $\tilde{N}_m = \xi_{\mathbf{m}}^{-1}(D_\lambda(0))$ are given in the Table below.

m	-2	-1	0	1	2	3
$D_\lambda(m)$	$(3,1)$	$(2,1),(3,2)$	$(1,1),(3,2)$	$(1,2)(2,3)$	$(1,3)$	$(1,4)$
$\tilde{N}_m$	6	3,8	$(1),5$	2,9	4	7

Note that the box $(1,1) \in D_\lambda(0)$ should contain the node $j = 1$, even if this node is not related to any Jucys-Murphy operator.

The table of contents of a given frame λ implies that the two extreme values of m, the maximal m_{max} and minimal m_{min}, are non-degenerate, due to standardness of λ. The value m_{max} determines the length

$$
\lambda_1 = m_{max} + 1 \tag{39}
$$

of the first row, and similarly, the value m_{min} yields

$$
(\lambda_{Tr})_1 = 1 - m_{min}, \tag{40}
$$

the length of the first column of λ. Deleting the sequence $1, 2, ..., m_{max}$ from a Jucys-Murphy vector $\mathbf{m}$, we are left with the sequence $\mathbf{m}'$ of components, corresponding to the Young diagram λ' obtained from λ by truncating its first row. Thus

$$
\lambda_2 = m'_{max} + 2, \tag{41}
$$

where m'_{max} is the maximal (non-degenerate) eigenvalue of $\mathbf{m}'$. Continuing this procedure, one derives from $\mathbf{m}$ the lengths of all rows of λ. In the same manner one derives the lengths of columns.

4. Interpretational remarks

The set of all Jucys-Murphy operators for given value of N has a natural quantum-mechanical interpretation for systems in which the symmetric group Σ_N is the group of physical symmetry. For $N > 2$, this group is nonabelian, so the elementary transpositions U_j, given by Eq. (11), do not commute mutually. Each of them is a physical observable, and thus can be measured, but they cannot be measured simultaneously. The Heisenberg principle of uncertainty is expressed here in terms of Young-Baxter relations

$$U_j U_{j+1} U_j = U_{j+1} U_j U_{j+1}, \quad j = 1, 2, ..., N - 1. \tag{42}$$

It points out that the non-commutativity is associated with only neighbour elementary transpositions U_j and U_{j+1}, since

$$U_j U_{j'} = U_{j'} U_j \text{ for } |j - j'| \geq 2. \tag{43}$$

In this context, the spectrum **m** of all Jucys-Murphy operators provides a maximal information on the statistics of the system, i.e. a maximal amount of observables which can be measured simultaneously. The measurement assumes a hierarchy of identical objects, manifested by the natural linear ordering in the set $\tilde{N}$. The eigenvalue $m_j(t)$ of the Jucys-Murphy operator $\hat{M}_j$ counts the number $c_j(t)$ of such objects $j' < j$ for which

$$(j, j')|\mathbf{m}> = |\mathbf{m}>, \tag{44}$$

diminished by the number $r_j(t)$ for objects with

$$(j, j')|\mathbf{m}> = -|\mathbf{m}>; \tag{45}$$

so that particles $j' < j$ with mixed symmetry under transposition are counted as zero (cf. Eq. (14)).

The case when identical objects are principally indistinguishable, favorises either bosonic or fermionic statistics, corresponding to the irrep $\lambda = \{N\}$ or $\{1^N\}$ of Σ_N, respectively. In these cases we have

$$(\mathbf{m})_{\text{bosonic}} = (1, 2, ..., N - 1) \tag{46}$$

and

$$(\mathbf{m})_{\text{fermionic}} = (-1, -2, ..., -N + 1), \tag{47}$$

so that the spectrum of Jucys-Murphy operators is trivially non-degenerate. Only such systems are found in Nature yet, but Ohnuki and Kamefuchi point out some further possibilities for parabosons and parafermions.

Much more prommissing, from the view of applications of non-trivial spectra of Jucys-Murphy operators, is the case of the Heisenberg model of magnetism. Here, each spin at a node $j \in \tilde{N}$ can be considered as a true copy of a standard object, but, at the same time, two nodes, j and $j' \neq j$, are, by definition of the model, completely distinguishable. Thus we are dealing with the model of N identical, but undoubtedly distinguishable physical objects - the single node spins. Clearly, the symmetric group Σ_N is the symmetry group of the Heisenberg Hamiltonian only in an exceptional case of the spherical model, when all spins interact pairwise with the same exchange integral. In BA, only the cyclic group $C_N \subset \Sigma_N$, serves as the translational symmetry group of the ring $\tilde{N}$. But Σ_N can be used as a dynamical group which permutes some energy levels, or some branches of energy bands, since it is well suited for a classification of states.

The RSK algorithm sketched in Sec.2 demonstrates that within the space $\mathcal{H}_\mu \subset \mathcal{H}$, spanned on an orbit 0_μ of the group Σ_N on the set $\tilde{n}^{\tilde{N}}$ of all magnetic configurations, the states related to a given irrep Δ^λ of Σ_N are classified in a complete way by the pair $(P(w), Q(w))$ of Weyl and Young tableaux, with $Q(w) \in SYT(\lambda)$ being a standard basis state of Δ^λ, and $P(w) \in WT(\lambda, \mu)$ playing the role of a repetition label for possibly different copies of Δ^λ in $\mathcal{H}_\mu$. But each $Q(w) = t$, being a standard Young tableau, is a common eigenfunction of the set of all Jucys-Murphy operators, and can be thus labeled by their spectrum. In this way, the Jucys-Murphy operators are good tools for classification of states of the Heisenberg model. In the case $s = 1/2$, or $n = 2$, when the Weyl tableau label $P(w)$ becomes redundand, these operators provide even complete classification of all states within a given number r of spin deviations from the ferromagnetic saturation. Clearly, the eigenstates of Jucys-Murphy operators are not the eigenstates of the Heisenberg Hamiltonian, so a diagonalization within each irrep Δ^λ is still needed, but the statistical properties of the wavefunctions are already fully determined.

References

1. T. Lulek, The Duality of Weyl and Bethe Ansatz for Heisenberg Chains, in SSPCM 1998, pp. 52-63.
2. B. Lulek and T. Lulek, Bethe Ansatz, Bloch Theorem and Rigged Configurations, in SSPCM 2000, pp. 298-310.
3. A. A. Jucys, On the Young Operators on the Symmetric Group, Lieturos Fizikos Rinkinys **6**, 163-80 (1966) (in Russian).
4. A. A. Jucys, Factorization of Young Projection Operators for the Symmetric Group, Lieturos Fizikos Rinkinys **11**, 5-10 (1971) (in Russian)
5. G. E. Murphy, The Idempotents of the Symmetric Group and Nakayama's conjecture, J. Algebra **81**, 258-65 (1983).
6. Cheng-Jiu Zhu and Jin-Quan Chen, A New Approach to Permutation Group

Representation II, J. Math. Phys. **24**, 2266-7 (1983).

7. Dian-Min Tong, Cheng-Jiu Zhu, and Zhong-QMa, Irreducible Representations of Braid Groups, J. Math. Phys. **33**, 2660-3 (1992).

8. Yoshio Ohnuki and Susumu Kamefudzi, Quantum Field Theory and Parastatistics, Univ. Tokyo Press, 1982.

9. A. Kerber, Applied Finite Group Actions, Springer-Verlag, Berlin 1998.

10. A. Lascoux, B. Leclerc, J.-Y. Thibon, "The plactic monoid" in: Algebraic combinatorics on words, ed., M. Lothaire, Cambridge University Press, 2002.

THE ANTISYMMETRIC COMPLEMENT OF THE RAREFIED BANDS OF TWO MAGNON STATES

R. OLCHAWA

Institute of Physics, Opole University, ul. Oleska 48
45-052 Opole, Poland, e-mail: rolch@uni.opole.pl

A. WAL

Instytute of Physics, University of Rzeszow ul. Rejtana 16A
35-959 Rzeszow, Poland

Some properties of the one-dimensional anisotropic Heisenberg Hamiltonian with the periodic boundary condition are pointed out. The antisymmetric two-magnon states by means of Bethe Ansatz are determined and compared with the symmetric ones. It is shown that the energy spectrum of the antisymmetric states together with the symmetric ones gives a regular band structure (i.e. without the rarefied bands).

1. Introduction

The model under consideration is the finite one–dimensional Heisenberg model of magnet defined by Hamiltonian

$$H = J \sum_{i=1}^{N} \left[\frac{1}{2} \left(S_i^+ S_{i+1}^- + S_i^- S_{i+1}^+ \right) + \varepsilon S_i^z S_{i+1}^z \right], \tag{1}$$

with the periodic boundary condition $N + 1 = 1$.

There are two types of approaches to the model: the first one based on the concept of magnetic configurations, such as Bethe Ansatz [1,2,3], the hierarchy of algebras [4,5] or a different kind of numerical calculations, and the second one based on the concept of magnons. In the second approach the Hamiltonian (1) is expressed in terms of magnon creation and annihilation operators, which are either boson or fermion operators. The widely used, in the spin wave theory, Holstein-Primakoff transformation [1,6] uses the boson operators, and transforms the Hamiltonian to the form of infinite series of products of the creation and annihilation operators. This infinite form is a source of problems when the exact eigenstates of the Hamiltonian are required. The fermion operators are used in the Wigner-Jordan transformation [1,8] and also arise from the single-band Hubbard model with a half-filled band in the limit of large on-site Coulomb repulsion. For the XY-model which corresponds to $\varepsilon = 0$ in (1), the Wigner-Jordan transformation leads to a system of free magnons.

It was shown that the energy spectrum of double excited states (i.e. states with two spin deviations) could be arranged into bands. [9,10] All these bands are doubly rarefied and only contain states for either odd or even vectors from the first Brillouin zone. In this article we are going to show that the spectrum of the antisymmetric two magnon excitations is described by the same dispersion law as for the symmetric ones and is doubly rarefied, too. However, there is a difference in distribution of the states over the Brillouin zone. As we shall see, the antisymmetric states fill all the gaps in the spectrum of the symmetric ones.

2. The antisymmetric two-magnon space

The Hamiltonian (1) could be easily diagonalized for one spin deviation. The eigenstates have the form

$$|k\rangle = \frac{1}{\sqrt{N}} \sum_{r=1}^{N} e^{ikr} |r\rangle, \tag{2}$$

where r is a position of the spin deviation, whereas k is a vector lying in the first Brillouin zone. A state $|k\rangle$ represents a plane wave, with the wave vector k. Such excitations are treated as quasiparticles and are called magnons. The dimension of the one-magnon space is equal to N. In order to create a space that contains two magnons we can use the product of the one-magnon space. The basis vectors in this space can be obtained in the form of the product of the states (2)

$$|k_1, k_2\rangle \equiv |k_1\rangle_1 |k_2\rangle_2 = \frac{1}{N} \sum_{r=1}^{N} \sum_{s=1}^{N} e^{i(k_1 r + k_2 s)} |r\rangle_1 |s\rangle_2. \tag{3}$$

It is easy to notice that such vectors contain configurations with the same positions $r = s$ of spin deviations, which is forbidden (for spin $s = 1/2$) from the physical point of view. However, the antisymmetric part of the space spanned by vectors

$$|k_1, k_2\rangle_a \equiv \frac{1}{\sqrt{2}} (|k_1, k_2\rangle - |k_2, k_1\rangle)$$

$$= \frac{1}{\sqrt{2N}} \sum_{r=1}^{N} \sum_{s=1}^{N} \left(e^{i(k_1 r + k_2 s)} - e^{i(k_2 r + k_1 s)} \right) |s\rangle_1 |r\rangle_2$$

$$= \frac{1}{N} \sum_{s=1}^{N} \sum_{r=s+1}^{N} \left(e^{i(k_1 r + k_2 s)} - e^{i(k_2 r + k_1 s)} \right) |s, r\rangle_a , \qquad (4)$$

where

$$|r, s\rangle_a = \frac{1}{\sqrt{2}} \left(|r\rangle_1 |s\rangle_2 - |s\rangle_1 |r\rangle_2 \right) , \qquad (5)$$

does not contain this unwanted terms. The states (5) are antisymmetric and can not be strictly treated as magnetic configurations but rather as states of two pseudoparticles (i.e. spin deviations).

In order to investigate the significance of the applied symmetry of states to the model, let us consider the action of the Hamiltonian (1) on a state $|r, s\rangle$ of two distinguishable pseudoparticels at positions r and s

$$H \, |r, s\rangle = a \, |r - 1, \ s\rangle + a \, |r, \ s + 1\rangle + b \, |r + 1, \ s\rangle + b \, |r, \ s - 1\rangle + c \, |r, \ s\rangle$$
$$= a \, |r_1, \ s_1\rangle + a \, |r_2, \ s_2\rangle + b \, |r_3, \ s_3\rangle + b \, |r_4, \ s_4\rangle + c \, |r, s\rangle , \qquad (6)$$

where $b = 0$ if r and s are the nearest neighbours. It follows that if $r < s$ then $r_i < s_i$, for $i = 1..4$. Thus, the pseudoparticles can never interchange their positions by a physical process. This property is characteristic of hard core particles in a one-dimensional open chain (see Figure 1). It fol-

Figure 1. Two hard core particles in one-dimensional open chain. These particles can not interchange their positions by a physical process.

lows from this that we do not need to apply any assumption about the interchange symmetry to construct the Hamiltonian matrix. However, the periodic boundary condition introduces a possibility for the interchange of the particle positions (Figure 2). If we consider the action of the Hamiltonian on a state $|r, N\rangle_a$, we notice that the second term in (6) gives $|r, N + 1\rangle_a = |r, 1\rangle_a = -|1, r\rangle_a$. So the matrix elements and - as a result - the energy spectrum would be different then in the case of symmetric states.

One can check that for odd number of particles the problem with the sign of the boundary states disappear.

3. The XY-model

The XY-model could be considered as a limit of the anisotropic Heisenberg model (1) for $\varepsilon \to 0$. According to the Wigner-Jordan transformation [8] the

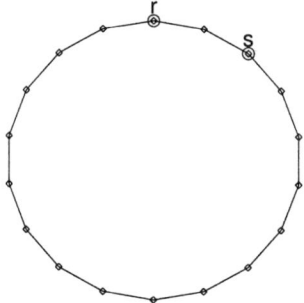

Figure 2.   Two hard core particles in one-dimensional chain with the periodic boundary condition. These particles have a possibility to interchange their positions.

XY-Hamiltonian can be expressed by the fermion annihilation and creation operators

$$c_k = (-1)^{\sigma_j} S_j^-,$$
$$c_k^\dagger = (-1)^{\sigma_j} S_j^+, \tag{7}$$

where σ_j denotes the number of spin deviations before j-th node. Using properties for spin operators S_j^- and S_j^+ one can easily prove the anticommutation relations for fermion operators

$$\{c_i, c_j^+\} = \delta_{ij}, \ \{c_i, c_j\} = \{c_i^+, c_j^+\} = 0. \tag{8}$$

The inverse relations to (7) allow us to write the hopping terms in the Hamiltonian in terms of fermion operators

$$S_i^+ S_{i+1}^- = c_i^+ c_{i+1},$$
$$S_i^- S_{i+1}^+ = -c_i c_{i+1}^+, \tag{9}$$

then the XY-Hamiltonian takes the form

$$H_{XY} = \frac{J}{2} \sum_{i=1}^{N} (c_i^+ c_{i+1} + c_{i+1}^+ c_i). \tag{10}$$

Some problem arises here if we introduce the periodic boundary conditions

$$c_N^+ c_{N+1} + c_{N+1}^+ c_N = c_N^+ c_1 + c_1^+ c_N. \tag{11}$$

For magnetic configurations with an even number of spin deviations the relations (9) are not fulfilled for the boundary case. [1] To illustrate this let us consider the action of $S_N^+ S_1$ and $c_N^+ c_1$ on a configuration $|1, n\rangle$ ($s < N$)

$$S_N^+ S_1^- |1, n\rangle = |N, n\rangle = |n, N\rangle$$
$$c_N^+ c_1 |1, n\rangle = c_N^+ |n\rangle = -|n, N\rangle \tag{12}$$

which yields

$$S_N^+ S_1^- = -c_N^+ c_1 \qquad (13)$$

So, if the configuration space is considered, the periodic boundary condition for the fermion operators (c-cycle boundary condition) is not equivalent to the periodic boundary condition for the spin operators. However, if we consider the antisymmetric states, then $|N, n\rangle_a = -|N, n\rangle_a$ and we get

$$S_N^+ S_1^- = c_N^+ c_1, \qquad (14)$$

thus, the Wigner-Jordan transformation becomes the exact transformation. In the Wigner-Jordan approach the Hamiltonian (10) is transformed to the diagonal form by the further transformation

$$a_k^+ = \frac{1}{\sqrt{N}} \sum_n e^{ikn} c_n^+,$$

$$a_k = \frac{1}{\sqrt{N}} \sum_n e^{-ikn} c_n. \qquad (15)$$

Finally, the Hamiltonian takes the form

$$H_{XY} = J \sum_k (\cos k)\, a_k^+ a_k. \qquad (16)$$

The energy spectrum of the Hamiltonian (16) for two-magnon excitation ($N = 20$) is presented in Figure 3. The relation between the symmetric and the antisymmetric eigenstates of the Hamiltonian (1) can be investigated

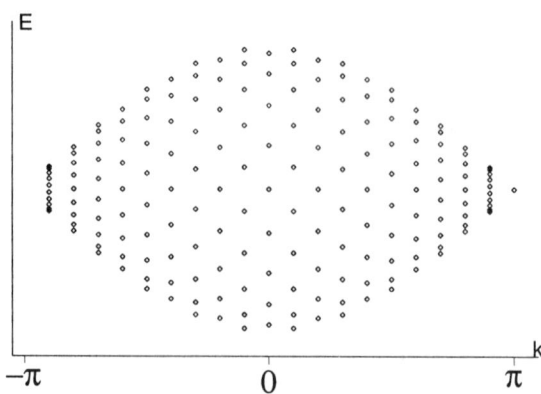

Figure 3. The energy spectrum of the antisymmetric two-magnon excitations for the finite XY model (for $N = 20$).

by Bethe Ansatz. The application of Bethe's and Hulthén's method to the case of two spin deviations leads to the assumption that all eigenvectors can be expressed in the following form [1,3]

$$|k_1, k_2\rangle_\zeta = \sum_{r<s} a_{r,s} |r, s\rangle_\zeta ,$$

$$a_{r,s} = e^{i(k_1 r + k_2 s + \frac{\phi}{2})} + e^{i(k_2 r + k_1 s - \frac{\phi}{2})} \tag{17}$$

where ζ was added to denote the symmetry of states, ϕ is a phase parameter. The quantum numbers k_1, k_2 can be complex, so they cannot be treated as the quasimomentum of magnons, in general. Substituting form (17) into the Hamiltonian (1) one obtains the following expression for energy levels

$$E = J \left(\cos k_1 + \cos k_2 + \varepsilon(\frac{N}{4} - 2) \right), \tag{18}$$

and the equation for the parameter ϕ

$$\cot \frac{\phi}{2} = \frac{\sin \frac{k_1 - k_2}{2}}{\frac{1}{\varepsilon} \cos \frac{k_1 + k_2}{2} - \cos \frac{k_1 - k_2}{2}}. \tag{19}$$

In the case of the XY-model $\varepsilon \to 0$ we get

$$E = J(\cos k_1 + \cos k_2) \tag{20}$$

and

$$\cot \frac{\phi}{2} = 0. \tag{21}$$

The periodic boundary condition for the symmetric states $a_{r,N+1} = a_{1,r}$ implies additional equations for k_1, k_2, and ϕ

$$Nk_1 - \phi = 2\pi\lambda_1,$$
$$Nk_2 + \phi = 2\pi\lambda_2. \tag{22}$$

Whereas for antisymmetric states the boundary condition of the form $a_{r,N+1} = -a_{1,r}$ gives equations

$$Nk_1 - \phi = \pi + 2\pi\lambda_1,$$
$$Nk_2 + \phi = \pi + 2\pi\lambda_2. \tag{23}$$

In both cases λ_1 and λ_2 are integers. The additional phase π which appears in (23) is a consequence of antisymmetry of the states.
The solution of equation (21) gives values for ϕ

$$\phi = \pi + 2\pi\alpha, \tag{24}$$

where α is an integer. Taking into account the conditions (22) and (23) we obtain allowed values for k_1 and k_2. In the case of the symmetric states we get

$$k_1 = \left(\beta_1 + \frac{1}{2}\right)\frac{2\pi}{N},$$

$$k_2 = \left(\beta_2 - \frac{1}{2}\right)\frac{2\pi}{N}, \tag{25}$$

and for the antisymmetric ones

$$k_1 = \beta_1\frac{2\pi}{N},$$

$$k_2 = \beta_2\frac{2\pi}{N}, \tag{26}$$

where β_1, β_2 are integers. It is easy to notice that in the case of anti-symmetric states, vectors k_1 and k_2 lie in the first Brillouin zone, what is in accordance with the solutions obtained by the Wigner-Jordan transformation. However, in the case of symmetric states the situation is quite different, the vectors (25) are not reciprocal lattice vectors and they cannot be treated as the quasimomentum of magnons. The spectrum of the symmetric states (for the same chain as in Figure 3) is presented in Figure 4. According to the idea presented in papers,[9,10] the obtained spectrum could

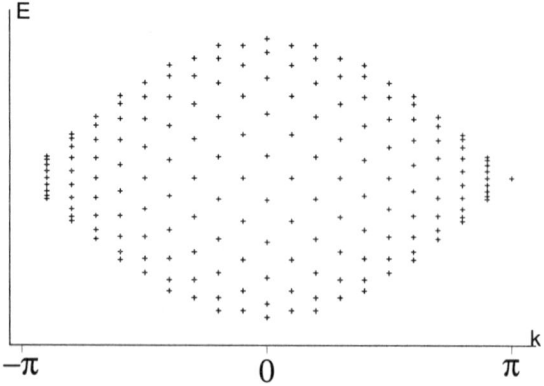

Figure 4. The energy spectrum of the symmetric two-magnon excitations for the finite XY model (for $N = 18$).

be arranged into bands by introducing the transformation

$$k = k_1 + k_2 = \kappa\frac{2\pi}{N},$$

$$q = k_1 - k_2 = \alpha \frac{2\pi}{N}, \qquad (27)$$

where $\kappa = \beta_1 + \beta_2$ and

$$\alpha = \beta_1 - \beta_2 + 1 = \begin{cases} 2, 4, 6, .., N - 2 & \text{for odd } k \\ 1, 3, 5, .., N - 1 & \text{for even } k \end{cases} \qquad (28)$$

for symmetric states, whereas for the antisymmetric ones

$$\alpha = \beta_1 - \beta_2 = \begin{cases} 1, 3, 5, .., N - 1 & \text{for odd } k \\ 2, 4, 6, .., N - 2 & \text{for even } k \end{cases} \qquad (29)$$

The energy spectrum (20) expressed by the new vectors can be rewritten in the form

$$E_\alpha(k) = 2J \cos \frac{q(\alpha)}{2} \cos \frac{k}{2}. \qquad (30)$$

The parameter α in (30) is interpreted as the band label. Relations (28,29) between parity of κ and α lead to the rarefied bands (i.e. bands that contain states not for all vectors from the first Brillouin zone).[11,12,13] It is easy to notice that the symmetric and antisymmetric energy spectrum complement each other and together form regular bands (Figure 5). For such a resultant spectrum the parameter $\alpha = 1, 2, ..N - 1$ for all k.

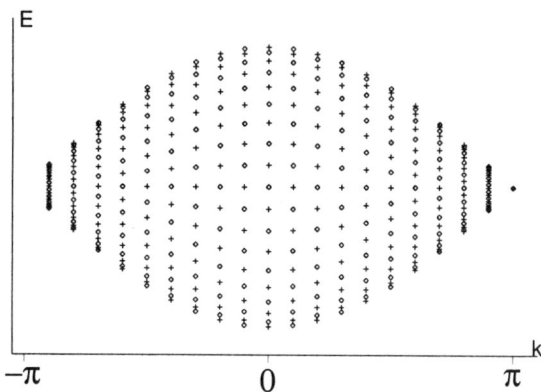

Figure 5.   The total energy spectrum of the symmetric two-magnon excitations and the antisymmetric ones for the finite XY model (for $N = 18$).

4. The antisymmetric states for $\varepsilon \neq 0$

The Hamiltonian (16) considered in the previous section described a system of free magnons which were fermions. Using the relation

$$S_n^z = c_n^+ c_n - \frac{1}{2}, \tag{31}$$

we can express the interaction of z-th components of spins in terms of the creation and annihilation operators

$$\sum_{n=1}^{N} S_n^z S_{n+1}^z = \frac{N}{4} - N_d + \sum_{n=1}^{N} c_n^+ c_{n+1}^+ c_{n+1} c_n = \frac{N}{4} - N_d +$$

$$+ \frac{1}{N} \sum_{k_1} \sum_{k_2} \sum_{k_2'} \sum_{k_1'} \delta(k_1 + k_2 - k_2' - k_1') e^{i(-k_2 + k_2')} a_{k_1}^+ a_{k_2}^+ a_{k_2'} a_{k_1'} \tag{32}$$

were N_d is the number of spin deviations. Adding this term into the Hamiltonian (16) we get the full Hamiltonian (1) expressed in terms of fermion operators. It is easy to notice that this term introduces the interaction between magnons, in which the total quasimomentum is preserved $k_1 + k_2 = k_2' + k_1' = k$. All the two-magnon eigenstates can be determined separately for any fixed k . Matrix elements of the Hamiltonian for two-magnons are given by the expression

$$H_{ij} = J \left(\cos k_1 + \cos k_2 + \varepsilon (\frac{N}{4} - 2) \right) \delta_{ij}$$

$$+ 2J\varepsilon \frac{\cos(k_2' - k_2) - \cos(k_2' - k_1)}{N}, \tag{33}$$

where i and j enumerate the basis vectors composed of pairs of vectors (k_1, k_2) from the first Brillouin zone such that $k_1 < k_2$ and $k_1 + k_2 = k$. The energy spectrum of the anisotropic Heisenberg Hamiltonian in the case of two magnon antisymmetric states together with the symmetric one is presented in Figure 6. It is easy to notice, comparing the energy spectrum obtained for free magnons (Figure 5) with the spectrum obtained for the full Hamiltonian (Figure 6), that the interaction between the magnons brought to form the bounded states that in the case of symmetric states.[9,10] Also as for the XY-model, obtained bands for the symmetric and antisymmetric states are doubly rarefied separately, but taken together yield a full band structure.

5. Conclusions

The periodic boundary condition applied to fermion operators leads to somewhat different energy spectrum from that in the case of symmetric

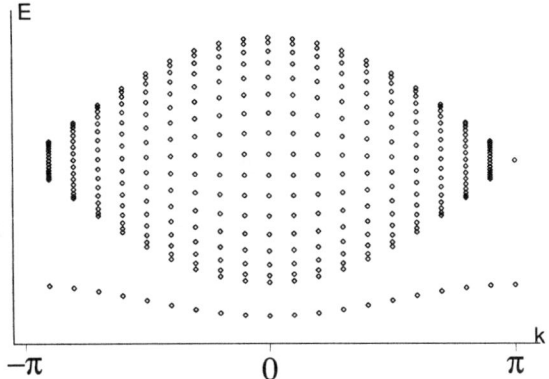

Figure 6. The energy spectrum of the antisymmetric two-magnon excitations for the finite Heisenberg model (for $N = 20$, $\varepsilon = 2$).

states. However in both cases the energy spectrum fulfil the same dispersion law, i.e. energy bands are described by the same functions $E_\alpha = E_\alpha(k)$, $\alpha = 1, 2, .., N - 1$. In the case of XY-model these functions take the form

$$E_\alpha(k) = 2J \cos \frac{\alpha\pi}{N} \cos \frac{k}{2}. \tag{34}$$

The difference in energy spectrum between symmetric and antisymmetric states results from the selection rule for k and α (28,29). It follows from the selection rule that a band α contains states either for even or odd vectors k only, depends on parity of α and considered symmetry. Both spectrums are mutually complementable spectra so, the resultant spectrum consists of full energy bands (i.e. contains states for all vectors k from the first Brillouin zone).

The same relation between symmetric and antisymmetric states occurs for the general case ($\varepsilon \neq 0$) of the anisotropy Heisenberg model. The energy spectrum obtained for this model, apart from the scattered state bands, which could be quite well (for $\varepsilon \simeq 1$) approximated by functions

$$E_\alpha(k) \simeq J \left(2 \cos \frac{\alpha\pi}{N - 1} \cos \frac{k}{2} + \varepsilon(\frac{N}{4} - 2) \right) \tag{35}$$

($a = 2, 3, .., N - 2$), also contains bounded state bands which very well satisfy the dispersion law

$$E(k) \simeq J \left(\frac{1 + \cos k}{2\varepsilon} + \varepsilon(\frac{N}{4} - 1) \right). \tag{36}$$

The above relations were obtained using the approach presented in previous papers. [9,10] The approximation for scattered states is based on the exact solution in the center of the Brillouin zone for $\varepsilon = 1$, whereas the bounded states approximation is obtained as a solution of equation (19), which for imaginery $q = k_1 - k_2 = ia$ can be written as

$$\coth\left(\frac{N}{2}\frac{a}{2}\right) = \frac{-\sinh\frac{a}{2}}{\frac{1}{\varepsilon}\cos\frac{k}{2} - \cosh\frac{a}{2}} \quad \text{for even } (\lambda_1 - \lambda_2), \qquad (37)$$

$$\tanh\left(\frac{N}{2}\frac{a}{2}\right) = \frac{-\sinh\frac{a}{2}}{\frac{1}{\varepsilon}\cos\frac{k}{2} - \cosh\frac{a}{2}} \quad \text{for odd } (\lambda_1 - \lambda_2), \qquad (38)$$

where the periodic boundary conditions (22) and (23) were applied. For big enough product Na

$$\coth\left(\frac{N}{2}\frac{a}{2}\right) \simeq \tanh\left(\frac{N}{2}\frac{a}{2}\right) \simeq 1 \qquad (39)$$

so the equations (37) take the form

$$\frac{1}{\varepsilon}\cos\frac{k}{2} = \cosh\frac{a}{2} - \sinh\frac{a}{2} = \exp\left(-\frac{a}{2}\right). \qquad (40)$$

The expresion for energy levels (18) can be rewritten in terms of a

$$E = J\left(2\cos\frac{k}{2}\cosh\frac{a}{2} + \varepsilon(\frac{N}{4} - 2)\right), \qquad (41)$$

which after expresing cosh in terms of exponential function and after substituting (40), gives the final result (36).

References

1. W.J. Caspers, *Spin Systems* (World Scientific, Singapore, 1989).
2. H.A. Bethe, *Z.Phys* **71** 205 (1931).
3. L. Hulthen, *Arkiv. Nat.Astron. Fys.* **26A** 1 (1938).
4. A.K. Theophilou, S. Thanos, *Physica B* **202**, 41 (1994).
5. S. Thanos, *Physica B*, **202**, 41 (1994).
6. T. Holsteisn and H. Primakoff, *Phys. Rev.*, **58**, 1098 (1940).
7. T.Lulek in *Symmetry and Structural Properties of Condensed Matter*, eds. T.Lulek, B.Lulek and A.Wal (World Scientific, Singapore, 1999).
8. P. Jordan and E. Wigner, *Z.Phys.*, **47**, 631 (1928).
9. R.Olchawa in *Symmetry and Structural Properties of Condensed Matter*, eds. T.Lulek, B.Lulek and A.Wal (World Scientific, Singapore, 1999)
10. R. Olchawa, *Physica B*, **291**, 29 (2000).
11. T. Lulek, *J.Physique*, **45**, 29 (1984).
12. W. Florek and T. Lulek, *J.Phys.*, **A 20**, 1921 (1987).
13. B. Lulek, *J.Phys.: Condens Matter*, **4**, 8737 (1992).

COMPLEX SOLUTIONS FOR THREE-PARTICLE SECTOR OF THE HEISENBERG MAGNET

A. WAL, M.LABUZ

Institute of Physics, University of Rzeszow
ul. Rejtana 16A, 35-950 Rzeszów, Poland.

R. OLCHAWA

Institute of Physics, University of Opole
Poland

The Bethe Ansatz equations are used to analyze three-particle sector of the spin 1/2 one-dimensional Heisenberg model. Fine structure of the complex solutions for this model was described for finite number N of spins.

1. Introduction

One-dimensional Heisenberg magnet has been a topic of research for many scientist. The research has initiated by Bethe,[1] Hulthen,[2] Clousoux[3] and continued by Yang and Yang,[4] Orbach,[5] Baxter[6] and others. The Hamiltonian of the system with interaction restricted to nearest neighbors is written:

$$H = \sum_{j=1}^{N} (4\mathbf{S}^j \mathbf{S}^{j+1} - 1), \tag{1}$$

where $\mathbf{S}^j$ is a spin operator acting on the j-th node of the chain consisting of N spins. The substitution proposed by Bethe, known as the Bethe Ansatz(BA):

$$\Psi = \sum_{1 \leq j_1 < j_2 ... < j_r \leq N} \left(\sum_P e^{i(\sum_{n=1}^{r} k_{P(n)} j_n + \frac{1}{2} \sum_{l<j} \phi_{P(l)P(j)})} \right) \Phi_{j_1, j_2, j_3, ... j_r} \tag{2}$$

leads to the set of equations called Bethe equations (BE):

$$N k_i = 2\pi \lambda_i + \sum_{j=1, j \neq i}^{r} \varphi_{ij}, \tag{3}$$

$$2\cot(\varphi_{ij}) = \cot(\frac{k_i}{2}) - \cot(\frac{k_j}{2}). \tag{4}$$

The number of equations in the set depends on the number of spin deviations r from the ferromagnetic state, in which all spins are parallel. Each
reversed spin can be treated as a pseudoparticle (magnon) characterized by
the pseudomomentum k_i which generally can be complex or even singular.
The phases φ_{ij} apear due to the interaction between pseudoparticles k_i and
k_j. The numbers $j_1, j_2, \ldots j_r$ in the equation (2) indicate the nodes in the
chain, in which the projection of spin is antiparallel to the ferromagnetic
state. In this language, $\Phi_{j_1, j_2, j_3, \ldots j_r}$ is the state with r reversed spins on
the position $j_1, j_2, \ldots j_r$. The sum over P in the same equation denotes all
permutation of the set $(1, 2, \ldots r)$. The total quasimomentum $k = \sum_{i=1}^{r} k_i$
should be real and is a good quantum number.

The solutions of the system (3-4) are parameterized by the integers λ_i
called winding numbers. To avoid equivalency in solutions these parameters
should satisfy the following equation

$$-\frac{N}{2} \leq \lambda_1 \leq \lambda_2 \leq \lambda_3 \ldots \lambda_r \leq \frac{N}{2}. \tag{5}$$

In this paper we restrict our consideration to the three deviations $r = 3$
and to the cases when two of pseudomomenta are complex. In the section 2
the equations (3-4) are written in explicit form for this case. We present, in
the section 3, the solutions for wide range of chain nodes N starting from
the large number $N \approx 1000$ and finish at $N \approx 10$.

2. Complex solutions for three deviation $r = 3$

To find the complex solutions (the solutions with complex values of pseudomomenta k_i, $i = 1, 2, 3$) for the three particle sector we made an assumption
that two of the pseudomomenta are complex conjugated, $k_i = k_j^*$. This
assumption is compatible with the string hypothesis.[7,8,9] We can fix the
indices for complex pair of pseudomomenta without loss of the generality,
because the equations are symmetric. Let us fix these indices as follows:

$$\begin{aligned}
k_1 &= x + Iy, \\
k_2 &= x - Iy,
\end{aligned} \tag{6}$$

where x and y are real and I is the imaginary unit. The remaining pseudomomentum k_3 is real, because sum of all psuedomomenta should be real.
The analysis of the equations (3) leads to the conclusion that the values
of phases should satisfy the following conditions: $\varphi_{13} = \varphi_{23}^*$ and φ_{12} is
complex. We can make the following substitutions:

$$\begin{aligned}
\varphi_{13} &= \alpha + I\beta, \\
\varphi_{23} &= \alpha - I\beta, \\
\varphi_{12} &= \gamma + I\delta,
\end{aligned} \tag{7}$$

with real values for α, β, γ. Using the substitutions (6) and (7) for equations (3) and (4) and writting equations separetly for the imaginary and real parts we obtain:

$$\sin(\alpha)\left[\sin\left(\frac{x}{2}\right)^2 + \sinh\left(\frac{y}{2}\right)^2\right] - \left[\sin\left(\frac{\alpha}{2}\right)^2 + \sinh\left(\frac{\beta}{2}\right)^2\right] \times$$

$$\left\{\frac{1}{2}\sin(x) - \cot\left(\frac{k_3}{2}\right)\left[\sin\left(\frac{x}{2}\right)^2 + \sinh\left(\frac{y}{2}\right)^2\right]\right\} = 0, \qquad (8)$$

$$\sinh(\beta)\left[\sin\left(\frac{x}{2}\right)^2 + \sinh\left(\frac{y}{2}\right)^2\right] - \frac{1}{2}\sinh(y)\left[\sin\left(\frac{\alpha}{2}\right)^2 + \sinh\left(\frac{\beta}{2}\right)^2\right] = 0 \quad (9)$$

$$\sinh(\delta)\left[\sin\left(\frac{x}{2}\right)^2 + \sinh\left(\frac{y}{2}\right)^2\right] - \sinh(y)\left[\sin\left(\frac{\gamma}{2}\right)^2 + \sinh\left(\frac{\delta}{2}\right)^2\right] = 0 \quad (10)$$

$$\sin(\gamma) = 0 \qquad (11)$$

$$Nx = 2\pi\lambda_1 + \gamma + \alpha \qquad (12)$$

$$Ny = \delta + \beta \qquad (13)$$

$$Nx = 2\pi\lambda_2 - \gamma + \alpha \qquad (14)$$

$$Nk_3 = 2\pi\lambda_3 - 2\alpha \qquad (15)$$

We are looking for the solution of these equations taking winding numbers λ_1, λ_2 and λ_3 as parameters. To avoid the equivalency in the solutions we should limit the real part values of pseudomomenta and phases to the range $(-\pi, \pi)$. The analysis of the system of equations, especially equations (11,12,14), leads to the conclusion that complex solutions can be possible only in the following two cases: I $\gamma = \pi$ and $\lambda_1 = \lambda_2 + 1$, II $\gamma = 0$ and $\lambda_1 = \lambda_2$

The system of equations (8 - 15) was solved for a wide range of node number N and for given parameters $\lambda_1, \lambda_2, \lambda_3$. This set of parameters λ_i was chosen in accordance with the equation (5). The two cases I and II are calculated separately.

Calculation was based on the assumption that the Bethe parameters k and φ manifest quasicontinuity with respect to N, i.e. they change in a quasicontinuous way during lowering (or increasing) N. Taking a solution of BE for given N as a starting point for numerical procedure we are able to find the new solution for $N' = N - \Delta N$, which was the new starting point for the next step $N'' = N' - \Delta N$. We repeated the steps until the solutions changed a type or vanished.

3. The solutions

3.1. $\lambda_1 = \lambda_2$, $\gamma = 0$

For this case the value of λ_3 parameter should be greater at least about 2 from the remaining parameters $\lambda_1 = \lambda_2$. The solutions of the equations (8 - 15) for this case exist in complex form for wide range of N starting from $N = 1000$. The values of both real and imaginary part of pseudomomenta and phases are small (in sense of module values) for such large number of nodes in spins' chain. By lowering N we observe that the dependency curves shape $k(N), \varphi(N)$ changes from flat form for large N to quickly decreasing or increasing form for small N. The left border of solutions range is provided by the values of $\{\lambda_i\}$ parameters.

The figure 1 presents the example of such solutions for the parameters $(\lambda_1, \lambda_2, \lambda_3) = (-1, -1, 1)$.

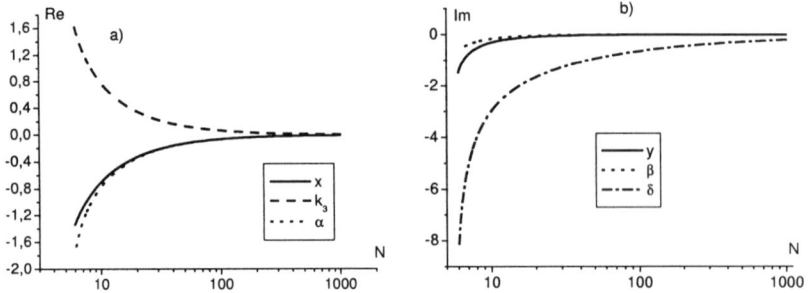

Figure 1. The solutions of BE for $\lambda_1 = -1, \lambda_2 = -1, \lambda_3 = 1$. The left picture a) presents the real parts of pseudomomenta and phases while the right picture b) the imaginary parts.

3.2. $\lambda_2 = \lambda_1 + 1$, $\gamma = \pi$

For this case the value of λ_3 parameter should be greater at least about 2 from the parameters λ_2. The solutions are different from those described in subsection 3.1. For the asymptotic solutions, $N \to \infty$, values of some Bethe parameters have limit 0: $\alpha \to 0$, $\beta \to 0$, and $\delta \to 0$. These properties can be explain by presenting the solutions of BE in graphic form. The number of unknowns in equations (8 - 15) was reduced to three $\{\alpha, \beta, \delta\}$ and these three equations were plotted in three dimensional space in such a way, that solutions of each equation were presented by a plane. The BE solutions are those points, they belong to all planes.

On Figure 2 the graphic form of solutions was plotted for two values of nodes in Heisenberg chain: part "a" corresponds to $N = 100$ and part "b" to $N = 10$. The points representing solutions for $N = 100$ are concentrated around the origin of the coordinate system α, β and δ. The calculations show that increasing N results in decreasing the distance between common points for all planes and the origin of coordinate system $(0,0,0)$, i.e. the solution tends to limit value 0. For small value of N there are solutions differ from 0 (see part "b" of the Figure 2).

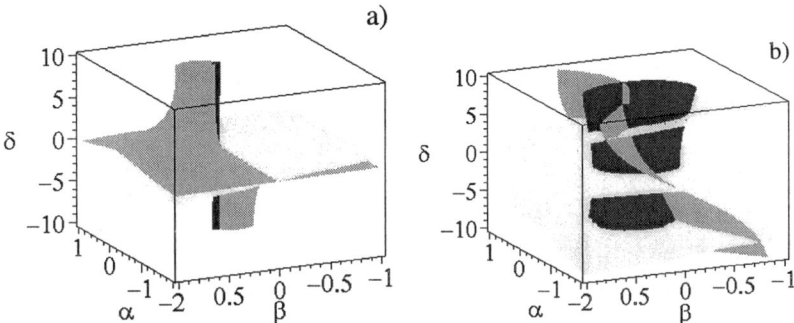

Figure 2. The solutions of BE in graphic form. The three colour planes correspond to three equations in terms of α, β, δ. The left part a): is ploted for $N = 100$, the right one b) for $N = 10$.

This properties of solutions lead to the conlcusion that complex solutions, in the sense we consider in this paper, exist only below some special chain length - the critical point N_{cr}. [10] Above that point the complex solution disappears. The value of N_{cr} depends on the values of equations parameters $\{\lambda_i\}$, for example the set $\lambda_1, \lambda_2, \lambda_3 = (-2, -1, 1)$ leads to $N_{\mathrm{cr}} = 23.188$, the other result in the following values: $\lambda_1, \lambda_2, \lambda_3 = (-2, -1, 2) \rightarrow N_{\mathrm{cr}} = 23.516$, $\lambda_1, \lambda_2, \lambda_3 = (-2, -1, 3) \rightarrow N_{\mathrm{cr}} = 23.661$. The quaiscontinuity of solutions with respect to N is now observed below this special point. The left border of such quasicontinuity is provided by values of parameters $\{\lambda_i\}$. On the Figure 3 we present an example of such type of solution.

4. Conclusions

The complex solutions of Bethe equations for three particle sector $r = 3$ was discussed in this paper in details. The asumption $k_1 = k_2^*$ allow us to determine the properties of the remaining parameters. The calculations show that there are two types of complex solution: I - for which $\lambda_1 = \lambda_2$

308

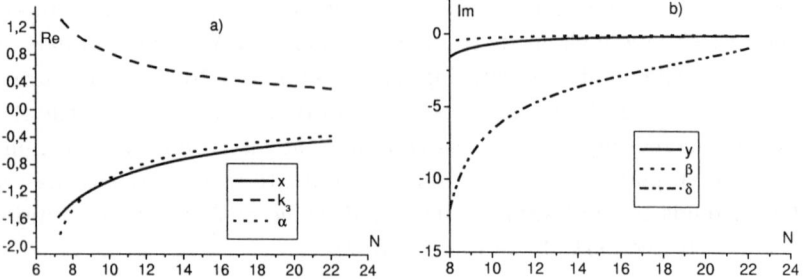

Figure 3. The solutions of BE for $\lambda_1 = -2, \lambda_2 = -1, \lambda_3 = 1$, $N_{cr} = 23.188$. The left picture a) presents the real parts of pseudomomenta and phases while the right picture b) the imaginary parts.

and II - $\lambda_2 = \lambda_1 + 1$. They manifest different properties. For first type I, if the solution exist, it is complex (in the sense of k and φ values). The second type is more intersting one. There is a special chain length N_{cr}, the critical point, and the complex solutions exist only below it. The value of N_{cr} is determined by the values of $\{\lambda_i\}$ parameters. The solutions exhibit quasicontinuous form with respect to N for both types, but for type II only for $N < N_{cr}$.

References

1. H. Bethe, *Z. Phys.* 71, 205-226 (1931).
2. L. Hulthen, *Arkiv Mat, Astron. Fysik* 26A, No 11, 1-106 (1938).
3. J. Des Cloizeaux and J. J. Pearson, *Phys. Rev.* 128, No 5, 2131-2135 (1958).
4. C. N. Yang, C. P. Yang, *Phys. Rev.* 150, No 1, 321-327 (1966); 150, No 1, 327-339 (1966).
5. R. Orbach, *Phys. Rev.*, **112** 309316 (1958).
6. R.J. Baxter, *Phys. Rev. Lett.*, **26**, 834 (1971).
7. M. Takahashi, *Prog. Theo. Phys.* **46**, 401 (1971).
8. S.V. Kerov, A.N. Kirillov and N.Yu. Reshetikhin, *Zapiski Nauchnykh Seminarov LOMI* **155** 50 (1986).
9. M. Kuzma, B. Lulek and T. Lulek, *Algebraic Combinatoric and Aplication* Proceedings of the Euroconference (ALCOMA), Goessweinstein 1999, Editors A. Betten, A. Kohnert, R. Laue and A. Wassermann (Springer 2001), p. 212.
10. W.J. Caspers, T. Lulek, B. Lulek, M. Kuzma and A. Wal, *Symmetry and Structural Properties of Condensed Matter* Proc. of the 7th International School on Theoretical Physics, Myczkowce 2002, Editors T. Lulek, B. Lulek, A. Wal (World Scientifi, Singapore).

MÖBIUS STRIP AS THE CONFIGURATION SPACE FOR THE CASE OF TWO MAGNONS IN BETHE ANSATZ

DOROTA JAKUBCZYK AND BARBARA LULEK

Institute of Physics, University of Rzeszów,
ul. Rejtana 16A, 35-310 Rzeszów, Poland
E-mail: dgol@univ.rzeszow.pl, barlulek@univ.rzeszow.pl

A finite Heisenberg magnetic ring with an arbitrary single-node spin and two spin deviations from the ferromagnetic saturation is considered as the system of two Bethe pseudoparticles. The set of all relevant magnetic configurations spans a surface which can be recognised as a Möbius strip. The dynamics of the system imposes the double twist of all regular orbits of the translation symmetry group.

1. Introduction

Bethe Ansatz solution of the eigenproblem of the isotropic Heisenberg Hamiltonian for a ring with N nodes [1] allows us to interpret some subspaces of the Hilbert space for the Heisenberg magnet as spaces of quantum states of a system of special kind of particles, referred hereafter to as Bethe pseudoparticles. They are spin deviations, treated as particles moving on the ring, with dynamics precisely defined by the Heisenberg Hamiltonian [8,9,10]. In this paper we aim to study the case of $r = 2$ spin deviations, with a particular attention paid to topological properties of the classical configuration space for such a system.

2. Kinematics and dynamics

Let

$$\tilde{N} = \{j = 1, 2, ..., N\} \tag{1}$$

be the set of all nodes of a magnetic ring. We refer hereafter to the set

$$Q = \{ |j_1, j_2 > \left| \begin{array}{l} 1 \leq j_1 < j_2 \leq N \text{ for } s = 1/2 \\ 1 \leq j_1 \leq j_2 \leq N \text{ for } s > 1/2 \end{array} \right. \}, \tag{2}$$

of all positions of the system of two spin deviations to as the classical configuration space of this system, where

$$|j_1, j_2> = \begin{cases} \hat{s}_{j_1}^- \hat{s}_{j_2}^- |0> & \text{for } j_1 < j_2, \\ 2^{-1} s^{-1/2} (2s-1)^{-1/2} (\hat{s}_{j_1}^-)^2 |0> & \text{for } j_1 = j_2 \ (s > 1/2), \end{cases} \quad (3)$$

with $\hat{s}_j^- = \hat{s}_j^x - \hat{s}_j^y$, etc., being spin operators at the node $j \in \tilde{N}$ and $|0> = |s, s, ..., s>$ is the set of all magnetic configurations of the magnet. Q is an orthonormal basis in the Hilbert space $\hat{H}$ of all quantum states of a magnet for which the z-projection of the total spin is equal $Ns - 2$. Kinematics and dynamics of the Heisenberg magnetic ring show up that the apparently evident cartesian structure of Q remains true only locally, whereas in the global scale it needs some essential topological departure from the picture of flat surface.

The set Q has, by its definition (4), a natural structure of a finite piece of a square lattice on the cartesian (j_1, j_2) - plane, or, more precisely, on the rectangular isosceles triangle

$$GT = \begin{cases} (1,2)(1,N)(N-1,N) & \text{for } s = 1/2, \\ (1,1)(1,N)(N,N) & \text{for } s > 1/2, \end{cases} \quad (4)$$

with the node (1,N) corresponding to the right angle, and the hypotenuse (1,2)(N-1,N) and (1,1)(N,N) for $s = 1/2$ and $s > 1/2$, respectively. We refer herefrom to GT as to the global triangle. Clearly,

$$GT(s > 1/2) = GT(1/2) \cup HC, \quad (5)$$

where

$$HC = \{(j,j)|j \in \tilde{N}\} \quad (6)$$

is the ,,hard core" orbit of Σ_N, corresponding to the Young subgroup with the two Bethe pseudoparticles at the same site (of.Fig.1). Let us discuss the structure, imposed on the set Q by the cyclic group C_N - the translation symmetry group of the system.

The configuration space Q forms an orbit of the action of the symmetric group Σ_N on the set of all magnetic configurations. Two orbits of the action of the symmetric group Σ_N on the set of all magnetic configurations decompose into complete sets of orbits of the subgroup C_N

$$R^{\Sigma_N : \Sigma^{\{N-2,2\}}} \downarrow C_N = \begin{cases} ((N-1)/2)R^{N:1} & \text{for N odd,} \\ (N/2-1)R^{N:1} + R^{N:2} & \text{for N even,} \end{cases} \quad (7)$$

$$R^{\Sigma_N : \Sigma^{\{N-1,1\}}} \downarrow C_N = R^{N:1}. \quad (8)$$

Here, $R^{\Sigma_N : \Sigma^\mu}$ denotes the transitive representation of a group Σ_N, with a stabiliser $\Sigma^\mu \subset \Sigma_N$. The action of the translation group C_N introduces

8.	18	28	38	48	58	68	78	88
7.	17	27	37	47	57	67	77	
6.	16	26	36	46	56	66		
5. $\uparrow j_2$	15	25	35	45	55			
4.	14	24	34	44				
3.	13	23	33					
2.	12	22						
1.	11					j_1		
	$\dot{1}$	$\dot{2}$	$\dot{3}$	$\dot{4}$	$\dot{5}$	$\dot{6}$	$\dot{7}$	$\dot{8}$

Figure 1. The classical configuration space Q for $N = 8$ and $s > 1/2$. The case $s = 1/2$ is obtained by removing the hypotenuse (11)(88) from the right isosceles triangle (11)(18)(88), which yields the triangle (12)(18)(78).

thus the structure of orbits in configuration space Q. In the cartesian coordinate system (j_1, j_2), these orbits are located on the lines, which are parallel to the main diagonal. .The hard-core orbit HC ($t = 0$) is regular and constitues the hypotenuse of GT ($s > 1/2$) (cf. Fig. 2).

Clearly, continuity requirements demand that each orbit t of the cyclic group C_N should consist of the set of consecutive points on a circle. It suggests to look at the presentation of the configuration space Q in the form of Eq.(2) or (4) as a *chart* of a continuous curved manifold $\bar{Q}$, such that each orbit t defines on $\bar{Q}$ a loop (homeomorphic to a circle), and that these loops do not intersect mutually.

The dynamics of the system is determined by the Heisenberg Hamiltonian

$$\hat{H}|\mathbf{j}>= \sum_{\mathbf{j'} \in Q_\mathbf{j}} (|\mathbf{j'} > -|\mathbf{j} >), \mathbf{j} \equiv (j_1, j_2) \in Q \tag{9}$$

where $Q_\mathbf{j}$ is the set of nearest neighbours of $\mathbf{j} \in Q$, including the fact of modular range of j_1 and j_2. This equation implies that the Hamiltonian $\hat{H}$ acts by short-distance jumps, from any magnetic configuration $\mathbf{j} \in Q$ to each of its nearest neighbours $\mathbf{j'} \in \mathbf{Q_j}$, along each locally available cartesian axis. We have

$$\hat{H}|j_1, j_2 >= |(\{j_1 + 1, j_2\}) > + |(\{j_1 - 1, j_2\}) > + |(\{j_1, j_2 + 1\}) > + \\ |(\{j_1, j_2 - 1\}) > - 4|j_1, j_2 > . \tag{10}$$

We can distinguish here two type of lines: cartesian and orbital lines. Both types of lines define a dissection of the surface $\bar{Q}$ into two-dimensional simplices: triangles with lexicographic order of points. The system of all these zero-, one- and two- dimensional simplices forms a *simplicial complex*

312

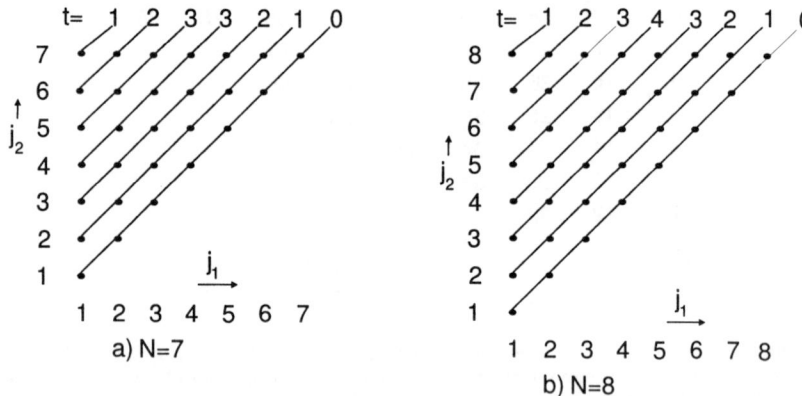

a) N=7

b) N=8

Figure 2. The structure of orbits of the translation group C_N on the configuration space Q for $N = 7$ (Fig. 2a) and $N = 8$ (Fig. 2b). The hard-core orbit $t = 0$ consists of a single line - the hypotenuse of the corresponding right triangle, whereas all regular orbits $t > 0$ are divided into two pieces. Fig. 2b shows also the doubly rarefied orbit $t = 4$ inside the triangle.

$\bar{\bar{Q}}$, or in other words, a triangulation of $\bar{Q}$ in the language of homotopy. Such a triangulation for the case $N = 8$ and $s > 1/2$ is presented in Fig. 3.

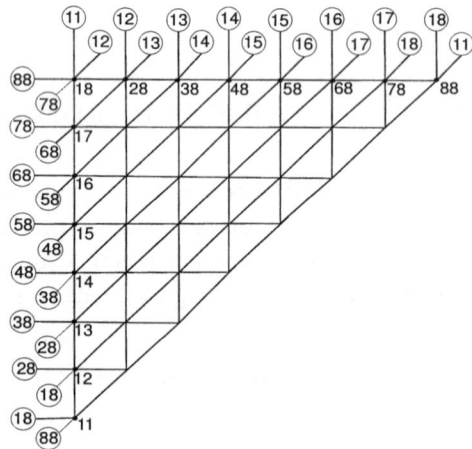

Figure 3. The triangulation $\bar{\bar{Q}}$ for $N = 8$, $s > 1/2$. Each vertical / horisontal line presents a cartesian loop, and each diagonal line - an orbital loop. Each line outgoing the GT should be connected with the appropriate encircled node.

Some topological properties of the surface $\bar{Q}$ are given by its first homotopy group $\pi_1(\bar{Q})$ which yields classes of non-contractible loops on $\bar{Q}$. By standard techniques, $\pi_1(\bar{Q})$ is isomorphic with the edge group $G(\bar{\bar{Q}}, L)$ of any of its triangulation $\bar{\bar{Q}}$, with $L \subset \bar{\bar{Q}}$ being any such a tree on $\bar{\bar{Q}}$ which encompasses all vertices, i.e. such that $Q \subset L$. We sketch now the construction of the edge group $G(\bar{\bar{Q}}, L)$, and therefore the homotopy group $\pi_1(\bar{Q})$, in some detail.

Let

$$L = \bigcup_{j'=2}^{N} \{(j, j') | 1 \le j < j'\} | \cup \{(1, j) | 2 \le j \le N\}, \tag{11}$$

that is, we choose the tree L as consisting of connected edges of nearest neighbours in (i) each horisontal axis $j' \in \{2, 3, ..., N\}$, (ii) the leftmost vertical axis ,, 1", in the map of Q, as given by Eq. (2) (Fig. 4).

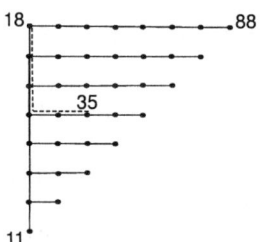

Figure 4. The tree L on the triangulation $\bar{\bar{Q}}$ for $N = 8$, $s > 1/2$. The nodes are labelled according to Fig. 1. Each node (j_1, j_2) has a uniquely defined path $v(j_1, j_2)$, which connects it with the initial node $(1, 8)$. E.g. the dashed line denotes the path $v(3, 5)$.

Let $\mathbf{j_0} = (1, N)$ be the ,, initial" vertex in the tree L, so that each vertex (j_1, j_2) has a uniquely defined path in $\bar{\bar{Q}}$

$$v(j_1, j_2) = (1, N)...(j_1, j_2) \tag{12}$$

which lies on the tree L, and connects the initial vertex $(1, N)$ with (j_1, j_2). With each edge $[\mathbf{j}, \mathbf{j'}]$ in $\bar{\bar{Q}}$ one associates a loop

$$\gamma(\mathbf{j}, \mathbf{j'}) = v(\mathbf{j})[\mathbf{j}, \mathbf{j'}]v(\mathbf{j'})^{-1}. \tag{13}$$

E.g. for N=8, and $(j_1, j_2) = (3, 5)$ we have

$$v(3, 5) = (18)(17)(16)(15)(25)(35). \tag{14}$$

Each loop corresponding to an edge, which lies on the tree ($[\mathbf{j}, \mathbf{j}'] \subset L$), is combinatorially trivial, so that the group $G(\bar{\bar{Q}}, L)$ is generated by all edges lying on $\bar{\bar{Q}} \setminus L$. Each triple $[\mathbf{j}_1, \mathbf{j}_2, \mathbf{j}_3]$ of edges, which forms a triangle (a two-dim simplex in $\bar{\bar{Q}}$) with $\mathbf{j}_1 < \mathbf{j}_2 < \mathbf{j}_3$ (for an arbitrarily chosen complete order in Q), should satisfy the relation

$$\gamma(\mathbf{j}_1, \mathbf{j}_2)\gamma(\mathbf{j}_2, \mathbf{j}_3) = \gamma(\mathbf{j}_1, \mathbf{j}_3). \tag{15}$$

(We use the lexicographic order in Q, but omit appropriate arrows on figures for reason of transparency).

Using these rules, one readily obtains the structure of the edge group $G(\bar{\bar{Q}}, L)$, and thus the homotopy group $\pi_1(\bar{\bar{Q}})$. E.g. when applying Eq. (15) to the triangle $(16)(17)(27)$ in Fig. 3 one gets $\gamma(16, 27) = 1$. In the same way, it follows that each edge $[\mathbf{j}, \mathbf{j}']$ inside the global triangle $(11)(18)(88)$ in Fig. 3 yields a trivial loop $\gamma(\mathbf{j}, \mathbf{j}') = 1$. Therefore, we have to consider only the edges outgoing from the boundary nodes in Fig. (cf. Fig. 5).

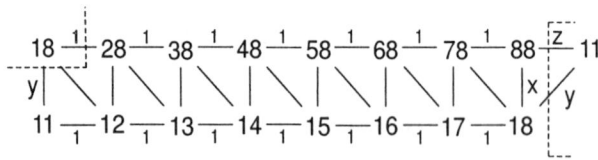

Figure 5. External simplices for the triangulation $\bar{\bar{Q}}$ from Fig. 3. Symbols x, y, z, and 1 denote the cartesian horisontal, cartesian vertical, orbital, and trivial loops, respectively. The upper right and left nodes (11 and 18, respectively), are repeated for convenience, but separated by dashed lines. Each line inside the above strip is equal to $x = y$, by virtue of Eq. (15) for consecutive two-dimensional simplices.

These edges are also mutually dependent, in virtue of several triangle conditions of Eq. (15), implied by triangulation $\bar{\bar{Q}}$, shown in Figs. 3 and 5. We introduce the notation

$$x = \gamma(18, 88), \quad y = \gamma(11, 18), \tag{16}$$

for cartesian loops, horisontal and vertical respectively, and

$$z = \gamma(11, 88) \tag{17}$$

for the orbital (hard-core) loop. It follows from Fig. 5 that these loops generate the homotopy group, and are mutually dependent. Also, all other loops can be expressed in terms of x, y, and z. In particular, the simplex $(12)(18)(28)$ implies

$$x = y, \tag{18}$$

which means that both cartesian loops are combinatorially (and thus homotopically) equivalent, whereas the simplex (11)(18)(88) yields

$$xy = x^2 = z. \tag{19}$$

We observe therefore that the orbital loop z is equivalent to the double repetition of a cartesian loop. Nb. the latter simplex, (11)(18)(88) should *not* be confused with the global triangle (11)...(18)...(88).

Finally, we obtain

$$\pi_1(\bar{Q}) \cong \mathcal{Z}, \tag{20}$$

that is, the first homotopy group of the considered configuration space $Q \subset \bar{Q}$ for the system of two Bethe pseudoparticles coincides with that for a circle. It is generated by a single loop, which can be chosen as any of the cartesian axes. Equivalently, the result (20) can be derived from Fig. 6 which presents the triangulation $\bar{Q}$ in more detail.

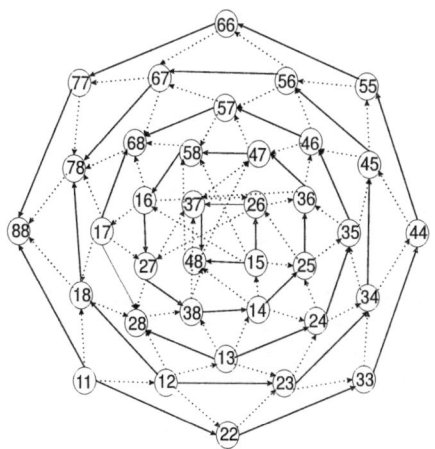

Figure 6. The triangulation $\bar{\bar{Q}}$ for $N = 8$, $s > 1/2$, in full detail. Dots denote the nodes $\mathbf{j} = (j_1, j_2)$ in accordance with Fig. 1. Solid lines denote the orbits of the translation group C_N, and dotted lines - the cartesian axes. The encircled parameters (j_1, j_2) at the end of each line denote the consecutive node to which the line should be connected in order to form eventually a loop.

For s=1/2 the general result (20) remains essentially the same, but cartesian axes x and y cease now to be loops, for the evident reason of a discontinuity implied by the absence of the diagonal orbit t=0 (HC of Eq. (6)). In this case, the group $\pi_1(\bar{Q})$ is also generated by a single loop, namely by any regular orbit, which is combinatorially equivalent to z.

Strictly speaking, the embedding $\bar{Q}$ cannot be a closed manifold, but rather a surface with a boundary. The boundary is a loop defined by the orbit $t = 0$ of hard core and $t = 1$ of the nearest neighbours, for $s > 1/2$ and $s = 1/2$, respectively, and the surface $\bar{Q}$ itself is a Möbius strip. It is well known that the homotopy of a Möbius strip is exactly the same as that of a circle, and thus is given by Eq. (20). A simple way of constructing of the Möbius strip for $N = 8$, $s > 1/2$, is presented in Fig. 7, which provides another chart of the configuration space Q. In this chart, horisontal lines correspond to orbits t, whereas the diagonal and antidiagonal lines are cartesian axes. The Möbius strip is obtained by gluing the vertical edges of the rectangle on Fig. 7 after making a twist.

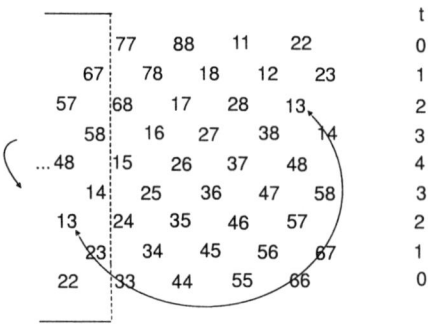

Figure 7. A rectangular chart of the configuration space Q for $N = 8$, $s > 1/2$. Horisontal lines correspond to orbits t of the cyclic group C_N, whereas diagonal and antidiagonal ones - to the cartesian axes. Twisting the vertical edge of the rectangle and appropriate identification as shown in the left-hand side yields the Möbius strip.

3. Discussion of paths in the configuration space

The only possible quantum transitions generated by the Heisenberg Hamiltonian occur between nearest neighbours in Q, along any cartesian axis. Thus a classical path correcponds to any cartesian axis or any zig-zacs from one axis to another. In particular, an orbit t cannot be, in general, a path since each elementary quantum transition is realised, according to Eq. (20) , between nearest orbits, t and $t = \pm 1$.

Eq. (20) admits, however, an interesting exception to the above rule. This is the case of N odd, and the orbit $t = (N-1)/2$ of furthermost neighbours, $\{(j, j + (N-1)/2)\}$. Let us consider, for example, the Möbius strip corresponding to $N = 7$ and $s = 1/2$ (cf. Figs. 2a and 8). Clearly, the path

$$(15)(25)(26)(36)(37)(47)(14)(15) \qquad (21)$$

lies entirely within the orbit t=3. All transitions involved in this path satisfy the condition (20) for the nearest neighbours just by virtue of the geometry of the Möbius strip, since the orbit t=2 becomes a double cover of circle. This structure spreads over next orbits, up to the t=1 of nearest neighbours, which is just the boundary of Möbius strip, and thus also a double cover circle. The same arguments apply also to the hard core orbit $t = 0$ for $s > 1/2$.

For N even, the situation changes only slightly, since here the furthermost orbit $t = N/2$ is doubly rarefied and admits only transitions to the next orbit, $t = N/2 - 1$, which is regular. Thus the latter orbit adapts to the rarefied one, again as a double cover (cf. Fig. 6). In this way, all regular orbits of the cyclic group C_N in the configuration space Q, embedded on the Möbius strip, acquire, by mechanical reasons imposed by the geometry, the structure of a double cover of a circle.

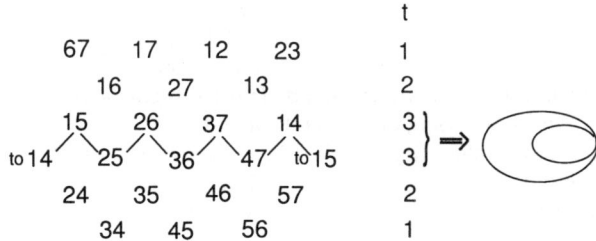

Figure 8. A rectangular chart of the configuration space Q for $N = 7$, $s = 1/2$. The lines show the physical path which lies within the furthermost orbit $t = 3$. On the corresponding Möbius strip, this path reproduces the orbit $t = 3$ as a double cover of a circle.

4. Conclusions

We have presented here some topological properties of the classical configuration space Q for the system of two Bethe pseudoparticles on a ring

318

consisting of N nodes. This space is consisting of a finite set of points $\mathbf{j} = (j_1, j_2)$ - possible locations of two pseudoparticles on the ring. The action of the translation group C_N - the symmetry group of the system - defines the structure of orbits. The Heisenberg Hamiltonian, which induces quantum jumps between nearest neighbours only, introduces the structure of neighbourhood, consistent with that of C_N - orbits. We have pointed out that both structures on Q can be interpreted as a triangulation $\bar{\bar{Q}}$ of a continuous surface $\bar{Q}$ which is locally flat, but globally reproduces a curvature imposed by loops in $\bar{\bar{Q}}$. Here we have shown that (i) the homotopy $\pi(\bar{Q}) \cong \mathcal{Z}$ coincides with that of a circle, i.e. is generated by a single loop, (ii) the surface $\bar{Q}$ itself is homeomorphic with a Möbius strip, (iii) each regular orbit of the translation group C_N corresponds, after embedding in $\bar{Q}$, to a loop which is a double cover of a circle. In this way, the geometry of the complement $\bar{Q}$ of the configuration space Q reproduces the dynamics of the system.

References

1. H. Bethe, Z. Physik **71**, 205-26(1931)(in German: English translation in: D.C. Mattis, The Many-Body Problem (World Sci., Singapore, 1993, pp. 689-716)
2. R. Baxter, Exactly Solvable Models in Statistical Mechanics (New York, Academic Press 1982)
3. C. N. Yang, Phys. Rev. Lett. **19**, 1312-15 (1961)
4. R. Baxter, Ann. Phys. (N.Y.) **70**, 193-228 (1972)
5. L. A. Takhtajan and L. D. Faddeev, LOMI **109**, 134-73 (1981)
6. H. J. de Vega, Int. J. Mod. Phys. **4**, 735-801 (1990)
7. E. K. Sklyanin, in: B.A.Kupersehmidt (Ed.) Integrable and Superintegrable Systems (World Sci., Singapore 1990), pp.8-33.
8. B. Lulek and T. Lulek, Rep. Math. Phys. **38**, 279-82 (1996)
9. B. Lulek and T. Lulek, in: T. Lulek, B. Lulek and A. Wal (Eds.) Symmetry and Structural Properties of Condensed Matter (World Sci., Singapore 1991) pp. 289-310.
10. B. Lulek and T. Lulek, Czech. J. Phys. **51**, 357-64 (2001)
11. S. V. Kerov, A. N. Kirillov, and N. Yu. Reshetikin, J. Sov. Math. **41**, 916-24 (1988).
12. T. Lulek, in: A. Betten, A. Kohnert,R. Laue and A. Wassermann (Eds.), Algebraic Combinatorics and Applications (Springer, Berlin 2000), pp. 261-72.
13. M. Gandin, La Function d'Onde de Bethe, Masson, Paris 1983.
14. B. Lulek, Acta Phys. Pol. **B22**, 371-88 (1992).
15. P.J.Hilton and S. Wylie, Homology Theory, Cambridge University Press, 1960.

THE REDUCED CONFIGURATION SPACE FOR THE SYSTEM OF BETHE PSEUDOPARTICLES

P. JAKUBCZYK, B. LULEK

Institute of Physics, University of Rzeszów,
ul. Rejtana 16A, 35-310 Rzeszów
Poland
E-mail: pjakub@univ.rzeszow.pl, barlulek@univ.rzeszow.pl

Exploitation of translational symmetry of a Heisenberg magnet leads to diminshing of size of the secular problem by the factor N (the number of magnetic nodes). The classical configuration space becames reduced . We want to present the structure of this reduced configurations space and dynamics in terms of spatial Fourier transforms on orbits of magnetic configurations. We describe also in a combinatorial way an algorithm, which allows to constuct immediately the reduced configuration space for an arbitrary r (the number of Bethe pseudoparticles) and N.

1. Introduction

We consider the one-dimensional Heisenberg magnetic ring, consisting of N nodes, with the single-node spin $s = 1/2$, outside closed shells [1]. In this case we have the sets:

$$\tilde{N} = \{j = 1, 2, \ldots, N\} \tag{1}$$

as the set of all nodes of the magnetic ring, and

$$\tilde{n} = \{0, 1\} \tag{2}$$

the single node standard basis for the spin $1/2$, where: "0" denotes spin up, the lack of a spin deviation, "1" - spin down, the spin deviation. The set $\tilde{n}$ corresponds to duoliteral alphabet with $0 < 1$ (completely ordered set).

Each mapping $f : \tilde{N} \to \tilde{n}$ from the set $\tilde{N}$ to $\tilde{n}$ can be presented in a form:

$$|f> = |m_1, m_2, \ldots, m_N > \text{ where } m_j \in \tilde{n}, j \in \tilde{N}, \tag{3}$$

and recognised as a magnetic configurations. The set of all magnetic configurations can be viewed as all words of the lenght N in the alphabet $\tilde{n}$.

The quantity

$$r = \sum_{i=1}^{N} m_i, \quad m_i \in \tilde{n} \tag{4}$$

denotes the number of spin deviations.

The set $\tilde{n}^{\tilde{N}} = \{f : \tilde{N} \to \tilde{n}\}$ of all such mappings provides an orthonormal basis, which spans the space $\mathcal{H}$ of all quantum states of the magnet $\mathcal{H} = \mathrm{lc}_C \tilde{n}^{\tilde{N}}$.

2. The action of Symmetric and Cyclic Group

Let Σ_N be the symmetric groups [2] on the set $\tilde{N}$. The action of symmetric group $P : \Sigma_N \times \tilde{n}^{\tilde{N}} \to \tilde{n}^{\tilde{N}}$ is defined as

$$P(\sigma) = \begin{pmatrix} f \\ f \cdot \sigma^{-1} \end{pmatrix}, \quad \sigma \in \Sigma_N, \ f \in \tilde{n}^{\tilde{N}}, \tag{5}$$

where σ^{-1} is the inverse to σ, $f : \tilde{N} \to \tilde{n}$, $\sigma^{-1} : \tilde{N} \to \tilde{N}$, $f \cdot \sigma^{-1} : \tilde{N} \to \tilde{n}$ is the composition of mappings f and σ^{-1}. The natural action P decomposes the set $\tilde{n}^{\tilde{N}}$ into orbits, labelled by the bipartition:

$$\mu = \{N - r, r\}, \quad r \in \{0, 1, 2, \ldots, E(\frac{N}{2})\}, \tag{6}$$

the function $E(x)$ returns integer part of the number x. The bipartition μ corresponds to the Young subgroup

$$\Sigma^{\mu} = \Sigma_{N-r} \times \Sigma_r, \tag{7}$$

i.e. Σ_{N-r} and Σ_r acting on the nodes occupied by the elements 0 and 1 respectively. The bipartition μ is a stabiliser on the corresponding orbit. We denote this orbit by $Q^{(r)}$ because all magnetic configurations belonging to $Q^{(r)}$ have just r spin deviations.

The linear (unitary) space

$$\mathcal{H}^{(r)} = \mathrm{lc}_C Q^{(r)} \tag{8}$$

spanned on each orbit $Q^{(r)}$ of Σ_N, is invariant under the action of the isoctropic Heisenberg Hamiltonian .

In order to make some links between Bethe Ansatz formulation and translational symmetry of the ring $\tilde{N}$, we consider here the structure of orbits of the translation group C_N on the classical configuration space $Q^{(r)}$ [3].

The corresponding action of the group C_N

$$B : C_N \times Q^{(r)} \to Q^{(r)}, \tag{9}$$

is specified by permutations:

$$B(j) = \begin{pmatrix} m_1, & m_2, & \cdots & , & m_N \\ m_{(1+j)modN}, & m_{(2+j)modN}, & \cdots, & m_{(N+j)modN} \end{pmatrix} \qquad (10)$$

and $j \in \tilde{N}$, $|m_1, m_2, \ldots, m_N > \in Q^{(r)}$.

The subgroup $C_N \subset \Sigma_N$ acting on the $Q^{(r)}$ decomposes it into complete orbits of the C_N according to:

$$Q^r \downarrow C_N = \sum_{\kappa \in K(N)} m(\mu, \kappa) R^{N:\kappa}, \qquad (11)$$

where $K(N)$ is the set of all divisors of the integer N, $m(\mu, \kappa)$ is the multiplicity of occurence of orbits of translational group of type κ in an orbit of symmetric group of type μ (cf. [4] for a detail formula for this multiplicity), $R^{N:\kappa}$ is the transitive representation of the group C_N with stabilizer κ. The set $Q^{(r)}$ decomposes into strata with epikernel κ.

3. Reduced Configurations Space (RCS)

Now we introduce the more detailed classification of the elements of strata, namely the orbits of translation group C_N. Let's introduce the vector:

$$\vec{t} = (t_1, t_2, \ldots, t_{r-1}) \qquad (12)$$

of integers $t_{j_\alpha} = (j_{\alpha+1} - j_\alpha)$, $j_\alpha \in \{0, 1, \ldots, N-1\}$ and j_α denote positions of succeeding ones in configuration $|f >= |m_1, m_2, \ldots, m_N >$, beginning from the left. The integer t_i is equal to distance beetwen nodes j_α and $j_{\alpha+1}$. Thus we can ascribe unambiguously to each magnetic configuration $f \in Q^{(r)}$ a vector $\vec{t}$. The set of such vectors $\vec{t}$ we denote by $T^{(r)}$. Let's treat the coordinates of vectors $\vec{t}$ as letter of a word. For example: The vector $\vec{t} = (1, 2, 2, 3)$ is equivalent to the word $t = 1223$.

Now we introduce some definitions on words, necessary to defining the reduced configuration space (RCS).

Two words t_1, t_2 are *conjugate* if:

$$t_1 \sim t_2 \Leftrightarrow \theta[t_1] = \theta[t_2] \qquad (13)$$

where $\theta[t]$ - denote the orbit of translation group, generated from the element t. The set of all conjugated vectors $\vec{t}$ we called the conjugacy class.

The Lyndon vector [5] is a primitive (i.e. $w = u^n$, $n \in \mathbb{N}$) vector which is minimal for the lexicographic order in its conjugacy class.

An integer $p \geq 1$ is a period of word $t = t_1, t_2, \ldots t_n$ where $t_i \in \mathbb{N}$ if $t_i = t_{i+p}$ for $i = \{1, 2, \ldots n - p\}$. The smallest period of t is called the *period* of t.

The set of all Lyndon vectors of the set $T^{(r)}$ we call the RCS or simply denote by T. Thus the RCS is the quotient:

$$T = T^{(r)}/ \sim \, . \tag{14}$$

And now, each magnetic configurations is uniquely presented as:

$$|f> = |\vec{t}, j>, \quad \text{where } \vec{t} \in T \tag{15}$$

and $j \in \{1, 2, \ldots, p\}$ where p denote the period of configuration $|f>$, $\vec{t}$ is the Lyndon vector for the configuration $|f>$. In this way we have obtained a new basis referred to as the basis of orbits. For example: $|f> = |011000000011> \Rightarrow$ period $p = 12$ thus $j \in \{1 \ldots 12\}$, but Lyndon vector is $\vec{t} = (1, 2, 1, 8)$ so, $j = 4$. Finally we receive: $|f> = |(1, 2, 1, 8), 4>$.

If we add to vector $\vec{t}$ the r-th coordinate $t_r = N - \sum_i^{r-1} t_i$ then $\vec{t} = (t_1, t_2, \ldots, t_r)$ where $\sum_{i=1}^r t_i = N$ then:

- For a vector $\vec{t}$, the coordinates t_i for which $i \in \{1, \ldots p\}$ where p is the period of $\vec{t}$, mark the succeeding twists of Yang-Baxter. An example is presented on Figure 1, where: a) for the vector $\vec{t} = (1, 2, 1, 8)$ period $p = 4 \Rightarrow i \in \{1, 2, 3, 4\}$ thus the coordinates $1, 2, 1, 8$ marks the twists of Yang-Baxter, b) for the vector $\vec{t} = (4, 4, 4)$ period $p = 1 \Rightarrow i \in \{1\}$ thus only first coordinate 4 mark the twist of Yang-Baxter.
- We can find the multiplicity κ of rarefied orbits:

$$\kappa = \frac{|\vec{t}|}{p}, \tag{16}$$

where $|\vec{t}|$ denotes the number of coordinates of the vector $\vec{t}$, p is the period of $\vec{t}$. For example: we have a magnetic configuration for which, we have N and $r = 4$ so, we can distinguish the cases:
- $p = 1 \Rightarrow \kappa = \frac{4}{1} = 4$, then $t_1 = t_2 = t_3 = t_4 = t \Rightarrow 4t = N \Rightarrow t = N/4$ therefore the vector of shape $\vec{t} = (N/4, N/4, N/4)$ describe fourfold rarefied orbits,
- $p = 2 \Rightarrow \kappa = \frac{4}{2} = 2$, then $t_1 = t_3, t_2 = t_4 \Rightarrow 2t_1 + 2t_2 = N \Rightarrow t_2 = \frac{(N - 2t_1)}{2}$ so the vectors of shape $\vec{t} = (t_1, \frac{(N - 2t_1)}{2}, t_1)$ and $t_1 \in \{1, 2, \ldots, \frac{N}{r} - 1\}$ describe twice rarefied orbits,
- and $p = 4 \Rightarrow \kappa = \frac{4}{1} = 4$ then $\vec{t} = (t_1, t_2, t_3)$ describes regular orbits. Recapitulating we have seen that T is the set of indices which indexes the orbits of cyclic groups.

For better understanding the structure of RCS we present the graphic representation of T for particularing cases (see Figure 2).

a)

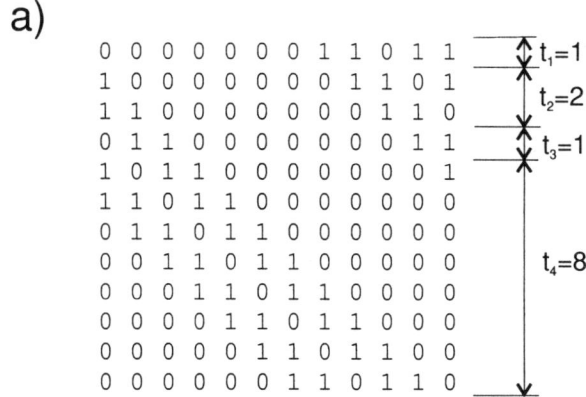

b)

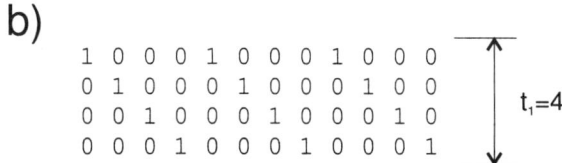

Figure 1. a) Presents the twists of Yang-Baxter for regular orbit $\vec{t} = (1, 2, 1, 8)$, b) presents the twist of Yang-Baxter for the threefold rarefied orbit $\vec{t} = (4, 4, 4)$

Every point on Figure 2 represents an element of the RCS, i.e. the vector. Parts a,b,c represent the one, two and three dimensional vectors, respectively. The numbers at arrows describe the ranges of variability of suitable coordinates. From this graphical representation we can easily read the vectors $\vec{t}$ which describe the rarefied orbits of C_N. These points are denoted in the pictures by boxes. We see that for a greater values of r we should use the $(r - 1)$ - dimensional space.

4. Algorithm

Exploitating of translation symmetry of the Heisenberg ring and generalizing the structure of T presented on the Figure 2, we can obtain the vectors of RCS immediately by an algorithm:

- Define the initial vector $\vec{t}$ as the primitive for which $\vec{t}_i = (\underbrace{11 \ldots 1}_{r-1})$, this vector can be seen as a number. The digits of the number are the coordinates of the vector.

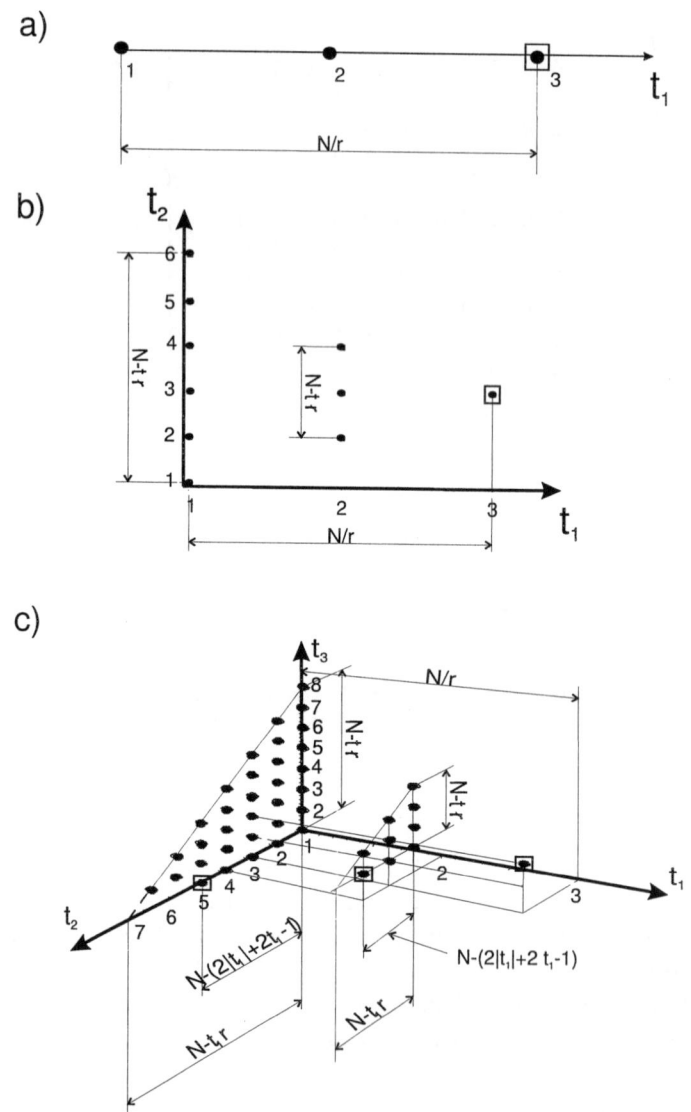

Figure 2. The graphic presentation of the RCS for a) $N = 6$ (the number of nodes of the magnetic ring) and $r = 2$ (the number of the spin deviation), b) $N = 9$ and $r = 3$, c) $N = 12$ and $r = 4$.

- We begin from the initial vector i.e. initial number $\vec{t}_i$, and obtain the next number (vector) $\vec{t}_{i+1}$ by increasing by one the number (vector) $\vec{t}_i$.

- In due course we have to obey the condition:

$$\sum_{i=1}^{r-1} t_i \leq N - 1 \text{ and } t_i \geq t_1, \ i \in \{2, \ldots, r-1\}. \tag{17}$$

- Next we add the r - th coordinate to vectors.
- The set of such vectors creates T^*.
- The set of all Lyndon vectors of the set T^* yields the RCS for N nodes and r spin deviations of the Heisenberg magnetic ring.

The RCS for $N = 12$ and $r = 6$ obtained by algorithm (using computer) is presented in Table 1.

Table 1. The RCS for $N = 12$ and $r = 6$.

(1, 1, 1, 1, 1, 7)	(1, 1, 1, 1, 2, 6)	(1, 1, 1, 1, 3, 5)	(1, 1, 1, 1, 4, 4)
(1, 1, 1, 1, 5, 3)	(1, 1, 1, 1, 6, 2)	(1, 1, 1, 2, 1, 6)	(1, 1, 1, 2, 2, 5)
(1, 1, 1, 2, 3, 4)	(1, 1, 1, 2, 4, 3)	(1, 1, 1, 2, 5, 2)	(1, 1, 1, 3, 1, 5)
(1, 1, 1, 3, 2, 4)	(1, 1, 1, 3, 3, 3)	(1, 1, 1, 3, 4, 2)	(1, 1, 1, 4, 1, 4)
(1, 1, 1, 4, 2, 3)	(1, 1, 1, 4, 3, 2)	(1, 1, 1, 5, 1, 3)	(1, 1, 1, 5, 2, 2)
(1, 1, 1, 6, 1, 2)	(1, 1, 2, 1, 1, 6)	(1, 1, 2, 1, 2, 5)	(1, 1, 2, 1, 3, 4)
(1, 1, 2, 1, 4, 3)	(1, 1, 2, 1, 5, 2)	(1, 1, 2, 2, 1, 5)	(1, 1, 2, 2, 2, 4)
(1, 1, 2, 2, 3, 3)	(1, 1, 2, 2, 4, 2)	(1, 1, 2, 3, 1, 4)	(1, 1, 2, 3, 2, 3)
(1, 1, 2, 3, 3, 2)	(1, 1, 2, 4, 1, 3)	(1, 1, 2, 4, 2, 2)	(1, 1, 2, 5, 1, 2)
(1, 1, 3, 1, 1, 5)	(1, 1, 3, 1, 2, 4)	(1, 1, 3, 1, 3, 3)	(1, 1, 3, 1, 4, 2)
(1, 1, 3, 2, 1, 4)	(1, 1, 3, 2, 2, 3)	(1, 1, 3, 2, 3, 2)	(1, 1, 3, 3, 1, 3)
(1, 1, 3, 3, 2, 2)	(1, 1, 3, 4, 1, 2)	(1, 1, 4, 1, 1, 4)	(1, 1, 4, 1, 2, 3)
(1, 1, 4, 1, 3, 2)	(1, 1, 4, 2, 1, 3)	(1, 1, 4, 2, 2, 2)	(1, 1, 4, 3, 1, 2)
(1, 1, 5, 1, 2, 2)	(1, 1, 5, 2, 1, 2)	(1, 2, 1, 2, 1, 5)	(1, 2, 1, 2, 2, 4)
(1, 2, 1, 2, 3, 3)	(1, 2, 1, 2, 4, 2)	(1, 2, 1, 3, 1, 4)	(1, 2, 1, 3, 2, 3)
(1, 2, 1, 3, 3, 2)	(1, 2, 1, 4, 1, 3)	(1, 2, 1, 4, 2, 2)	(1, 2, 2, 1, 2, 4)
(1, 2, 2, 1, 3, 3)	(1, 2, 2, 2, 1, 4)	(1, 2, 2, 2, 2, 3)	(1, 2, 2, 2, 2, 3)
(1, 2, 2, 2, 3, 2)	(1, 2, 2, 3, 1, 3)	(1, 2, 2, 3, 2, 2)	(1, 2, 3, 1, 2, 3)
(1, 2, 3, 1, 3, 2)	(1, 2, 3, 2, 1, 3)	(1, 2, 3, 2, 2, 2)	(1, 3, 1, 3, 1, 3)
(1, 3, 1, 3, 2, 2)	(1, 3, 2, 1, 3, 2)	(1, 3, 2, 2, 2, 2)	(2, 2, 2, 2, 2, 2)

5. Dynamics

We consider dynamics of RCS in terms of spatial Fourier transforms on orbits of magnetic configurations.

Each orbit $\vec{t} \in T$ spans a space which is invariant under C_N, namely a carrier space of the transitive representation $R^{N:\kappa}$ with $\kappa = N/|\vec{t}|$. Decomposition of $R^{N:\kappa}$ into irreps of C_N has the form

$$R^{N:\kappa} = \sum_{k \in B/\kappa} \oplus \Gamma_k, \tag{18}$$

where $B \equiv B/1 = \{k = 0, \pm 1, \pm 2, \ldots, \begin{cases} \pm(N/2 - 1), N/2 \text{ for } N \text{ even} \\ \pm(N-1)/2, \quad \text{for } N \text{ odd} \end{cases}\}$ is the Brillouin zone for the ring $\tilde{n}$, $B/\kappa = \{k \in B \mid k/\kappa \in B\} \subset B$, i.e. the subset of k's in B, which are divisible by κ is the κ - tuply rarefied Brillouin zone, and Γ_k is an irrep of C_N, specified by

$$\Gamma_k(j) = \exp(\frac{2\pi i k j}{N}), j \in \tilde{N}. \tag{19}$$

The decomposition (18) yields an orthonormal basis in the space $H^{(t)}$ of quantum states of the system of r Bethe pseudoparticles in a form

$$|k, \vec{t}> = \frac{1}{\sqrt{N/\kappa}} \sum_{j=1}^{N/\kappa} e^{2\pi i k j/N} |\vec{t}, j>. \tag{20}$$

We refer hereafter to $\{|k, \vec{t}> | \vec{t} \in T, k \in B/\kappa\}$ as to the **basis of wavelets** [6].

The Hamiltonian matrix in the wavelet of basis for the space $H^{(r)}$ becomes block-diagonalized due to the symmetry of C_N, each block $H^{(r)}(k)$ is labelled by the quasimomentum k from the Brillouin zone B and correspond to the reduced quantum space, labelled by elements $\vec{t}$ of the T.

6. Conclusions

We have presented the notion of Reduced Configuration Space, their properties and structure. We introduced the algorithm which allows us construct the Reduced Configuration Space, avoiding the action of symmetric and cyclic groups and using only combinatorics. We have seen that such space is used in the method of basis of wavelets which diminishes the size of the secular problem.

References

1. H. Bethe, Z. Physik **71**, 205 (1931) (in German; English translation in: D.C.Mattis, The Many-Body Problem, World Sci., Singapore 1993, pp. 689-716).
2. G. James, A. Kerber, The representation of Symmetric Group in: G. C. Rota (Eds.), Encyclopedia of Mathematics and its Applications, **16**, 1981.
3. B. Lulek and T. Lulek, Bethe Ansatz, Bloch Theorem, and Rigged Configurations, in B. Lulek, T. Lulek and A. Wal (Eds), Symmetry and Structural Properties of Condensed Matter, Vol. **6**, World Sci., Singapore 2001, pp. 298-310.
4. B. Lulek, Acta Phys. Pol. **B22**, 371(1992).
5. M. Lothaire, Combinatorcs on Words, Cambridge University Press,(1983).
6. A. Wal, T. Lulek, B. Lulek and E. Kozak, Int. J. Mod. Phys. **B13**, 3307-21 (1999).

RELATION BETWEEN THE CHARGE OF STANDARD YOUNG TABLEAU OF LASCOUX AND SCHÜTZENBERGER AND QUASIMOMENTUM

P. KRÓL, T. LULEK

Institute of Physics, University of Rzeszów
Al. T. Rejtana 16A, 35-310 Rzeszów,
Poland

An irreducible representation $\Delta^{\{\lambda\}}$ of the symmetric group Σ_N under restriction to the cyclic group C_N becomes reducible and can be decomposed to one-dimensional irreducible representations Γ_k of the cyclic group, where the index k can be identified with quasimomentum in the Brillouin zone [5,6]. The irreducible representation $\Delta\{\lambda\}$ corresponds to the standard Young diagram with the shape λ [3]. Lascoux and Schützenberger defined the charge of Young tableau as the rank of a poset structure on the set of tableaux with the same weight [1,2]. We expect that the set of charges of Young tableaux with the shape λ corresponds to the set of indices of representations Γ_k and can be used for a description of the distribution of quantum states of one-dimensional Heisenberg magnet over the Brillouin zone.

1. Introduction

1.1. *Magnetic configurations.*

We consider Heisenberg magnetic ring [5-7] consisting of N nodes. Let

$$\tilde{N} = \{j = 1,2,..., N\} \tag{1}$$

be the set of nodes.

We assume that each node $j \in \tilde{N}$ is occupied by a spin s. Let

$$\tilde{n} = \{i = 1,2,...,n\} \tag{2}$$

be the set of single-node projections -s, -$s+1$, ...,+s, where $n = 2s + 1$. Any mapping $f : \tilde{N} \to \tilde{n}$ is called a magnetic configuration. All magnetic configurations form a set

$$\tilde{n}^{\tilde{N}} = \{f : \tilde{N} \to \tilde{n}\}, \tag{3}$$

which spans the n^N –dimensional linear unitary space L of quantum states of the magnet.

1.2. Permutational representation of the symmetric group acting on the set of magnetic configurations.

The action of the symmetric group Σ_N on the set $\widetilde{n}^{\widetilde{N}}$ of all magnetic configurations [5,6] is defined by the formula

$$P(\sigma) = \begin{pmatrix} f \\ f \circ \sigma^{-1} \end{pmatrix}, f \in \widetilde{n}^{\widetilde{N}}, \sigma \in \Sigma_N. \tag{4}$$

This formula determines the permutational representation P of the symmetric group Σ_N. We will briefly describe some decompositions of this representation leading to the irreducible representations of the translation group.

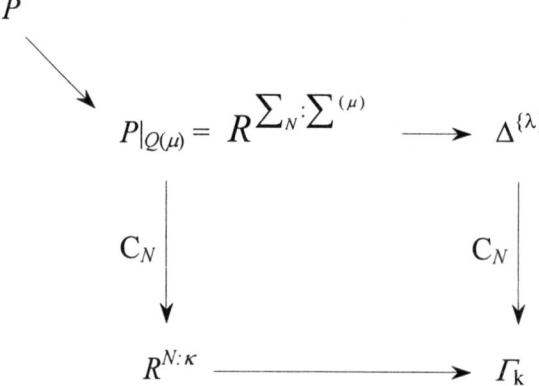

Figure 1. Two ways of decomposition of the permutational representation of the symmetric group to irreducible representations of the cyclic group.

2. Decomposition of the transitive representations.

Under the action of the group Σ_N the set $\widetilde{n}^{\widetilde{N}}$ decomposes into orbits $Q^{(\mu)}$

$$\widetilde{n}^{\widetilde{N}} = \bigcup_{|\mu|=N} Q^{(\mu)} \tag{5}$$

i.e. the sets of all configurations with the same number of different single-node projections of the spin s [5-7]. Here μ is an (improper) partition of N ($|\mu| = N$)

and is called the type of orbit

$$\mu = (\mu_1, \mu_2, ..., \mu_n),$$ (6)

where the part μ_i denotes the number of nodes with the spin i.

The restriction of a representation P to a single orbit $Q^{(\mu)}$ is a transitive representation of the symmetric group Σ_N acting on the orbit $Q^{(\mu)}$. We may note decomposition

$$P = \sum_{\mu|=N} R^{\Sigma_N : \Sigma^{(\mu)}}.$$ (7)

Under restriction to the cyclic group the transitive representation of the symmetric group decomposes to transitive representations of the cyclic group C_N [6].

$$R^{\Sigma_N : \Sigma^{(\mu)}} \downarrow C_N = \sum_{\kappa \in K(N)} m(\mu, \kappa) R^{N : \kappa},$$ (8)

where $K(N)$ is the lattice of all divisors of the integer N and $m(\mu, \kappa)$ can be found [6] as the number of a κ-tuply rarefied orbits of the cyclic group in the orbit $Q^{(\mu)}$ of the symmetric group.

The last step leading to irreducible representation of a group C_N is a decomposition of a transitive representation of the cyclic group

$$R^{N : \kappa} = \sum_{k \in B} m(\kappa, k) \Gamma_k,$$ (9)

where $m(\kappa, k)$ is given by formula

$$m(\kappa, k) = \begin{cases} 1 & \text{when lcd } (\kappa, k) = \kappa, \\ 0 & \text{in other cases,} \end{cases}$$ (10)

and B is the set of labels of all irreducible representations of the group C_N

$$B = \left\{ k = 0, \pm 1, \pm 2, ..., \begin{cases} \pm(N/2 - 1), N/2 & \text{for } N \text{ even} \\ \pm(N-1)/2 & \text{for } N \text{ odd} \end{cases} \right\}$$ (11)

This set is referred to as the Brillouin zone in condensed matter physic and the label k means a quasimomentum [5-7].

The irreducible representations of the cyclic group are one-dimensional

$$\Gamma_k(j) = e^{2\pi i k j / N}, \qquad j \in \tilde{N}, \ k \in B. \tag{12}$$

The transitive representation of the symmetric group is reducible and can be decomposed to the irreducible representations of the group Σ_N

$$R^{\Sigma_N : \Sigma^{(\mu)}} = \sum_{\lambda | - N} K_{\lambda \mu} \Delta^{\{\lambda\}}, \tag{13}$$

where the numbers $K_{\lambda \mu}$ denote the elements of the Kostka matrix K [3] and are equal to the number of the Young tableaux with the shape λ and weight μ.

3. The restriction of the irreducible representation of the symmetric group under the cyclic group.

In order to find out about how the irreducible representation of the symmetric group Σ_N decomposes under restriction to the cyclic group C_N to irreducible representations of the group C_N we will connect the formulas (8), (9) and (13) in to equation

$$\sum_{\lambda | - N} K_{\lambda \mu} \Delta^{\{\lambda\}} \downarrow C_N = \sum_{\kappa \in K(N), k \in B} m(\mu, \kappa) m(\kappa, k) \Gamma_k. \tag{14}$$

Rewriting this equation for all partitions μ we will get a set of equations, which is easy to solve due to the fact that the Kostka matrix K is upper triangular. The solution will provide numbers $m(\lambda, k)$ in the decomposition

$$\Delta^{\{\lambda\}} \downarrow C_N = \sum_{k \in B} m(\lambda, k) \Gamma_k. \tag{15}$$

Example 1.
Let us calculate the decomposition of a representation $\Delta^{\{2, 1, 1\}}$ under restriction to the cyclic group C_4 to the irreducible representations of this group.
The set of equations (4) will appear

$$\begin{cases} \Delta^{\{4\}} \downarrow C_4 = \Gamma_0 \\ \left(\Delta^{\{4\}} + \Delta^{\{3,1\}}\right) \downarrow C_4 = \Gamma_{-1} + \Gamma_0 + \Gamma_1 + \Gamma_2 \\ \left(\Delta^{\{4\}} + \Delta^{\{3,1\}} + \Delta^{\{2,2\}}\right) \downarrow C_4 = \Gamma_{-1} + 2\Gamma_0 + \Gamma_1 + 2\Gamma_2 \\ \left(\Delta^{\{4\}} + 2\Delta^{\{3,1\}} + \Delta^{\{2,2\}} + \Delta^{\{2,1,1\}}\right) \downarrow C_4 = 3\Gamma_{-1} + 3\Gamma_0 + 3\Gamma_1 + 3\Gamma_2 \end{cases} \tag{16}$$

The solution of this set of equations shows decompositions

$$\Delta^{\{4\}} \downarrow C_4^{\cdot} = \Gamma_0$$

$$\Delta^{\{3,1\}} \downarrow C_4 = \Gamma_{-1} + \Gamma_1 + \Gamma_2$$

$$\Delta^{\{2,2\}} \downarrow C_4 = \Gamma_0 + \Gamma_2$$

$$\Delta^{\{2,1,1\}} \downarrow C_4 = \Gamma_{-1} + \Gamma_0 + \Gamma_1.$$

(17)

4. Cyclage.

We will show another possibility to determine the decomposition described above but this time we will consider the poset structure on standard Young tableaux, which is obtained by an operation of cyclage [1,2]. The definition of the cyclage rests on the circular permutation on a word.

Let us recall some important definitions and then We will explain how to construct the graph representing the poset structure on standard Young Tableaux.

We shall treat the set of nodes, as a totally ordered alphabet A.

$$A = \tilde{N} = \{1,2,...,N\}$$

(18)

This set generates a free monoid A^* [1,2]. Let $w \in A^*$ be a word

$$w = a_1, a_2, \ldots , a_k. \quad \text{,where } a_i \in A .$$

(19)

and $w' \in A^*$ be a word obtained from a word w by moving the fist letter to the end of the word

$$w' = a_2, \ldots , a_k, a_1, \quad , a_i \in A .$$

(20)

We denote by ζ the bijection defined by

$$\zeta(w) = w'.$$

(21)

Using insertion algorithm (known also as Schensted's algorithm [2]) we may associate to each word $w \in A^*$ a Young tableau $t = P(w)$.

Let $\mathrm{Tab}(\cdot, \mu)$ denote the set of all Young tableaux with the weight μ. We may now define the cyclage [1,2] as the mapping

$$C : \text{Tab}(\cdot,\mu) \setminus \text{Row}(\mu) \rightarrow \text{Tab}(\cdot,\mu) \tag{22}$$

by the formula

$$C(t) = P(\zeta(w)), \tag{23}$$

where t is a tableau $t = P(w)$.

Shortly the operation of cyclage transforms the tableau t into the tableau $t' = P(\zeta(w))$.

Example 2.

The word $w = 4312$ by the insertion algorithm leads to the tableau t.

$$P(4312) = \begin{array}{|c|c|} \hline 1 & 2 \\ \hline 3 \\ \cline{1-1} 4 \\ \cline{1-1} \end{array} = t$$

By the operation of cyclage acting on the tableau t we obtain a tableau t'.

$$C(t) = P(\zeta(4312)) = P(3124) = \begin{array}{|c|c|c|} \hline 1 & 2 & 4 \\ \hline 3 \\ \cline{1-1} \end{array} = t'$$

5. The poset structure on standard Young tableaux.

We are ready now to use the map C to define a graph structure [1,2] on the set Tab $(\cdot,\mu)$ of Young tableaux. The vertices of this graph are the Young tableaux and the edges are defined by

$$t \longrightarrow t' \Leftrightarrow C(t) = t' \text{ and } a_1 > 1, \tag{24}$$

where a_1 is the first letter in a word w corresponding to a tableau t.

If the weight μ of Young tableaux is equal to 1^N then we obtain the graph representing the ranked poset structure on the set of standard Young tableaux with the row and column tableaux as the minimal and maximal element respectively.

The cocharge co(t) of a tableau t is defined as its rank in the poset $\text{Tab}(\cdot,\mu)$, that is the number of cyclages needed to transform t into the row tableau [1,2].

Alternatively, one can compute the cocharge of a Young tableau t whose weight is a partition $\mu = (1^N)$ using the following algorithm [1,2,4].

1. Read the word $w = (a_1, a_2, \ldots , a_N)$ from the tableau t by reading t from left to right and bottom to top.
2. Label each letter of w according to the following rules:
   - start at the letter $a_i = 1$ and label this letter with 0.
   - after labelling a_r with number k, label the next letter $a_j = a_r + 1$ with

number $k+1$ if it lies to the left of a_r or with number k if it lies to the right of a_r

3. The cocharge of t is the sum of all labels.

The complementary statistic it is the charge[1,2,4] of a table t.

$$c(t) = m(\mu) - co(t),\qquad(25)$$

where by $m(\mu)$ is denoted the maximal value of the cocharge on $\mathrm{Tab}(\cdot,\mu)$

$$m(\mu) = \frac{1}{2}(N(N-1)), \quad \text{for} \quad \mu = (1^N).\qquad(26)$$

Example 3.

Reading a tableau

$$t = \begin{array}{|c|c|}\hline 1 & 3 \\\hline 2 & 4 \\\hline\end{array}$$

we obtain the word $w = 2\ 4\ 1\ 3$, which we label $2_1\ 4_2\ 1_0\ 3_1$. The sum of labels is 4. Thus the cocharge of tableau t is equal to 4, $c(t) = 4$. This tableau is one of the vertices of the graph obtained for $\mu = (1,1,1,1)$. This is illustrated in Figure 2.

6. The Young tableaux and the irreducible representations of a cyclic group.

The irreducible representation $\Delta^{\{\lambda\}}$ of a symmetric group corresponds to a set of standard Young tableaux with the shape λ and the dimension of this representation is equal to the number of these tableaux [3]. Thus for representation $\Delta^{\{2,\ 1,\ 1\}}$ we have three tableaux

$$t_1 = \begin{array}{|c|c|}\hline 1 & 2 \\\hline 3 \\\hline 4 \\\hline\end{array} \qquad t_2 = \begin{array}{|c|c|}\hline 1 & 3 \\\hline 2 \\\hline 4 \\\hline\end{array} \qquad t_3 = \begin{array}{|c|c|}\hline 1 & 4 \\\hline 2 \\\hline 3 \\\hline\end{array}\ .\qquad(27)$$

$$\dim \Delta^{\{2,1,1\}} = 3$$

We find those tableaux in the graph (Figure 2.) at the levels labelled with the cocharges $c(t_1) = 3$, $c(t_2) = 4$ and $c(t_3) = 5$.

If we reduce the cocharges to the integers in the range given by the Brillouin zone we obtain the levels of the graph labelled with $k \in B$ (which are also the labels of the irreducible representations of the cyclic group C_N) Now we may

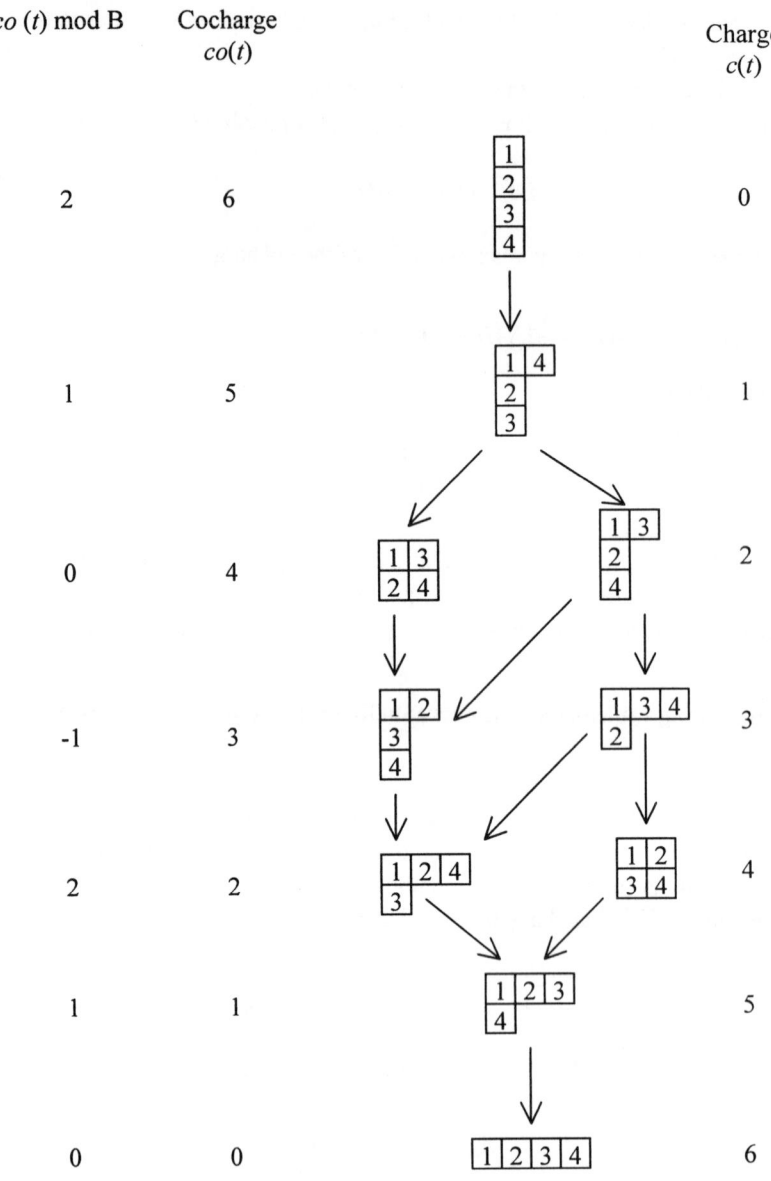

Figure 2. The ranked poset structure on the set of standard Young tableaux with the weight $\mu = (1,1,1,1)$. The cocharges reduced to the integers in the range given by the Brillouin zone label the levels of the graph by $k \in B$.

notice (Figure 2.) that the positions of the tableaux (27) correspond to the decomposition of the representation $\Delta^{\{2,\,1,\,1\}}$ under restriction to the group C_4 to the irreducible representations Γ_k

$$\Delta^{\{2,1,1\}} \downarrow C_4 = \Gamma_{-1} + \Gamma_0 + \Gamma_1. \tag{28}$$

Exactly the same decomposition we have obtained earlier (Eqs. (17)) by solving the set of equations (16). Checking the positions of the tableaux with the shapes (4), (3,1), (2,2) and obtaining the decompositions

$$\Delta^{\{4\}} \downarrow C_4 = \Gamma_0$$
$$\Delta^{\{3,1\}} \downarrow C_4 = \Gamma_{-1} + \Gamma_1 + \Gamma_2 \tag{29}$$
$$\Delta^{\{2,2\}} \downarrow C_4 = \Gamma_0 + \Gamma_2$$

which are the same as described by Eqs (17) we become convinced that for $N = 4$ the decomposition we are interested in satisfies relation given by conjecture

$$\Delta^{\{\lambda\}} \downarrow C_N = \sum_{t \in \mathrm{SYT}(\lambda)} \Gamma_{co(t)\,\mathrm{mod}\,B}. \tag{30}$$

This conjecture suggests that there is a relation between a set of cocharges of all standard Young tableaux with the shape λ and the set of quasimomenta of the quantum states represented by these Young tableaux.
Let

$$Co(\lambda) = \{co(t)|t \in \mathrm{SYT}(\lambda)\} \tag{31}$$

be the set of charges of all standard Young tableaux with the shape λ and

$$G(\lambda) = \{k = co(t) \,\mathrm{mod}\, B | t \in \mathrm{SYT}(\lambda)\} \tag{32}$$

be the set of quasimomenta k. We denote by g a map

$$g : Co(\lambda) \rightarrow F(\lambda) \tag{33}$$

$$g(co(t)) = co(t) \,\mathrm{mod}\, B = k$$

This map describes the relation between the set of cocharges and the set of quasimomenta.

7. Conclusions

We have noticed that the set of cocharges of Young tableaux with the shape λ corresponds to the set of indices of representations Γ_k. This is described by map (33). The application of this map to the stet of standard Young tableaux representing the quantum states of a Heisenberg magnet gives the distribution of quantum states over the Brillouin zone.

References

1. J. Désarménien, B. Leclerc, J.-Y. Thibon, *Hall-Littlewood functions and Kostka-Foulkes polynomials in representation theory*, Sèminaire Lotharingien de Combinatoire, **B32c** (1994).
2. A. Lascoux, B. Leclerc and J.-Y. Thibon *The plactic monoid*, in: *Algebraic combinatorics on words*, ed., M. Lothaire, Cambridge University Press, 2002.
3. B.G. Wybourne, *Symmetric Functions and Their Application to Problems in Physics*, in: *Symmetry and Structural Properties of Condensed Matter*, eds., T. Lulek, A. Wal, and B. Lulek; World Sci., Singapore (1998).
4. S. Dasmahapatra and O. Foda, *Strings, paths and standard tableaux*, Int. J. Mod. Phys. **A13**, 501-522 (1998).
5. B. Lulek, *Density of states in the Brillouin zone for a finite magnetic linear chain with a single impurity*, Acta Phys. Pol. **A74**, 453-63 (1988).
6. B. Lulek, *Recepta Weyla orbity nieregularne i łamanie symetrii translacyjenj jednowymiarowego magnetyka Heisenberga*, Uniwersytet im. A. Mickiewicza, Poznań. 1993.
7. A. Wal, T. Lulek, B. Lulek and E. Kozak, *Int. J. Mod. Phys.* **B13**, 3307 (1999).

N-DEPENDENCE OF THE SPECTRUM OF THE ONE-DIMENSIONAL HEISENBERG HAMILTONIAN IN TERMS OF SPECTRAL PARAMETERS

M. ŁABUZ[1] AND A. WAL[2]

Institute of Physics, Rzeszow University
Rejtana 16 a, 35 – 310 Rzeszow, Poland
[1]*labuz@univ.rzeszow.pl,* [2]*wal@univ.rzeszow.pl*

We evaluate solutions of The Heisenberg Hamiltonian as functions of the number N nodes for a specified set of winding numbers $\{n_1, n_2, n_3\}$. We plot the spectrum of solutions starting from the asymptotic case, up to $N = 6$.

1. Introduction

We investigate one-dimensional Heisenberg chain containing identical particles with spin 1/2 with an interaction only between neighbour particles in the chain. For such system the Hamiltonian has the form

$$\hat{H} = \sum_{n=1}^{N} (4\mathbf{S}^n \mathbf{S}^{n+1} - 1) \tag{1}$$

with boundary conditions

$$\mathbf{S}^N \equiv \mathbf{S}^1, \tag{2}$$

$\mathbf{S}^n$ indicates the spin vector at the position n in the chain.

The focus is on three-magnon excitations from ferromagnetic vacuum for finite chain. Vacuum state is the state with all spins oriented in the same direction and every deviation from this system is considered as pseudoparticle. Three-magnon excitations are equivalent to the chain with three spin deviations (or pseudoparticles).

We concentrate on Algebraic Bethe Ansatz (ABA), but also approach of coordinate substitution is used (CBA). We pay our attention especially on String Hypothesis (SH) and its appearance in solutions for fixed set of winding numbers $\{n_1, n_2, n_3\}$ and for fixed N. String Hypothesis was formulated for the first time by M. Takahashi. The paper published in 1971 contains form of hypothesis and wide mathematical discussion. There are also other papers, in which SH was used for Heisenberg chain. Essler was considering two-magnon excitations and discovered deficiencies in the String Hypothesis. For some cases, he obtained new pairs of real solutions instead of those in complex forms.

2. Co-ordinate Bethe Ansatz

Stationary state for the system being discussed following Bethe can be written in the form:

$$\Psi = \sum_{1 \le n_1 \le n_2 \ldots \le n_r \le N} a(n_1, n_2, \ldots, n_r) \, |n_1 n_2 \ldots n_r>, \tag{3}$$

where

$$a(n_1, n_2, \ldots, n_r) = \sum_P \exp\left(i \left(\sum_{l=1}^{r} k_{P(l)} n_l + \frac{1}{2} \sum_{j<l} \phi_{P(j)P(l)} \right) \right). \tag{4}$$

Here k_l are pseudomomenta, $\phi_{l,j}$ are the phases between interacting pseudoparticles and P is a permutation of the positions of r pseudoparticles to assure, that the pseudoparticles are indistinguishable. Finally $|n_1 n_2 \ldots n_r>$ denotes the magnetic configuration that corresponds to localisation of r inversions on proper nodes of the chain.

The energy of eigensate (3) is given by

$$E = -\sum_{l=1}^{r} (1 - \cos k_l). \tag{5}$$

Pseudomomenta and phases are related with each other by the reflection condition:

$$2 \cot \frac{\phi_{l,j}}{2} = \cot \frac{k_l}{2} - \cot \frac{k_j}{2} \,, \qquad \phi_{l,j} = -\phi_{j,l}. \tag{6}$$

Furthermore these parameters should also satisfy the boundary conditions:

$$N k_l = 2\pi n_l + \sum_{j \ne l} \phi_{l,j} \,, \qquad l = 1, 2, \ldots, r. \tag{7}$$

Here N is the length of the chain and n_l are winding numbers fulfilling the condition:

$$-\frac{N}{2} \le n_1 \le n_2 \ldots \le \frac{N}{2} \,. \tag{8}$$

3. Algebraic Bethe Ansatz

The equations (6,7) - called Bethe equations - contain two types of parameters: pseudomomenta and phases. It turns out that the problem of Heisenberg Hamiltonian can be solved using only one type of parameter - spectral parameter λ. We can make the substitution

$$\lambda_l = \cot\left(\frac{k_l}{2}\right), \tag{9}$$

which is equivalent to

$$\exp(ik_l) = \frac{\lambda_l + i}{\lambda_l - i} = e(\lambda_l). \tag{10}$$

In this parameterisation we find, that Bethe function can be written as

$$\Psi = \tag{11}$$

$$\sum_{1\leq n_1\leq\ldots\leq n_r} \sum_{P} \prod_{l=1}^{n_l} \left(e(\lambda_{P(l)})\right)^{n_l} \prod_{\substack{1<m<n<r \\ P(m)<P(n)}} e\left(\frac{\lambda_{P(m)} - \lambda_{P(n)}}{2}\right) |n_1 n_2 \ldots n_r>.$$

The Bethe equation takes the following form called Algebraic Bethe Ansatz:

$$\left(\frac{\lambda_l + i}{\lambda_l - i}\right)^N = \prod_{m\neq l}^{r} \frac{\lambda_l - \lambda_m + 2i}{\lambda_l - \lambda_m - 2i}, \tag{12}$$

or equivalently:

$$[e(\lambda_l)]^N = \prod_{\substack{m=1 \\ m\neq l}} \left[e\left(\frac{\lambda_l - \lambda_m}{2}\right)\right]. \tag{13}$$

The energy of state (12) is given by

$$E(\lambda_1, \ldots, \lambda_r) = \sum_{l=1}^{r} \frac{-2}{\lambda_l^2 + 1} . \tag{14}$$

The solutions of Bethe equation (3) have the special form. It was noticed by Takahashi and called String Hypothesis. According to it, we should expect that for the limit $N \to \infty$ (for fixed r) solutions $(\lambda_1, \ldots, \lambda_r)$ consist of sets of strings:

$$\lambda_m^{lv} = \lambda^{lv} + i(l + 1 - 2m) + O(e^{-\delta}), \qquad m = 1, 2, \ldots, l. \tag{15}$$

where $l \geq 1$ gives the length of the string, v labels different strings of the same length, m specifies the imaginary part of λ and $\delta > 0$.

4. Spectral parameters

System of equations (3) seems to be a task that is difficult to solve, especially for large chains of length N. To find values of spectral parameters λ in wide range of N we used formulas of CBA and relation between pseudomomenta k and λ. Procedure of calculations starts from solution of Bethe equations in asymptotic case. To get correct results we used asymptotic formulas, which allowed us to find values of k [10]. Next we applied numerical procedure of computations which let us investigate the change of spectral parameters λ as a function of chain length N starting from $N = 1000$ till $N = 6$. We noticed that the character of solutions does not change drastically and values of parameters evidence quasi-continuous form.

System of equations (7) is parameterised by winding numbers satisfying condition (8). From all of sets $\{n_1, n_2, n_3\}$ there have been chosen those, which guaranteed the complex form of solutions for pseudomomenta k and thus spectral parameters λ. As the representative cases, in this paper there have been selected the solutions, for which the character corresponds the best with the predictions of String Hypothesis. For three overturned spins we should expect:

    a) one string of length 3,
    b) one string of length 2 and one of length 1,
    c) three strings of length 1.

According to formula (15), complex part of string solutions should take the form:

    a) $-2i, 0, 2i$
    b) $-i, i; 0$

In calculations we have chosen sets $\{n_1, n_2, n_3\}$ for which we should expect strings of a) and b) type. Furthermore we restricted sets of winding numbers to those for which the character of solutions remain identical there are no changes from complex to real type of solution.

Finally we have chosen three representative sets of winding numbers: $\{-1,-1,-1\}$, $\{-1,1,1\}$, $\{-2,1,1\}$ and we plotted the change of λ in function of N in wide range of chain length.

5. Results

Results obtained during computations are presented in Figures 1, 2 and 3.

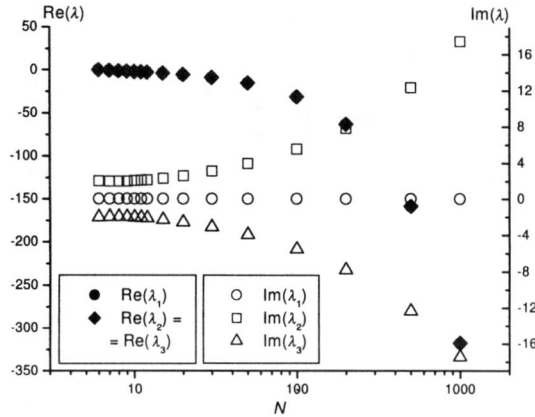

Figure 1.   Change of λ vs N for the set of winding numbers $\{-1,-1,-1\}$.

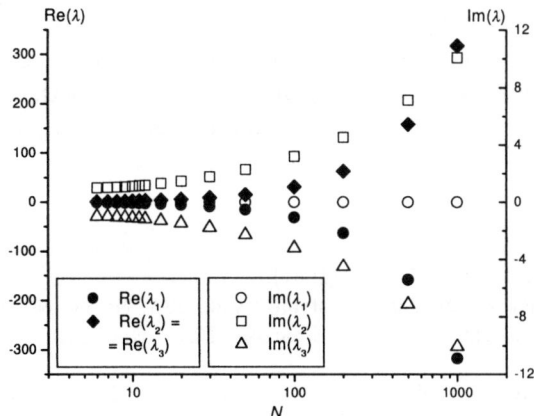

Figure 2.   Change of λ vs N for the set of winding numbers $\{-1,1,1\}$.

6. Conclusions

The method used in the paper allows us to follow the changes of spectral parameters λ for fixed set of winding numbers $\{n_1, n_2, n_3\}$ in wide range of chain length N - even for very long chains. But the way of computations disables to get all the results of ABA immediately for selected N and r.

342

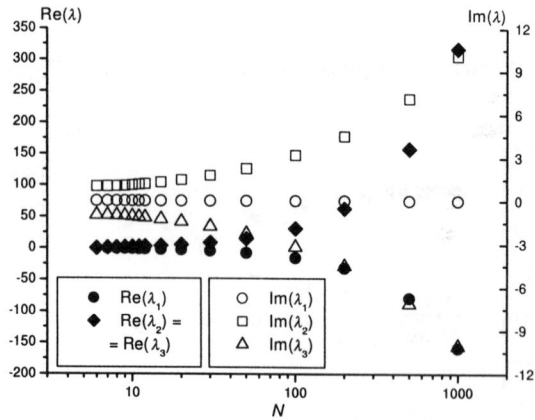

Figure 3.   Change of λ vs N for the set of winding numbers $\{-2,1,1\}$.

For chosen sets $\{n_1, n_2, n_3\}$ it was possible to plot the change of λ in function of N for $N \in (6, 1000)$. Analysing obtained results, one can see, that complex part of solutions (especially for large number of N) is not of the form included in (15). The larger the number of nodes is, the larger divergence from expected results may be observed. Complex values of spectral parameters increase exponentially together with chain length N. Unexpectedly in few cases, String Hypothesis is fulfilled for small N (e.g. $N = 6$).

References

1. H. Bethe, Z. Phys. 71, 205 (1931).
2. L. Hulthen, Archiv Math, Astron. Fysik, 26A, No 11, 1-106 (1938).
3. J. Des Cloizeaux and J. J. Pearson, Phys. Rev. 128, No 5, 2131-2135 (1958).
4. C. N. Yang, C. P. Yang, Phys. Rev. 150, No 1, 321-327 (1966); 150, No 1, 327-339 (1966).
5. M. Takahashi, Progress of Theoretical Physics, Vol. 46, No. 2 (1971).
6. L. D. Faddeev, L. A. Takhtadzhyan, Zap. Nauch. Sem. LOMI, 109, 134 (1981).
7. W. J. Caspers, "Spin Systems", World Scientific, Singapore, (1989).
8. F. H. L. Essler, V. E. Korepin, K. Schoutens, J. Phys: Math. Gen., 25, 4115-4126 (1992).
9. W. J. Caspers, P. K. H. Gragert, T. Lulek, A. Wal in *Symmetry and Structural Properties of Condensed Matter*, Proc. of the 6th International School of Theoretical Physics, eds. T. Lulek, B. Lulek, A. Wal (World Scientific, Singapore 2001) p. 285.
10. W. J. Caspers, to be published.

Part C

ENERGY BAND STRUCTURE IN SOLIDS

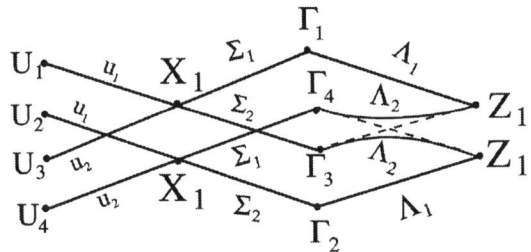

LOCALIZED FUNCTIONS AND THE ELEMENTARY ENERGY BANDS IN CRYSTALS WITH STRONGLY ANISOTROPIC STRUCTURE

M. SZNAJDER, D.M. BERCHA,

Institute of Physics, University of Rzeszów,
Rejtana 16a, 35-310 Rzeszów, Poland

C.C. TOVSTYUK

State University "Lvivska Polytechnika",
Bandera Street 12, 290046 Lviv, Ukraine

This paper is intended to show that the localized functions method enables to obtain the elementary energy bands in energy spectra of strongly anisotropic crystals. The elementary energy bands reflecting the Davydov splitting can exist in the energy spectra of such crystals in a more complex form than that following from the space symmetry group of a crystal. The localized functions method is applied to equivalent representations describing energy states in As_2S_3 crystal and to non-equivalent representations in the case of SbSI crystal.

1. Introduction

Recently it has been demonstrated[1] that the elementary energy bands[2,3] can be traced in numerically calculated energy spectra of crystals (electronic and phonon spectra). Three methods to predict them in energy spectra were discussed: the induction procedure on the basis of a given Wyckoff position in the unit cell, the compatibility relations for characters of irreducible representations and the empty lattice approximation. Since the elementary energy bands should be regarded as the smallest building parts of energy spectra of crystals one can utilize them as a useful tool for the prediction and verification of numerical calculations. This is especially important in the case of complicated crystals, i.e. with a large number of atoms in the unit cell, as well as strongly anisotropic ones whose band structure topology is also complicated. It is known that the band structure of an anisotropic crystal with at least two translationally non-equivalent weakly interacting structural units in the unit call should display the Davydov splitting. It is expected that the elementary energy bands should also enable to reflect this phenomenon.

2. The localized functions method

The localized functions method can be applied to investigate the Davydov splitting at any $\vec{k}$ point of the Brillouin zone (BZ). The localized orbitals (in particular the Wannier functions) are created by means of the Bloch functions. An arbitrary Bloch basis function of an irreducible representation of the space symmetry group of a crystal can be presented as a linear combination of localized functions by means of the projection operator:

$$\Psi^{\alpha j}_{i,\vec{k}_0}(\vec{r}) = \frac{f_{\alpha,\vec{k}_0}}{Nn} \sum_g \tau^{\alpha}_{ij,\vec{k}_0}(g)\, \hat{g}\, \varphi(\vec{r}) = \frac{f_{\alpha,\vec{k}_0}}{Nn} \sum_g \tau^{\alpha}_{ij,\vec{k}_0}(g)\, \varphi_g(\vec{r}). \qquad (1)$$

In this equation g is an element of the wave vector group $G_{\vec{k}_0}$, $\tau^{\alpha}_{ij,\vec{k}_0}$ is a matrix element of the α representation, $f_{\alpha,\vec{k}_0}$ denotes a dimension of the α irreducible representation of the wave vector group $G_{\vec{k}_0}$, n is a number of elements of $G_{\vec{k}_0}$, j denotes a representation equivalent to the α-representation, N means a number of unit cells in a crystal. By $\varphi(\vec{r})$ we understand a localized function in a given position of the unit cell: $(e_1, e_2, , ..., e_r, e_n)$. Further this function can be conveniently denoted as follows: $\varphi(\vec{r}) \equiv \varphi(1,1,0)$ where the right-hand mark means a number of the unit cell, the middle mark is a number of a position, the left-hand mark is a number of a basis function. Since an element of the space group of the crystal g can include a translation by vector $\vec{a}$, therefore, an action of this element on $\varphi(\vec{r})$ function can be presented as: $g\,\varphi(\vec{r}) \equiv g\varphi(1,1,0) \equiv \varphi(1,r,\vec{a})$ where $g \equiv \{E/\vec{a}\}g_{r1}$, and a matrix element: $\tau^{\alpha}_{ij,\vec{k}_0}(g) \equiv \tau^{\alpha}_{ij,\vec{k}_0}(g_{r1})\, e^{i\vec{k}_0\vec{a}}$. Hence, Eq. (1) can be rewritten as follows:

$$\Psi^{\alpha j}_{i,\vec{k}_0}(\vec{r}) = \frac{f_{\alpha,\vec{k}_0}}{Nn} \sum_{g_{r1},\vec{a}} \tau^{\alpha}_{ij,\vec{k}_0}(g_{r1}) e^{i\vec{k}_0\vec{a}}\, \varphi(1,r,\vec{a}) =$$

$$= \frac{f_{\alpha,\vec{k}_0}}{n} \sum_{g_{r1}} \tau^{\alpha}_{ij,\vec{k}_0}(g_{r1}) \cdot \frac{1}{N} \sum_{\vec{a}} e^{i\vec{k}_0\vec{a}}\, \varphi(1,r,\vec{a}) = \qquad (2)$$

$$= \frac{f_{\alpha,\vec{k}_0}}{n} \sum_{g_{r1}} \tau^{\alpha}_{ij,\vec{k}_0}(g_{r1}) \cdot \varphi(1,r,\vec{k}_0),$$

where $\varphi(1,r,\vec{k}_0) \equiv \frac{1}{N} \sum_{\vec{a}} e^{i\vec{k}_0\vec{a}}\, \varphi(1,r,\vec{a})$ is an extended function presented in a representation which is analogous to the **kq** one, introduced by Zak.[4] It can be easily shown that the extended function is transformed under the action of an element $g_{s1} = \{h_{s1}/\vec{\alpha}_{s1}\} \in G_{\vec{k}_0}$ into extended function built from the function localized in another position and a phase factor can appear:

$$g_{s1}\varphi(1,r,\vec{k}_0) = \{h_{s1}/\vec{\alpha}_{s1}\}\frac{1}{N}\sum_{\vec{a}} e^{i\vec{k}_0\vec{a}}\,\varphi(1,r,\vec{a}) =$$

$$= \{h_{s1}/\vec{\alpha}_{s1}\}\frac{1}{N}\sum_{\vec{a}} e^{i\vec{k}_0\vec{a}}\,\{h_{r1}/\vec{\alpha}_{r1}\}\varphi(1,1,\vec{a}) =$$

$$= \frac{1}{N}\sum_{\vec{a}} e^{i\vec{k}_0\vec{a}}\,\underbrace{\{h_{s1}/\vec{\alpha}_{s1}\}\{h_{r1}/\vec{\alpha}_{r1}\}}_{\{E/\vec{a}'_{sr}\}g_{f1}}\,\varphi(1,1,\vec{a}) = \tag{3}$$

$$= \frac{1}{N}\sum_{\vec{a}} e^{i\vec{k}_0\vec{a}}\{E/\vec{a}'_{sr}\}g_{f1}\varphi(1,1,\vec{a}) = \frac{1}{N}\sum_{\vec{a}} e^{i\vec{k}_0\vec{a}}\varphi(1,f,\vec{a}+\vec{a}'_{sr}).$$

$$g_{s1}\varphi(1,r,\vec{k}_0) = \frac{1}{N}e^{-i\vec{k}_0\vec{a}'_{sr}}\sum_{\vec{a}''} e^{i\vec{k}_0\vec{a}''}\,\varphi(1,f,\vec{a}''),$$

where $\vec{a}'' = \vec{a} + \vec{a}'_{sr}$. Finally:

$$g_{s1}\varphi(1,r,\vec{k}_0) = e^{-i\vec{k}_0\vec{a}'_{sr}} \cdot \varphi(1,f,\vec{k}_0). \tag{4}$$

Phase factors and new positions follow from the multiplication table for elements of the wave vector group.

An element of the secular matrix which determines the energy spectra of a crystal can be written as a matrix element of the Hamiltonian in the basis of $\Psi^{\alpha j}_{i\vec{k}_0}$ functions:

$$D^{\alpha}_{iji'\,j'}(\vec{k}_0) = \int \Psi^{\alpha j*}_{i\,\vec{k}_0}(\vec{r})\,\hat{H}\,\Psi^{\alpha j'}_{i'\,\vec{k}_0}(\vec{r})\,dV =$$

$$= \frac{f^2_{\alpha,\vec{k}_0}}{n^2}\sum_{g_r,g'_r} T^{*\alpha}_{ij,\vec{k}_0}(g_{r1})\,T^{\alpha}_{i'j',\vec{k}_0}(g_{r'1})\int \varphi^*(1,r,\vec{k}_0)\hat{H}\varphi(1,r',\vec{k}_0)dV. \tag{5}$$

It can be used to calculate energy states in a given $\vec{k}_0$ point. In Eq. (5) the Hamiltonian $\hat{H} = -\frac{\hbar^2}{2m}\nabla^2 + V(\vec{r})$ is invariant with respect to crystal symmetry transformations and it is Hermitian, $V(\vec{r})$ denotes the crystal potential, while a matrix element $D^{\alpha}_{iji'\,j'}$ is invariant with respect to the coordinate transformations. The secular matrix that is built from the above matrix elements can be written for a given crystal, this will be presented in the next section.

3. Elementary energy bands for strongly anisotropic crystals

Consider the layered As$_2$S$_3$ crystal (orpiment) from the monoclinic system, described by the C^5_{2h} space symmetry group. There are two translationally non-equivalent layers in a unit cell of the crystal. The symmetry group of a single layer is C_s^2. Since the interaction between the non-equivalent layers is weak one should observe the Davydov splitting in the energy spectra of the orpiment crystal. We check if the splitting exists at the edge of the BZ, namely in Z ($\vec{k} = 1/2\vec{b}_3$) and X ($\vec{k} = 1/2\vec{b}_1$) points. The wave vector groups at those points have only one representation, therefore, all the energy states at those points are described by the equivalent representations. The Bloch basis functions for points Z and X according to Eq. (2) are the following:

Z point:

$$\Psi_{11} = \frac{2}{4N} \sum_{\vec{a}} e^{i\vec{k}_0 \cdot \vec{a}} \left(\varphi(1,1,\vec{a}) + \varphi(1,4,\vec{a}) \right) = \frac{1}{2} \left(\varphi(1,1,\vec{k}_0) + \varphi(1,4,\vec{k}_0) \right),$$

$$\Psi_{12} = \frac{-2}{4N} i \sum_{\vec{a}} e^{i\vec{k}_0 \cdot \vec{a}} \left(\varphi(1,2,\vec{a}) - \varphi(1,3,\vec{a}) \right) = -\frac{1}{2} i \left(\varphi(1,2,\vec{k}_0) - \varphi(1,3,\vec{k}_0) \right),$$

$$\Psi_{21} = \frac{-2}{4N} i \sum_{\vec{a}} e^{i\vec{k}_0 \cdot \vec{a}} \left(\varphi(1,2,\vec{a}) + \varphi(1,3,\vec{a}) \right) = -\frac{1}{2} i \left(\varphi(1,2,\vec{k}_0) + \varphi(1,3,\vec{k}_0) \right),$$

$$\Psi_{22} = \frac{2}{4N} \sum_{\vec{a}} e^{i\vec{k}_0 \cdot \vec{a}} \left(\varphi(1,1,\vec{a}) - \varphi(1,4,\vec{a}) \right) = \frac{1}{2} \left(\varphi(1,1,\vec{k}_0) - \varphi(1,4,\vec{k}_0) \right).$$

Positons e_1, e_2, e_3, e_4 where the fuctions $\varphi(\vec{r})$ are localized were obtained by the action of elements from $G_{\vec{k}_0}$ on the chosen initial position e_1:

$$g_{11}e_1 = e_1; \quad g_{11} = \{E/0\}, \qquad g_{31}e_1 = e_3; \quad g_{31} = \left\{ I/\tfrac{\vec{a}_1+\vec{a}_3}{2} \right\},$$
$$g_{21}e_1 = e_2; \quad g_{21} = \left\{ C_{2z}/\tfrac{\vec{a}_3}{2} \right\}, \qquad g_{41}e_1 = e_4; \quad g_{41} = \left\{ \sigma_z/\tfrac{\vec{a}_1}{2} \right\}.$$

X point:

$$\Psi_{11} = \frac{2}{4N} \sum_{\vec{a}} e^{i\vec{k}_0 \cdot \vec{a}} \left(\varphi(1,1,\vec{a}) - i\varphi(1,4,\vec{a}) \right) = \frac{1}{2} \left(\varphi(1,1,\vec{k}_0) - i\varphi(1,4,\vec{k}_0) \right),$$

$$\Psi_{12} = \frac{2}{4N} \sum_{\vec{a}} e^{i\vec{k}_0 \cdot \vec{a}} \left(i\varphi(1,2,\vec{a}) + \varphi(1,3,\vec{a}) \right) = \frac{1}{2} i \left(\varphi(1,2,\vec{k}_0) - i\varphi(1,3,\vec{k}_0) \right),$$

$$\Psi_{21} = \frac{2}{4N} \sum_{\vec{a}} e^{i\vec{k}_0 \cdot \vec{a}} \left(-i\varphi(1,2,\vec{a}) + \varphi(1,3,\vec{a}) \right) = \frac{-1}{2} i \left(\varphi(1,2,\vec{k}_0) + i\varphi(1,3,\vec{k}_0) \right),$$

$$\Psi_{22} = \frac{2}{4N} \sum_{\vec{a}} e^{i\vec{k}_0 \cdot \vec{a}} \left(\varphi(1,1,\vec{a}) + i\varphi(1,4,\vec{a}) \right) = \frac{1}{2} \left(\varphi(1,1,\vec{k}_0) + i\varphi(1,4,\vec{k}_0) \right),$$

Since the localized functions are the basis of a regular representation therefore, due to the Burnside theorem, to every two-dimensional representation corresponds another equivalent representation. Hence, the secular matrix D is built from four Bloch basis functions and the matrix elements are the following:

$$D_{ijkl} \equiv \int \Psi_{ij}^* \hat{H} \Psi_{kl} dV.$$

After calculations the following matrices were obtained:

Z point :   **X** point :

$$D = \begin{pmatrix} A+B & K & 0 & 0 \\ K^* & A-B & 0 & 0 \\ 0 & 0 & A+B & K \\ 0 & 0 & K^* & A-B \end{pmatrix}, \quad D = \begin{pmatrix} A & L^* & 0 & 0 \\ L & A & 0 & 0 \\ 0 & 0 & A & L^* \\ 0 & 0 & L & A \end{pmatrix}. \quad (6)$$

where

$$A \sim \langle \varphi(1,1,\vec{k}_0) | \hat{H} | \varphi(1,1,\vec{k}_0) \rangle,$$
$$B \sim \langle \varphi(1,1,\vec{k}_0) | \hat{H} | \varphi(1,4,\vec{k}_0) \rangle,$$
$$K \sim \langle \varphi(1,1,\vec{k}_0) | \hat{H} | \varphi(1,3,\vec{k}_0) \rangle,$$
$$L \sim \langle \varphi(1,4,\vec{k}_0) | \hat{H} | \varphi(1,3,\vec{k}_0) \rangle + \langle \varphi(1,1,\vec{k}_0) | \hat{H} | \varphi(1,3,\vec{k}_0) \rangle$$

Elements A and B are larger then K and L since they are built from localized functions describing atoms in the same layer of the crystal. Solutions of the corresponding secular equations for matrices defined by Eq. (6) are the following:

$$\textbf{Z} \text{ point}: E = A \pm \sqrt{B^2 + KK^*}, \quad \textbf{X} \text{ point}: E = A \pm \sqrt{LL^*}. \quad (7)$$

Hence, one can expect a large splitting between energy states at Z point, resulting from the interlayer interaction and, a small splitting at X point, the Davydov one. In order to draw a scheme of the energy spectrum of As$_2$S$_3$ crystal between X, Γ and Z points it is necessary to estimate a splitting in the center of the BZ, i.e. in the Γ point. Using the Davydov method, we reduced irreducible representations of the symmetry group of a single layer (in the Γ point) into irreducible representations of the symmetry group of the crystal (Γ point):

$$\Gamma_1' \longrightarrow \Gamma_1, \Gamma_4; \quad \Gamma_2' \longrightarrow \Gamma_2, \Gamma_3.$$

These pairs describe the energy states in the Γ point where the Davydov splitting can be observed. Hence, a scheme of the energy spectrum of the As$_2$S$_3$ crystal presented in Fig. 1.

One can observe there four connected bands. We established additionally using the compatibility relations that the elementary energy bands for the symmetry group of the orpiment crystal C_{2h}^5 consist of 2 bands. They are presented in Fig. 2a. However, the representations at the Γ point in Fig. 2a are not the Davydov dublets. In order to model the predicted Davydov splitting at the Γ point one should overlap the elementary enregy bands in the way presented in Fig. 2b. The last scheme can be compared to Fig. 1 (directions X-Γ-Z).

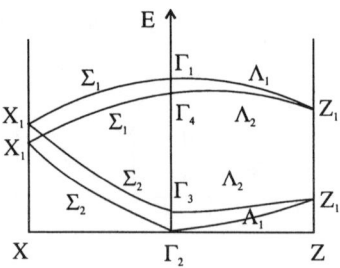

Figure 1. A scheme of the energy spectrum of the As_2S_3 crystal.

Fig. 1 revelas an advantage of the localized functions method - it enables to determine in the energy spectra of strongly anisotropic crystals the physically composed elementary energy bands. They exist in such spectra in a complex form that reflect physical properties of the crystal. Numerical calculations of the band structure of As_2Se_3 crystal[5] (that is isomorphic to As_2S_3 one) confirm an existence of the overlapping elementary energy bands presented in Fig. 1. At the calculated band structure one can observe the predicted in Fig. 1 unavoidable degeneracy of bands in the $\Gamma - X$ direction.

The localized function method can be used to obtain the elementary energy bands in the case of strongly anisotropic crystals of higher symmetry, e.g. the orthorhombic one. For the orthorhombic system two dimensional, non-equivalent representations can describe the energy states of points from the BZ's edges. We have analyzed a chain SbSI crystal described by D_{2h}^{16} space symmetry group. Recently, our numerical calculations[1] have confirmed an existence of the elementary energy bands in the band structure of the SbSI crystal. Our analysis shows, as should be expected, that the Davy-

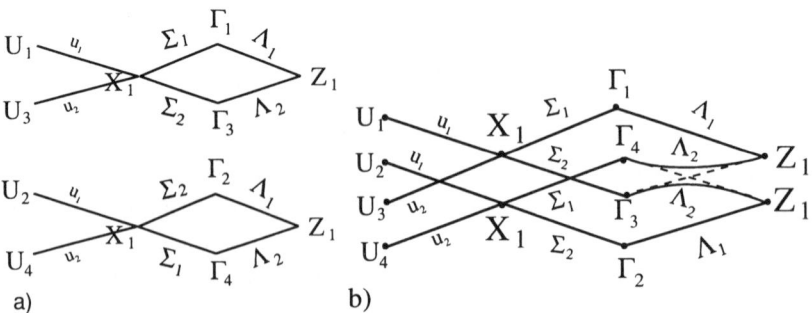

Figure 2. a) Elementary energy bands of As_2S_3 crystal. b) Two overlapping elementary energy bands of As_2S_3.

dov splitting exists between the energy states described by non-equivalent two-dimensional representations at the BZ's edges.

4. Conclusions

The localized functions method allows to complete the group-theoretical analysis with information about the energy distances between bands in high-symmetry points of the BZ. It enables to estimate the splitting between bands quantitatively, moreover, involving the compatibility relations one can predict the topology of bands. The elementary energy bands can be obtained by the localized functions method in the energy spectra of strongly anisotropic crystals. They exist in such spectra in a complex, overlapping form that reflects the Davdyov splittings.

References

1. D.M. Bercha, M. Sznajder, K.Z. Rushchanskii, in *Symmetry and Structural Properties of Condensed Matter*, eds.: T. Lulek, B. Lulek, A. Wal (World Scientific, Singapore, 2000), pp. 376-382.
2. L. Michel and J. Zak, *Phys. Rev.* **B59**, 5998 (1999).
3. D.M. Bercha and O.B. Mitin, *Fiz. Tekh. Polup.* **27**, 508 (1987).
4. J. Zak, *Phys. Rev. Lett.* **45**, 1025 (1980).
5. E. Tarnow, A. Antonelli, and J.D. Joannopoulos, *Phys. Rev.* **B34**, 4059 (1986).

HARMONIC OSCILLATORS REVISITED: RELATIVISTIC FORMULATIONS, CONFINED PARTICLES, AND SEVERAL OTHER ASPECTS

JACEK KARWOWSKI

Instytut Fizyki, Uniwersytet Mikołaja Kopernika,
Grudziądzka 5, PL-87-100 Toruń, Poland
E-mail: jka@phys.uni.torun.pl

The eigenvalue problems corresponding to one and two particles confined in harmonic oscillator potentials are analyzed. In particular Klein-Gordon and Dirac equations, which in the non-relativistic limit transform, respectively, into Lévy-Leblond and Schrödinger equations for harmonic oscillators, are constructed. Differences between the equations describing spin-0 and spin-$\frac{1}{2}$ particles are briefly discussed. In particular the eigenvalue spectra of different kinds of harmonic oscillators are compared. Besides, some properties of a system of two identical particles interacting with a Coulomb force and bound by a Hook's law potential (harmonium) are discussed. Usefulness of the harmonic-oscillator-type potentials in modeling of spatially confined systems is emphasized.

1. Introduction

Studies on quantum objects confined by external potentials constitute a substantial part of nuclear, atomic, molecular, and solid state physics. In particular, bound states of an N-electron atom are usually obtained by solving an eigenvalue problem of a Hamiltonian describing N electrons confined by the external potential generated by a positive charge. Molecular vibrations or behavior of quantum dots can be modeled by confining the pertinent particles in a harmonic-oscillator-type potential. Confinement by different kinds of power-law (i.e. r^n) potentials is frequently used to modeling nucleons and other systems of elementary particles. Behavior of atoms embedded in plasmas or in neutral media, such as liquid helium, may be modeled by introducing an exponential (Yukawa-type) confinement. Consequently, models of confinement which lead to analytical solutions of the corresponding Hamiltonian eigenvalue problem are of a particular interest and importance. Among them, problems known as "hydrogen atom", i.e. confinement of an electron by a $-Z/r$ potential and "harmonic oscillator" when the confinement is described by a Hooke's law potential $\omega^2 r^2/2$ constitute essential parts of all textbooks of quantum mechanics.

In this paper we are concerned with some aspects of the harmonic oscillator problem. Due to its formal simplicity, it is used as the most basic pedagogical example of exactly solvable quantum mechanical problems. However it also offers a possibility of modeling a large class of physical phenomena. The Schrödinger equation for an electron in a uniform magnetic field confined by a harmonic oscillator type potential was solved already in 1928 by Fock[1] and two years later again by Darwin[2]. Since then numerous studies on a variety of applications of the harmonic-oscillator-type potentials have been published. Among very recent studies on models based on harmonic oscillator one should mention a paper on symplectic models of n-particle systems confined in a harmonic oscillator potential[3] and another one on rotation-vibration spectra of diatomic molecules[4]. A variety of exactly and quasi-exactly solvable models have been discussed during the last decade[5,6,7]. Many other examples may be found in the Ushveridze monograph[8] and in a review by Jaskólski[9]. Also in modeling quantum dots particular attention has been paid to the confinement by the harmonic oscillator potential [10,11].

The eigenvalue problem of a Hamiltonian describing two interacting electrons confined by an external harmonic oscillator potential proved to be also analytically solvable[12]. The motivation for studies on this system, known as *harmonium*, apart of purely academic curiosity, resulted from an attempt to understand in more detail the electron correlation problem. A very recent paper on this subject is also concerned with various aspects of electron correlation[13]. A detailed study on the spectrum of harmonium is subject of another paper presented at this School[14]. Another example of exactly (or rather quasi-exactly) solvable problem is a system of two electrons interacting by a Coulomb repulsion potential with an additive linear term and bound by a central force represented by a harmonic oscillator potential[15,16]. An application of the Heun equation[17] to this class of problems may open new possibilities of finding their exact solutions.

Relativistic generalizations of the harmonic oscillator appear to be neither trivial nor unique. First of all, the mode of a generalization depends upon the definition of the relativistic oscillator. The first studies, performed in the early thirties by Nikolsky[18] and Postępska[19] were concerned with a Dirac equation for an electron in the field of a scalar quadratic potential. The resulting eigenvalue problem reduces to a quartic equation with no bound solutions. The non-relativistic discrete energy levels correspond in this oscillator to resonances. Another approach, recently formulated by Toyama and Nogami[20], aims at finding a Dirac system with an infinite number of bound states whose energies are all equally spaced. Such a system has been constructed using the inverse scattering method[21,20]. An approach, leading to the so called Dirac oscillator, is based on a construction of the Dirac equation which is exactly solvable and in the non-relativistic limit gives the Schrödinger harmonic oscillator equation[22,23,24,25,26].

In this paper we construct relativistic oscillators which have another feature of the non-relativistic harmonic oscillator: the Hamiltonian is invariant with respect to a canonical transformation interchanging coordinates and momenta. Thus, we are looking for oscillators which are described by the same Hamiltonian in both coordinate and momentum representations. A particular attention is given to the spin of the confined particle. We discuss an electron described by the Dirac equation and, in the non-relativistic limit by the Lévy-Leblond equation[27], and a spin 0 particle described by the Klein-Gordon and by the Schrödinger equations. Usefulness of non-Hermitian vector potentials to construction of some of the models is demonstrated in an analysis of the Lévy-Leblond and Dirac equations. Besides, non-relativistic models describing two interacting particles confined in harmonic oscillator potentials are briefly discussed.

Atomic units are used in this paper. The velocity of light $c \approx 137$. The mass m of the particle is explicitly given in all equations.

2. One particle

Let us assume that a particle moves in an external field described by a stationary model potential which, in general, may be composed of a vector potential $\boldsymbol{A}(\boldsymbol{r})$ and a scalar potential $V(\boldsymbol{r})$. We are interested in energies of the bound states of the particle. The model potentials have to be neither of electromagnetic origin nor Hermitian but we assume that they depend only on the coordinates $\boldsymbol{r}$ of the particle. We also assume that the resulting Hamiltonian is Hermitian. The origin of the energy scale is at the ionization limit in both non-relativistic and relativistic models. Thus, the total relativistic energy of the particle is equal to $E + \mathrm{m}c^2$.

A spin-less particle is described by the Klein-Gordon equation

$$\left[\frac{1}{2\mathrm{m}}(\boldsymbol{p} - \boldsymbol{A}^\dagger)(\boldsymbol{p} - \boldsymbol{A}) + V - E - \frac{(V - E)^2}{2\mathrm{m}c^2} \right] \Psi_K = 0, \qquad (1)$$

where all symbols have their usual meaning. In the non-relativistic limit the Klein-Gordon equation transforms to the Schrödinger equation

$$\left[\frac{1}{2\mathrm{m}}(\boldsymbol{p} - \boldsymbol{A}^\dagger)(\boldsymbol{p} - \boldsymbol{A}) + V - E \right] \Psi_S = 0. \qquad (2)$$

A spin-$\frac{1}{2}$ particle is described by the Dirac equation which, in the standard (Dirac-Pauli) representation, reads

$$\begin{bmatrix} (V-E)\boldsymbol{I}, & c\boldsymbol{\sigma}\cdot(\boldsymbol{p}-\boldsymbol{A}^{\dagger}) \\ c\boldsymbol{\sigma}\cdot(\boldsymbol{p}-\boldsymbol{A}), & (V-E-2mc^2)\boldsymbol{I} \end{bmatrix} \begin{bmatrix} \Psi_{\mathcal{D}}^{L} \\ \Psi_{\mathcal{D}}^{S} \end{bmatrix} = 0, \tag{3}$$

where $\boldsymbol{I}$ is a 2×2 unit matrix, $\boldsymbol{\sigma}$ are the Pauli spin matrices and $\Psi_{\mathcal{D}}^{L}/\Psi_{\mathcal{D}}^{S}$ are traditionally called the large/small components of the wavefunction. The non-relativistic limit of Eq. (3)

$$\begin{bmatrix} (V-E)\boldsymbol{I}, & c\boldsymbol{\sigma}\cdot(\boldsymbol{p}-\boldsymbol{A}^{\dagger}) \\ c\boldsymbol{\sigma}\cdot(\boldsymbol{p}-\boldsymbol{A}), & -2mc^2\boldsymbol{I} \end{bmatrix} \begin{bmatrix} \Psi_{\mathcal{L}}^{L} \\ \Psi_{\mathcal{L}}^{S} \end{bmatrix} = 0 \tag{4}$$

is known as the Lévy-Leblond equation[27]. The elimination of $\Psi_{\mathcal{L}}^{S}$ from Eq. (4) gives

$$\left\{ \left[\frac{1}{2m}(\boldsymbol{p}-\boldsymbol{A}^{\dagger})(\boldsymbol{p}-\boldsymbol{A}) + V - E \right]\boldsymbol{I} - \frac{\boldsymbol{\sigma}}{2m}[(\nabla\times\boldsymbol{A})+\boldsymbol{M}] \right\} \Psi_{\mathcal{L}}^{L} = 0, \tag{5}$$

where

$$\boldsymbol{M} = \left(\boldsymbol{A}^{\dagger}-\boldsymbol{A}\right)\times\nabla - i\left(\boldsymbol{A}^{\dagger}\times\boldsymbol{A}\right) \tag{6}$$

and $\Psi_{\mathcal{L}}^{L}$ is a two-component Pauli spinor. If $\boldsymbol{A}$ is Hermitian then $\boldsymbol{M}=0$ and Eq. (5) takes the standard form of the non-relativistic Schrödinger-Pauli equation. Let us note that Eq. (5) rather than Eq. (2) should be interpreted as the non-relativistic limit of the Dirac equation.

We define a harmonic oscillator as a quantum system for which the Hamiltonian remains invariant under transformation

$$p_1 \leftrightarrow a_1 x_1, \quad p_2 \leftrightarrow a_2 x_2, \quad p_3 \leftrightarrow a_3 x_3, \tag{7}$$

where a_1, a_2, a_3 are constants. As one can easily see, Eqs. (1) – (4) correspond the harmonic oscillators if

$$\boldsymbol{A} = ik_a \boldsymbol{r}, \tag{8}$$

where $\boldsymbol{r} = \{x_1, x_2, x_3\}$ and $V = V_0 = const.$ Introducing operators

$$\mathbf{a} = \frac{1}{\sqrt{2k_a}}\left(k_a \boldsymbol{r} + i\boldsymbol{p}\right), \quad \mathbf{a}^{\dagger} = \frac{1}{\sqrt{2k_a}}\left(k_a \boldsymbol{r} - i\boldsymbol{p}\right) \tag{9}$$

we get

$$\left[\mathbf{a}_j, \mathbf{a}_k^{\dagger}\right] = \delta_{jk}, \tag{10}$$

$$\frac{1}{2m}(\boldsymbol{p}-\boldsymbol{A}^{\dagger})(\boldsymbol{p}-\boldsymbol{A}) = \omega\,\mathbf{a}^{\dagger}\mathbf{a} = \frac{p^2}{2m} + \frac{m\omega^2}{2}r^2 - \frac{3k_a}{2m} \tag{11}$$

and

$$M = 2ik_a \left[\mathbf{a}^\dagger \times \mathbf{a} \right] = 2k_a \mathbf{L}, \tag{12}$$

where $\omega = |k_a|/\mathrm{m}$ and $\mathbf{L} = \mathbf{r} \times \mathbf{p}$ is the orbital angular momentum.

Another option, valid for Eq. (2) and Eq. (4), is

$$A = ik_a \mathbf{r}, \quad V = k_v r^2, \tag{13}$$

where k_a and k_v are constants. However, scalar potentials which produce either Klein-Gordon or Dirac equations invariant with respect to transformation (7), if different from a constant, have to be energy-dependent. For example, in the case of the Klein-Gordon equation, such a pseudo-potential is real in a finite range of r only and reads

$$V_E = (\mathrm{mc}^2 + E) \left[1 - \sqrt{1 - k_v r^2 \frac{2\mathrm{mc}^2}{(\mathrm{mc}^2 + E)^2}} \right]. \tag{14}$$

In the non-relativistic limit we get

$$V_E = k_v r^2 - \frac{kr^2}{\mathrm{mc}^2} (E + \tfrac{1}{2} kr^2) + \mathcal{O}\left[(r\alpha)^4 \right]. \tag{15}$$

A less obvious solution, valid for two-dimensional problems, is

$$A = \frac{1}{2} [\mathbf{B} \times \mathbf{r}], \tag{16}$$

where $\mathbf{r} = 1/\sqrt{2k_a}(\mathbf{a}^\dagger + \mathbf{a})$ and $\mathbf{B} = \nabla \times \mathbf{A}$ may be interpreted as the external magnetic field. In this case $M = 0$ and, assuming $\mathbf{B} = \{0, 0, B\}$,

$$(\mathbf{p} - \mathbf{A})^2 = p^2 + \frac{1}{4} B^2 \rho^2 - \mathbf{B} \cdot \mathbf{L}, \tag{17}$$

where $\rho^2 = x_1^2 + x_2^2$. In the non-relativistic case [Eqs. (2) and (4)] we can take additionally $V = k_v x_3^2$ in order to get the equations invariant with respect to the three-dimensional $\mathbf{r} \leftrightarrow \mathbf{p}$ transformation.

As it is seen, harmonic oscillator Hamiltonians defined in this section are expressible in terms of linear combinations of $a_j^\dagger a_k$ and, thus, are elements of $SU(3)$ algebra. Consequently, the corresponding eigenvalue problems are all exactly solvable and the spectra may be generated using appropriate ladder operators.

2.1. Spin-0 particle

In the case described by Eq. (8), taking for simplicity $V = 0$, both Klein-Gordon and Schrödinger equation may be written as

$$\left(\frac{p^2}{2\mathrm{m}} + \frac{\mathrm{m}\omega^2}{2} r^2 - \varepsilon \right) \Psi = 0, \tag{18}$$

where $\omega = |k_a|/\mathrm{m}$ and

$$\varepsilon = \begin{cases} E + 3k_a/(2\mathrm{m}), & \text{(Schrödinger)}, \\ E + 3k_a/(2\mathrm{m}) + E^2/(2\mathrm{m}c^2), & \text{(Klein-Gordon)}. \end{cases} \qquad (19)$$

As it is well known, $\varepsilon = (2n + \ell + \frac{3}{2})\omega$. Therefore in the Schrödinger case

$$E \equiv E_{n\ell}^{\mathcal{S}} = (2n + \ell + \frac{3}{2}\delta)\omega, \qquad (20)$$

and in the Klein-Gordon one

$$E_{n\ell}^{\mathcal{K}} = \mathrm{m}c^2 \left[\sqrt{1 + \frac{2E_{n\ell}^{\mathcal{S}}}{\mathrm{m}c^2}} - 1 \right], \qquad (21)$$

where $n, \ell = 0, 1, 2, \ldots$, with ℓ corresponding to the angular momentum and $\delta = 1 - k_a/|k_a| = 0, 2$. In the Klein-Gordon case the energy levels are not equally spaced (though the eigenvalues ε are). The distance between two consecutive levels is equal to

$$E_{N+1}^{\mathcal{K}} - E_N^{\mathcal{K}} = \omega - (4n + 2\ell + 3\delta + 1)\frac{\alpha^2\omega^2}{2\mathrm{m}} + \mathcal{O}(\alpha^4), \qquad (22)$$

where $N = 2n + \ell$. Then, it is always smaller than in the Schrödinger case and decreases with increasing energy.

In the case of the axial oscillator [Eq. (16) with $V = 0$], again both Klein-Gordon and Schrödinger equations may be written as

$$\left[\frac{p^2}{2\mathrm{m}} + \frac{\mathrm{m}\omega^2}{2}\rho^2 - \xi L_z - \varepsilon \right] \Psi = 0 \qquad (23)$$

where $z \equiv x_3$, $\xi = k_a/\mathrm{m} = B/(2\mathrm{m})$ and

$$\varepsilon = \begin{cases} E, & \text{(Schrödinger)}, \\ E + E^2/(2\mathrm{m}c^2), & \text{(Klein-Gordon)}. \end{cases} \qquad (24)$$

The Hamiltonian in Eq. (23) commutes with L_z. Therefore its eigenfunctions may be expressed as

$$\Psi(\rho, \phi) = \psi_{nm}(\rho)\Phi_m(\phi) \qquad (25)$$

with $L_z\Phi_m = m\Phi_m$ and

$$\varepsilon_{nm} = (2n + |m| + 1)\omega + m\xi. \qquad (26)$$

Again, relations between Schrödinger and Klein-Gordon energy levels are given by Eq. (21). Let us note that in the non-relativistic case also scalar potentials $V = k_v r^2$ or $V = k_v z^2$ may be added to the Hamiltonian.

2.2. Spin 1/2 particle

The case of a spin 1/2 particle is much more interesting. As it was already mentioned, in the non-relativistic case it is described by the Lévy-Leblond rather than by the Schrödinger equation. Both Lévy-Leblond and Dirac equations may be separated to the large- and small-component equations. The equation for the large component of the wavefunction reads

$$\left[\left(\frac{p^2}{2m} + \frac{m\omega^2}{2} r^2 - \varepsilon \right) I - \xi \boldsymbol{\sigma} \cdot \boldsymbol{L} \right] \Psi^L = 0, \tag{27}$$

where in the Dirac case $\xi = \frac{|k_a|}{k_a} \omega$ and in the Lévy-Leblond one, when a scalar potential is introduced, ξ may be considered as an independent parameter. The eigenvalue ε is defined as in Eq. (19) except that "(Schrödinger)" has to be replaced by "(Lévy-Leblond)" and "(Klein-Gordon)" by "(Dirac)". The angular and spinor parts of the large component may easily be obtained from the requirement that it is an eigenfunction of the angular momentum operators J^2, L^2 and S^2. Since

$$\boldsymbol{\sigma} \cdot \boldsymbol{L} = 2\boldsymbol{S} \cdot \boldsymbol{L} = J^2 - L^2 - S^2, \tag{28}$$

where $\boldsymbol{S} = \frac{1}{2}\boldsymbol{\sigma}$ is the spin operator, and $\boldsymbol{J} = \boldsymbol{L} + \boldsymbol{S}$, the scalar product $\boldsymbol{\sigma} \cdot \boldsymbol{L}$ in the radial equation may be replaced by the corresponding eigenvalue $[j(j+1) - \ell(\ell+1) - 3/4]I$ matrix. Consequently, both components of the spinor describing the large component of the wavefunction are associated with the same radial function (eigenfunction of the spherical harmonic oscillator). The small components of the Dirac bi-spinor may easily be obtained from the large one using Eq. (3).

The eigenvalues, in the Lévy-Leblond case, are equal to

$$\varepsilon_{n\ell j} = \left(2n + \ell + \tfrac{3}{2} \right) \omega + \begin{cases} - \left(\ell + \tfrac{3}{2} \right) \xi, & \text{if } j = \ell + \tfrac{1}{2}, \\ + \left(\ell - \tfrac{1}{2} \right) \xi, & \text{if } j = \ell - \tfrac{1}{2}. \end{cases} \tag{29}$$

If $|\xi| \ll \omega$, the last equation describes the spectrum of the Schrödinger spherical harmonic oscillator perturbed by the spin-orbit splitting with the spin-orbit parameter equal to ξ. A comparison of the non-relativistic boson and fermion harmonic oscillator spectra is given in Figure 1.

In the Dirac case the values of ξ are restricted to $\pm\omega$. Consequently, the eigenvalues are equal to

$$\varepsilon_{n\ell j} = \begin{cases} (2n + \tfrac{3}{2})\omega, & \text{if } j = \ell + \tfrac{1}{2}, \\ (2n + 2\ell + +\tfrac{5}{2})\omega, & \text{if } j = \ell - \tfrac{1}{2}, \end{cases} \tag{30}$$

if $k_a > 0$ and

$$\varepsilon_{n\ell j} = \begin{cases} (2n + 2\ell + \tfrac{3}{2})\omega, & \text{if } j = \ell + \tfrac{1}{2}, \\ (2n + \tfrac{1}{2})\omega, & \text{if } j = \ell - \tfrac{1}{2}, \end{cases} \tag{31}$$

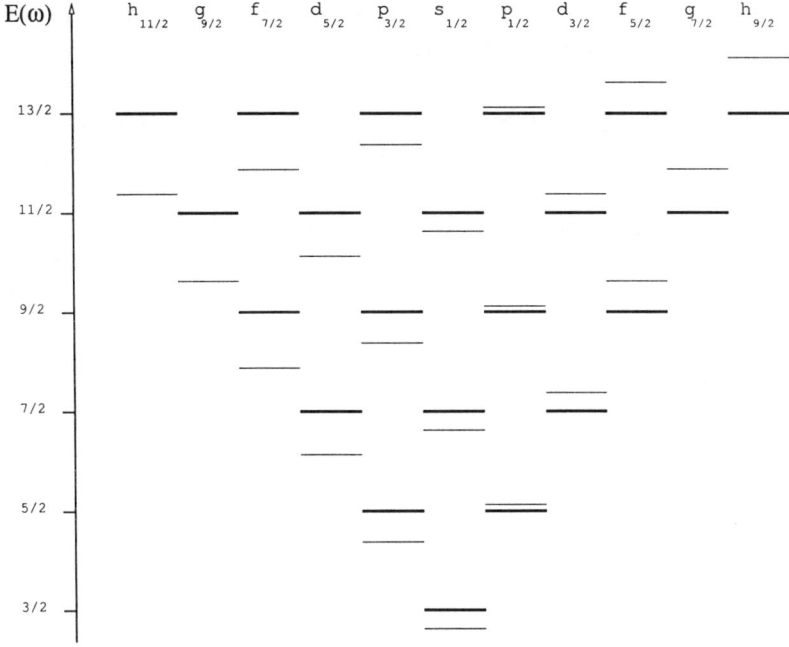

Figure 1.   A comparison between harmonic oscillator spectra of non-relativistic particles with spin 0 (thick lines) and spin 1/2 (thin lines).

if $k_a < 0$. In both cases [Eqs. (30) and (31)] the spectra are highly degenerate. If $j = \ell + \frac{1}{2}$ and $k_a > 0$ or $j = \ell - \frac{1}{2}$ and $k_a < 0$ then the energies are ℓ-independent, i.e. each energy level is infinitely degenerate. The second branch of the spectrum ($j = \ell - \frac{1}{2}$ and $k_a > 0$ or $j = \ell + \frac{1}{2}$ and $k_a < 0$) is structurally similar to the Schrödinger case.

3. Two particles

Recent advances in semiconductor technology allowed the construction of new quantum systems, sometimes referred to as *quantum dots* or *artificial atoms*. A quantum dot may be modeled by a number of electrons confined in a potential well. There exists some experimental evidence that the confining potential in real quantum dots is very close to the parabolic one[10]. Therefore two interacting electrons confined in a harmonic oscillator potential offer the simplest model of a quantum dot[15]. It is most interesting that the 6-dimensional eigenvalue problem of two interacting particles confined by a parabolic potential is separable and that the equations describing 5 degrees of freedom of this system are exactly solvable (they may be reduced to the angular momentum eigenvalue problem and to a spherical harmonic oscillator). The remaining one-dimensional interaction-dependent Schrödinger equation in several cases of practical importance is quasi-exactly solvable[12,16,28]. In other cases it may be easily solved numerically to an arbitrary accuracy[14].

Let us consider a system of two identical particles confined by a harmonic oscillator potential with interaction described by a scalar potential $V = V(r_{12})$, where $r_{12} = |r_1 - r_2|$ is the distance between the particles. The system is described by the Hamiltonian

$$H(r_1, r_2) = \frac{1}{2} \left(p_1^2 + p_2^2 \right) + \frac{\omega^2}{2} \left(r_1^2 + r_2^2 \right) + V(r_{12}), \tag{32}$$

where, for simplicity, we set $m = 1$. Introducing the center-of-mas and the relative coordinates, respectively

$$\mathbf{R} = \tfrac{1}{2}(\mathbf{r}_1 + \mathbf{r}_2) \quad \text{and} \quad \mathbf{r} = \mathbf{r}_1 - \mathbf{r}_2, \tag{33}$$

we get

$$H(r_1, r_2) = H_{\mathbf{R}}(\mathbf{R}) + H_{\mathbf{r}}(\mathbf{r}), \tag{34}$$

where

$$H_{\mathbf{R}}(\mathbf{R}) = -\tfrac{1}{4}\nabla_{\mathbf{R}}^2 + \omega^2 \mathbf{R}^2 \tag{35}$$

and

$$H_{\mathbf{r}}(\mathbf{r}) = -\nabla_{\mathbf{r}}^2 + \tfrac{1}{4}\omega^2 \mathbf{r}^2 + V(r), \tag{36}$$

where $r = |\mathbf{r}|$.

The eigenfunction of H may be factorized as

$$\Psi(1, 2) = \Xi(\mathbf{R})_{nlm} \Phi(\mathbf{r})_{\nu\lambda\mu} \Theta_{sm_s}, \tag{37}$$

where Θ_{sm_s} is a two-electron spin function (either antisymmetric singlet with $s = m_s = 0$ or symmetric triplet with $s = 1$ and $m_s = 0, \pm 1$) while $\Xi(\mathbf{R})_{nlm}$ and $\Phi(\mathbf{r})_{\nu\lambda\mu}$ are eigenfunctions of H_R and H_r, respectively. Consequently,

$$\Xi(\mathbf{R})_{nlm} = \frac{1}{R} \chi(R)_{nl} Y_{lm}(\hat{\mathbf{R}}) \tag{38}$$

and

$$\Phi(\mathbf{r})_{\nu\lambda\mu} = \frac{1}{r} \phi(r)_{\nu\lambda} Y_{\lambda\mu}(\hat{\mathbf{r}}). \tag{39}$$

If the eigenvalues corresponding to $\Xi(\mathbf{R})_{nlm}$ and $\Phi(\mathbf{r})_{\nu\lambda\mu}$ are, respectively, equal to E_{nl}^R and $E_{\nu\lambda}^r$, then the total energy of the system is

$$E_{nl,\nu\lambda} = E_{nl}^R + E_{\nu\lambda}^r. \tag{40}$$

The radial function $\chi(R)$ fulfills the eigenvalue equation of the spherical harmonic oscillator

$$\left[-\frac{1}{4}\frac{d^2}{dR^2} + \frac{l(l+1)}{4R^2} + \omega^2 R^2 \right] \chi(R)_{nl} = E_{nl}^R \, \chi(R)_{nl}. \tag{41}$$

Consequently

$$\chi(R)_{nl} \sim R^{l+1} e^{-\omega R^2/2} F\left[-\tfrac{1}{2}(n-l), \tfrac{1}{2}(2l+3), \omega\, R^2 \right], \tag{42}$$

where F is the confluent hypergeometric function.

The dipole moment operators $\mathbf{D} \sim \mathbf{r}_1 + \mathbf{r}_2 \sim \mathbf{R}$. Therefore the dipole transition moment does not depend upon a specific form of $\Phi(\mathbf{r})_{\nu\lambda\mu}$. The dipole transitions between two functions (37) defined by sets of quantum numbers $\{nlm, \nu\lambda\mu, sm_s\}$ and $\{n'l'm', \nu'\lambda'\mu', s'm_s'\}$ are forbidden unless $\{\nu\lambda\mu, sm_s\} = \{\nu'\lambda'\mu', s'm_s'\}$. The values of the transition moments and the selection rules for nlm quantum numbers correspond to those of a spherical harmonic oscillator.

As we see, the only function which has to be determined is the radial part of Φ. As one can easily check it fulfills the following radial equation

$$\left[-\frac{d^2}{dr^2} + \frac{\lambda(\lambda+1)}{r^2} + \frac{1}{4}\omega^2 r^2 + V(r) \right] \phi(r)_{\nu\lambda} = E_{\nu\lambda}^r \, \phi(r)_{\nu\lambda}. \tag{43}$$

Since Ψ has to be antisymmetric with respect to the transposition of the particles and $\Xi(\mathbf{R})$ is always symmetric, $\Phi(\mathbf{r})$ has to be even for triplets and odd for singlets, i.e.

$$\Phi(-\mathbf{r})_{\nu\lambda\mu} = (-1)^s \, \Phi(\mathbf{r})_{\nu\lambda\mu}. \tag{44}$$

Eq. (43) may easily be solved numerically to an arbitrary accuracy[14] for all well behaving potentials $V(r)$. However, in some cases, it may also be solved analytically. In particular, if $V(r) = ar^2 + b/r^2$, where a and b are constants, then Eq. (43) may be reduced to Eq. (41) with modified values of ω and l.

Less trivial and more interesting are two other cases. In the first one, studied in detail by Ghosh and Samanta[16,28],

$$V(r) = \frac{1}{r} - \beta\omega r. \tag{45}$$

Introducing reduced variable $\rho = r\sqrt{\omega/2}$, one may easily show that the asymptotically correct solution is

$$\phi(\rho) = \rho^{l+1}e^{-\rho(\rho-\alpha)/2}\,W(\rho), \tag{46}$$

where $\alpha = 2\beta\sqrt{2/\omega}$, $W(0) = const$ and

$$\lim_{\rho\to\infty} e^{-\rho^2}\,W(\rho) = 0.$$

The energy is equal to

$$E_{\nu\lambda} = \omega\left(\nu + \lambda + \tfrac{3}{2}\right) - \beta^2. \tag{47}$$

However, Eq. (47) is not valid for a *given* β. A power-series expansion of $W(\rho)$ reduces to a polynomial for certain values of β only. As one can easily check, β has to be a root of a $(\nu+1)$-th order algebraic equation. In particular, for $\nu = 0$

$$\beta = \frac{1}{2(l+1)}$$

and the node-less radial wave functions are equal to

$$\phi(r)_{0\lambda} = A\,\rho^{l+1}e^{-\rho(\rho-\alpha)/2}, \tag{48}$$

where A is a normalization constant. For larger ν the corresponding expressions become rather cumbersome.

In the second case

$$V(r) \sim \frac{1}{r}. \tag{49}$$

The corresponding system is known as *harmonium* and, probably, is the only system of confined particles interacting with the Coulomb force for which analytical solutions of the Hamiltonian eigenvalue problem are known. In this case the asymptotic solution reads

$$\phi(\rho) = \rho^{l+1}e^{-\rho^2/2}\,W(\rho). \tag{50}$$

The most interesting property of the resulting equation for W is that its solution is a polynomial in ρ if ω is a root of a hypergeometric function:

$$F(\lambda, \nu, 2\lambda + 2\nu + 1, \tfrac{\omega}{2}) = 0 \tag{51}$$

where $\nu \geq 2$ and $\lambda \geq 0$. If Eq. (51) is fulfilled then the condition for $W(\rho)$ to be a polynomial gives the restriction for the eigenvalues $E_{\nu\lambda}^r$ in the form:

$$F(\lambda, \nu, \tfrac{2}{\omega} E_{\nu\lambda}^r, \tfrac{\omega}{2}) = 0. \tag{52}$$

From here, for given ν and λ, one gets

$$E_{\nu\lambda}^r = \omega(\nu, \lambda)\left(\nu + \lambda + \tfrac{1}{2}\right), \tag{53}$$

where $\omega(\nu, \lambda)$ is the solution of Eq. (51) for the pertinent values of ν and λ. In particular, $\omega(2, \lambda) = 1/(2\lambda + 2)$, $\omega(3, \lambda) = 1/(8\lambda + 10)$, etc. One can demonstrate[12,14] that in the limit of small ω

$$E^r = 3\left(\frac{\omega}{4}\right)^{2/3} + \sqrt{3}\omega\left(m + \tfrac{1}{2}\right), \tag{54}$$

where $m = 0, 1, 2, \ldots$. In the limit of large ω, the Coulomb term may be neglected and

$$E^r = \omega\left(m + \tfrac{3}{2}\right). \tag{55}$$

The last two examples of exactly solvable two-particle Schrödinger equations are of a particular importance for studies of properties of systems of N interacting particles. The knowledge of the exact structure of the electron correlation energy and the form of the wavefunction in the vicinity of $r_{12} = 0$ is essential in designing approximate methods of solving N-electron Hamiltonian eigenvalue problems in many areas of physics and in quantum chemistry.

There are many open problems in theory of two interacting particles confined by a harmonic oscillator potential. An explicit introduction of spin and, consequently, of the spin-spin, spin-orbit and orbit-orbit interactions, would prevent an exact separation of the two-particle equation. However by representing the spin-dependent operators in a basis of eigenfunctions of the exactly separable spin-less Hamiltonian may lead to a high-quality description of the spin-orbit effects in low orders of the perturbation expansion.

The original motivation for studies on the eigenfunctions of harmonium was an exploration of the electron correlation effects on both energies and wave-functions[29]. However there are still many questions to be answered. This includes the exact behavior of the intracule density and of the Coulomb hole as functions of the Hamiltonian parameters, particularly in the excited states and the influence of the electron correlation effects on dipole-forbidden transition probabilities between different two-particle states.

Studies on the relativistic harmonium may contribute to understanding of the very difficult problem of using explicitly correlated wave-functions in variational solving a two-particle Dirac-Coulomb equation. Finally, searching for more $V(r_{12})$ potentials leading to exactly solvable Schrödinger equations is certainly worth of effort.

Acknowledgments

I am most grateful to the Organizers of SSPCM'2002, especially to Barbara and Tadeusz Lulek for their warm hospitality at Myczkowce. This work was supported by the Polish State Committee for Scientific Research (KBN) under project No. 5 P03B 119 21. Helpful remarks of Karol A. Penson are highly appreciated.

References

1. V. Fock, *Bemerkung zur Quantenlung des harmonischen Oszillators im Magnetfeld*, Z. Phys. **47**, 446-448 (1928).
2. C. G. Darwin, *The diamagnetism of the free electron*, Proc. Cambridge Philos. Soc. **27**, 86-90 (1930).
3. K. Grudziński and B. G. Wybourne, *Symplectic models of n-particle systems*, Rep. Math. Phys. **38**, 251-266 (1996).
4. D. J. Rowe and C. Bahri, *Rotation-vibration spectra of diatomic molecules and nuclei with Davidson interactions*, J. Phys. A: Math. Gen. **31**, 4947-4961 (1998).
5. M. Znojil, *Harmonic oscillations in quasi-relativistic regime*, J. Phys. A: Math. Gen. **29**, 2905-2917 (1996).
6. M. Bednar, J. Ndimubandi and A. G. Nikitin, *On connection between the two-body Dirac oscillator and Kemmer oscillators*, Can. J. Phys. **75**, 283-290 (1997).
7. S. N. Datta and A. Misra, *Exact solution of the relativistic dynamics of a spin-$\frac{1}{2}$ particle moving in a homogeneous magnetic field*, Int. J. Quantum Chem. **82**, 209-217 (2001).
8. A. G. Ushveridze, *Quasi-Exactly Solvable Models in Quantum Mechanics*, Institute of Physics Publishing, Bristol 1994.
9. W. Jaskólski, *Confined many-electron systems*, Phys. Rep. **271**, 1-66 (1996).
10. P. A. Maksym and T. Chakraborty, *Quantum dots in a magnetic field: Role of electron-electron interactions*, Phys. Rev. Lett. **65**, 108-111 (1990).
11. D Bielińska-Wąż, J. Karwowski and G. H. F. Diercksen, *Spectra of confined two-electron atoms*, J. Phys. B: At. Mol. Opt. Phys. **34**, 1987-2000 (2001).
12. M. Taut, *Two electrons in an external oscillator potential: Particular analytic solutions of a Coulomb correlation problem*, Phys. Rev. A **48**, 3561-3566 (1993).
13. J. Cioslowski and K. Pernal, *The ground state of harmonium*, J. Chem. Phys. **113**, 8434-43 (2000).
14. L. Cyrnek, *The energy spectrum of harmonium*, This volume.
15. U. Merkt, J. Huser and M. Wagner, *Energy spectra of two electrons in a harmonic quantum dot*, Phys. Rev. B **43**, 7320-7323 (1991).
16. S. K. Ghosh and A. Samanta, *Study of correlation effects in an exactly solvable model two-electron system*, J. Chem. Phys. **94**, 517-522 (1991).
17. W. Lay, K. Bay and S. Yu Slavyanov, *Asymptotic and numeric eigenvalues of the double confluent Heun equation*, J. Phys. A: Math. Gen. **31**, 8521-8531 (1998).
18. K. Nikolsky, *Das Oszillatorproblem nach Diracschen Theorie*, Z. Phys. **62**, 677-681 (1930).
19. I. Postępska, *Harmonischer Oszillator nach der Diracschen Wellengleichung*, Acta Phys. Polon. **4**, 269-280 (1935).

20. F. M. Toyama and Y. Nogami, *Harmonic oscillators in relativistic quantum mechanics*, Phys. Rev. **A 59**, 1056-1062 (1999).

21. Y. Nogami and F. M. Toyama, *Supersymmetry aspects of the Dirac equation in one dimension with a Lorentz scalar potential*, Phys. Rev. **A 47**, 1708-1714 (1993).

22. P. A. Cook, *Relativistic harmonic oscillators with intrinsic spin structure*, Lettere al Nuovo Cimento, **1**, 419-426 (1971).

23. M. Moreno and A. Zentella, *Covariance, CPT and the Foldy-Wouthuysen transformation for the Dirac oscillator*, J. Phys. A: Math. Gen. **22**, L821-L825 (1989).

24. M. Moshinsky and A. Szczepaniak, *The Dirac oscillator*, J. Phys. A: Math. Gen. **22**, L817-L820 (1989).

25. R. M. Mir-Kasimov, $SU_q(1,1)$ *and the relativistic oscillator*, J. Phys. A: Math. Gen. **24**, 4283-4302 (1991).

26. M. Moshinsky and A. del Sol Mesa, *The Dirac oscillator of arbitrary spin*, J. Phys. A: Math. Gen. **29**, 4217-4236 (1996).

27. J. M. Lévy-Leblond, *Non-relativistic particles and wave equations*, Commun. Math. Phys. **6**, 286-311 (1967).

28. A. Samanta and S. K. Ghosh, *Correlation in exactly solvable two-particle quantum system*, Phys. Rev. **A 42**, 1178-1183 (1990).

29. N. R. Kestner and O. Sinanoglŭ, *Study of electron correlation in helium-like systems using an exactly soluble model*, Phys. Rev. **128**, 2687-2693 (1962).

BOUNDS TO VARIATIONAL DIRAC ENERGIES

GRZEGORZ PESTKA

Instytut Fizyki, Uniwersytet Mikołaja Kopernika,
Grudziądzka 5, PL-87-100 Toruń, Poland
E-mail: gp@phys.uni.torun.pl

Conditions under which the Dirac Hamiltonian eigenvalue problem algebraized in a finite dimensional space gives upper bounds to the exact eigenvalues have been formulated. The theorems presented generalize of the well known kinetic balance condition. The theorems lead to computational methods for which the convergence pattern is both predictable and correct. Additionally, it has been shown that though the correct boundary conditions dramatically improve the quality of the approximations, they are not necessary to determine the bound properties of the eigenvalues.

1. Introduction

In this work we describe conditions under which eigenvalues $\tilde{E}^{\mathrm{r}}$ of a matrix which represents Dirac equation

$$\begin{pmatrix} V & c\,\boldsymbol{\sigma}\cdot\hat{\boldsymbol{p}} \\ c\,\boldsymbol{\sigma}\cdot\hat{\boldsymbol{p}} & V - 2mc^2 \end{pmatrix} \begin{pmatrix} \Psi^{\mathrm{l}} \\ \Psi^{\mathrm{s}} \end{pmatrix} = E^{\mathrm{r}} \begin{pmatrix} \Psi^{\mathrm{l}} \\ \Psi^{\mathrm{s}} \end{pmatrix} \tag{1}$$

in the finite dimensional model space

$$\mathcal{H} = \mathcal{H}^{\mathrm{l}} \otimes \mathcal{H}^{\mathrm{s}},$$

composed of N_{l}-dimensional large component space $\mathcal{H}^{\mathrm{l}}$ and N_{s}-dimensional small component space $\mathcal{H}^{\mathrm{s}}$, are upper bounds to the corresponding eigenvalues E^{r} of the Dirac operator. The projection of Eq. (1) onto the model space leads to the algebraic Dirac equation

$$\begin{pmatrix} V^{\mathrm{l}} & c\,T \\ c\,T^{\dagger} & V^{\mathrm{s}} - 2mc^2\,S^{\mathrm{s}} \end{pmatrix} \begin{pmatrix} C^{\mathrm{l}} \\ C^{\mathrm{s}} \end{pmatrix} = \tilde{E}^{\mathrm{r}} \begin{pmatrix} S^{\mathrm{l}} & 0 \\ 0 & S^{\mathrm{s}} \end{pmatrix} \begin{pmatrix} C^{\mathrm{l}} \\ C^{\mathrm{s}} \end{pmatrix}. \tag{2}$$

The resulting approximations to the large and small components of the wavefuntion are, respectively,

$$\Psi^{\mathrm{l}} \approx \sum_{p=1}^{N_{\mathrm{l}}} C_p^{\mathrm{l}} \varphi_p^{\mathrm{l}}, \quad \Psi^{\mathrm{s}} \approx \sum_{q=1}^{N_{\mathrm{s}}} C_q^{\mathrm{s}} \varphi_q^{\mathrm{s}},$$

where φ_p^{l} and φ_q^{s} are the basis functions in, respectively, $\mathcal{H}^{\mathrm{l}}$ and $\mathcal{H}^{\mathrm{s}}$.

2. Relations between large and small component spaces

2.1. *Lévy-Leblond equation*

The non-relativistic counterpart of Dirac equation (1), known as Lévy-Leblond (LL) equation,[1] may be written as[2]

$$
\begin{pmatrix} V & c\,\boldsymbol{\sigma}\cdot\hat{\boldsymbol{p}} \\ c\,\boldsymbol{\sigma}\cdot\hat{\boldsymbol{p}} & -2mc^2 \end{pmatrix}
\begin{pmatrix} \Psi^{\mathrm{l}} \\ \Psi^{\mathrm{s}} \end{pmatrix}
= E^{\mathrm{nr}}
\begin{pmatrix} 1 & 0 \\ 0 & 0 \end{pmatrix}
\begin{pmatrix} \Psi^{\mathrm{l}} \\ \Psi^{\mathrm{s}} \end{pmatrix}.
\tag{3}
$$

As one can easily see, the elimination of the small component Ψ^{s} reduces Eq. (3) to the Schrödinger equation. The algebraic LL equation obtained by the projection of Eq. (3) onto $\mathcal{H}$ reads

$$
\begin{pmatrix} \boldsymbol{V}^{\mathrm{l}} & c\,\boldsymbol{T} \\ c\,\boldsymbol{T}^{\dagger} & -2mc^2\,\boldsymbol{S}^{\mathrm{s}} \end{pmatrix}
\begin{pmatrix} \boldsymbol{C}^{\mathrm{l}} \\ \boldsymbol{C}^{\mathrm{s}} \end{pmatrix}
= \tilde{E}^{\mathrm{L}}
\begin{pmatrix} \boldsymbol{S}^{\mathrm{l}} & 0 \\ 0 & 0 \end{pmatrix}
\begin{pmatrix} \boldsymbol{C}^{\mathrm{l}} \\ \boldsymbol{C}^{\mathrm{s}} \end{pmatrix}.
\tag{4}
$$

As one can easily see, it may be reduced to the algebraic Schrödinger equation

$$
\left(\frac{1}{2m}\,\boldsymbol{P}^2 + \boldsymbol{V} \right) \boldsymbol{C}^{\mathrm{l}} = \tilde{E}^{\mathrm{S}}\,\boldsymbol{S}^{\mathrm{l}}\boldsymbol{C}^{\mathrm{l}},
\tag{5}
$$

obtained by the projection of the Schrödinger equation onto $\mathcal{H}^{\mathrm{l}}$, if

$$
\boldsymbol{P}^2 = \boldsymbol{T}^{\dagger}\,(\boldsymbol{S}^{\mathrm{s}})^{-1}\,\boldsymbol{T}.
\tag{6}
$$

It was noticed by Dyall et al.[3] that Eq. (6) holds if the kinetic balance condition

$$
\boldsymbol{\sigma}\cdot\hat{\boldsymbol{p}}\ \mathcal{H}^{\mathrm{l}} \subset \mathcal{H}^{\mathrm{s}}
\tag{7}
$$

is fulfilled. Consequently, spectra derived from Eqs. (4) and (5) are the same if condition (7) holds. Moreover, under condition (7) $\tilde{E}^{\mathrm{L}}$ does not depend on a specific choice of $\mathcal{H}^{\mathrm{s}}$. In particular, enlarging the dimension of $\mathcal{H}^{\mathrm{s}}$ has no influence on $\tilde{E}^{\mathrm{L}}$. Consequently, according to Hylleraas-Undheim-MacDonald theorem,[4,5] for a given energy level, $\tilde{E}^{\mathrm{L}} \geq E^{\mathrm{nr}}$.

2.2. *Dirac equation*

The behaviour of Dirac energies is similar though slightly more complicated. Let us divide the spectrum of the Dirac Hamiltonian to two parts. The first part consists of the *positive spectrum* describing electronic states; in the non-relativistic limit it converges to the spectrum of the Schrödinger equation. The second part consists of the *negative spectrum* connected with the positron states; in the non-relativistic limit it approaches $-\infty$. The same partition is applied to the spectrum of the algebraic Dirac equation

(2): there are N_l positive spectrum eigenstates and N_s negative spectrum eigenstates.

Let us assume that the atomic (rigorous) balance condition[6]

$$\bigcup_{i=1}^{N_\mathrm{l}} \left(\hat{W}_i\, \boldsymbol{\sigma}\cdot\hat{\boldsymbol{p}}\, \mathcal{H}^\mathrm{l} \right) \subset \mathcal{H}^\mathrm{s} \tag{8}$$

is fulfilled, where $\hat{W}_i = \left(\tilde{E}_i^\mathrm{r} + 2mc^2 - V \right)^{-1}$, and $\tilde{E}_i^\mathrm{r}$ are the positive energy spectrum solutions of the algebraic Dirac equation (2), and let $\mathcal{H}^+$ be a space spanned by N_l eigenvectors $\tilde{\Psi}_i = \sum_{p=1}^{N_\mathrm{l}} C_{i,p}^\mathrm{l}\varphi_p^\mathrm{l} + \sum_{q=1}^{N_\mathrm{s}} C_{i,q}^\mathrm{s}\varphi_q^\mathrm{s}$ of the positive spectrum obtained from Eq. (2). Under these assumptions

Lemma: *The eigenvalues $\tilde{E}_i^\mathrm{r}$ and the space $\mathcal{H}^+$ are independent of the small component space $\mathcal{H}^\mathrm{s}$.*

Proof: Let $\mathcal{H}_i^+$ be the one-dimensional space spanned by $\tilde{\Psi}_i = \begin{pmatrix} \tilde{\Psi}_i^{\,\mathrm{l}} \\ \tilde{\Psi}_i^{\,\mathrm{s}} \end{pmatrix}$ and let $\tilde{E}_i^\mathrm{r}$ be the corresponding eigenvalue. To the function $\tilde{\Psi}_i$ corresponds eigenvalue equation

$$\begin{pmatrix} \boldsymbol{V}_i^\mathrm{l} & c\,\boldsymbol{T}_i \\ c\,\boldsymbol{T}_i^\dagger & \boldsymbol{V}_i^\mathrm{s} - 2mc^2\,\boldsymbol{S}_i^\mathrm{s} \end{pmatrix} \begin{pmatrix} \boldsymbol{C}_i^\mathrm{l} \\ \boldsymbol{C}_i^\mathrm{s} \end{pmatrix} = \tilde{E}_i^\mathrm{r} \begin{pmatrix} \boldsymbol{S}_i^\mathrm{l} & 0 \\ 0 & \boldsymbol{S}_i^\mathrm{s} \end{pmatrix} \begin{pmatrix} \boldsymbol{C}_i^\mathrm{l} \\ \boldsymbol{C}_i^\mathrm{s} \end{pmatrix}, \tag{9}$$

where $\boldsymbol{V}_i^{\mathrm{l/s}}$, $\boldsymbol{S}_i^{\mathrm{l/s}}$, $\boldsymbol{T}_i$ are calculated using $\tilde{\Psi}_i^{\,\mathrm{l}}$ and $\tilde{\Psi}_i^{\,\mathrm{s}}$. After elimination of the small component coefficient $\boldsymbol{C}_i^\mathrm{s}$ Eq. (9) reduces to

$$\left[\boldsymbol{V}_i^\mathrm{l} + c^2 \boldsymbol{T}_i \left((2mc^2 + \tilde{E}_i)\boldsymbol{S}_i^\mathrm{s} - \boldsymbol{V}_i^\mathrm{s} \right)^{-1} \boldsymbol{T}_i^\dagger \right] \boldsymbol{C}_i^\mathrm{l} = \tilde{E}_i^\mathrm{r} \boldsymbol{S}_i^\mathrm{l} \boldsymbol{C}_i^\mathrm{l}.$$

It is easy to see that

$$\left\langle \tilde{\Psi}_i^{\,\mathrm{l}} \,\middle|\, \boldsymbol{\sigma}\cdot\hat{\boldsymbol{p}}\,\hat{W}_i\,\boldsymbol{\sigma}\cdot\hat{\boldsymbol{p}} \,\middle|\, \tilde{\Psi}_i^{\,\mathrm{l}} \right\rangle = \boldsymbol{T}_i^\dagger \left((2mc^2 + \tilde{E}_i)\boldsymbol{S}_i^\mathrm{s} - \boldsymbol{V}_i^\mathrm{s} \right)^{-1} \boldsymbol{T}_i \tag{10}$$

if

$$\hat{W}_i\,\boldsymbol{\sigma}\cdot\hat{\boldsymbol{p}}\,\mathcal{H}_i^\mathrm{l} \subset \mathcal{H}_i^\mathrm{s}, \tag{11}$$

where $\mathcal{H}_i^\mathrm{l}$ and $\mathcal{H}_i^\mathrm{s}$ are spanned, respectively, by $\tilde{\Psi}_i^{\,\mathrm{l}}$ and $\tilde{\Psi}_i^{\,\mathrm{s}}$ (c.f. the kinetic balance condition (7) and Eq. (6)). Since the LHS of Eq. (10) does not depend on the small component $\tilde{\Psi}_i^{\,\mathrm{s}}$, we can conclude that enlarging the $\mathcal{H}_i^\mathrm{s}$ space in inclusion (11) has influence neither on $\tilde{E}_i^\mathrm{r}$ nor on $\tilde{\Psi}_i$ and $\mathcal{H}_i^+$. If we consider all positive spectrum solutions of Eq. (2) then this result leads to the condition (8). This conclusion proves the Lemma. $\qquad\square$

More details are given in a forthcoming article.[7] In particular it is shown that if conditions of the *Lemma* are fulfilled and the potential V is negative definite then the following *Theorem* holds:

Theorem *Let us assume that:*

- *The negative energy spectrum of the algebraic Dirac equation (2) is bounded from above by E_{sep},*
- *The ground state energy E_{g} of the Dirac Hamiltonian is greater then E_{sep},*
- *The rigorous kinetic balance condition (8) is fulfilled.*

Then the energies derived from Eq. (2) corresponding to the positive spectrum bound states approximate monotonically from above the exact energies of the positive spectrum bound states of the Dirac equation (1). The remaining energies are either in the positive or in the negative continuum. □

It can be easily checked that the assumptions of the theorem are fulfilled for the Coulomb central potential $V = -Z/r$, $1 \le Z \le 137$. For this potential the approximate calculations are usually carried out in either Slater or Gauss type bases and the fulfilment of the atomic balance condition (8) is, in most cases, impossible. However it turns out[7] that its simplified form, i.e., the *asymptotic balance condition*

$$\boldsymbol{\sigma} \cdot \hat{\boldsymbol{p}} \, \mathcal{H}^{\mathrm{l}} \cup r \boldsymbol{\sigma} \cdot \hat{\boldsymbol{p}} \, \mathcal{H}^{\mathrm{l}} \subset \mathcal{H}^{\mathrm{s}} \tag{12}$$

obtained by combining of the long range limit $r \to +\infty$ (kinetic balance condition (7)) and the short range limit $r \to 0$ of (8), is sufficient for getting the upper bounds to the exact Dirac energies. This analysis shows that the correct asymptotic relation between large and small components of the basis functions is most essential for obtaining bound properties of the approximate energies. An imposition of the correct boundary conditions to the trial functions, though very essential for the convergence rate, is of a secondary importance.

3. Numerical examples

Numerical illustrations of the Theorem have been constructed for the Coulomb potential $V = -Z/r$, with $Z = 92$, using functional bases in which the radial parts of both components can be expressed as

$$\chi_{n,k} = r^{n+\gamma-2} e^{-\lambda r^k} \qquad n = 1, 2, \ldots, N_{\mathrm{l/s}}.$$

In particular, we used the Slater ($k = 1$) and the Gaussian ($k = 2$) bases. The spin-angular parts of the bases have limited to $\kappa = -1, 1$, i.e. to $S_{1/2}$ and $P_{1/2}$ species. In the Fig. 1 the approximate ground state ($\kappa = -1$, $1S_{1/2}$) LL $\left(E_{\mathrm{g}}^{\mathrm{L}}\right)$ and Dirac $\left(E_{\mathrm{g}}^{\mathrm{r}}\right)$ energies are plotted versus dimension N_{s}

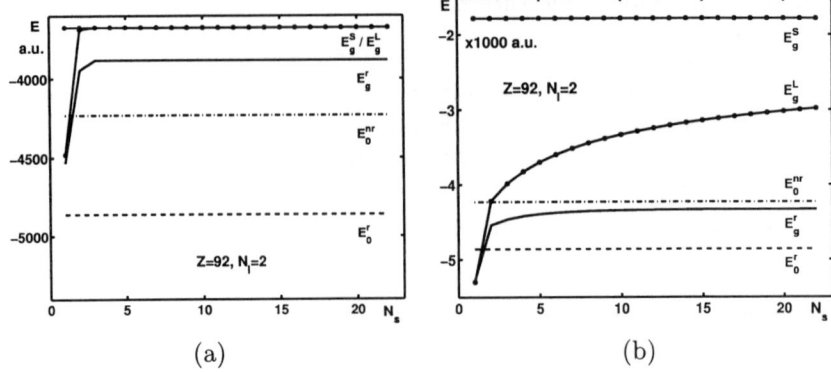

Figure 1. Convergence pattern of LL E_g^L and Dirac E_g^r ground state energies in the case of Gauss (a) and Slater (b) bases.

of the small component space $\mathcal{H}^s$, while the large component space $\mathcal{H}^l$ is spanned by $N_l = 2$ basis functions. For a better exemplification of the behaviour of solutions of Eq. (2) treated as approximations to the corresponding solutions of Eq. (1), the values of the nonlinear parameters γ and λ are different from the optimal ones. In the exact wavefunctions $\gamma = \sqrt{1 - (Z/c)^2} \approx 0.74$, $\lambda = Z = 92$. In the present calculations for the Slater basis $\gamma = 0.55$ and $\lambda = 30$, while for the Gaussian basis $\gamma = 1$ and $\lambda = 2500$. In Fig. 1 the horizontal dashed and dash-dotted lines represent, respectively, the exact relativistic (E_0^r) and non-relativistic (E_0^{nr}) energies. One can see the convergence of the LL energy E_g^L to the Schrödinger energy E_g^S (the horizontal solid line with dots) obtained in $\mathcal{H}^l$. In the same figure plots of the Dirac energies E_g^r are also displayed. Fig. 1a illustrates the behaviour of the energy levels calculated in the Gaussian basis. The kinetic balance condition is fulfilled for $N_s > 2$. Consequently, the LL energy E_g^L is equal to Schrödinger energy E_g^S if $N_s > 2$. The atomic balance condition (8) is never fulfilled, but due to the fulfilment of the kinetic balance (7) and of the more restrictive asymptotic balance condition (12), a similar stabilization can be observed also for the Dirac energy E_g^r. Fig. 1b shows the behaviour of the energy levels calculated in the Slater basis. In this case neither kinetic (7) nor atomic (8) balance condition is fulfilled. However, the larger is the dimension N_s of the small component space, the better approximation to these conditions is obtained. In consequence all energies in Fig. 1b monotonically increase and converge to the values which would be obtained if the relevant balance condition was fulfilled. In particular, for N_s sufficiently large, the upper bounds to the exact energies are obtained.

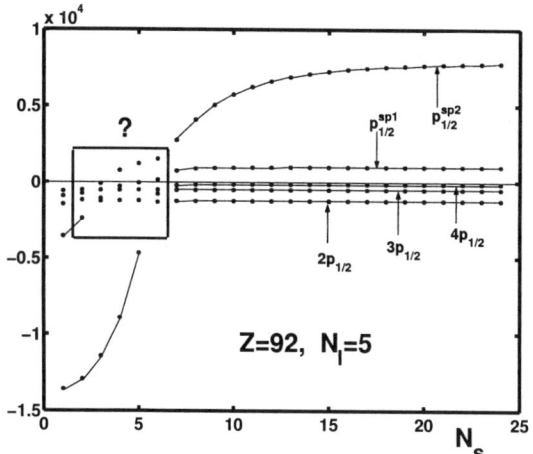

Figure 2. Approximate spectrum of Dirac Hamiltonian for $\kappa = 1$ ($P_{1/2}$) symmetry.

In Fig. 2 spectrum of the algebraic Dirac equation (2) in the Slater basis ($\gamma = 1$, $\lambda = Z/2 = 46$) for $\kappa = 1$ ($P_{1/2}$ symmetry) as a function of N_s while $N_l = 5$ is displayed. For N_s large enough (i.e. if the balance conditions (8, 12) are relatively well satisfied), the sequence of the eigenvalues is the same as of those in Eq. (1). However this is not the case for small values of N_s: if $N_s < 7$ (the region marked by a box) the sequence of the eigenvalues derived from Eq. (2) not in every case is correct and their direct assignment to the eigenvalues of the Dirac Hamiltonian is impossible.

4. Conclusions

Conditions have been formulated under which the eigenvalues of the algebraic LL/Dirac equations are the upper bounds to the corresponding solutions of the LL/Dirac equations. The fulfilment of these conditions ensures the correct global behaviour (sequence of the energy levels, complete basis limit and non-relativistic limit) of the approximate spectra. As a consequence the approximate (derived from the algebraic equations) energies may be assigned in a simple way to the exact ones. This makes possible a construction black-box-type computational programs..

Acknowledgements

The author is grateful to Prof. Jacek Karwowski for his encouragement and for many discussions and comments on subjects related to this paper. This work was supported by the Polish State Committee for Scientific Research, Grant No. 5 P03B 119 21.

References

1. J. M. Lévy-Leblond, *Non-relativistic particles and wave equations, Commun. Math. Phys.* **6**, 286-311 (1967).
2. A. Rutkowski, *Relativistic perturbation theory: I. A new perturbation approach to the Dirac equation, J. Phys. B: At. Mol. Phys.* **19**, 149-158 (1986).
3. K. G. Dyall, I. P. Grant and S. Wilson, *The Dirac equation in the algebraic approximation: I. Criteria for the choice of basis functions and minimum basis set calculations for hydrogenic atoms, J. Phys. B.: At. Mol. Phys.* **17**, L45-L50 (1984).
4. E. A. Hylleraas and B. Undheim, *Numerische Berechnung der 2 S-Terme von Ortho- und Par-Helium, Z. Phys.* **65**, 759-772 (1930).
5. J. K. L. MacDonald, *Successive Approximations by the Rayleigh-Ritz Variation Method, Phys. Rev.* **43**, 830-833 (1933).
6. W. Kutzelnigg, *Relativistic one-electron Hamiltonians 'for electrons only' and the variational treatment of the Dirac equation, Chem. Phys.* **225**, 203-222 (1997).
7. G. Pestka, *in preparation.*

THE ENERGY SPECTRUM OF HARMONIUM

LECH CYRNEK

Instytut Fizyki, Uniwersytet Mikołaja Kopernika,
Grudziądzka 5, PL-87-100 Toruń, Poland
E-mail: lcyrnek@phys.uni.torun.pl

Spectrum of harmonium, i.e. of a system of two electrons interacting by a Coulomb force and confined in a harmonic oscillator is analyzed. The dependence of the energy levels on the strength of the confining potential (the coupling constant ω in the harmonic oscillator potential) is determined in the whole range of ω. In particular the pattern of degeneracies of the energy levels in the limit of $\omega \to 0$ is determined.

1. Introduction

Harmonium is a system consisting of two electrons interacting with Coulomb potential confined by harmonic-oscillator potential

$$H = -\frac{1}{2}\Delta_1 + \frac{1}{2}\omega^2 \mathbf{r}_1^2 - \frac{1}{2}\Delta_2 + \frac{1}{2}\omega^2 \mathbf{r}_2^2 + \frac{1}{|\mathbf{r}_1 - \mathbf{r}_2|}. \tag{1}$$

It is probably the only system of two particles interacting by Coulomb forces for which analytical solutions of the Hamiltonian eigenvalue problem are known[1]. Harmonium was used by many authors as a model system for studying the electron correlation effects[2,3,4,5]. It is also a prototype for the simplest three-dimensional quantum dot[6]. However, in most cases, the analysis was limited to either the ground state or to several exited states. In this paper, probably for the first time, the complete spectrum of harmonium has been determined.

2. The eigenvalue equation

In order to solve the eigenvalue problem of Hamiltonian (1), we use the following transformation

$$\mathbf{r} = \mathbf{r}_1 - \mathbf{r}_2, \quad \mathbf{R} = \frac{1}{2}(\mathbf{r}_1 + \mathbf{r}_2), \tag{2}$$

where $\mathbf{R}$ is the center of the mass vector and $\mathbf{r}$ is the relative motion vector $\mathbf{r}$. Due to this transformation Hamiltonian (1) becomes

$$H = H_\mathbf{r} + H_\mathbf{R}, \tag{3}$$

where

$$H_\mathbf{r} = -\Delta_\mathbf{r} + \frac{1}{4}\omega^2\mathbf{r}^2 + \frac{1}{r} \tag{4}$$

describes relative motion of the electrons and

$$H_\mathbf{R} = -\frac{1}{4}\Delta_\mathbf{R} + \omega^2\mathbf{R}^2. \tag{5}$$

describes the motion of the center of mass of the electrons. The eigenvalue problem $H\Psi = E\Psi$, after substitution

$$\Psi(1,2) = \varphi(\mathbf{r})\zeta(\mathbf{R})\chi(s_1, s_2), \tag{6}$$

where $\chi(s_1, s_2)$ is a two-electron spin function, separates to

$$H_\mathbf{r}\varphi(\mathbf{r}) = \varepsilon\varphi(\mathbf{r}) \quad \text{and} \quad H_\mathbf{R}\zeta(\mathbf{R}) = \eta\zeta(\mathbf{R}), \tag{7}$$

with the total energy equal to $E = \eta + \varepsilon$.

The eigenvalue equation of $H_\mathbf{R}$ describes a spherical harmonic-oscillator. For this system the analytical solutions are well known. The energy spectrum is given by

$$\eta_{n'l'} = \omega\left(2n' + l' + \frac{3}{2}\right), \tag{8}$$

where n' and l' are spherical harmonic-oscillator quantum numbers.

The eigenvalue equation of $H_\mathbf{r}$ is very similar to the spherical harmonic-oscillator equation, but it contains an additional term $1/r$ which strongly influences the behavior of the eigenfunctions at $r \approx 0$. After substitution

$$\varphi(\mathbf{r}) = \frac{u(r)}{r}Y_{lm}(\Omega_\mathbf{r}) \tag{9}$$

the radial equation reads

$$\left[-\frac{d^2}{dr^2} + \frac{1}{4}\omega^2 r^2 + \frac{1}{r} + \frac{l(l+1)}{r^2}\right]u_{nl}(r) = \varepsilon_{nl}u_{nl}(r) \tag{10}$$

Eq. (10) has analytical solutions only for a discrete (but infinite) set of ω values[1]. In order to get its solutions for arbitrary ω, we integrate this equation numerically on a grid of points. At the limit of $\omega \to \infty$, term $1/r$ may be neglected and we get the spectrum of a spherical harmonic oscillator with equidistant energy levels separated by ω. At the limit of $\omega \to 0$ the electronic repulsion dominates. As a consequence the maxima

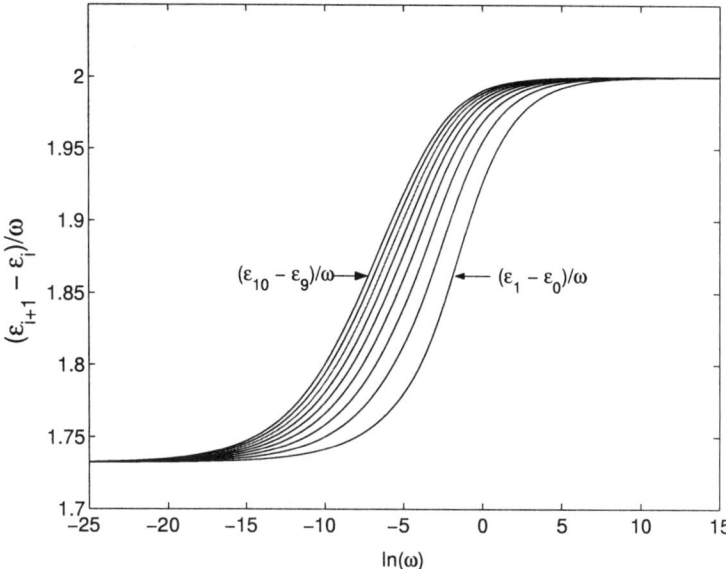

Figure 1. Differences between adjacent eigenvalues of H_r, for $l = 0$, scaled by ω^1, ϵ/ω, versus $\ln\omega$. The consecutive curves (from left to right) correspond to $(\epsilon_{2,0} - \epsilon_{1,0})/\omega$, $(\epsilon_{3,0} - \epsilon_{2,0})/\omega$, ..., $(\epsilon_{10,0} - \epsilon_{9,0})/\omega$. It is seen that all differences at the limit $\omega \to 0$ converge to $\sqrt{3}$ and at the limit $\omega \to \infty$ converge to 2.

of the density $|u(r)|^2$, for all states, are shifted toward large values of r (the probability of distances between electrons becomes small)[1]. Consequently, the term $l(l + 1)/r^2$ may be neglected. Thus, the energy levels for small ω are l-independent. As one can show[1] the distance between energy levels, for small ω, approaches $\sqrt{3}\omega$ and the energy of the ground state approaches

$$V_0 = 3\left(\frac{\omega}{4}\right)^{(2/3)}.$$
(11)

The energy spectrum of H_r is shown in Figs. (1) and (2).

3. The energy spectrum of harmonium

By combining spectra of H_R and H_r we can get the energy spectrum of the whole system. It is displayed in Fig. (3).

At large ω limit the interaction term may be neglected and the spectrum is the same as that of two non-interacting electrons confined in a harmonic oscillator potential. The ground state, ^1S, is not degenerate. The degree of

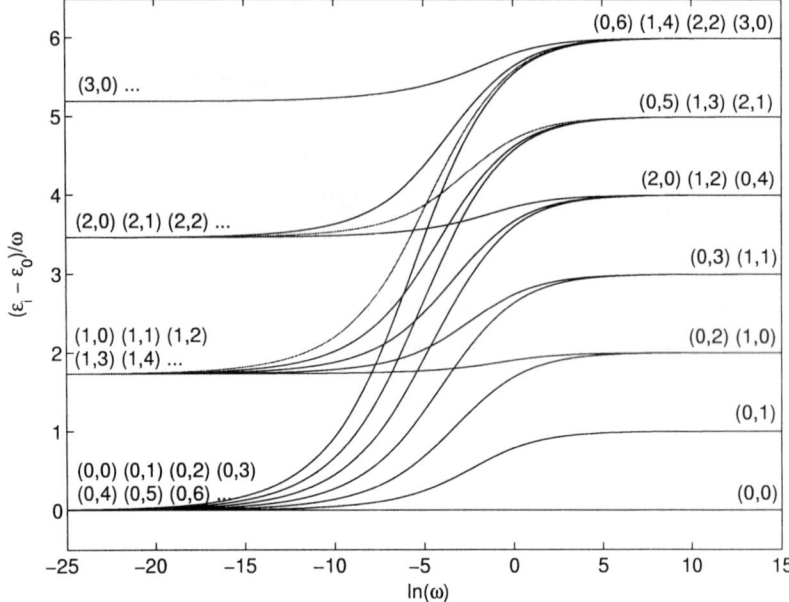

Figure 2. Excitation energies (relative to the ground state) in the spectrum of $H_{\mathbf{r}}$. At each level pairs of quantum numbers (n, l) are given.

degeneracy of excited states is always finite and increases with increasing energy. The differences between adjacent energy levels is equal to ω.

At small ω limit the degree of degeneracy of each energy level, including the ground state, is infinite. Each energy level consists of a bundle of infinite number of energy levels, some of them with arbitrarily high energy at the large ω limit. This feature of the spectrum of harmonium may explain instabilities recently detected by Cioslowski and Pernal in basis set calculations of the ground state of harmonium[2]. There are two progressions of the energy levels at small ω limit, each composed of equally spaced levels: in one energy levels differ by ω and in the other one – by $\sqrt{3}\omega$.

Acknowledgments

This work was supported by the Polish State Committee for Scientific Research (KBN) under project No. 5 P03B 119 21. Helpful discussions with J. Karwowski are highly appreciated.

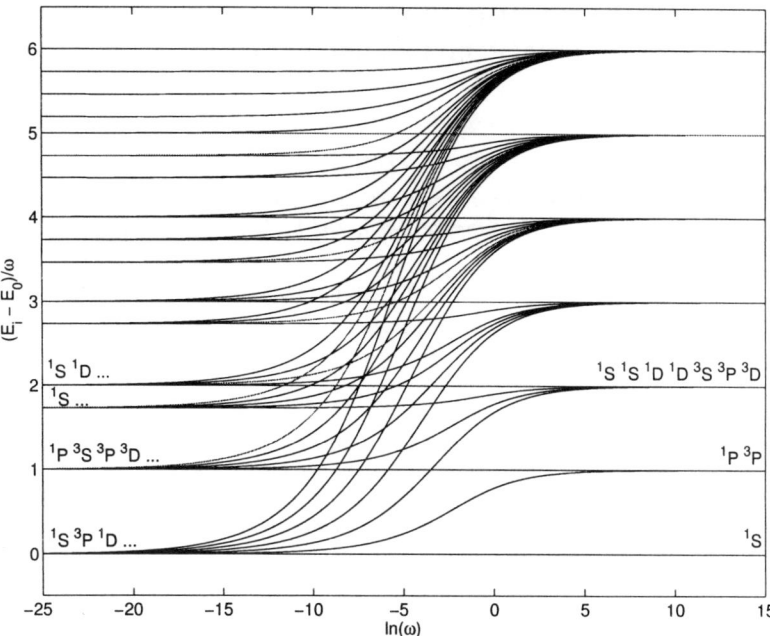

Figure 3. Excitation energies (relative to the ground state) of harmonium, scaled by ω^{-1}, versus $\ln \omega$.

References

1. M. Taut, *Two electrons in an external oscillator potential: Particular analytic solutions of a Coulomb correlation problem*, Phys. Rev. A **48**, 3561-3566 (1993).
2. J. Cioslowski and K. Pernal, *The ground state of harmonium*, J. Chem. Phys. **113**, 8434-43 (2000).
3. S. K. Ghosh and A. Samanta, *Study of correlation effects in an exactly solvable model two-electron system*, J. Chem. Phys. **94**, 517-522 (1991).
4. A. Samanta and S. K. Ghosh, *Correlation in exactly solvable two-particle quantum system*, Phys. Rev. **A 42**, 1178-1183 (1990).
5. N. R. Kestner and O. Sinanoglǔ, *Study of electron correlation in helium-like systems using an exactly soluble model*, Phys. Rev. **128**, 2687-2693 (1962).
6. U. Merkt, J. Huser and M. Wagner, *Energy spectra of two electrons in a harmonic quantum dot*, Phys. Rev. **B 43**, 7320-7323 (1991).

SUPERSPACE GROUPS FOR INCOMMENSURATE COMPOSITE STRUCTURES

P. ZEINER

Institute for Theoretical Physics & CMS, TU Wien,
Wiedner Hauptsraße 8-10, 1040 Vienna, Austria

T. JANSSEN

Institute for theoretical Physics, University of Nijmegen,
Toernooiveld, 6525 ED Nijmegen, The Netherlands

We generalize the notion of superspace groups for the needs of incommensurate composite structures. We motivate two different kinds of definitions and discuss to which systems they can be applied. We illustrate their main properties for the special case of a composite structure consisting of two subsystems.

1. Introduction

It is well known that there is an intimate relation between quasi periodic structures in three dimensions and periodic structures in dimensions $n > 3$. Quasiperiodic structures can be embedded in spaces of higher dimension[1] (so-called superspaces) and conversely quasiperiodic structures can be constructed from periodic structures in higher dimensions, e.g. by the well known cut-and-projection scheme[2] for quasicrystals or by considering appropriate three-dimensional cuts of the superspace.[1] Thus the symmetry of quasiperiodic structures can be characterized by space groups in higher dimensions. However, the three-dimensional physical space, seen as a subspace of the superspace, plays a distinguished role which has to be taken into account. Thus the notion of superspace groups was introduced by A. Janner and T. Janssen some decades ago.[3]

Their concept has been developed for (in)commensurately modulated structures and is based on the existence of a periodic average structure. However, this is a feature that is characteristic for modulated structures, whereas other quasiperiodic structures such as incommensurate composite structures or quasicrystals do not have a periodic average structure in general. Thus the original notion of superspace groups cannot be applied to quasicrystals and only with some restrictions to incommensurate composite structures.

In fact, if one wants to apply the usual notion of superspace groups to incommensurate composite structures one needs a periodic average structure. Since an incommensurate composite structure by definition consists of a finite number of mutually incommensurate modulated structures, one can choose the average structure of one of these subsystems. However, the choice of the subsystem is arbitrary and assigns a distinguished role to a particular subsystem. Moreover different choices of subsystems (and thus different periodic average structures) will lead to different (and inequivalent) superspace groups in general. This is of course not acceptable for the crystallographer, since each physical system should be characterized by a unique symmetry group.

In order to remedy this situation we generalize the notion of superspace groups for the needs of composite systems. The important feature is that all subsystems are treated in the same way and no one plays a distinguished role. This generalization will provide us in a natural way with a definition of equivalence of superspace groups for composite systems.

We organize the paper as follows. We first recall the usual definition of superspace groups and their main properties. However, the usual setting is a bit too special for our purposes and thus we add a section on a non-standard presentation of superspace groups. Finally we discuss the superspace groups for composite systems.

2. Incommensurately modulated systems

As mentioned in the introduction, modulated structures in n dimensions with a d-dimensional modulation can be embedded in $n + d$-dimensional superspace in such a way, that the corresponding structure in superspace, the so-called supercrystal is a periodic structure in the $n + d$-dimensional superspace.[4,5,3,1] The symmetry group of the supercrystal is thus a $n + d$-dimensional space group. It is important to mention that modulated systems can be seen as a periodic basis structure with an (in)commensurate modulation. As a consequence the diffraction pattern consists of two types of reflection peaks, main peaks due to the periodic basis structure and satellites corresponding to the modulation. This distinction imposes some restrictions on the possible space groups which leads to the definition of the so-called superspace groups,[3] which are discussed briefly in the following section. For more details we refer to the work of A. Janner and T. Janssen.[1]

2.1. *Superspace groups*

The important feature of the $n + d$-dimensional superspace V_S is the fact that it contains a distinguished n-dimensional subspace V_E, called the position or external space. The orthogonal complement of V_E is called internal space and denoted by V_I (it is of course d-dimensional), and hence we have $V_S = V_E \oplus V_I$. It is this distinction of external and internal spaces V_E and V_I that takes into account the distinction of a basic structure and a modulation or main reflections and satellites, respectively. The main point is that all space group operations have to respect this distinction into V_E and V_I. In order to formulate this feature in a mathematical way, we have to recall some notions. Let $E(n+d)$, $E(n)$ and $E(d)$ be the Euclidean groups of the spaces V_S, V_E and V_I, respectively. Furthermore let $T(n + d)$, $T(n)$ and $T(d)$ be the corresponding subgroups of translations and $O(n + d)$, $O(n)$ and $O(d)$ the corresponding orthogonal groups.

A set Λ in an n-dimensional Euclidean space V is called a lattice if there is a discrete subgroup T_Λ of the translation group $T(n)$ such that (1) T_Λ acts transitively on Λ, (2) T_Λ is isomorphic to $\mathbb{Z}^n$ and (3) the (real) linear space spanned by Λ is V. The group T_Λ is called the lattice group corresponding to the lattice Λ.

With these definitions at hand we can define the notion of superspace group:

Definition 2.1. An (n,d)-dimensional superspace group G is a subgroup of $E(n) \times E(d)$ such that

(S1) $G \cap T(n + d) =: T_\Sigma$ is an $(n + d)$-dimensional lattice group and
(S2) $G \cap T(d) =: T_D$ is a d-dimensional lattice group.

An alternative but equivalent definition of superspace groups is the following one[3]:

Definition 2.2. An (n,d)-dimensional superspace group is a subgroup of $E(n+d)$ such that the conditions (S1) and (S2) are satisfied and $G \cap T(d)$ is an invariant subgroup of G.

For superspace groups a new equivalence criterion is needed, one which takes the additional structure due to the distinguished subspace into account[3]:

Definition 2.3. Two (n, d)-dimensional superspace groups G and G' are equivalent if and only if there exists a group isomorphism $\chi : G \to G'$ such that $\chi(G \cap T(d)) = G' \cap T(d)$.

2.2. *Lattices*

A very important property of the $n + d$-dimensional lattice Σ is that it admits a standard basis, i.e. a basis $a_1, \ldots, a_{n+d}$ such that $a_{n+1}, \ldots, a_{n+d}$ form a basis of the d-dimensional lattice D. An explicit expression of such a standard basis will be given immediately. Let us first remark that the orthogonal projection of Σ into V_E is an n-dimensional lattice $\Lambda = \pi_E \Sigma$, where we have introduced the orthogonal projection $\pi_E : V_S \rightarrow V_E$. In the following we identify the subspaces V_E and V_I with $\mathbb{R}^n$ and $\mathbb{R}^d$, respectively, and write the elements of V_S as $x = (\mathbf{x}, \mathbf{y})$ with $\mathbf{x} \in V_E$ and $\mathbf{y} \in V_I$. Correspondingly we identify the lattices Λ and D with the corresponding lattices in $\mathbb{R}^n$ and $\mathbb{R}^d$, respectively. Note that Λ is the lattice of the n-dimensional periodic basis structure. A standard basis of Σ is now given by

$$a_i = (\mathbf{a}_i, -\mathbf{b}_{d+i}) \qquad \mathbf{a}_i \in \Lambda, \mathbf{b}_{d+i} \in V_I \qquad i = 1, \ldots, n \qquad (1a)$$

$$a_{n+j} = (\mathbf{0}, \mathbf{b}_j) \qquad \mathbf{b}_j \in D \qquad j = 1, \ldots, d. \qquad (1b)$$

In the same way as for ordinary space groups one defines reciprocal lattices of the lattices Σ, Λ, and D, which are denoted by Σ^*, Λ^*, and D^*, respectively. The reciprocal basis corresponding to the standard basis (1), which we therefore call a reciprocal standard basis reads

$$a_i^* = (\mathbf{a}_i^*, \mathbf{0}) \qquad\qquad i = 1, \ldots, n \qquad (2a)$$

$$a_{n+j}^* = (\mathbf{a}_{n+j}^*, \mathbf{b}_j^*) \qquad\qquad j = 1, \ldots, d, \qquad (2b)$$

which satisfies the orthogonality relation

$$a_i^* \cdot a_j^* = 2\pi \delta_{ij} \qquad\qquad i, j = 1, \ldots, n + d. \qquad (3)$$

In addition, the vectors $\mathbf{a}_i^*$ and $\mathbf{b}_j^*$ form a basis of Λ^* and D^*, respectively, and we have

$$\mathbf{a}_i^* \cdot \mathbf{a}_k^* = 2\pi \delta_{ik} \qquad\qquad i, k = 1, \ldots, n \qquad (4)$$

$$\mathbf{b}_j^* \cdot \mathbf{b}_k^* = 2\pi \delta_{jk} \qquad\qquad j, k = 1, \ldots, d. \qquad (5)$$

The vectors $\mathbf{b}_{d+i}$ and $\mathbf{a}_{n+j}^*$ can be expressed in term of the basis vectors $\mathbf{b}_j$ and $\mathbf{a}_i^*$, respectively:

$$\mathbf{b}_{d+i} = \sum_{j=1}^{n} \sigma_{ji} \mathbf{b}_j \qquad\qquad (6a)$$

$$\mathbf{a}_{n+j}^* = \sum_{i=1}^{d} \sigma_{ji} \mathbf{a}_i^*. \qquad\qquad (6b)$$

Note in passing that the lattice Λ^* describes the main reflections in the diffraction pattern, whereas the vectors a^*_{n+j} describe the (in)commensurate modulation of the system, which gives rise to the satellite reflections. As a warning we mention that $\Lambda^* = (\pi_E \Sigma)^* \neq \pi_E \Sigma^*$, where the latter is the set of all reflections in the diffraction pattern.

2.3. *Point group*

Since the superspace group G is a subgroup of $E(n) \times E(d)$ by definition, the point group of the superspace group is a subgroup of $O(n) \times O(d)$. In fact the point group P is a subdirect product of the point groups $P_E \subset O(n)$ and $P_I \subset O(d)$, which leave the corresponding lattices invariant: $P\Sigma = \Sigma$, $P_E \Lambda = \Lambda$, $P_I D = D$. Corresponding equations hold for the reciprocal lattices. An important result is that the integral matrix representation of P that describes the action of P on the lattice Σ is reducible. The matrix $\Gamma(R)$ corresponding to the point group element $R = (R_E, R_I)$ reads explicitly

$$\Gamma(R) = \begin{pmatrix} \Gamma_E(R) & 0 \\ \Gamma_M(R) & \Gamma_I(R) \end{pmatrix}, \tag{7}$$

$$Ra_i = \sum_{j=1}^{n+d} \Gamma(R)_{ji} a_j. \tag{8}$$

The matrices $\Gamma_E(R)$ and $\Gamma_I(R)$ describe the action of R_E and R_I on the basis $a_i, i = 1, \ldots, n$ of Λ and the basis $b_j, j = 1, \ldots, d$ of D, respectively:

$$Ra_i = \sum_{j=1}^{n} \Gamma_E(R)_{ji} a_j \tag{9}$$

$$Rb_i = \sum_{j=1}^{d} \Gamma_I(R)_{ji} b_j \tag{10}$$

$\Gamma_M(R)$ can be expressed in terms of the matrices $\Gamma_E(R)$, $\Gamma_I(R)$, and the $d \times n$ connecting matrix σ of Eq. (6):

$$\Gamma_M(R) = \sigma \Gamma_E(R) - \Gamma_I(R)\sigma = \sigma_r \Gamma_E(R) - \Gamma_I(R)\sigma_r. \tag{11}$$

In the second equation we have split the connecting matrix σ into two parts, an invariant part σ_i and a rational part σ_r:

$$\sigma = \sigma_i + \sigma_r \tag{12}$$

$$\sigma_r = \sigma + \sum_{R \in P} \Gamma_I(R)\sigma \Gamma_E(R)^{-1} \tag{13}$$

$$\sigma_i = \Gamma_I(R)\sigma_i \Gamma_E(R)^{-1} \qquad \text{for all } R \in P. \tag{14}$$

Note that the matrices σ, σ_i and σ_r depend on the actually chosen standard basis.

2.4. Generalizations

For our purpose the definition of superspace groups as given above is too restrictive. We thus generalize this definition. However, this does not lead to new types of superspace groups. It just gives a more general setting for the original types of superspace groups which is needed for the description of composite systems. This new definition reflects the fact that a modulated system can be embedded into a higher dimensional space in various ways.

Instead of decomposing the superspace V_S in the physical subspace V_E and its orthogonal complement, we may write V_S as the direct sum $V_S = V_E \oplus V_I$, where V_I is now a d-dimensional subspace not necessarily orthogonal to V_E, but such that $V_E \cap V_I = \{0\}$. The Euclidean and orthogonal group of the superspace are again denoted by $E(n+d)$, and $O(n+d)$, respectively. For the Euclidean groups of V_E and V_I we use the notation $E_E(n)$, $E_I(d)$ and $O_E(n)$, $O_I(d)$, respectively. Note that $E_E(n) \times E_I(d)$ is no subgroup of $E(n+d)$ in general, since $O_E(n) \times O_I(d) \subset O(n+d)$ if and only if V_E and V_I are orthogonal.

Definition 2.4. An (n,d)-dimensional superspace group G is a subgroup of $E_E(n) \times E_I(d) \cap E(n+d)$ such that

(S1) $G \cap T(n+d) =: T_\Sigma$ is an $(n+d)$-dimensional lattice group and
(S2) $G \cap T_I(d) =: T_D$ is a d-dimensional lattice group.

It is immediate that this definition reduces to the definition 2.1 in case that V_E and V_I are orthogonal. Note that there is an essential difference in the two definitions. In the standard case we have a pair of distinguished subspaces of the superspace V_S, namely V_E and its orthogonal complement V_I. Any point group element $R \in P$ leaves these two subspaces invariant: $RV_E = V_E$ and $RV_I = V_I$. In the generalized case, however, we have in general two pairs of distinguished subspaces, namely V_E, V_I and their orthogonal complements $V_E^\perp$, $V_I^\perp$, which are all invariant under the action of a point group element R. As a consequence, the point group P of the superspace group G is not only a subgroup of $O_E(n) \times O_I(d) \cap O(n+d)$, but also a subgroup of $O_E(n) \times O_I(d) \cap O_I^\perp(n) \times O_E^\perp(d) \cap O(n+d)$, where $O_I^\perp(n)$ and $O_E^\perp(d)$ denote the orthogonal groups corresponding to the subspaces $V_E^\perp$ and $V_I^\perp$, respectively.

This fact is also reflected in the following equivalent definition of the superspace groups, which is an analog to definition 2.2:

Definition 2.5. An (n,d)-dimensional superspace group is a subgroup of $E(n + d)$ such that the conditions (S1) and (S2) are satisfied and

(S3) $G \cap T_I(d) = T_D$ is an invariant subgroup of G,
(S4) the point group P of G leaves V_E invariant.

In the case that V_E and V_I are orthogonal condition (S4) is an immediate consequence of (S3).

Due to the fact that we can decompose the superspace V_S in two ways, we have two different kinds of projections onto V_E. By π_E we denote the unique (non orthogonal) projection with the properties $\pi_E V_E = V_E$ and $\pi_E V_I = \{0\}$, the orthogonal projection onto V_E shall be Π_E. Analogously we define the non orthogonal projection π_I by $\pi_I V_I = V_I$ and $\pi_I V_E = \{0\}$ and the orthogonal projection Π_I.

2.4.1. *Lattices*

It is no problem to show that there exists a standard basis of Σ also in the general case. In fact the proof involves only the fact that T_D is an invariant subgroup and does not make use of the fact that V_E and V_I are orthogonal. Similarly many theorems on the lattices Σ, D and $\Lambda =: \pi_E \Sigma$ and there reciprocal lattices remain true. However, one has to be careful and choose the right projection. For instance, $\Lambda = \pi_E \Sigma$ is an n-dimensional lattice, whereas $\Pi_E \Sigma$ is in general no lattice (it is not isomorphic to $\mathbb{Z}^n$). Analogously $D^* := \pi_I^* \Sigma^*$ is a d-dimensional lattice, where π_I^* is the adjoint of the projection π_I. Note that π_I^* projects V_S onto $V_E^\perp \neq V_I$ and correspondingly π_E^* projects V_S onto $V_I^\perp$. Thus D^* is a subset of $V_E^\perp$, whereas D is a subset of V_I. Hence D^* is reciprocal to D only in the sense that the bases $\{\pi_I^* a_{n+i}, i = 1, \ldots, d\}$ of D^* and $\{a_{n+i}, i = 1, \ldots, d\}$ of D satisfy the relation

$$\pi_I^* a_{n+i}^* \cdot a_{n+j} = 2\pi \delta_{ij} \qquad i, j = 1, \ldots, d. \tag{15}$$

A similar remark has to be made about the n-dimensional lattice $\Lambda^* = V_I^\perp \cap \Sigma^*$, which is reciprocal to Λ only in the sense that

$$a_i^* \cdot \pi_E a_j = 2\pi \delta_{ij} \qquad i, j = 1, \ldots, n. \tag{16}$$

Since V_E and V_I are not orthogonal any more, the explicit form of the vectors of a standard basis is more complicated. The vectors of $x \in V_E$ and $y \in V_I$ can be written in the form $x = (\mathbf{x}, \mathbf{0})$ with $\mathbf{x} \in \mathbb{R}^n$ and $y = (\hat{\Pi}_E \mathbf{y}, \mathbf{y})$ with $\mathbf{y} \in \mathbb{R}^d$. Here $\hat{\Pi}_E$ is a $d \times n$ matrix such that $\Pi_E y = \hat{\Pi}_E \mathbf{y}$. The basis vectors of a standard basis of the lattice Σ can now be written as follows

$$a_i = (\mathbf{a}_i - \hat{\Pi}_E \mathbf{b}_{d+i}, -\mathbf{b}_{d+i}) \quad \mathbf{a}_i \in \mathbb{R}^n, \mathbf{b}_{d+i} \in \mathbb{R}^d \quad i = 1, \ldots, n \quad (17a)$$

$$a_{n+j} = (\hat{\Pi}_E \mathbf{b}_j, \mathbf{b}_j) \qquad\qquad \mathbf{b}_j \in \mathbb{R}^d \qquad\qquad j = 1, \ldots, d. \quad (17b)$$

Here the vectors $\mathbf{a}_i, i = 1, \ldots, n$ span a lattice in $\mathbb{R}^n$ which we denote by $\hat{\Lambda}$. Similarly the vectors $\mathbf{b}_j$ span a lattice in $\mathbb{R}^d$ which we call $\hat{D}$. In fact $\hat{\Lambda}$ is just the n-dimensional lattice of the periodic average structure. The reciprocal standard basis reads

$$a_i^* = (\mathbf{a}_i^*, -\hat{\Pi}_E^T \mathbf{a}_i^*) \qquad\qquad i = 1, \ldots, n \qquad (18a)$$

$$a_{n+j}^* = (\mathbf{a}_{n+j}^*, \mathbf{b}_j^* - \hat{\Pi}_E^T \mathbf{a}_{n+j}^*) \qquad j = 1, \ldots, d, \qquad (18b)$$

where $\mathbf{a}_i^*$ and $\mathbf{b}_j^*$ are the basis vectors of the reciprocal lattices $\hat{\Lambda}^*$ and $\hat{D}^*$, respectively. The relation between $\mathbf{a}_{n+j}^*$ and $\mathbf{b}_{d+i}$ and the basis vectors $\mathbf{a}_i^*$ and $\mathbf{b}_j$, respectively, is again established by the connecting matrix σ

$$\mathbf{b}_{d+i} = \sum_{j=1}^n \sigma_{ji} \mathbf{b}_j \qquad\qquad (19a)$$

$$\mathbf{a}_{n+j}^* = \sum_{i=1}^d \sigma_{ji} \mathbf{a}_i^*. \qquad\qquad (19b)$$

If we introduce the natural isomorphisms

$$\phi_E : V_E \to \mathbb{R}^n, (\mathbf{x}, \mathbf{0}) \to \mathbf{x} \qquad \phi_I : V_I \to \mathbb{R}^d, (\hat{\Pi}_E \mathbf{y}, \mathbf{y}) \to \mathbf{y} \qquad (20)$$

we can sum up the relation between the lattices Λ, $\hat{\Lambda}$, D, $\hat{D}$ and their reciprocal lattices in the following table

$$\Lambda = \pi_E \Sigma \longleftrightarrow \hat{\Lambda} = \phi_E(\Lambda) \qquad\qquad \pi_E a_i = (\mathbf{a}_i, \mathbf{0}) \longleftrightarrow \mathbf{a}_i \qquad (21)$$

$$\Lambda^* = V_I^\perp \cap \Sigma^* \longleftrightarrow \hat{\Lambda}^* = (\hat{\Lambda})^* \qquad a_i^* = (\mathbf{a}_i^*, -\hat{\Pi}_E^T \mathbf{a}_i^*) \longleftrightarrow \mathbf{a}_i^* \qquad (22)$$

$$D = V_I \cap \Sigma \longleftrightarrow \hat{D} = \phi_I(D) \qquad\qquad a_{n+j} = (\hat{\Pi}_E \mathbf{b}_j, \mathbf{b}_j) \longleftrightarrow \mathbf{b}_j \qquad (23)$$

$$D^* := \pi_I^* \Sigma^* \longleftrightarrow \hat{D}^* = (\hat{D})^* \qquad\qquad \pi_I^* a_{n+j}^* = (\mathbf{0}, \mathbf{b}_j^*) \longleftrightarrow \mathbf{b}_j^*. \qquad (24)$$

Finally we want to remark that the orthogonal projection of Σ^* onto V_E describes the diffraction peaks of the modulated structure in question. It is important to note that this pattern $\Pi_E \Sigma^*$ is independent of the matrix $\hat{\Pi}_E$. This is of course necessary, since the way of embedding the physical structure into the superspace may not affect the diffraction pattern. Furthermore Eq. (18) gives the most general way of embedding the diffraction pattern into a $n + d$-dimensional superspace such that all a_i^* are linearly independent. The embedding is thus fixed by the (linearly independent) vectors $\mathbf{b}_j^*$ and the matrix $\hat{\Pi}_E$. However, note that only the vectors $\mathbf{b}_j^*$

need to be linearly independent, whereas the set $\mathbf{b}_j^* - \hat{\Pi}_E^T \mathbf{a}_{n+j}^*$, $j = 1, \ldots, d$ need not to be linearly independent and therefore does not necessarily span $\hat{V}_I$. Furthermore the matrix $\hat{\Pi}_E$ has to satisfy certain restrictions arising from the point group symmetry of the superspace group, which is discussed below.

2.4.2. *Point group*

We have already mentioned above that the point group P of a superspace group G is a subgroup of $O_E(n) \times O_I(d) \cap O(n + d)$ by its definition. As an immediate consequence P is also a subgroup of $O_E(n) \times O_E^\perp(d)$, $O_I^\perp(n) \times O_I(d)$ and $O_I^\perp(n) \times O_E^\perp(d) \cap O(n + d)$. Thus every point group element $R \in P$ can be written as $R = (R_E, R_I)$ where $R_E \in O_E(n)$ and $R_I \in O_I(d)$, but also as $R = (R_E, R_E^\perp)$ with $R_E \in O_E(n)$ and $R_E^\perp \in O_E^\perp(d)$. Here R_E, R_I and $R_E^\perp$ describe the action of R on the subspaces V_E, V_I and $V_E^\perp$, respectively. The corresponding point groups, i.e. the point groups induced from P on these subspaces shall be denoted by P_E, P_I and $P_E^\perp$, respectively, the corresponding homomorphisms from P into these groups shall be denoted by φ_E, φ_I and $\varphi_E^\perp$. P is not only a subdirect product of P_E and P_I, but also a subdirect product of P_E and $P_E^\perp$. Furthermore we can define the following normal subgroups

$$N_E := \varphi_E(\ker \varphi_I) = \varphi_E(\ker \varphi_E^\perp) \lhd P_E \tag{25a}$$

$$N_I := \varphi_I(\ker \varphi_E) \lhd P_I \qquad\qquad N_E^\perp := \varphi_E^\perp(\ker \varphi_E) \lhd P_E^\perp, \tag{25b}$$

where Def. (25a) is well defined due to $\ker \varphi_I = \ker \varphi_E^\perp$. From a basic theorem on subdirect products[6] we infer the following isomorphisms

$$P_E/N_E \simeq P_I/N_I \simeq P_E^\perp/N_E^\perp. \tag{26}$$

Moreover we have the important isomorphism $P_I \simeq P_E^\perp$.

A very important consequence of the definition of P is that there is a connection between the structure of the point group and the relative orientation of V_E and V_I. Let $\hat{R}_E$ and $\hat{R}_I$ be the orthogonal $n \times n$ and $d \times d$ matrices corresponding to the operations R_E and R_I with respect to an orthonormal basis of V_E and V_I, respectively. Then the following condition must hold true

$$\hat{\Pi}_E \hat{R}_I = \hat{R}_E \hat{\Pi}_E, \tag{27}$$

which puts a restriction on the possible combinations of R_E and R_I for a given matrix $\hat{\Pi}_E$. Thus, in contrast to the standard case not every subdirect product of point groups P_E and P_I is a point group of a superspace group. On the other hand, given a point group P of a superspace group, Eq. (27) puts a restriction on the relative orientation of V_E and V_I. These restrictions have to be taken into account when one is embedding the physical structure in the superspace, otherwise one could lose some (or even all) symmetry elements. Note in passing that $\hat{R}_I$ is not only an orthogonal matrix but also leaves the metric tensor $g_I = E_d + \hat{\Pi}_E^T \hat{\Pi}_E$ invariant, where E_d is the d-dimensional unit matrix: $\hat{R}_I^T g_I \hat{R}_I = g_I$.

There exists an important connection between (in)commensurability and the matrix $\hat{\Pi}_E$. If Eq. (27) has the only solution $\hat{\Pi}_E = 0$, then the system is commensurate. Conversely incommensurability implies the existence of a non-zero solution of Eq. (27).

One of the main results on the point group for superspace groups remains unchanged in the generalized case: The integral matrix representation of P that describes the action of P on the lattice Σ is reducible. The matrix $\Gamma(R)$ corresponding to the point group element $R = (R_E, R_I)$ reads as before

$$\Gamma(R) = \begin{pmatrix} \Gamma_E(R) & 0 \\ \Gamma_M(R) & \Gamma_I(R) \end{pmatrix}, \tag{28}$$

$$Ra_i = \sum_{j=1}^{n+d} \Gamma(R)_{ji} a_j, \tag{29}$$

where the matrices $\Gamma_E(R) = \Gamma_E(R_E)$ and $\Gamma_I(R) = \Gamma_I(R_I)$ again depend only on R_E and R_I, respectively. As before they describe the action of R_E and R_I on the basis $a_i, i = 1, \ldots, n$ of $\hat{\Lambda}$ and the basis $b_j, j = 1, \ldots, d$ of $\hat{D}$, respectively:

$$R\mathbf{a}_i = \sum_{j=1}^{n} \Gamma_E(R_E)_{ji} \mathbf{a}_j \tag{30}$$

$$R\mathbf{b}_i = \sum_{j=1}^{d} \Gamma_I(R_I)_{ji} \mathbf{b}_j \tag{31}$$

$\Gamma_M(R)$ can again be expressed in terms of the matrices $\Gamma_E(R_E)$, $\Gamma_I(R_I)$, and the $d \times n$ connecting matrix σ of Eq. (6):

$$\Gamma_M(R) = \sigma \Gamma_E(R_E) - \Gamma_I(R_I)\sigma. \tag{32}$$

2.4.3. Equivalence of superspace groups

The concept of equivalence of superspace groups has to be adapted for the generalized definition of superspace groups. It is important that this generalization is compatible with the old one, i.e. that no new classes of superspace groups emerge or that inequivalent superspace groups are equivalent under the new definition. We define equivalence as follows (where we assume that V_E is the same for both superspace groups)

Definition 2.6. Two (n, d)-dimensional superspace groups G and G' are equivalent if and only if there exists a group isomorphism $\chi : G \to G'$ such that $\chi(G \cap T_I(d)) = G' \cap T_I'(d)$.

One immediately realizes that this reduces to the standard definition if $V_I = V_I' = V_E^\perp$, since $T_I(d) = T_I'(d)$ holds in that case.

3. Incommensurate composite systems

Incommensurate composite systems consist of several (incommensurately modulated) subsystems. We exclude the case of commensurate composite systems and assume in addition that the atomic surfaces are continuous and topologically equivalent to $\mathbb{R}^d$. This rules out any discontinuous modulations and all composite systems that are not fully incommensurate. Otherwise the different subsystems are not well defined and different choices of subsystems are possible.[7] It might be even not clear whether to describe the system as a composite system or a modulated system.[8,9,10]

3.1. Superspace groups — Definition

The subsystem of a composite system may be equivalent or not. In the first case there may be symmetry transformations that map a subsystem onto another one, in the latter this possibility is excluded. We consider the case of inequivalent subsystems first. In this case we have the following requirements for the definition of the superspace group for a composite system with inequivalent subsystems: the superspace group must be a superspace group in the sense of the previous chapter for each subsystem, and no element of the superspace group may transform one subsystem into another.

Let assume we have m subsystems of dimension $n + d$. Then to each of the m subsystems there corresponds an internal space $V_{I\nu}$ and a decomposition $V_S = V_E \oplus V_{I\nu}$. The corresponding Euclidean groups and translation groups are denoted by $E_{I\nu}$ and $T_{I\nu}$, respectively. We assume that all $V_{I\nu}$ are different. Otherwise two subsystems would have the same periodic average structure and could be described as a single subsystem.

We can now formulate the definition of a superspace group for a composite system with m inequivalent subsystems, which is a direct generalization of definition 2.5

Definition 3.1. An (n, d)–dimensional superspace group G is a subgroup of $E(n + d)$ such that

(C1) $G \cap T(n + d) =: T_\Sigma$ is an $(n + d)$–dimensional lattice group,

(C2) $G \cap T_\nu(d) =: T_{D_\nu}$ are d–dimensional lattice groups for all $\nu = 1, \ldots, m$

(C3) $G \cap T_\nu(d) = T_{D_\nu}$ are invariant subgroups of G for all $\nu = 1, \ldots, m$

(C4) the point group P of G leaves V_E invariant.

We conclude immediately by means of the remark in the previous chapter that this definition implies that the superspace group G is a subgroup of the direct product $E_E \times E_{I\nu}$ for all $\nu = 1, \ldots, m$. Furthermore it follows that G is a superspace group for each of the (modulated) subsystems. However, we mention as a warning that this need not be the full superspace group of the particular subsystem under consideration, but may be just a subgroup of the latter (because the symmetry group of a subsystem may be larger than the symmetry of the whole system).

Let us call the superspace groups of definition 3.1 superspace groups of type I.

Next we consider the case that all or some of the subsystems of the composite system are equivalent. Then subsystems may be mapped onto each other by a symmetry transformation. Thus we generalize the definition 3.1 as follows

Definition 3.2. An (n, d)–dimensional superspace group G is a subgroup of $E(n + d)$ such that

(C1) $G \cap T(n + d) =: T_\Sigma$ is an $(n + d)$–dimensional lattice group,

(C2) $G \cap T_\nu(d) =: T_{D_\nu}$ are d–dimensional lattice groups for all $\nu = 1, \ldots, m$

(C3') $g T_{D_\nu} g^{-1} = T_{D_{\pi_g(\nu)}}$, where π_g is a permutation of $\nu = 1, \ldots, m$

(C4) the point group P of G leaves V_E invariant.

Let us call them superspace groups of type II if not all π_g in (C3') are trivial, otherwise this definition reduces to definition 3.1. Note that any superspace group of type II has a unique maximal normal subgroup of type I, i.e. the subgroup of all symmetry operations that leave all subsystems fixed.

In this setting the following definition of equivalence emerges naturally (of course only composite systems with the same number of subsystems can be equivalent):

Definition 3.3. Two (n, d)–dimensional superspace groups G and G' are equivalent if and only if there exists a group isomorphism $\chi : G \to G'$ such that $\chi(G \cap T_\nu(d)) = G' \cap T_{\pi(\nu)}(d)$, where π is an appropriate permutation.

3.2. Lattices

For each of the subsystems ν there exists a standard basis $a_i^\nu, i = 1, \ldots, n+d$ of Σ such that $a_i^\nu, i = n+1, \ldots, n+d$ is a basis of the lattice D_ν. For any ν we can define the sublattices Λ_ν and D_ν^*, which correspond to the periodic average lattice and the lattice of main reflections of the subsystem ν, and we can apply all our knowledge from the previous chapter.

For composite systems consisting of two subsystems we can find an additional lattice basis, which we shall call standard basis of Σ in the following. Let $D_{12} = D_1 \cap D_2$ and s its dimension. Then we can find a basis of Σ with the following properties: $a_{n+1}, \ldots, a_{n+d} \in D_1$ and $a_{n-s+1}, \ldots, a_{n-s+d} \in D_2$, and thus $a_{n-s+1}, \ldots, a_n \in D_{12}$. This basis has the nice property that it is simultaneously an extension of a basis for D_1 and D_2. This basis will prove very useful for discussing the structure of the point group for $m = 2$. A generalization for arbitrary m is in general impossible.

3.3. Point group

For any standard basis corresponding to a subsystem ν we can immediately write down the corresponding matrices $\Gamma^\nu(R)$

$$\Gamma^\nu(R) = \begin{pmatrix} \Gamma_E^\nu(R) & 0 \\ \Gamma_M^\nu(R) & \Gamma_I^\nu(R) \end{pmatrix}, \tag{33}$$

$$Ra_i = \sum_{j=1}^{n+d} \Gamma(R)_{ji}^\nu a_j \tag{34}$$

and go on as in the previous chapter.

For $m = 2$ we have an additional basis at hand, namely the standard basis. Let us assume $D_{12} = \{0\}$ for simplicity. Then the rotation matrices $\Gamma(R)$ with respect to this basis can be further reduced

$$\Gamma(R) = \begin{pmatrix} \Gamma_0(R) & 0 & 0 \\ \Gamma_{M2}(R) & \Gamma_2(R) & 0 \\ \Gamma_{M1}(R) & 0 & \Gamma_1(R) \end{pmatrix}, \qquad \Gamma_{M\nu}(R) = \tau_\nu \Gamma_0(R) - \Gamma_\nu(R)\tau_\nu,$$

$$\tag{35}$$

where $\Gamma_0(R)$, $\Gamma_{M\nu}(R)$ and $\Gamma_\nu(R)$ are $(n-d) \times (n-d)$, $d \times (n-d)$ and $d \times d$ integer matrices. The connection with Eq. (33) is given by

$$\Gamma_E^1(R) = \begin{pmatrix} \Gamma_0(R) & 0 \\ \Gamma_{M2}(R) & \Gamma_2(R) \end{pmatrix} \qquad \Gamma_E^2(R) = \begin{pmatrix} \Gamma_0(R) & 0 \\ \Gamma_{M1}(R) & \Gamma_1(R) \end{pmatrix} \qquad (36)$$

$$\Gamma_M^\nu(R) = \big(\Gamma_{M\nu}(R), 0\big) \qquad\qquad \Gamma_I^\nu(R) = \Gamma_\nu(R) \qquad\qquad (37)$$

In addition we have the following relation for the connecting matrices

$$\sigma_\nu = \big(\tau_\nu, \rho_\nu\big) \qquad (38)$$

where ρ_ν is a $d \times d$ matrix that is in general no integer matrix. In fact one can prove[10] that ρ_ν is invertible and $\rho_1 = \rho_2^{-1}$. From this one can deduce that Γ_1 and Γ_2 are equivalent representations of the point group P. In fact the corresponding intertwining matrix is just ρ_1. Moreover Γ_E^1 and Γ_E^2 are equivalent representations, too. Thus the two subsystems must have the same point group symmetry, and in fact their symmetry group must belong to the same geometric crystal class. However, they may belong to different arithmetic crystal classes. In fact, the last assertion is not restricted to the case $m = 2$. For general m it holds true that the symmetry groups of the subsystems must belong to the same geometric crystal class but not necessarily to the same arithmetic class.

Finally we want to mention that one can show that to each arithmetic class of superspace groups there exists a finite group of integer matrices of the form (35). On the other hand, there exists a unique arithmetic class of superspace groups to each finite group of integer matrices of the form (35), if Γ_1 and Γ_2 are equivalent representations and an additional compatibility relation for $\Gamma_{M\nu}$ is satisfied. Thus we can prove a one-to-one correspondence between arithmetic crystal classes and equivalence classes of certain finite groups of integer matrices.[10]

Let us make some remarks on the case of superspace groups of type II. These superspace groups G contain a normal subgroup H of index 2 which is a superspace group of type I, whose point group elements can thus be written in the form (35). Let us denote the point groups of these superspace groups with P_G and P_H, respectively. Then we can write the rotation matrices as follows

$$\Gamma(R) = \begin{pmatrix} \Gamma_0(R) & 0 \\ \Gamma_M(R) & \Gamma_{12}(R) \end{pmatrix}, \qquad \Gamma_M(R) = \tau\Gamma_0(R) - \Gamma_{12}(R)\tau, \qquad (39)$$

where all matrices involved are integer matrices (except τ) and the representation Γ_{12} of P_G is obtained from the integer representation Γ_1 of P_H by induction (or alternatively from Γ_2). This representation looks similar to Eq. (7), but let us point out an important difference: $\Gamma_E(R)$ is an $n \times n$ matrix whereas $\Gamma_0(R)$ is a $(n-d) \times (n-d)$ matrix. Correspondingly $\Gamma_I(R)$ and $\Gamma_{12}(R)$ have different dimensions, too.

Again one can prove that there is a one-to-one correspondence of the arithmetic classes of superspace groups of type II with certain equivalence classes of finite groups of integer matrices. This can be generalized easily for the case that D_1 and D_2 have more than the zero translation in common, too.[10]

4. Conclusions and outlook

We have presented a generalization of the notion of superspace groups for composite systems. We have mentioned some basic properties of this new kind of superspace groups for general m and discussed some of the main properties for $m = 2$. In particular the case of general m has to be treated in more detail and shall be discussed elsewhere.

Acknowledgments

One of the authors (P. Z.) acknowledges financial support by the Austrian Academy of Sciences (APART-program).

References

1. T. Janssen. The symmetry of quasiperiodic systems. *Acta Cryst. A*, 47:243–255, 1991.
2. M. Duneau and A. Katz. Quasiperiodic patterns. *Phys. Rev. Lett.*, 54:2688–2691, 1985.
3. A. Janner and T. Janssen. Superspace groups. *Physica*, 99:47–76, 1979.
4. P. M. de Wolff. The pseudo-symmetry of modulated crystal structures. *Acta Cryst. A*, 30:777–785, 1974.
5. A. Janner and T. Janssen. Symmetry of periodically distorted crystals. *Phys. Rev. B*, 15:643–658, 1977.
6. M. Hall Jr. *The Theory of Groups*, chapter 5.5. McMillan, New York, 1959.
7. J. Darriet, L. Elcoro, A. El Abed, E. Gaudin, and J. M. Perez-Mato. Crystal structure of $Ba_{12}Co_{11}O_{33}$. re-investigation using the superspace approach of orthorhombic oxides $A_{1+x}(A'_xB_{1-x})O_3$ based on $[A_8O_{24}]$ and $[A_8A'_2O_{18}]$ layers. *Chemical Materials*, to be published.
8. J. Etrillard, P. Bourges, and C. T. Lin. Incommensurate composite structure of the superconductor $Bi_2Sr_2CaCu_2O_{8+\delta}$. *Phys. Rev. B*, 62:150–153, 2000.
9. D. Grebille, H. Leligny, and O. Pérez. Comment on "incommensurate composite structure of the superconductor $Bi_2Sr_2CaCu_2O_{8+\delta}$". *Phys. Rev. B*, 64:106501, 2001.
10. P. Zeiner and T. Janssen. to be published.

POINT SYMMETRY OF THE YTTRIUM-IRON GARNETS

R. J. WOJCIECHOWSKI*, A. LEHMANN-SZWEYKOWSKA*, T. LULEK**,
J. BARNAŚ*, P.E. WIGEN***

* Faculty of Physics, Adam Mickiewicz University,
ul. Umultowska 85, 61-614 Poznań, Poland

** The University of Rzeszów Institute of Physics,
ul. Rejtana 16a, 35-959 Rzeszów, Poland

*** Department of Physics, Ohio State University,
Columbus, Ohio, USA

We investigate an influence of the point symmetry on the microscopic properties of a simple nanostructure consisting of two hybridising iron-oxygen octahedral and tetrahedral clusters. Within the framework of the standard three-band Hubbard model, energy levels with their eigen-states are found for the clusters. The eigen-states are labeled by the irreducible representations of the appropriate point symmetry groups. The results serve next as a point of departure both for estimating superexchange integrals between the clusters, and, also, for determining a microscopic origin of a strong magnetic anisotropy of the charge transfer between them.

1. Introduction

We investigate an influence of the point symmetry of building units of the crystal unit cell of yttrium-iron garnet (YIG) on the magnetic superexchange and the charge transfer. The building units of YIG are 12 tetrahedral (TMO_4) and 8 (TMO_6) octahedral iron-oxygen $p - d$ hybridising clusters[1-4]. In the tetrahedral (d) cluster, the central iron site is surrounded by the four oxygen nearest neighbours, whereas the octahedral (a) cluster, apart from its central iron, consists of six anion sites. The point symmetry group of the tetrahedron is S_4, and that of the octahedron is S_6[5]. The clusters of either type can be labeled by their respective central iron sites. The overall symmetry of the garnet is cubic and its space symmetry is given by the cubic space group O_h^{10}. Hybridising clusters of the same type are separated from one another, whereas the nearest-neighbour octahedron and tetrahedron share an oxygen corner, which is important since a crucial role in the charge transfer is played by the $p - d$ hybridisation. As known, the cations are three-valent iron ions, Fe^{3+}, and the anions - oxygen ions, O^{2-}. It has also to be mentioned that 12 sites of the third (dodecahedral) type are populated by three-valent yttrium ions with their electronic shells closed up. For the sake of our further consideration, it seems to be more convenient to express an electronic structure of the system in terms of holes rather than in terms of the electrons. Thus the system consists of holes of two types: $3d$ holes localized at the iron ions and $2p$ holes which populate the anions. The simplest possible Hamiltonian, which can be used as a point of departure to account for the magnetic and transport properties of the YIG derivatives, is a three-band Hubbard model [6,7]:

The Hamiltonian consists of the following terms:

$$H = \sum_{\alpha,m_s} E_a n_\alpha + \sum_{\beta,m_s} E_d n_\beta + U \sum n_{\alpha,m_s} n_{\alpha,-m_s} + \sum_{\gamma,m_s,i} E_\gamma n_{\gamma,m_s,i} + H_{p-d}(1)$$

The summation in the first and second term is over all five $3d$ orbital states (α, β) and also over the spin (m_s). E_a and E_d are octahedral and tetrahedral energies of the $3d$ holes, respectively. The Hubbard on-site repulsion is between apposite-spin $3d$ holes, being in the same orbital state. The summation in the fourth term of the Hamiltonian runs over all the oxygen ions, which are the nearest-neighbours of both the iron sites (i). Moreover, it runs over all the three $2p$ orbital states (γ) and the two possible spins (m_s). The hybridization term H_{p-d} takes the following standard form:

$$H_{p-d} = \sum_{i,\alpha,\gamma,m_s,m_{s'}} \delta_{m_s,m_{s'}} (V_a(\alpha, \gamma) d_{\alpha,m_s}^\dagger c_{i,\gamma,m_{s'}} + hc) +$$

$$\sum_{i,\beta,\gamma,m_s,m_{s'}} \delta_{m_s,m_{s'}} (V_d(\beta, \gamma) d_{\beta,m_s}^\dagger c_{i,\gamma,m_{s'}} + hc) \qquad (2)$$

The first term of H_{p-d} describes the hybridisation between the $3d$ holes of the octahedral iron with $2p$ holes localized at all its six oxygen neighbours. The second term gives an analogous expression to the tetrahedral $3d$ holes and the $2p$ holes at their four oxygen neighbours.

Under the influence of the largest (cubic) contribution to the crystal field, five orbital $3d$ states are separated into t_{2g} triplet and e_g doublet[8]. The ground state of the hole in the tetrahedral position, is the triplet (E_d), and in the case of the octahedral site, it is the doublet (E_a. A separation between the cubic energy levels is given by respective parameters Δ_{cf}^d and Δ_{cf}^a. The $2p$ orbital states are not split in the cubic crystal field, forming t_{1u} triplet. In the first step of our procedure, this Hamiltonian is projected separately onto the tetrahedron and octahedron, neglecting both the p-p hopping and also, the occurrence of the mutual oxygen corner between each pair of the nearest-neighbour (nn) tetrahedral and octahedral clusters. The group-theoretical analysis provides information on all the symmetry-permitted hybridising $3d$ and $2p$ states[5], which can be then used as proper initial states in the the perturbation procedure, with the $p-d$ hybridisation considered as a perturbation. The overall cubic symmetry of the crystal is broken in the nearest-neighbour iron-oxygen distance. What we finally obtain, is a set of four eigen-values with their respective eigen-states for each one-dimensional irreducible representation of either point symmetry group (S_4 and S_6), attached to each of the hybridising clusters[5]. In the next step, the inter-cluster $p-d$ hybridization which results from the occurrence of the mutual oxygen corner of the nearest tetrahedral and octahedral clusters, is taken into account, which gives rise to the magnetic superexchange interaction between the clusters and inter-sublattice hole's motion. A possible intra-sublattice hole motion i.e. that between clusters of the same symmetry, must be mediated by the $p-d$ hybridisation combined with the $p-p$ hopping to the next nearest neighbours.

As might have been expected, the Hubbard on-site repulsive interaction between holes with the same orbital states and opposite spins, finally results in an occurrence of the magnetic superexchange interactions between the clusters. The problem can be discussed, separately, for each irreducible representation, in terms of the difference in energies, corresponding to the parallel and anti-parallel mutual orientations of the spins, attached to the respective orbital states.

The paper is organised as follows. In the second section, the energy levels and their respective eigen-states, labeled by the irreducible representations of the respective symmetry groups, are thoroughly discussed. Also, the question of the hole localization on either the iron or oxygen sites of the cluster, is analyzed on the basis of the wave-functions of the cluster's orbital eigen-states.

In the third section, magnetic superexchange interactions between two two $S = 1/2$ spins localized at their respective d and a clusters , are introduced and analyzed with special emphasis put on the question of a role of the point symmetry of the clusters. Magnitude of the superexchange integral between the iron-oxygen clusters is also estimated.

The fourth section deals with the hole's transfer between the tetrahedral and octahedral clusters. The $p-d$ spin-factorized transition matrix elements for different irreducible representations are found. Also, the charge transfer probabilities are discussed in terms of the different representations and as functions of the model parameters.

2. Single-hole energy levels and eigen-states of the hybridising clusters

In a first step of our procedure, the truncated Hubbard Hamiltonian (1) is projected onto a single iron-oxygen cluster of either type. It turns out, that, due to the occurrence of the on-site Coulomb repulsion, the matrix of the projected Hamiltonian is naturally factorized with respect to the spin. Thus, for clusters of both types, the problem is actually reduced to diagonalisation of the two 2x2 submatrices for each irreducible representation, which can be easily done. Wave-functions of the hybridising clusters are assumed in an LCAO MO-like form i.e. as linear combinations of proper single-hole $3d$ and $2p$ states with their coefficients determined in a diagonalisation procedure. As already stated before, in the nearest neighbour distance, the overall cubic symmetry is broken and the point symmetry of an octahedron is S_6, and that of a tetrahedron is S_4. Within the framework of the group theory, we can precisely determine the original proper symmetry-permitted single-hole states[5]. For the obvious physical reasons, those must be linear combinations of the $3d$ and $2p$ states, which guarantee non-vanishing $p - d$ hybridising matrix element. The states can be determined as bases of these irreducible representations of the respective point symmetry groups, which, if the point groups are extended to the space group of the crystal, can induce the same irreducible representation of the space group O_h^{10}. For the octahedral cluster, it is the irrep $a_u^{S_6}$, which labels all five sets of four eigen-states. At the iron site, the cubic doublet $e_g^{O_h}$ is not actually split as the two irreps of the group S_6, which label the $3d$ wave functions obtained from the doublet, are mutually complex conjugate. Those are $e_g^{S_6}$ and $(e_g^{S_6})*$, respectively. The cubic triplet $t_{2g}^{O_h}$ is split into a singlet, labeled by the irrep $a_g^{S_6}$, and the doublet labeled again by $e_g^{S_6}$ and its complex conjugate. Wave functions both at the iron center of the cluster and at each of its oxygen sites, can be obtained by the well-known projection procedure. The wave function of all the octahedral oxygen sites $|P_a(\nu) >$ is given as a linear combination of the single-hole ν states, where ν is $e_u^{S_6}$, $(e_u^{S_6})*$ and $a_u^{S_6}$, respectively, at each site with half of the functions having the opposite sign of their phase factors. Of course, all five Cartesian products of the appropriate $3d$ and $2p$ representations are equal to $a_u^{S_6}$.

In a case of the tetrahedron, the ground state cubic triplet t_{2g} is split into the $b_{3d}^{S_4}$ singlet and the doublet whose states are labeled by the complex representation $e_{3d}^{S_4}$ and its conjugate, respectively. The excited cubic doublet is split into the $a_{3d}^{S_4}$ and $b_{3d}^{S_4}$ singlets. The $2p$ t_{1u} triplet states are split into $b_{2p}^{S_4}$ singlet and the doublet, which corresponds to the irrep $e_{2p}^{S_4}$ and its complex conjugate. Again, like in the case of the octahedron, both the $3d$ and $2p$ wave-functions are obtained by using the projection procedure. The appropriate Cartesian products of the $3d$ and $2p$ states are equal either to the irrep b^{S_4}, or to the irrep e^{S_4}. Analogously to the previous case, the oxygen wave function of each tetrahedron $|P_d >$ is a linear combination of the four single-site $b^{S_4}_{2p}$, or e^{S_4} states with half of the sites having the opposite phase factor to that of the others.

After some very simple algebra, from one 2x2 submatrix corresponding to the spin $m_s(a)$ or $m'_s(d)$, we obtain two eigen-energies: E^l_1 and E^l_3 ($l = a, d$) with their respective eigen-states. As concerns the eigen-states, in the diagonalisation procedure there are determined four coefficients, which indicate contributions of the appropriate $3d$ state (c^1_{1l} and c^3_{1l} with $l = a, d$ to the eigen-states. And analogously, four coefficients are obtained which indicate contributions of the oxygen states $|P_l >$ ($l = a, d$) (c^1_{3l} and c^3_{3l}, where $l = a, d$) to these eigen-states, respectively.

The second 2x2 submatrix of the Hamiltonian, which corresponds to the opposite value of the spin i.e. $-m_s$ for an octahedral cluster (a) and $-m'_s$ for a tetrahedral cluster(d), gives also two eigen-energies for each cluster :E^l_2 and E^l_4($l = a, d$). The $3d$ contribution to the eigen-states are determined by the coefficients: c^2_{2l} and c^4_{2l}, whereas the $2p$ contributions are given by c^2_{4l} and c^4_{4l} with $l = a, d$ for clusters of both types.

The eight energies are solutions of the four respective square master equations and like coefficients in the eigen-states, they are expressed in terms of the model parameters which are: the cubic crystal-field ground-state energies of the single-hole $3d$ localized at the octahedral (E_a) and tetrahedral (E_d) iron site, respectively, the energy distance between the ground and excited states at the octahedral (Δ^a_{cf}), and tetrahedral (Δ^d_{cf}) iron sites. Also, the on-site Coulomb repulsion parameter (U), energy of the $2p$ single hole localized at oxygen site (E_p), energy cost of the transfer between iron sites and their oxygen nearest neighbours ($\Delta_a = E_p - E_a$ and $\Delta_d = E_p - E_d$). An analysis of the cubic crystal-field for the two iron positions is a textbook problem[8]. Since we have decided to discuss the problem in terms of the holes rather than electrons, we have e_g doublet as a ground state of the $3d$ holes on a octahedral iron site, and t_{2g} triplet on a tetrahedral. The energy level E_a is lower than E_d. Assuming mutually related values of the respective ground states, one obtains the energy distances to the excited states expressed by means of only one parameter i.e.$10Dq$. In Figs 1a , there are presented two lowest-lying energy levels of the set of the four energy levels with the $3d$ contribution to their eigen-functions labeled by $e^{S_6}_g$, originated from e_g (solid line), and those of the four energy levels labeled by $a_g^{S_6}$, originated from the triplet t_{2g} (dashed line), respectively, as functions of the two basic hybridization parameters V_a and V_d. In Fig.1b, analogously, we present two of the energy levels with the $3d$ contribution to their eigen-states,labeled by $b_{3d}^{S_4}$, originated from t_{2g} (dashed line), and the two energies with the eigen-states labeled by $b_{3d}^{S_4}$ from the doublet e_g (solid line). For the sake of simplicity, we have assumed V_a and V_d being equal to each other. Mutual differences in the clusters' energy levels, corresponding to the different irreps of the point symmetry groups, are determined by the respective values of the $p - d$ hybridization matrix elements between appropriate $3d$ and $2p$ states. All the matrix elements can be expressed in terms of elementary two-center integrals of covalency,which are assumed to be: V_a between the basic $3d$ states (x^2, y^2, z^2) and $2p$ states (x, y, z), and V_d between xy, yz, zx and x, y, z states. In pure YIG, at $T = 0K$, five lowest-lying energy levels are occupied at either cluster. On the basis of the Hund's rule, they all have the same spin state. In the case of valence-uncompensated doping, extra holes are induced, which occupy the sixth in magnitude energy level with the attached spin opposite to that of the lowest orbital state. A probability of localization of the particular holes - either at the iron sites or sharing the surrounding oxygen sites - can be obtained as modulo-square wave-function coefficients. As seen in Figs 2a and 2b, where the probabilities of the localisation vs the hybridisation parameters, are given for the two energy levels of those in Figs 1a and 1b, respectively. The degree of localisation at the central iron site is prevailing for the orbital ground state, whereas in its excited state, the hole shows a tendency to be shared by the surrounding oxygen sites. Of course, its spin is opposite to that of the ground-state hole. A degree of either iron or oxygen localisation, is strongly dependent upon a strength of the $p - d$ hybridisation, which can explain some experimental reports on an occurrence of Fe^{+4} in Ca-doped YIG.

The results obtained within the framework of the linear combination of atomic orbital (LCAO MO) method, can reproduce those proposed by Zhang and Rice[9] for the $Cu - O$ clusters in superconducting materials. Considering a two-hole space at one cluster of either type, with its basis consisting of two of the cluster's four single-particle energy levels with mutually opposite spins, we obtain the resultant cluster spin being a Zhang-Rice singlet with the proper energy value.

3. Superexchange interactions between a pair of the hybridising clusters

As mentioned before, each pair of the nearest clusters belonging to the different sublattices, share a common oxygen site which enables their mutual communication due to the $p - d$ hybridisation. The $p - d$ Hamiltonian is projected onto such pair consisting of the octahedral and tetrahedral clusters. The crucial question, which arises here, is whether there is a non-zero matrix element of the $p - d$ hybridisation between states of the respective irreps of S_4 and S_6. The occurrence of non-zero matrix element enables both a superexchange coupling of the spins attached to these states, and also, a possible hole transfer. Our analysis shows that matrix elements of the $p - d$ hybridization take non-zero values between all respective states of the two clusters. We perform a diagonalisation of the projected Hamiltonian (1) separately for each pair of the states. In order to estimate a strength of the superexchange coupling between the spins attached to the respective orbital states, we consider two cases: i) the clusters have the same spin direction of their ground states (ferromagnetic order) and ii) the clusters have opposite spins of their ground states (antiferromagnetic order). In both cases different off-diagonal transition matrix elements of the $p - d$ Hamiltonian are obtained. As an example, two off-diagonal matrix elements are presented below in an explicit form:

i) for ferromagnetic order

$$A_{13}^{ad} = (A_{31}^{da})^* = c_{1a}^1 c_{3d}^3 a V_a + c_{3a}^1 c_{1d}^3 d^* V_d \tag{3}$$

and analogously

ii) for antiferromagnetic order

$$A_{12}^{ad} = (A_{21}^{da})^* = c_{1a}^1 c_{4d}^2 a V_a + c_{3a}^1 c_{2d}^2 d^* V_d \tag{4}$$

where the coefficients a and d take a specific form different for each pair of the hybridising states. After some simple algebra two different sets of the clusters' single-particle energy levels together with their eigen-states are found. We discuss both pure YIG and, also, its derivatives. They differ in a number of holes per a hybridising cluster. In pure YIG, the population is five holes per each cluster. It means that always the lowest-lying energy level of the four, which correspond to the same Cartesian product of the irreps in the $3d$ and $2p$ spaces, is populated at $T = 0K$. And, as a further conclusion, it means that two of the eight two-cluster energy levels are occupied and the higher of the two is the Fermi level. Let us consider a contribution to the total ground-state energy of the two-cluster system, which is a sum of the two lowest-lying single-hole energy levels.

As might have been expected[10,11] in pure YIG, the "antiferromagnetic" ground-state energy level lies below the "ferromagnetic" one for a wide range of the Hubbard repulsion. It roughly means that a pair consisting

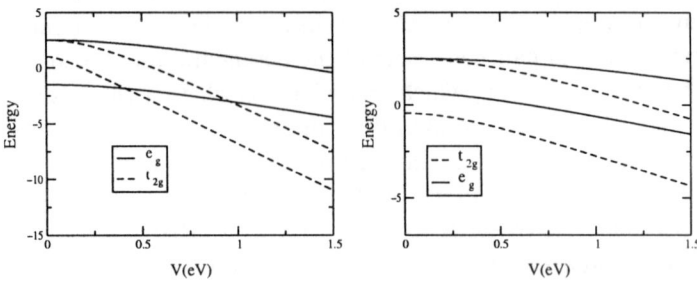

Figure 1.   The energy levels for the octahedral (a) and tetrahedral (b) clusters for $U = 8eV$, $E_a = -1.5eV$, $E_d = -0.44eV$, $E_p = 2.5eV$.

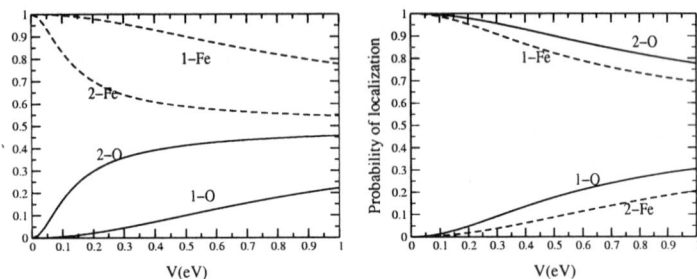

Figure 2.   The probabilities of localization either at the iron and oxygen sites for the energy levels given in Fig.1

of the octahedral and tetrahedral clusters, favours mutually opposite directions of their respective spins. Thus the resultant spin of the two clusters is equal to zero. Information on a difference between the ferromagnetic and antiferromagnetic two-cluster energy enables us to estimate actual values of the superexchange integrals as functions of the model parameters. The estimation is based on a very simple analysis giving rise to the following formula for a respective contribution to the superexchange integral J_{ad}:

$$J_{ad} = (\frac{2}{3})(E_f - E_{af}),$$

(5)

where E_f and E_{af} stand for specific energies, corresponding to the ferromagnetic and antiferromagnetic mutual order of the $S = 1/2$ spins, attached to the specific orbital states of the two clusters. In this approach, the superexchange integral consists of contributions, corresponding to the pairs of spins of all the orbital states[7]. Our results seem to remain in good agreement with the common knowledge on the subject. They show rather weak dependence on the Hubbard U. It is not surprising, however, that we observe the contributions to the superexchange integral to be significantly influenced by the orbital states of the interacting spins. Besides, the dependence on the $p-d$ hybridisation parameters is much more pronounced than that on the parameter U. In Fig. 3, the result is shown graphically: as an example, we present the superexchange integrals for two pairs of spins attached to the ground (line 1) and first excited (line 2) energy levels of the octahedral and tetrahedral clusters.

As stated before, valence-uncompensated doping induces extra-holes into the system. If we assume that three holes may get located at the pair of two hybridising clusters, the third energy level gets populated. According to the same line, we then obtain that a ferromagnetic rather then antiferromagnetic order of the ground-state spins will be favoured, which means that an extra hole can influence the resultant spin of the system in two ways: by inducing an extra $S = 1/2$ spin, and also, by changing the superexchange interaction integral.

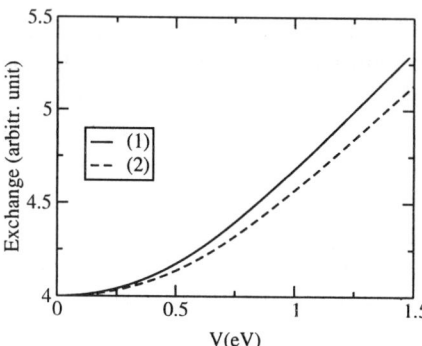

Figure 3. Contribution to the exchange parameter between the octahedral and tetrahedral clusters for the same parameters as those in Fig.1

4. The anisotropy of magnetoresistance

In order to find a microscopic origin of a strong anisotropy, shown by the magnetoresistance in Ca:doped YIG, we investigate again a simple system of two hybridizing iron-oxygen clusters with tetrahedral and octahedral symmetries, respectively. By applying a group-theoretical approach to the diagonalization of the Hubbard model with the Zeeman term included, we obtain energy levels and the orbital eigen-states for both the clusters. This is the point of departure for calculating the total probability of the charge transfer between the clusters[10]. And, finally, the magnetoresistance, as a function of the external field, is found at room temperatures. As an origin of the anisotropy in the magnetoresistance, we suggest the symmetry of the orbital eigen-states of the clusters, which finds its reflection in a form of the orbital contribution to the magnetic moments of the clusters.

In order to study the influence of the external magnetic field on the transport properties of the system, we extend the initial Hamiltonian (1) by a Zeeman term[11,12]. An important point is that magnetic moments of single-fermion states of any cluster consist of both the orbital and spin-like contributions. The structure of the electronic states of the clusters can be considerably alternated under the influence of the external magnetic field applied in different directions.

Our numerical calculations have been performed for the specific nanostructure which consisted of the octahedral cluster with its central iron located at the beginning of the Cartesian coordinate system i.e. at $(0,0,0)$ and one of its six nearest tetrahedral neighbours, whose central iron site was located at the point $(1/4, -1/8, 0)$. For the sake of simplicity, we consider only one of the possible irreducible representations at each cluster. In practice, it means that only four energy levels are attached to either of the clusters. Two of the orbital eigen-states at either cluster have their spin opposite to the remaining ones. In the pure system, only one of the four eigen-states of either cluster is occupied. In a doped system, an extra hole is localized at one of the clusters, reducing its resultant spin. We consider the external magnetic field applied along the three following, mutually perpendicular axes: along the axis between the central irons of the clusters i.e. along $< 2, -1, 0 >$, along the axis $< 1, 2, 0 >$ and, finally, along $< 0, 0, 1 >$[9].Due to the respective orbital contributions to the magnetic moments of the clusters[3], a strong anisotropy is shown by the magnetoresistance. In our model, no spin-orbit coupling is considered. And therefore, the spin contribution to the magnetic moment is "harmless" with respect to the effect of the field. Thus, the strong anisotropy of the magnetoresistance must be an orbital effect. The total probability of the hopping transfer between the clusters is defined as follows[10]:

$$T_{m_s,m_{s'}}(H) = \sum_{p,q} |A_{p,q}|^2 \delta_{m_s,m'_s}(P(E_a(p)(1 - P(E_d(q)) + P(E_d(q)) \atop (1 - P(E_a(q)))) \tag{6}$$

where $|A_{p,q}|^2$ are the modulo-squared matrix elements of the $p-d$ hybridisation Hamiltonian between the eigen-states of the cluster $a(p)$ and the cluster $d(q)$ i.e. the probabilities of all the single hops between these clusters with a conservation of the spins of the involved states. The temperature popula-

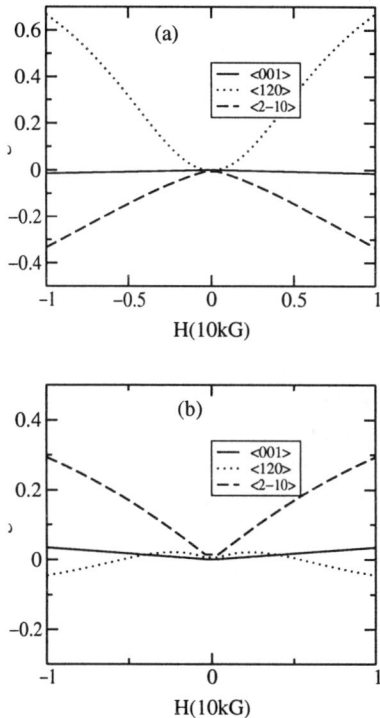

Figure 4. The relative magnetoresistance as a function of the external magnetic field in the three different directions at T=300K for U=5eV and $V_a=V_d=0.5$eV. (a) corresponds to the parallel (↑↑) and (b) to the antiparallel (↑↓) mutual orientation of the spins of the clusters.

tions of the eigen-energy levels i.e. $E_a(p)$ and $E_d(q)$, are determined by the Boltzmann distribution functions, $P(E_a(p))$ and $P(E_d(q))$, respectively. In order to investigate the magnetoresistance in the doped system, we consider two different situations; i) the ground-state spins of the clusters are parallel to each other, ii) the ground-state spins are mutually antiparallel, which can imply an excitation of the spin-wave. In both cases the relative magnetoresistance can be defined in the following standard manner:

$$M_{\uparrow\uparrow(\uparrow\downarrow)}H) = (T_{\uparrow\uparrow(\uparrow\downarrow)}(H) - T_{\uparrow\uparrow(\uparrow\downarrow)}(0))/T_{\uparrow\uparrow(\uparrow\downarrow)}(H)) \qquad (7)$$

where H is an external magnetic field. The examples of our results are presented in Fig.4. Values of the microscopic parameters of the model are taken from literature on the transition metal oxides.The diagrams in Fig.4 show the relative magnetoresistance, as a function of the external magnetic field, applied in three different directions. The most profound change with the magnitude of the field is observed when the field is applied along the axis joining the central iron ions of the clusters i.e. the field is along the direction of the possible hops of the hole. The change in the magnetoresistance is rather weak if the field is perpendicular to the plane (x, y), in which both central ions are located. It seems to be quite interesting that the results obtained with two different mutual orientations of the spins of the clusters i.e. parallel and antiparallel, show a tendency to be mutually reversed. Within the framework of our oversimplified model, we achieve a satisfactory agreement with the experimental data[13]. It does not seem possible, however, to reproduce the data quantitatively, since the latter are taken from thin-film samples, which requires averaging over all the directions in the crystal. Nevertheless, two important goals are reached: it is rather obvious that the magnetic anisotropy of the hopping transport in Ca:YIG can be attributed to the orbital contribution to the magnetic moment. This contribution is strongly direction-dependent, as the cubic symmetry of garnets is broken at the nearest-nieghbourhood distance. An important role is also played by the spins, which determine the possible transitions between the eigen-states of the clusters.

Acknowledgments

This work was supported by the Committee for Scientific Research under contract PBZ-KBN-044/P03/2001.

References

1. L.Klein and A.Aharony, Phys.Rev.B 45, 9915-9925 (1992)
2. H.Donnenberg, S.Toebben and A.Birkholz, J.Phys.:Condens.Matter 9(1997) 6359-6370

3. S.Ishihara, J.Inoue and S.Maekawa, Phys.Rev.B 55(1997) 8280-8286
4. A.Lehmann-Szweykowska, R.J.Wojciechowski and G.A.Gehring, Acta Phys.Polon.A 97(2000) 563-566
5. A.Lehmann-Szweykowska, T.Lulek and M.M.Kaczmarek, J.Phys.:Condens.Matter 13(2001) 3607-3621
6. J.Hubbard, Proc.Roy.Soc.London Ser. A 276 (1963) 238
7. A.Oles, Phys.Rev.B 28 (1983) 327
8. J.S. Griffith, "The Theory of Transition-Metal Ions", Cambridge University Press, 1971
9. F.C.Zhang and T.Rice, Phys.Rev.B 37(1988) 3759-3761
10. C.Srinitiwarong and G.A.Gehring, J.Phys.: Condens.Matter 13(2001) 7987-7998
11. R.J.Wojciechowski, A.Lehmann-Szweykowska, P.E.Wigen, and J.Barnaś, phys.stat. sol b (in press)
12. A.Lehmann-Szweykowska, R.J.Wojciechowski J.Barnaś, and P.E.Wigen, Molecular Low Dimensional and Nanostructured Materials for Advanced Applications, 297, Kluver Academic Publishers 2002
13. P.Wigen and Donglei Li, private communication

TRANSPORT PROPERTIES OF DOMAIN WALLS IN FERROMAGNETS

J. BARNAŚ

Department of Physics, Adam Mickiewicz University, ul. Umultowska 85,
61-614 Poznań, Poland
E-mail: barnas@amu.edu.pl

V. K. DUGAEV

Institute for Problems of Materials Science, Vilde 5, 58001 Chernovtsy, Ukraine
Max-Planck-Insitut für Mikrostukturphysik, Weinberg 2, 06120 Halle, Germany
E-mail: dugaev@yahoo.co.uk

The influence of domain walls on transport properties of ferromagnetic materials is analyzed theoretically and the results are compared with recent experiments. In the case of diffusive transport through a thick domain wall, the semiclassical approximation is applied and a local spin transformation is performed, which replaces the system with a domain wall by the corresponding system without domain wall but with an additional gauge field. Due to a redistribution of single-particle electron states at the wall, one obtains either negative or positive contributions to resistivity. On the other hand, suppression of the weak localization corrections to conductivity by the gauge field created by the wall leads to an increase in the low-temperature conductivity in the presence of a domain wall. In the case of narrow domain walls the semiclassical approximation is not valid. Instead of this one can use an approach based on scattering matrix. In this particular case, the domain wall induces a large positive contribution to the resistivity. The corresponding magnetoresistance in nanostructures with sharp domain walls can be large, in accordance with recent experiments.

1. Introduction

It is well know for long time that magnetic domain walls (DWs) in a ferromagnetic metal influence its electronic transport properties by producing an additional contribution to electrical resistivity. Since DWs give rise to electron scattering,[1,2] one could expect that this contribution is positive. This expectation was also supported by early experiments. It was only very recently when a single DW contribution to electrical resistivity could be extracted in a controllable way from the overall resistance.[3,4,5,6] Surprisingly, it turned out that the resistance of a system with DWs in some cases was smaller than in the absence of DWs,[3,4] whereas in other cases it was larger.[7,8,9] This intriguing observation led to considerable theoretical interest in electronic transport through DWs.[10,11,12,13,14] The interest is additionally stimulated by possible applications of the associated magnetoresistance in magnetoelectronics devices. This is because creation and destruction of DWs can be controlled by a weak magnetic field. The corresponding magnetoresistance can be then either positive or negative.

Recent experiments on magnetic point contacts showed that constrained DWs formed at the very contact between ferromagnetic wires produce an unexpectedly large contribution to electrical resistivity, and consequently lead to large negative magnetoresistance.[15] The characteristic feature of DWs in point contact geometry is their very small width (a few angstroms),[16,17] which is much smaller than the DWs width in bulk materials, thin films, or in wires.

In the following we will describe theoretically basic features of the electronic transport through DWs, and will present explanation of the above described experimental observations. Two limits will be analyzed in detail – the limit of thick DW, when electronic transport through the wall is diffusive, and the limit of narrow DW, when the transport is ballistic. In the former case the theoretical treatment is based on a semiclassical approach, which is valid for $k_{F\uparrow(\downarrow)}D \gg 1$, where $k_{F\uparrow}$ and $k_{F\downarrow}$ are the Fermi wavevectors corresponding to the two spin channels, and D is a characteristic length of the magnetization variation (DW width).[18] In such a case DW can lead to redistribution of single-electron quasiparticles, and this can lead either to positive or to negative contribution to resistivity. Another mechanism which leads to negative contribution is based on the suppression of weak localization (WL) corrections to conductivity by DWs.[10] At sufficiently low temperatures quantum interference effects in a magnetically uniform system (without DWs) lead to an increase in the resistivity due to enhanced back scattering.[19,20] Creation of DWs destroys the interference effects and therefore diminishes resistivity of the system.

When, however, the DW width D is of atomic size, like in some nanoconstrictions,[16] the condition of semiclassical behavior is not fulfilled. In that case, one has to use a different approach, like for instance the one based on the scattering matrix and Landauer formalism.

2. Diffusive transport through a thick domain wall

2.1. *Model*

Assume a simplified model of a ferromagnetic metal, in which conduction electrons with a parabolic energy spectrum interact with a nonuniform magnetization that smoothly varies across a certain DW. Assume also that the electrons are scattered by defects with the corresponding scattering potential being independent of the spin orientation (in a general case this potential can be spin dependent). When the domain wall is sufficiently thick, $D \gg l$, where l is the electron mean free path, electronic transport across the wall is diffusive.

The single-particle Hamiltonian describing conduction electrons locally exchange-coupled to the magnetization $\mathbf{M(r)}$ takes the form

$$H_0 = -\frac{1}{2m}\frac{\partial^2}{\partial \mathbf{r}^2} - J\boldsymbol{\sigma} \cdot \mathbf{M(r)}, \qquad (1)$$

where J is the exchange parameter, $\boldsymbol{\sigma} = (\sigma_x, \sigma_y, \sigma_z)$ are the Pauli matrices, and the unit system with $\hbar = 1$ is used.

The domain wall is characterized by a magnetization profile $\mathbf{M(r)}$. For the sake of simplicity we assume $|\mathbf{M(r)}| = M_0 = \text{const}$. Thus, we can write

$$J\mathbf{M(r)} = M\mathbf{n(r)}, \qquad (2)$$

where $\mathbf{n(r)}$ is a unit vector field specific for a particular type of DWs (to be defined later), and $M = JM_0$ is measured in energy units.

In order to control the charge density of the electron gas, we include the Coulomb electron-electron interaction in the mean-field approximation *via* the term

$$H_{int} = e\phi(z), \qquad (3)$$

where e is the electron charge ($e < 0$) and the field $\phi(z)$ is the mean-field electrostatic potential in the presence of the wall, which obeys the equation

$$\frac{d^2\phi(z)}{dz^2} = -4\pi e\left(\langle \psi^\dagger \psi \rangle - n_0\right), \qquad (4)$$

with $\langle ..... \rangle$ denoting the ground state average, n_0 being the electron gas density in the absence of DW, and ψ and $\psi^\dagger$ denoting the spinor field operators. The potential $\phi(z)$ has to be calculated self-consistently, which assures that the total charge accumulated at the wall vanishes, though the charge neutrality may be violated locally. The total Hamiltonian H of the system can be then written as

$$H = H_0 + H_{int}, \qquad (5)$$

where H_0 and H_{int} are given by Eq.(1) and Eq.(3), respectively.

2.2. *Gauge transformation*

The key point of the approach is a local unitary transformation

$$\psi \to T(\mathbf{r})\,\psi, \quad T^\dagger(\mathbf{r})\,T(\mathbf{r}) = \breve{1}, \tag{6}$$

where $\breve{1}$ is the 2×2 unit matrix. $T(\mathbf{r})$ transforms the problem of electrons in a system with nonuniform magnetization to an equivalent problem of electrons in a system with uniform magnetization, but with an additional gauge field.[10,18] In other words, $T(\mathbf{r})$ transforms the second term in Eq. (1) as

$$\boldsymbol{\sigma} \cdot \mathbf{n}(\mathbf{r}) \to \sigma_z, \tag{7}$$

or equivalently

$$T^\dagger(\mathbf{r})\,\boldsymbol{\sigma} \cdot \mathbf{n}(\mathbf{r})\,T(\mathbf{r}) = \sigma_z. \tag{8}$$

Explicit form of $T(\mathbf{r})$ is given by[21]

$$T(\mathbf{r}) = \frac{1}{\sqrt{2}} \left(\breve{1}\sqrt{1 + n_z(\mathbf{r})} + i\,\frac{n_y(\mathbf{r})\,\sigma_x - n_x(\mathbf{r})\,\sigma_y}{\sqrt{1 + n_z(\mathbf{r})}} \right). \tag{9}$$

Generally, the above transformation can be applied not only to simple DWs, but also to other types of topological excitations in ferromagnetic systems, for instance to helicoidal waves, skyrmions, and others.

Upon applying the transformation (6) to the kinetic part of the Hamiltonian (1) one obtains

$$\frac{\partial^2}{\partial \mathbf{r}^2} \to \left(\frac{\partial}{\partial \mathbf{r}} + \mathbf{A}(\mathbf{r}) \right)^2, \tag{10}$$

where the non-Abelian gauge field $\mathbf{A}(\mathbf{r})$ is given by

$$\mathbf{A}(\mathbf{r}) = T^\dagger(\mathbf{r})\,\frac{\partial}{\partial \mathbf{r}}\,T(\mathbf{r}). \tag{11}$$

According to Eq. (9), the gauge field $\mathbf{A}(\mathbf{r})$ is a matrix in the spin space.

Assume now a more specific DW in a bulk system, which is translationally invariant in the x-y plane: $\mathbf{M}(\mathbf{r}) \to \mathbf{M}(z)$ and $\mathbf{n}(\mathbf{r}) \to \mathbf{n}(z)$. For a simple DW with $\mathbf{M}(z)$ lying in the plane normal to the wall one can parameterize the vector $\mathbf{n}(z)$ as

$$\mathbf{n}(z) = (\,\sin\varphi(z),\ 0,\ \cos\varphi(z)\,), \tag{12}$$

where the phase $\varphi(z)$ determines the type of DWs. The transformation (9) is then reduced to

$$T(z) = \frac{1}{\sqrt{2}} \left(\breve{1}\sqrt{1 + \cos\varphi(z)} - i\sigma_y\,\frac{\sin\varphi(z)}{\sqrt{1 + \cos\varphi(z)}} \right), \tag{13}$$

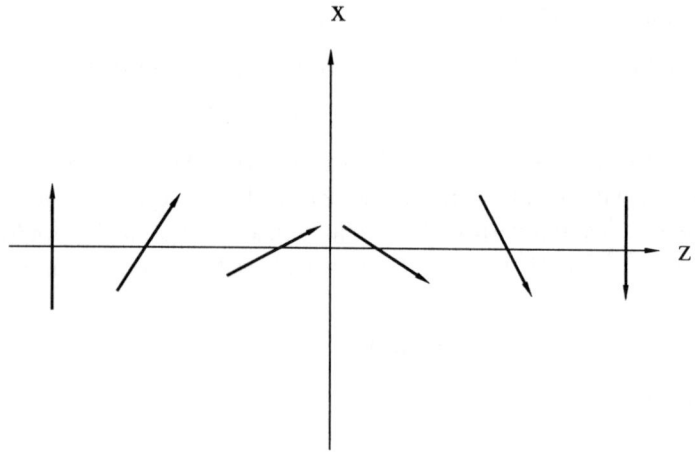

Figure 1. Variation of the magnetization in the domain wall.

and the gauge field acquires the simple form

$$\mathbf{A}(z) = \left(0,\, 0,\, -\frac{i}{2}\, \sigma_y\, \varphi'(z) \right),$$ (14)

where $\varphi'(z) \equiv \partial \varphi(z)/\partial z$.

Taking into account the above formulas one can write the full transformed Hamiltonian in the form

$$H = -\frac{1}{2m}\frac{\partial^2}{\partial \mathbf{r}^2} - M\sigma_z + e\phi(z) + \frac{m\beta^2(z)}{2} + i\sigma_y\frac{\beta'(z)}{2} + i\sigma_y\beta(z)\frac{\partial}{\partial z},$$ (15)

where

$$\beta(z) = \frac{\varphi'(z)}{2m}.$$ (16)

When $k_{F\uparrow(\downarrow)}D \gg 1$, the perturbation due to DW is weak and the semiclassical approximation is well justified. The last three terms on the right hand side of Eq. (15) can be then treated as a small perturbation.

If one assumes the domain wall in the form of a kink shown schematically in Fig. 1, then

$$\varphi(z) = -\frac{\pi}{2}\, \tanh{(z/L)}$$ (17)

with $L = D/2$, and the parameter $\beta(z)$ is given by

$$\beta(z) = -\frac{\pi}{4mL \cosh^2(z/L)}. \tag{18}$$

2.3. Local conductivity

The general formula for the local conductivity (without localization corrections and for electric field applied along the axis z) has the following form

$$\sigma_{zz} = \frac{e^2}{2\pi m^2} \operatorname{Tr} \int \frac{d^3 k}{(2\pi)^3} (k_z - m\beta\sigma_y) G_{\mathbf{k}}^R (k_z - m\beta\sigma_y) G_{\mathbf{k}}^A, \tag{19}$$

where the gauge potential $\mathbf{A}(z)$ given by Eq. (14) is taken into account, and the retarded (R) and advanced (A) Green functions are both evaluated at the Fermi level,

$$G_{\mathbf{k}}^{R,A} = \frac{-\varepsilon_{\mathbf{k}} - M\sigma_z - k_z\beta(z)\sigma_y + \mu_r(z)}{[-\varepsilon_{\mathbf{k}\uparrow}(z) + \mu_r(z) \pm i/2\tau_\uparrow(z)][-\varepsilon_{\mathbf{k}\downarrow}(z) + \mu_r(z) \pm i/2\tau_\downarrow(z)]}. \tag{20}$$

Here, $\varepsilon_{\mathbf{k}} = (q^2 + k_z^2)/2m$ with $q^2 = k_x^2 + k_y^2$, $\mu_r(z) = \mu - m\,\beta^2(z)/2 - e\phi(z)$ with μ denoting the chemical potential, and

$$\varepsilon_{\mathbf{k}\uparrow(\downarrow)}(z) = \varepsilon_{\mathbf{k}} \mp \left[M^2 + k_z^2\beta^2(z)\right]^{1/2}, \tag{21}$$

where the upper (lower) sign refers to $\uparrow$ ($\downarrow$). The quasi-particle energies $\varepsilon_{\mathbf{k}\uparrow(\downarrow)}(z)$ are the eigenstates of the whole Hamiltonian (poles of the Green functions). They correspond to pure spin states only outside the wall, whereas inside the wall they have no pure spin-up (spin-down) form because of spin mixing by the wall. Finally, $\tau_\uparrow(z)$ and $\tau_\downarrow(z)$ in Eq.(20) are the relaxation times, which for impurity scattering potential V_0 independent of the electron spin have the form

$$\frac{1}{\tau_{\uparrow(\downarrow)}(z)} = \frac{m V_0^2}{2\pi} \left[k_{F\uparrow}(z) + k_{F\downarrow}(z) \pm \frac{M}{\beta(z)} \operatorname{arcsinh} \frac{k_{F\uparrow}(z)\,\beta(z)}{M} \mp \right.$$

$$\left. \mp \frac{M}{\beta(z)} \operatorname{arcsinh} \frac{k_{F\downarrow}(z)\,\beta(z)}{M} \right], \tag{22}$$

where $k_{F\uparrow(\downarrow)}(z)$ are the appropriate Fermi wavevectors,

$$k_{F\uparrow(\downarrow)}^2(z) = 2m\mu_r(z) + 2m^2\beta^2(z) \pm$$

$$\pm 2m \left[2m\mu_r(z)\beta^2(z) + m^2\beta^4(z) + M^2\right]^{1/2}. \tag{23}$$

The difference in scattering times is due to a difference in the density of states at the Fermi level for $\uparrow$ and $\downarrow$ states.

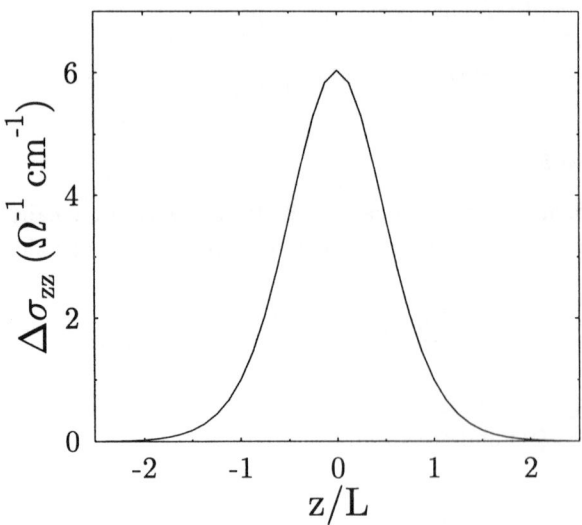

Figure 2. Domain wall contribution to local conductivity, calculated for $L = 50$ Å, Fermi energies $E_{F\uparrow} = 3$ eV and $E_{F\downarrow} = 2.5$ eV, and for impurity scattering potential leading to the bulk conductivity (without domain wall) $\sigma = 0.67 \cdot 10^5$ Ohm^{-1} cm^{-1}.

The local conductivity σ_{zz} is a smoothly varying function of z, $\sigma_{zz} = \sigma_{zz}(z)$,

$$\sigma_{zz}(z) = \frac{e^2}{2\pi^2 m} \sum_{\sigma=\uparrow,\downarrow} \tau_\sigma(z) \left(\frac{k_{F\sigma}^3(z)}{3} + m^2\beta^2(z)\, k_{F\sigma}(z) - \right.$$

$$\left. - m^2 M\beta(z) \arctan \frac{k_{F\sigma}(z)\beta(z)}{M} \right). \tag{24}$$

Given the conductivity $\sigma_{zz}(z)$, the average resistivity of a sample of length d with a domain wall can be found as

$$\rho = \frac{1}{d} \int \frac{dz}{\sigma_{zz}(z)}. \tag{25}$$

The DW contribution to the local conductivity, $\Delta\sigma_{zz}$, is shown in Fig.2 as a function of z. It is evident that this contribution is positive, i.e., the conductivity is enhanced within the wall. The enhancement shown in Fig.2 is not large, but it could be larger when one would assume appropriate spin asymmetry of the impurity scattering potential. It should be noted, however, that by taking opposite spin asymmetry for the impurity scattering potential, the enhancement can be diminished or even can change sign, i.e., the conductivity within the wall can be lower than outside the wall. Thus, this model can account for both signs of the magnetoresistance associated with DWs. This sign depends on the spin asymmetry of impurity scattering potential.

2.4. Weak localization effects

It is well known that external magnetic field suppresses the WL corrections to conductivity. More specifically, the vector potential associated with the magnetic field produces an additional phase shift which destroys the quantum interference effects responsible for these corrections. From the discussion of the two preceding subsections follows, that the DW effect on electrons in a ferromagnetic metal can be described in terms of the spin dependent gauge potential $\mathbf{A}(\mathbf{r})$. One may then expect, that this gauge potential has a similar influence on the quantum corrections to conductivity as the vector potential associated with an external magnetic field. This problem was analyzed by Tatara and Fukuyama,[10,14] and also by Lyanda-Geller et al.[22]

The quantum corrections to conductivity due to WL are important at low temperatures, while at higher temperatures they are suppressed by inelastic scattering processes. They are usually accounted for theoretically in terms of cooperons,[19,20] i.e., propagators which describe propagation in space and time of the superconductive density fluctuations. In nonmagnetic systems there are singlet and triplet cooperons, which contribute with opposite signs. There is, however, an essential difference between WL effects in nonmagnetic and ferromagnetic systems, even if there are no DWs in the latter case. This difference is due to internal magnetic field and associated vector potential, which diminishes or even suppresses in some cases the WL corrections in ferromagnets. If, however, the corrections are not suppressed, then the magnetization usually suppresses the singlet cooperon. As a consequence, there is no weak antilocalization effect in ferromagnetic systems with strong spin-orbit scattering, contrary to nonmagnetic ones. Accordingly, the magnetoresistance related to WL in ferromagnets is always negative.[23]

The cooperon $C(\mathbf{r}, \mathbf{r}')$ can be found from relevant ladder-type diagrams.[19,20,22,23] The corresponding localization correction to the local conductivity is then related to the cooperon via the expression

$$\delta\sigma = -2D\frac{e^2}{\pi}C(\mathbf{r}, \mathbf{r}), \tag{26}$$

where D is the diffusion constant The key question is then to find the *cooperon* from the appropriate integral or differential equations.

The effect of DW on quantum corrections is related to a specific mechanism of the suppression of cooperons by the gauge field associated with the domain walls. Accordingly, the resistance of a system without DWs is larger than in their presence because DWs destroy the negative corrections to conductivity. Associated magnetoresistance is therefore negative. Weak localization correction to conductivity in ferromagnets in the presence of DWs was analyzed by Tatara and Fukuyama[10,14] as well as by Lyanda-Geller *et al*,[22] but in the case of quasi-one-dimensional wires. When the lateral dimension $d_\perp$ obeys the condition $d_\perp \gg l \gg k_F^{-1}$, then electron motion is three dimensional. However, when additionally $l_\phi \gg d_\perp$, where l_ϕ is the phase coherence length, the system behaves like quasi-one-dimensional from the point of view of WL effects. Under this condition Lyanda-Geller *et al*[22] found the localization correction

$$\delta\sigma^{(1)} = -\frac{e^2}{4\pi}(l_\phi + l_w) \tag{27}$$

where $l_w^{-2} = D^{-1}(\tau_w^{-1} + \tau_\phi^{-1})$. Here τ_ϕ is phase coherence time in the absence of DW, and $1/\tau_w = D/4M^2\tau^2 Ld$. Thus, the wall contributes to phase decoherence and one can note that $\delta\sigma^{(1)}$ in the presence of the wall is smaller than in its absence. Accordingly, conductivity of a system with DW is larger than that of a system without the wall. The parameters τ_ϕ, l_ϕ, and k_F, which enter Eq.(27), were assumed to be independent of the spin orientation. In a general case, however, they are spin dependent.

The above considerations apply to DWs which are uniform in their planes. When, however, this condition is not fulfilled, then there is an additional factor leading to suppression of the WL corrections, which is related to the Berry phase.[22] This contribution is described quantitatively by the additional dephasing length l_M related to the fact that the magnetization (which is coupled to electron spin) encircles a nonzero solid angle Ω for electrons completing a self-crossing trajectory. It is worth to note that $\Omega = 0$ for DWs which are uniform in their planes.

3. Transport through an atomic-size domain wall

3.1. *Scattering states*

Let us consider again the Hamiltonian (1) describing electrons in a spatially inhomogeneous magnetization $\mathbf{M}(\mathbf{r})$. For a very narrow constrained DW one may consider only a few channels for electronic transport. A limiting situation is when there is only a single transport channel. In such a one-dimensional case the Hamiltonian (1) can be rewritten as

$$H = -\frac{1}{2m}\frac{d^2}{dz^2} - JM_z(z)\,\sigma_z - JM_x(z)\,\sigma_x. \tag{28}$$

We will make use of the scattering states taken in the form

$$\chi_{R\uparrow k}(z) = \begin{cases} \begin{pmatrix} e^{ik_\uparrow z} + r_{R\uparrow}\,e^{-ik_\uparrow z} \\ r_{R\uparrow}^f\,e^{-ik_\downarrow z} \end{pmatrix}, & z \ll -L \\[2mm] \begin{pmatrix} t_{R\uparrow}\,e^{ik_\downarrow z} \\ t_{R\uparrow}^f\,e^{ik_\uparrow z} \end{pmatrix}, & z \gg L \end{cases} \tag{29}$$

where $k_{\uparrow(\downarrow)} = \sqrt{2m(E \pm M)}$, and E is the electron energy. This state describes the spin-up electron wave incident from $-\infty$ and partly reflected and transmitted into the spin-up and spin-down channels. The coefficients $t_{R\uparrow}$ and $t_{R\uparrow}^f$ are the transmission amplitudes without and with spin reversal, respectively, whereas $r_{R\uparrow}$ and $r_{R\uparrow}^f$ are the relevant reflection amplitudes. The analogous form have the scattering states related to the spin down wave incident from left to right (labeled with $R \downarrow k$), as well as the scattering states related to electron waves incident on DW from the right.

Integrating the Schrödinger equation $H\psi = E\psi$ with the Hamiltonian (28) from $-\delta$ to $+\delta$ in the vicinity of $z = 0$ (where the domain wall is located), and assuming $L \ll \delta \ll k_{\uparrow(\downarrow)}^{-1}$, one obtains

$$-\frac{1}{2m}\left(\left.\frac{d\chi_n}{dz}\right|_{+\delta} - \left.\frac{d\chi_n}{dz}\right|_{-\delta}\right) - \lambda\,\sigma_x\,\chi_n(0) = 0, \tag{30}$$

where n is the electron state index ($n \equiv R(L) \uparrow (\downarrow) k$) and λ is a factor defined as

$$\lambda \simeq \int_{-\infty}^{\infty} dz\, JM_x(z) \simeq ML \tag{31}$$

Equation (30) has the form of a spin-dependent condition for transmission through a δ-like potential barrier located at $z = 0$.

Taking into account the scattering states (29) and the condition (30), in combination with the continuity condition for the wave functions, one finds the following expressions for the transmission amplitudes:

$$t_{R\uparrow(\downarrow)} = t_{L\downarrow(\uparrow)} = \frac{2v_{\uparrow(\downarrow)}(v_\uparrow + v_\downarrow)}{(v_\uparrow + v_\downarrow)^2 + 4\lambda^2}, \tag{32}$$

$$t_{R\uparrow(\downarrow)}^f = t_{L\downarrow(\uparrow)}^f = \frac{4i\lambda\,v_{\uparrow(\downarrow)}}{(v_\uparrow + v_\downarrow)^2 + 4\lambda^2}, \tag{33}$$

where $v_{\uparrow(\downarrow)} = k_{\uparrow(\downarrow)}/m$.

According to Eq.(33), the magnitude of the spin-flip transmission coefficient can be estimated as

$$|t^f|^2 \sim \left(\frac{\lambda v}{v^2 + \lambda^2}\right)^2 \sim \left(\frac{M\varepsilon_0}{\varepsilon_F \varepsilon_0 + M^2}\right)^2 (k_F L)^2, \tag{34}$$

where $\varepsilon_F = k_F^2/2m$, and $\varepsilon_0 = 1/mL^2$. For $k_F L \ll 1$ one finds $\varepsilon_0 \gg \varepsilon_F$. Taking $\varepsilon_F \sim M$, once obtains

$$|t^f|^2 \sim \left(\frac{M}{\varepsilon_F} k_F L\right)^2 \ll 1. \tag{35}$$

Thus, a sharp domain wall can be considered as an effective barrier for the spin-flip transmission.

It should be noted that the conservation of flow has the following form

$$v_\uparrow \left(1 - |r_{R\uparrow}|^2\right) - v_\downarrow \left|r_{R\uparrow}^f\right|^2 = v_\downarrow |t_{R\uparrow}|^2 + v_\uparrow \left|t_{R\uparrow}^f\right|^2, \tag{36}$$

and analogous equations hold also for the other scattering states.

3.2. Resistance of the domain wall

To calculate the conductivity we start from the current operator

$$\hat{j}(z) = e\,\psi^\dagger(z)\,\hat{v}\,\psi(z). \tag{37}$$

Expanding $\psi(z)$ in the scattering states (30), and performing quantum-mechanical averaging, one obtains the following formula for the current

$$j(z) = -ie \sum_n \int \frac{d\varepsilon}{2\pi}\, e^{i\varepsilon\delta}\, G_n(\varepsilon)\, \chi_n^\dagger(z)\, \hat{v}\, \chi_n(z), \tag{38}$$

where n is the index of scattering states. The matrix elements of the velocity operator $\hat{v} = -(i/m)\,\partial_z$ can be calculated in the basis of the scattering states, and one obtains

$$v_{R\uparrow(\downarrow)} \equiv \langle R\uparrow(\downarrow)k | \hat{v} | R\uparrow(\downarrow)k\rangle = v_{\downarrow(\uparrow)} |t_{R\uparrow(\downarrow)}|^2 + v_{\uparrow(\downarrow)} \left|t_{R\uparrow(\downarrow)}^f\right|^2, \tag{39}$$

and similar expressions for the other states.

The retarded Green function $G_n(\varepsilon)$ in Eq. (38) is diagonal in the basis of scattering states. Assuming that the transmission of electrons through the barrier is small, one can take the chemical potential constant $\mu = \mu_R$ for $z < 0$, and $\mu = \mu_L$ for $z > 0$. This corresponds to the voltage drop $U = (\mu_R - \mu_L)/e$ across the barrier. The Green function $G_{R\uparrow k}(\varepsilon)$ acquires then the following simple form

$$G_{R\uparrow k}(\varepsilon) = \frac{1}{\varepsilon - \varepsilon_{R\uparrow}(k) + \mu_R + i\delta}, \tag{40}$$

where $\varepsilon_{R\uparrow}(k) = k^2/2m - M$. The other components of the Green function have a similar form.

After integrating over ε, one finds

$$
j(z) = e \int \frac{dk}{2\pi} \left\{ v_\uparrow \, \chi_{R\uparrow k}^\dagger(z) \, \chi_{R\uparrow k}(z) \, \theta \left[\mu_R - \varepsilon_{R\uparrow}(k) \right] \right.
$$
$$
+ v_\downarrow \, \chi_{R\downarrow k}^\dagger(z) \, \chi_{R\downarrow k}(z) \, \theta \left[\mu_R - \varepsilon_{R\downarrow}(k) \right]
$$
$$
- v_\uparrow \, \chi_{L\uparrow k}^\dagger(z) \, \chi_{L\uparrow k}(z) \, \theta \left[\mu_L - \varepsilon_{L\uparrow}(k) \right]
$$
$$
\left. - v_\downarrow \, \chi_{L\downarrow k}^\dagger(z) \, \chi_{L\downarrow k}(z) \, \theta \left[\mu_L - \varepsilon_{L\downarrow}(k) \right] \right\} . \tag{41}
$$

In view of the conservation of charge, the current does not depend on z, and therefore can be calculated for $z = 0$. Moreover, the total current from the states $\varepsilon_{R\uparrow(\downarrow)}(k), \varepsilon_{L\uparrow(\downarrow)}(k) \le \mu_L$ vanishes and only the states obeying the condition $\mu_L < \varepsilon_{R\uparrow(\downarrow)}(k) < \mu_R$ contribute to the current. The conductance G can be then found as a linear response to small perturbation (in the limit of $U \to 0$), and one finds

$$
G = \frac{e^2}{2\pi} \left(\frac{v_\downarrow}{v_\uparrow} \left| t_{R\uparrow} \right|^2 + \left| t_{R\uparrow}^f \right|^2 + \frac{v_\uparrow}{v_\downarrow} \left| t_{R\downarrow} \right|^2 + \left| t_{R\downarrow}^f \right|^2 \right), \tag{42}
$$

where all the velocities and transmission coefficients are taken at the Fermi level.

Finally, using Eqs (32) and (33), one can write the conductance in the form

$$
G = \frac{4e^2}{\pi} \frac{v_\uparrow v_\downarrow \left(v_\uparrow + v_\downarrow \right)^2 + 2\lambda^2 \left(v_\uparrow^2 + v_\downarrow^2 \right)}{\left[\left(v_\uparrow + v_\downarrow \right)^2 + 4\lambda^2 \right]^2} \tag{43}
$$

In the limit of $v_\uparrow = v_\downarrow$ and $\lambda \to 0$ one obtains the conductance of a single spin-degenerate channel, $G_0 = e^2/\pi$.

The dependence of G/G_0 on the wall parameter L is shown in Fig. 3 for different values of the parameter M. One can note that the conductance in the presence of a domain wall is generally much smaller than in the absence of the wall. Accordingly, the associated magnetoresistance can be very large (more than 100%, which corresponds to $G/G_0 < 0.5$)), in agreement with experimental observations. It is also worth to note that resistance of an abrupt domain wall is not so large as the resistance of a thicker domain wall (provided the conditions assumed for the model are fulfilled).

4. Summary

We have presented theoretical description of the domain wall contribution to electrical resistivity of metallic ferromagnets. Two limiting cases were

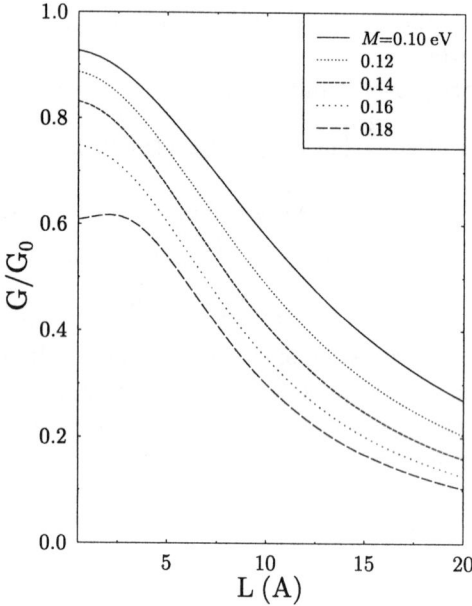

Figure 3. Relative conductance in the presence of a domain wall calculated as a function of L for $E_F = 0.2$ eV and for different values of the parameter M as indicated. For these parameters k_F is about 7 Å.

analyzed in details - the case of a thick domain wall with diffusive electron transport across the wall, and the limit of atomic-size and constrained domain wall, which effectively could be described by a one-dimensional model. These two possibilities are not the only ones. In very pure systems electronic transport across thick domain wall can be ballistic, despite the fact that the domain wall itself may be considered quasi-classically.[24] Apart from this, transport in real nanoconstrictions involves more channels and should be described by a more general theory. Anyway, such an approach may be useful particularly in the cases of point contacts based on new semiconductors heavily doped with magnetic impurities, like ferromagnetic GaMnAs or related compounds.

Acknowledgments

V.K.D. is grateful to Mianowski Foundation for the Science Fellowship and to L.A. Turski for numerous discussions. This work is supported by the Polish Committee for Scientific Research through the Grants 5 P03B 091 20 and PBZ/KBN/044/P03/2001.

References

1. G.G. Cabrera and L.M. Falicov, *Phys. Status Solidi B* **61**, 539 (1974); **62**, 217 (1974).
2. L. Berger, *J. Appl. Phys.* **49**, 2156 (1978).
3. K. Hong and N. Giordano, *J. Phys. Condens. Matter* **13**, L401 (1998).
4. U. Rüdiger, J. Yu, S. Zhang, A.D. Kent, and S.S.P. Parkin, *Phys. Rev. Lett.* **80**, 5639 (1998).
5. A.D. Kent, U. Rüdiger, J. Yu, L. Thomas, and S.S.P. Parkin, *J. Appl. Phys.* **85**, 5243 (1999).
6. A.D. Kent, J. Yu, U. Rüdiger, and S.S.P. Parkin, *J. Phys. Condens. Matter* **13**, R461 (2001).
7. J.F. Gregg, W. Allen, K. Ounadjela, M. Viret, M. Hehn, S.M. Thompson, and J.M.D. Coey, *Phys. Rev. Lett.* **77**, 1580 (1996).
8. N. Garcia, M. Muñoz, and Y.W. Zhao, *Phys. Rev. Lett.* **82**, 2923 (1999).
9. U. Ebels, A. Radulescu, Y. Henry, L. Piraux, and K. Ounadjela, *Phys. Rev. Lett.* **84**, 983 (2000).
10. G. Tatara and H. Fukuyama, *Phys. Rev. Lett.* **78**, 3773 (1997).
11. P.M. Levy and S. Zhang, *Phys. Rev. Lett.* **79**, 5110 (1997).
12. R.P. van Gorkom, A. Brataas, and G.E.W. Bauer, *Phys. Rev. Lett.* **83**, 4401 (1999).
13. P.A.E. Jonkers, S.J. Pickering, H. De Raedt, and G. Tatara, *Phys. Rev. B* **60**, 15970 (1999).
14. G. Tatara, *Int. J. Mod. Phys. B* **15**, 321 (2001).
15. H.D. Chopra and S.Z. Hua, *Phys. Rev. B* **66**, 020403(R) (2002).
16. P. Bruno, *Phys. Rev. Lett.* **83**, 2425 (1999).
17. H. Imamura, N. Kobayashi, S. Takahashi, and S. Maekawa, *Phys. Rev. Lett.* **84**, 1003 (2000).
18. V.K. Dugaev, J. Barnaś, A. Łusakowski, L.A. Turski, *Phys. Rev. B* **65**, 224419 (2002).
19. B.L. Altshuler, A.G. Aronov, D.E. Khmelnitskii, and A.I. Larkin, in *Quantum Theory of Solids*, ed. I.M. Lifshits (Mir, Moscow, 1982), pp. 130-237.
20. P.A. Lee and T.V. Ramakrishnan, *Rev. Mod. Phys.* **57**, 287 (1985).
21. V.I. Falko and S.V. Iordanskii, *Phys. Rev. Lett.* **82**, 402 (1999).
22. Yu. Lyanda-Geller, I.L. Aleiner, and P.M. Goldbart, *Phys. Rev. Lett.* **81** 3215 (1998).
23. V.K. Dugaev, P. Bruno, and J. Barnaś, *Phys. Rev. B* **64**, 144423 (2001).
24. L.R. Tagirov, B.P. Vodopyanov, and K.B. Efetov, *Phys. Rev. B* **63**, 104428 (2001).

NEW APPROACH TO THE SPIN-ORBIT SCATTERING OF ELECTRONS IN DISORDERED METALLIC SYSTEM

B. J. SPISAK AND A. PAJA

Department of Solid State Physics, Faculty of Physics and Nuclear Techniques,
University of Mining and Metallurgy,
Al. Mickiewicza 30, 30-059 Krakow, Poland
E-mail: spisak@novell.ftj.agh.edu.pl
E-mail: paja@novell.ftj.agh.edu.pl

We have developed a set of transport equations for electrons which are moving in disordered metallic systems, explicitly taking into account the electron spin. We have used the quantum mechanical phase-space distribution (the Wigner distribution function) for calculation of the transport coefficients as an alternative to the commonly applied Kubo-Greenwood formula. The results allow us to investigate the role of spin-orbit interaction in the region of weak localisation.

1. Introduction

Theory of the electron transport in disordered media is still an interesting problem. The existence of disorder induces weak localisation (WL) which causes high resistivity of amorphous alloys and other disordered materials. WL arises due to quantum interference (QI) of electron waves going along closed paths in oposite directions and coming back to initial position [1]. Various factors as thermal vibrations, external magnetic fields, some kinds of impurities may destroy this subtle phenomenon by partial decoherence of the electron waves, diminishing electron localisation and therefore increasing the conductivity. Correct transport theory should explain this surprising effect.

It is well known fact that heavy metal impurities in light metal matrix introduce strong spin-orbit scattering. Experimental data for resistivity of a-CaAl alloys doped with Au and Ag clearly show a substantial decrease of resistivity with increasing content of impurities [2]. The theory based on the Kubo-Greenwood formula, given by Fukuyama and Hoshino [3], cannot be applied here because it is based on the earlier works of Kawabata [4, 5] which was developed for heavily doped semiconductors where $k_F l \gg 1$ (k_F - the Fermi wave vector, l - mean free path of electron). It could be probably applied for amorphous alloys of low resistivity but it is not the case because undoped $Ca_{70}Al_{30}$ amorphous alloy has the resistivity of 310 $\mu\Omega$cm which gives $k_F l \approx 1$. Thus we propose below a new approach based on the "$2k_F$-scattering" developed by Morgan et al. [6] for highly resistive alloys, taking into account the spin-orbit scattering.

2. The theoretical model

In the presented description of electron transport properties we assume nearly-free-electron model and neglect electron-electron interaction effects. We consider the motion of electrons with the effective mass m and spin one-half in three-dimensional disordered system, where the scattering centres have no magnetic moments. Such a system can be described by the one-particle Hamiltonian

$$\widehat{H} = \frac{\widehat{\mathbf{p}}^2}{2m} + U_0(\mathbf{R}) + U_{SO}(\mathbf{R}), \tag{1}$$

where $\widehat{\mathbf{p}}$ is the momentum operator, $U_0(\mathbf{R}){=}\sum_{i=1}^{N} u_a(\mathbf{R}{-}\mathbf{r}_i)$ is the ordinary part of the potential, which is defined as a superposition of single atomic potentials located at random positions $\mathbf{r}_i$, $U_{SO}(\mathbf{R}){=}a_{SO}\widehat{\sigma}\cdot[\nabla U_0(\mathbf{R})\times\widehat{\mathbf{p}}]$ is the spin-orbit part, $\widehat{\sigma}$ - the Pauli spin matrices, $a_{SO} = \hbar/(2mc)^2$ and $c, \hbar$ have their usual meanings.

The role of spin-orbit interaction increases with the atomic number as Z^4 [7] and therefore this effect should be clearly observed in metals where electrons are scattered by heavy ions placed in a matrix of lighter ones. This is important point in the context of quantum interference effects and their influence on the diffusive motion of an electron. Since the scattering of the electron spin causes additional phase shift of the electronic wave function, the QI is partially destroyed and it changes the quantum corrections to the classical Drude conductivity.

The linearised von Neumann equation in the Wigner representation [8], for the system described by the Hamiltonian in the form (1), being under influence of the external electric field $\mathbf{E}$ which is switched on in the adiabatic manner with the infinitesimal constant ε is given by the formula

$$[\frac{\hbar\mathbf{K}}{m}\cdot\nabla+\varepsilon]\rho_{s's}(\mathbf{R},\mathbf{K})+\frac{2}{\hbar}\sum_{\mathbf{q}}\mathrm{Im}\{U_0(\mathbf{q})e^{i\mathbf{q}\cdot\mathbf{R}}\}$$

$$\times\rho_{s's}(\mathbf{R},\mathbf{K}+\frac{1}{2}\mathbf{q})+i\frac{b_{SO}}{2\hbar}\sum_{s''}\sum_{\mathbf{q}}[\widehat{\Lambda}(\mathbf{q},\mathbf{K})\rho_{s's''}(\mathbf{R},\mathbf{K}+\frac{1}{2}\mathbf{q})\cdot\sigma_{s''s}$$

$$-\widehat{\Lambda}^*(\mathbf{q},\mathbf{K})\cdot\sigma_{s's''}\rho_{s''s}(\mathbf{R},\mathbf{K}+\frac{1}{2}\mathbf{q})$$

$$=-e\mathbf{E}\cdot\sum_{s''}[\frac{\hbar\mathbf{K}}{m}\delta_{s's''}+v^{SO}_{s's''}(\mathbf{K})]\phi_{s''s}(\mathbf{R},\mathbf{K})\,,\quad(2)$$

where $b_{SO}=\hbar a_{SO}$, s denotes the z-component of the spin, $\phi_{s''s}(\mathbf{R},\mathbf{K})$ is the Wigner representation of the derivative of the Fermi-Dirac distribution, $U_0(\mathbf{q})$ - the Fourier transform of the static potential in the solid, and $\widehat{\Lambda}(\mathbf{q},\mathbf{K})$ is given by

$$\widehat{\Lambda}(\mathbf{q},\mathbf{K})=U_0(\mathbf{q})e^{i\mathbf{q}\cdot\mathbf{R}}[\mathbf{q}\times(\nabla-2i\mathbf{K})].\quad(3)$$

The term in the square brackets in the right-hand side of equation (2) represents the generalised velocity, and $\rho_{ss'}(\mathbf{R},\mathbf{K})$ is the Wigner matrix defined by the Hermitian matrix 2×2. The space average of the diagonal elements of this matrix gives the probability of finding the electron in the state $\mathbf{K}$ with the spin up (α) or down (β), respectively, and off-diagonal elements describe the fluctuations from the average values. The equation (2) is equivalent to a set of four equations for the elements of the Wigner matrix, but the Hermitian property of the Wigner matrix means that this set can be reduced to two equations. In the next step, we are using the projection operator method based on the Zwanzig and Mori idea [9, 10] for the description of the transport properties of the system. In the result we obtain the set of two equations for the average part of the diagonal and off-diagonal elements of the Wigner matrix. We ignore the equation for the off-diagonal elements of the Wigner matrix because the current involving the usual velocity does not depend on the off-diagonal elements. The equation for the diagonal elements has the form [11]:

$$\epsilon\rho^A_{\alpha\alpha}(\mathbf{K})+\sum_{\mathbf{K}'}\{[T_0(\mathbf{K},\mathbf{K}')+T_{\alpha\beta,\beta\alpha}(\mathbf{K},\mathbf{K}')][\rho^A_{\alpha\alpha}(\mathbf{K}-\rho^A_{\alpha\alpha}(\mathbf{K}')]\}$$

$$=-e\mathbf{E}\cdot v_{\alpha\alpha}(\mathbf{K})[1+\gamma_0(\mathbf{K})]\phi^A_{\alpha\alpha}(\mathbf{K})\,,\quad(4)$$

where, generally speaking, $T(\mathbf{K},\mathbf{K}')$ are the terms which describe the scattering of electrons, and $\gamma_0(\mathbf{K})$ modifies the injection of electrons into the system taking into account the fluctuations in the Wigner representation of the Fermi-Dirac distribution function.

The operator $T_0(\mathbf{K},\mathbf{K}')$ is responsible for the ordinary scattering. The operator defined as $T_{\alpha\beta,\beta\alpha}(\mathbf{K},\mathbf{K}')$ is the spin-flip scattering part.

We now use an effective medium approximation to determine the analytical form of the generalised scattering kernel. The main idea of this approximation is based on the following picture. The disordered system is replaced by a medium, the properties of which are determined by two quantities: $T(\mathbf{K}, \mathbf{K}')$ and $\gamma(\mathbf{K})$. A scatterer is embedded in this medium and $T(\mathbf{K}, \mathbf{K}')$ and $\gamma(\mathbf{K})$ must be determined self-consistently. After some rearrangement we obtain the form of $T(\mathbf{K}, \mathbf{K}')$

$$
\begin{aligned}
T(\mathbf{K}, \mathbf{K}') \approx{} & \frac{1}{\hbar^2} \frac{N}{V^2} \sum_{\mathbf{Q}} u_a(\mathbf{Q}) \tilde{u}_a(-\mathbf{Q})[(1 - \delta_{\mathbf{K}, \mathbf{K}'}) \\
& \times \{1 + b_{SO}^2[(\mathbf{K} \times \mathbf{Q}) \cdot \sigma_{\beta\alpha}][(\mathbf{K}' \times \mathbf{Q}) \cdot \sigma_{\alpha\beta}]\} \\
& \qquad \times G(\mathbf{Q}, \mathbf{K} - \frac{1}{2}\mathbf{Q}, \mathbf{K}' + \frac{1}{2}\mathbf{Q}) \\
& + \{1 + b_{SO}^2[(\mathbf{K} \times \mathbf{Q}) \cdot \sigma_{\alpha\beta}][(\mathbf{K}' \times \mathbf{Q}) \cdot \sigma_{\beta\alpha}]\} \\
& \qquad \times G(\mathbf{Q}, \mathbf{K} + \frac{1}{2}\mathbf{Q}, \mathbf{K}' - \frac{1}{2}\mathbf{Q}) \\
& - \{1 + b_{SO}^2[(\mathbf{K} \times \mathbf{Q}) \cdot \sigma_{\beta\alpha}][(\mathbf{K}' \times \mathbf{Q}) \cdot \sigma_{\alpha\beta}]\} \\
& \qquad \times G(\mathbf{Q}, \mathbf{K} + \frac{1}{2}\mathbf{Q}, \mathbf{K}' - \frac{1}{2}\mathbf{Q}) + \\
& - \{1 + b_{SO}^2[(\mathbf{K} \times \mathbf{Q}) \cdot \sigma_{\alpha\beta}][(\mathbf{K}' \times \mathbf{Q}) \cdot \sigma_{\beta\alpha}]\} \\
& \qquad \times G(\mathbf{Q}, \mathbf{K} - \frac{1}{2}\mathbf{Q}, \mathbf{K}' + \frac{1}{2}\mathbf{Q})],
\end{aligned} \tag{5}
$$

where $\tilde{u}_a(\mathbf{R}' - \mathbf{r}_i)$ is a modification of the single ionic potential, V is the volume of the system, and G is the Green function which can be determined by the Dyson equation.

3. The quantum interference effect

To investigate the quantum interference effect it is convenient to separate the generalised scattering kernel into three parts, namely

$$
T(\mathbf{K}, \mathbf{K}') = T_1(\mathbf{K}, \mathbf{K}') + T_2(\mathbf{K}, \mathbf{K}') + T_3(\mathbf{K}, \mathbf{K}'), \tag{6}
$$

where $T_1(\mathbf{K}, \mathbf{K}')$ is the normal Boltzmann equation scattering obtained by perturbation theory, $T_2(\mathbf{K}, \mathbf{K}')$ and $T_3(\mathbf{K}, \mathbf{K}')$ describe the multiple scattering peaked in the backward direction and the same peaked in the forward direction, respectively. As it was shown in [6], the quantum interference effects are represented by $T_2(\mathbf{K}, \mathbf{K}')$. When the spin-orbit interaction is included $T_2(\mathbf{K}, \mathbf{K}')$ takes the form [11]:

$$T_2(\mathbf{K}, \mathbf{K}') \approx \frac{1}{\hbar^2} \frac{N}{V^2} \sum_{\mathbf{Q}} u_a(\mathbf{Q}) \tilde{u}_a(-\mathbf{Q}) [(1 - \delta_{\mathbf{K} - \mathbf{K}', \mathbf{Q}})$$

$$\times \{1 + b_{SO}^2 [(\mathbf{K} \times \mathbf{Q}) \cdot \sigma_{\beta\alpha}] [(\mathbf{K}' \times \mathbf{Q}) \cdot \sigma_{\alpha\beta}]\}$$

$$\times G(\mathbf{Q}, \mathbf{K} - \frac{1}{2}\mathbf{Q}, \mathbf{K}' + \frac{1}{2}\mathbf{Q}) + (1 - \delta_{\mathbf{K}' - \mathbf{K}, \mathbf{Q}})$$

$$\{1 + b_{SO}^2 [(\mathbf{K} \times \mathbf{Q}) \cdot \sigma_{\alpha\beta}] [(\mathbf{K}' \times \mathbf{Q}) \cdot \sigma_{\beta\alpha}]\}$$

$$\times G(\mathbf{Q}, \mathbf{K} + \frac{1}{2}\mathbf{Q}, \mathbf{K}' - \frac{1}{2}\mathbf{Q}). \tag{7}$$

This can be converted to a self-consistent equation using the Green function technique, which was proposed in [6], but its structure is a little different, because the spin-orbit interaction is included. If we neglect the third part of scattering kernel in the formula (6), and next we sum over $\mathbf{K}'$ with the weight factor $1 - cos\theta$, where θ is the angle between incident and outgoing wave vectors, then after some reckoning we obtain the analytical formula for the inverse transport relaxation time

$$\tau_{tr}^{-1} = \frac{1 + \frac{1}{2}\frac{\tau_{FZ}}{\tau_T} Y^2 F_{MHP}^{1/2}(Y)}{1 - \frac{3}{64}\frac{\tau_T}{\tau_{EL}} F_{MHS}^{1/2}(Y) F_{MHP}^{1/2}(Y)}, \tag{8}$$

where τ_{FZ} is the Faber-Ziman relaxation time, τ_{EL} is the elastic relaxation time, τ_T is the total relaxation time

$$\tau_T^{-1} = \tau_{EL}^{-1} + \tau_{SO}^{-1}, \tag{9}$$

and the functions $F_{MHS}(Y)$ and $F_{MHP}(Y)$ were defined in [12] [a].

4. Conclusions

We presented a new approach to the calculations of the transport properties of disordered systems based on the Wigner representation. We obtained a set of transport equations which depends on electron spin and includes the spin-orbit interaction.

The final formula for transport relaxation time can be compared with experimental data by the use of the formula for the reduced resistivity

$$\rho_{red}^{th} = \frac{[n(x)\tau_{tr}(x)]^{-1} - [n(0)\tau_{tr}(0)]^{-1}}{[n(0)\tau_{tr}(0)]^{-1}}, \tag{10}$$

[a]In equation (37) of [12] the term $+\sqrt{2}((1 - Y^2)^{1/2} + 1)^{1/2}$ should be replaced by $-\sqrt{2}((1 + Y^2)^{1/2} + 1)^{1/2}$ and numerical factor of 2 should be added before the logarithm.

where $n(x)$ is the atomic concentration of conduction electrons for heavy atoms content x.

We applied our results for the interpretation of experimental data of the resistivity for $Ca_{70}Al_{30-x}Au_x$ and we obtained quite good agreement with experiment [11].

Acknowledgments

The authors thank prof. G. J. Morgan (Department of Physics and Astronomy, University of Leeds, United Kingdom) for fruitful discussions. The work was supported in part by the Polish Committee for Scientific Research under grant No. 5P03B 026 20.

References

1. J. S. Dugdale, *Contemporary Physics* **28**, 547 (1987).
2. A. Sahnoune, J. O. Ström-Olsen, H. E. Fischer, *Phys. Rev.* **B46**, 10035 (1992).
3. H. Fukuyama, K. Hoshino, *J. Phys. Soc. Japan* **50**, 2131 (1981).
4. A. Kawabata, *Solid State Commun.* **34**, 431 (1980).
5. A. Kawabata, *J. Phys. Soc. Japan* **49**, 628 (1980).
6. G. J. Morgan, M. A. Howson, K. Šaub, *J. Phys. F: Metal Phys* **15**, 2157 (1985).
7. L. D. Landau, E. M. Lifshitz, *Quantum Mechanics*, PWN, Warsaw, 1979 (in Polish).
8. M. Hillery, R. F. O'Connell, M. O. Scully, E. P. Wigner *Physics Reports* **106**, 121 (1984).
9. R. Zwanzig, *Phys. Rev.* **124**, 983 (1961).
10. H. Mori, *Prog. Theor. Phys.* **34**, 399 (1965).
11. B. J. Spisak, *Influence of the spin-orbit interaction on the electrical conductivity of disordered system*, Thesis, Krakow, 2002 (in Polish).
12. A. Paja, G. J. Morgan, *phys. stat. sol. (b)* **206**, 701 (1998).

SYMMETRIES OF LCAO BAND STRUCTURE CALCULATIONS IN CRYSTALLINE SOLIDS

W. OBERMAYR

Institute for Electronic Engineering, FH Joanneum,
Werk-VI-Strasse 46, A-8605 Kapfenberg, Austria
E-mail: werner.obermayr@fh-joanneum.at

The linear combination of atomic orbitals (LCAO) expansion of the crystal wave function is one of the oldest methods[2] of calculating electron band structures. Despite of its mathematical transparency it was seldom used in the early years of quantum mechanics, because the determination of the matrix elements of the crystal potential was almost impossible. But nowadays the computers are so powerful that such calculations can be performed. In this context the concept of symmetry is basic for several reasons: So the symmetry of the crystal lattice leads to the model of tight-binding (nearest neighbor approximation), to the method of performing band structure calculations via the least-squares scheme, and it permits a minimization of the in general very high numerical effort. In this work we discuss some results in context with these subjects.

1. Introduction

In this work we describe a method for performing LCAO band structure calculations in crystalline solids: We apply the symmetries of a crystal to derive parameters which are very well suited for tight-binding models and furthermore lead to a mathematically well defined and efficient least-squares scheme for the calculation of the matrix of the crystal potential.

Explicit numerical results for potassium are compared with results of augmented plane wave (APW) and full-potential linearized augmented plane wave (FPLAPW) results.

2. Determination of the Electronic Band Structure via the LCAO Method

2.1. *Method of Calculation*

The Schrödinger equation for a one-electron wave function in a crystal is given by

$$H \psi = E \psi \quad \text{with} \quad H \equiv -1/2 \nabla^2 + V, \tag{1}$$

426

where we use atomic units, V is a (effective, one-particle) crystal potential, and H is the Hamiltonian of the crystal. V has the symmetry of the crystal lattice.

In the following we denote the space group of the crystal lattice by $\mathcal{G}$, its translation group by $\mathcal{T}$, and its point group by $\mathcal{G}_0$. The operators of $\mathcal{G}$ are denoted by $\{\alpha|a\}$ and we define their action on a coordinate vector $\mathbf{x}$ according to Seitz[8] by $\{\alpha|a\}\,\mathbf{x} = \alpha\,\mathbf{x} + \mathbf{a}$. Furthermore, we denote by $\{\alpha|a\}^{-1}$ the inverse operator of $\{\alpha|a\}$, and $\{\epsilon|0\}$ is the identity of the space group. The operators $P_{\{\alpha|a\}}$, which act upon a function $f(\mathbf{r})$ via the elements of the space group by

$$P_{\{\alpha|a\}}\, f(\mathbf{r}) = f(\{\alpha|a\}^{-1}\,\mathbf{r}) \tag{2}$$

form a group which is isomorphic to the space group $\mathcal{G}$.

$\mathbf{R}$ is a vector of the Bravais lattice in question and $\mathbf{s}$ a relative position vector.

In the next step we perform an LCAO expansion $\psi_{\mathbf{k}}^{LCAO}(\mathbf{r})$ of the crystal wave function $\psi_{\mathbf{k}}(\mathbf{r})$

$$\psi_{\mathbf{k}}^{LCAO}(\mathbf{r}) = \sum_{s=s_1}^{s_N} \sum_{l=0}^{l_s} \sum_{n=1}^{n_{ls}} \sum_{h=1}^{h_l} \psi_{nl,s;\,\mathbf{k}}^{(\nu_{lh})}(\mathbf{r}) \cdot c_{nl,s;\,\mathbf{k}}^{(\nu_{lh})} \tag{3}$$

with

$$\psi_{nl,s;\,\mathbf{k}}^{(\nu_{lh})}(\mathbf{r}) = \sum_{\mathbf{R}} e^{i\mathbf{k}\cdot(\mathbf{s}+\mathbf{R})}\, \chi_{nl,s}^{(\nu_{lh})}(\mathbf{r} - [\mathbf{s}+\mathbf{R}]). \tag{4}$$

In our case there are N atoms in the Wigner-Seitz cell situated at the relative positions $\mathbf{s}_1, \ldots, \mathbf{s}_N$. The $\psi_{nl,s;\,\mathbf{k}}^{(\nu_{lh})}(\mathbf{r})$ are Bloch wave functions obtained from the atomic-like orbitals $\chi_{nl,s}^{(\nu_{lh})}(\mathbf{r} - \mathbf{s})$ centered at the relative location $\mathbf{s}$. The basis set $\{\chi(\mathbf{r})\}_s$ associated with the atom at $\mathbf{s}$ consists of orbitals $\chi_{nl,s}^{(\nu_{lh})}(\mathbf{r})$ with angular momentums $l = 0, \ldots, l_s$ and n_{ls} radial parts with l. The index h runs over the number of irreducible representations ν_{lh} of $\mathcal{G}_0$ compatible with l. The atomic-like orbitals are

$$\chi_{nl}^{(\nu)}(\mathbf{r}) = \chi_{nl}(r)\,\mathbf{Y}_l^{(\nu)}(\mathbf{r}), \tag{5}$$

where $\chi_{nl}(r)$ and $\mathbf{Y}_l^{(\nu)}(\mathbf{r})$ are the radial and spherical part of the orbital, respectively. The column $\mathbf{Y}_l^{(\nu)}(\mathbf{r})$ forms a set of basis functions for the ν-th unitary irreducible representations $\mathcal{D}^{(\nu)}$ of the point group $\mathcal{G}_0$, thus:

$$P_{\{\alpha|0\}}\,\tilde{\chi}_l^{(\nu)}(\mathbf{r}) = \tilde{\chi}_l^{(\nu)}(\mathbf{r})\,D^{(\nu)}(\alpha), \qquad \alpha \in \mathcal{G}_0. \tag{6}$$

The ˜-sign stands for transposition. The dimension of $\mathcal{D}^{(\nu)}$ is d_ν.

With the Eqs. (2) and (6) we find for $\psi_{nl,s;\,\mathbf{k}}^{(\nu_{lh})}(\mathbf{r})$ the symmetries:

$$\psi_{nl,s;\,\mathbf{k}}^{(\nu_{lh})}(\mathbf{r}) = \psi_{nl,\{\epsilon|\mathbf{R}\}\,s;\,\mathbf{k}}^{(\nu_{lh})}(\mathbf{r}), \qquad \{\epsilon|\mathbf{R}\} \in \mathcal{T}, \tag{7}$$

$$P_{\{\alpha|\mathbf{a}\}}\,\tilde{\psi}_{nl,s;\,\mathbf{k}}^{(\nu_{lh})}(\mathbf{r}) = e^{-i\alpha\mathbf{k}\cdot\mathbf{a}}\,\tilde{\psi}_{nl,\{\alpha|\mathbf{a}\}s;\,\alpha\mathbf{k}}^{(\nu_{lh})}(\mathbf{r})\,D^{(\nu_{lh})}(\alpha), \tag{8}$$

for $\{\alpha|\mathbf{a}\} \in \mathcal{G}$.

Now we determine the coefficients $c_{nl,s;\,\mathbf{k}}^{(\nu_{lh})}$ of the LCAO ansatz $\psi_{\mathbf{k}}^{LCAO}(\mathbf{r})$ for the crystal wave function via the minimum property of the ground state energy. Variation of the expectation value of the energy leads to the *generalized eigenvalue problem*

$$H_{\mathbf{k}}\,\mathbf{c}_{\mathbf{k}} = E_{\mathbf{k}}\,S_{\mathbf{k}}\,\mathbf{c}_{\mathbf{k}}. \tag{9}$$

$H_{\mathbf{k}}$ is the matrix of the Hamiltonian H, $S_{\mathbf{k}}$ the overlap matrix, $\mathbf{c}_{\mathbf{k}}$ represents the coefficients of $\psi_{\mathbf{k}}^{LCAO}(\mathbf{r})$, and $E_{\mathbf{k}}$ is the energy associated with the wave function $\psi_{\mathbf{k}}^{LCAO}(\mathbf{r})$.

The matrices $H_{\mathbf{k}}$ and $S_{\mathbf{k}}$ have many properties in common. In order to derive them simultaneously, we write $M_{\mathbf{k}}$ for $H_{\mathbf{k}}$ and $S_{\mathbf{k}}$.

Obviously, the matrix $M_{\mathbf{k}}$ has block structure, where the block in the row (s, l, n, h) and column (s', l', n', h') is given by the matrix

$$M_{nl,s;\,n'l',s';\,\mathbf{k}}^{(\nu_{lh})(\nu_{l'h'})} = \left\langle \psi_{nl,s;\,\mathbf{k}}^{(\nu_{lh})} \,|\, M \,|\, \tilde{\psi}_{n'l',s';\,\mathbf{k}}^{(\nu_{l'h'})} \right\rangle \tag{10}$$

with $M = H$ or $\mathbf{1}$. The multiplication of $\psi_{nl,s;\,\mathbf{k}}^{(\nu_{lh})}$ and $\tilde{\psi}_{n'l',s';\,\mathbf{k}}^{(\nu_{l'h'})}$ is performed in the dyadic way, $\langle\,|\,|\,\rangle$ is applied to each matrix element.

For the establishment of the *generalized eigenvalue problem* (9) the matrix elements (10) have to be determined. This can be done efficiently by the application of a least-squares scheme, which we describe in the following.

At first it is our aim to find a minimum number of parameters having the property, that all matrix elements (10) can be represented by linear combinations in term of them. For this purpose we take into account the invariance of M with respect to the operators of the space group in the integrals (10): With the symmetries (7) and (8) of $\psi_{nl,s;\,\mathbf{k}}^{(\nu_{lh})}(\mathbf{r})$ we find

$$M_{\mathbf{s};\,\mathbf{s}';\,\mathbf{k}}^{(\nu)(\nu')} = \tilde{D}^{(\nu)}(\alpha)\,M_{\mathcal{T}_{\{\alpha|\mathbf{a}\}}\mathbf{s};\,\mathcal{T}_{\{\alpha|\mathbf{a}\}}\mathbf{s}';\,\alpha\mathbf{k}}^{(\nu)(\nu')}\,D^{(\nu')}(\alpha) \tag{11}$$

for $\alpha \in \mathcal{G}_0$, where we denote by $\mathcal{T}\mathbf{s}$ the set of relative positions, which differ from $\mathbf{s}$ by primitive translations only.

Obviously, the relations (11) combine matrices with different sets of relative positions. In order to find these matrices we determine the *stabilizer* $\mathcal{G}_{\mathbf{ss}'}/\mathcal{T}$ and the *orbit* of $\mathcal{G}/\mathcal{T}$ on $\mathbf{s}$ and $\mathbf{s}'$. $\mathcal{G}_{\mathbf{ss}'}/\mathcal{T}$ is a subgroup of $\mathcal{G}/\mathcal{T}$. The rotational parts of $\mathcal{G}_{\mathbf{ss}'}/\mathcal{T}$ form the point group $\mathcal{G}_{\mathbf{ss}'0}$ and these groups are isomorphic. We denote the elements of $\mathcal{G}_{\mathbf{ss}'}/\mathcal{T}$ by $\mathcal{T}_{\{\beta|\mathbf{b}\}}$.

Now we decompose $\mathcal{G}/\mathcal{T}$ into its left cosets with respect to $\mathcal{G}_{ss'}/\mathcal{T}$ and find representatives which we denote by $\mathcal{T}_{\{\tilde{\beta}|\tilde{\mathbf{b}}\}}$. So the *orbit* of $\mathcal{G}/\mathcal{T}$ on s and s' consists of the different pairs of relative locations

$$\mathcal{T}_{\{\tilde{\beta}|\tilde{\mathbf{b}}\}}\mathbf{s}, \qquad \mathcal{T}_{\{\tilde{\beta}|\tilde{\mathbf{b}}\}}\mathbf{s}'. \tag{12}$$

Now we rewrite the Eqs. (11) in the form:

$$M^{(\nu)(\nu')}_{\mathbf{s};\,\mathbf{s}';\,\mathbf{k}} = \tilde{D}^{(\nu)}(\beta)\, M^{(\nu)(\nu')}_{\mathbf{s};\,\mathbf{s}';\,\beta\mathbf{k}}\, D^{(\nu')}(\beta) \tag{13}$$

$$M^{(\nu)(\nu')}_{\mathbf{s};\,\mathbf{s}';\,\mathbf{k}} = \tilde{D}^{(\nu)}(\check{\beta})\, M^{(\nu)(\nu')}_{\mathcal{T}_{\{\tilde{\beta}|\tilde{\mathbf{b}}\}}\mathbf{s};\,\mathcal{T}_{\{\tilde{\beta}|\tilde{\mathbf{b}}\}}\mathbf{s}';\,\check{\beta}\mathbf{k}}\, D^{(\nu')}(\check{\beta}) \tag{14}$$

for all $\beta \in \mathcal{G}_{ss'0}$ and $\mathcal{T}_{\{\tilde{\beta}|\tilde{\mathbf{b}}\}}$. Obviously, the matrices $M^{(\nu)(\nu')}_{\mathcal{T}_{\{\tilde{\beta}|\tilde{\mathbf{b}}\}}\mathbf{s};\,\mathcal{T}_{\{\tilde{\beta}|\tilde{\mathbf{b}}\}}\mathbf{s}';\,\mathbf{k}}$ can be represented in terms of the $M^{(\nu)(\nu')}_{\mathbf{s};\,\mathbf{s}';\,\mathbf{k}}$: This means that it is sufficient to choose the independent parameters in such a manner, that the matrix $M^{(\nu)(\nu')}_{\mathbf{s};\,\mathbf{s}';\,\mathbf{k}}$ can be represented by linear combinations in terms of them.

For this purpose we define operators R_β which act upon a matrix element via the elements β of $\mathcal{G}_{ss'0}$ by

$$R_\beta\, M^{(\nu)(\nu')}_{\mathbf{s};\,\mathbf{s}';\,\mathbf{k};\,rs} = M^{(\nu)(\nu')}_{\mathbf{s};\,\mathbf{s}';\,\beta^{-1}\mathbf{k};\,rs},$$

Next we determine a unitary matrix $U^{\nu,\,\nu'}$ which transforms the *Kronecker product* $\bar{D}^{(\nu)}(\beta) \otimes D^{(\nu')}(\beta)$ into its *Clebsch-Gordan series*, that is into the direct sum of irreducible representations $D^{(\mu)}_{\mathcal{G}_{ss'0}}$ of $\mathcal{G}_{ss'0}$. According to Eq. (13) a linear transformation $U^{\nu,\,\nu'}$ of the matrix $M^{(\nu)(\nu')}_{\mathbf{s};\,\mathbf{s}';\,\mathbf{k}}$ leads to $P^{(\nu)(\nu')}_{\mathbf{s};\,\mathbf{s}';\,\mathbf{k}}$, which consist of sets of elements

$$P^{(\nu)(\nu');\,(\mu)}_{\mathbf{s};\,\mathbf{s}';\,\mathbf{k};\,1}, \ldots, P^{(\nu)(\nu');\,(\mu)}_{\mathbf{s};\,\mathbf{s}';\,\mathbf{k};\,d_{\mathcal{G}_{ss'0;\,\mu}}},$$

which form basis for $D^{(\mu)}_{\mathcal{G}_{ss'0}}$ of $\mathcal{G}_{ss'0}$. In order to find our independent parameters we take one element from each set of partner functions, e.g. $P^{(\nu)(\nu');\,(\mu)}_{\mathbf{s};\,\mathbf{s}';\,\mathbf{k};\,r}$ and represent its partners in terms of it[6,7]:

$$R_\beta\, P^{(\nu)(\nu');\,(\mu)}_{\mathbf{s};\,\mathbf{s}';\,\mathbf{k};\,r} = \sum_{\beta'\in\mathcal{G}_{ss'0}} \lambda^{\mu;\,\beta'}_{r;\,\beta}\, [R_{\beta'}\, P^{(\nu)(\nu');\,(\mu)}_{\mathbf{s};\,\mathbf{s}';\,\mathbf{k};\,r}]_{\text{ind}}, \tag{15}$$

$$P^{(\nu)(\nu');\,(\mu)}_{\mathbf{s};\,\mathbf{s}';\,\mathbf{k};\,s} = \frac{d_{\mathcal{G}_{ss'0;\,\mu}}}{g_{ss'0}} \sum_{\beta\in\mathcal{G}_{ss'0}} \bar{D}^{(\mu)}_{\mathcal{G}_{ss'0;\,sr}}(\beta)\, R_\beta\, P^{(\nu)(\nu');\,(\mu)}_{\mathbf{s};\,\mathbf{s}';\,\mathbf{k};\,r}, \qquad (s \neq r). \tag{16}$$

So all elements of $P^{(\nu)(\nu')}_{\mathbf{s};\,\mathbf{s}';\,\mathbf{k}}$ are represented by linear combinations of $d_{\mathcal{G}_{ss'0;\,\mu}}$ independent parameters $[R_{\beta'}\, P^{(\nu)(\nu');\,(\mu)}_{\mathbf{s};\,\mathbf{s}';\,\mathbf{k};\,r}]_{\text{ind}}$. In the following we determine the elements $P^{(\nu)(\nu');\,(\mu)}_{\mathbf{s};\,\mathbf{s}';\,\mathbf{k};\,r}$.

At first we observe, that the matrix $M^{(\nu_{lh})(\nu_{l'h'})}_{nl,s;\,n'l',s';\,\mathbf{k}}$ of the definition (10) can be expressed by a Fourier series with respect to the Bloch vector $\mathbf{k}$,

$$M^{(\nu_{lh})(\nu_{l'h'})}_{nl,s;\,n'l',s';\,\mathbf{k}} = e^{-i\mathbf{k}\cdot(\mathbf{s}-\mathbf{s}')} \sum_{\mathbf{R}} e^{i\mathbf{k}\cdot\mathbf{R}}\, M^{(\nu_{lh})(\nu_{l'h'})}_{nl,s;\,n'l',s';\,\mathbf{R}}, \qquad (17)$$

where the Fourier coefficients are given by

$$M^{(\nu_{lh})(\nu_{l'h'})}_{nl,s;\,n'l',s';\,\mathbf{R}} = \int_{(V_c)} \bar{\chi}^{(\nu_{lh})}_{nl,s}(\mathbf{r}-\mathbf{s})\, M\, \tilde{\chi}^{(\nu_{l'h'})}_{n'l',s'}(\mathbf{r}-[\mathbf{s}'+\mathbf{R}])\, d^3 r. \qquad (18)$$

$M = H$, or in our case, the crystal potential V of the Eq. (1). We denote the volume of the crystal by V_c.

Obviously the Fourier coefficients $M^{(\nu)(\nu')}_{s;\,s';\,\mathbf{R}}$ satisfy the symmetries

$$M^{(\nu)(\nu')}_{\{\epsilon|\mathbf{R}'\}s;\,s';\,\mathbf{R}} = M^{(\nu)(\nu')}_{s;\,\{\epsilon|-\mathbf{R}'\}s';\,\mathbf{R}} = M^{(\nu)(\nu')}_{s;\,s';\,\{\epsilon|-\mathbf{R}'\}\mathbf{R}},$$

$$M^{(\nu)(\nu')}_{s;\,s';\,\mathbf{R}} = \tilde{D}^{(\nu)}(\alpha)\, M^{(\nu)(\nu')}_{\{\alpha|a\}s;\,\{\alpha|a\}s';\,\alpha\mathbf{R}}\, D^{(\nu')}(\alpha). \qquad (19)$$

These integrals are invariant with respect to simultaneous primitive translations of $\mathbf{s}$ and $\mathbf{s}'$ only.

Next we determine the influence of the operators $\{\beta|\mathbf{b}\}$ of $\mathcal{T}_{\{\beta|\mathbf{b}\}}$ on the Fourier coefficients. Obviously there exist translational parts $\mathbf{b}^s$ and $\mathbf{b}^{s'}$ associated with each rotational part β such that

$$\{\beta|\mathbf{b}^s\}s = \{\beta|\mathbf{v}_\beta + \mathbf{R}^s_\beta\}s = s,$$

$$\{\beta|\mathbf{b}^{s'}\}s' = \{\beta|\mathbf{v}_\beta + \mathbf{R}^{s'}_\beta\}s' = s'.$$

Consequently,

$$\{\beta|\mathbf{R}^s_\beta - \mathbf{R}^{s'}_\beta\}\,(\mathbf{s}-\mathbf{s}') = (\mathbf{s}-\mathbf{s}') \qquad (20)$$

for all $\beta \in \mathcal{G}_{ss'0}$. We abbreviate these operators by

$$\beta^{\mathbf{s}-\mathbf{s}'} = \{\beta|\mathbf{R}^s_\beta - \mathbf{R}^{s'}_\beta\},$$

since they leave the difference $\mathbf{s}-\mathbf{s}'$ of the relative locations invariant. They form a group which we denote by $\mathcal{G}^{\mathbf{s}-\mathbf{s}'}_{ss'}$. It is isomorphic to $\mathcal{G}_{ss'0}$. Therefore a set of irreducible unitary representations of $\mathcal{G}^{\mathbf{s}-\mathbf{s}'}_{ss'}$ is given by $\mathcal{D}_{\mathcal{G}^{\mathbf{s}-\mathbf{s}'}_{ss'}} \sim \mathcal{D}_{\mathcal{G}_{ss'0}}$.

Now we can write the Eq.(19) as

$$M^{(\nu)(\nu')}_{s;\,s';\,\mathbf{R}} = \tilde{D}^{(\nu)}(\beta)\, M^{(\nu)(\nu')}_{s;\,s';\,\beta^{\mathbf{s}-\mathbf{s}'}\mathbf{R}}\, D^{(\nu')}(\beta), \qquad (21)$$

$$M^{(\nu)(\nu')}_{s;\,s';\,\mathbf{R}} = \tilde{D}^{(\nu)}(\breve{\beta})\, M^{(\nu)(\nu')}_{\{\breve{\beta}|\mathbf{v}_{\breve{\beta}}\}s;\,\{\breve{\beta}|\mathbf{v}_{\breve{\beta}}\}s';\,\breve{\beta}\mathbf{R}}\, D^{(\nu')}(\breve{\beta}). \qquad (22)$$

These equations for the Fourier coefficients of the matrix elements correspond to the Eqs. (13) and (14) for the matrix elements.

In order to find appropriate independent parameters for the Fourier coefficients, we now proceed in the same way as in the case of the $\mathbf{k}$ dependent matrix elements. We define operators R'_β acting upon a Fourier coefficient via the elements $\beta^{\mathbf{s}-\mathbf{s}'}$ of $\mathcal{G}^{\mathbf{s}-\mathbf{s}'}_{\mathbf{ss}'}$ by

$$R'_\beta \, M^{(\nu)(\nu')}_{\mathbf{s};\,\mathbf{s}';\,\mathbf{R};\,rs} = M^{(\nu)(\nu')}_{\mathbf{s};\,\mathbf{s}';\,(\beta^{\mathbf{s}-\mathbf{s}'})^{-1}\mathbf{R};\,rs}. \tag{23}$$

In order to find the independent parameters we perform the linear transformation $U^{\nu,\,\nu'}$ of $M^{(\nu)(\nu')}_{\mathbf{s};\,\mathbf{s}';\,\mathbf{R}}$ to obtain $P^{(\nu)(\nu')}_{\mathbf{s};\,\mathbf{s}';\,\mathbf{R}}$. They consist of sets of elements $P^{(\nu)(\nu');\,(\mu)}_{\mathbf{s};\,\mathbf{s}';\,\mathbf{R}}$, which form basis for the irreducible representations $\mathcal{D}^{(\mu)}_{\mathcal{G}^{\mathbf{s}-\mathbf{s}'}_{\mathbf{ss}'}}$ of $\mathcal{G}^{\mathbf{s}-\mathbf{s}'}_{\mathbf{ss}'}$. So we can apply the same method as we used before in order to express the Fourier coefficients in terms of independent parameters:

$$R'_\beta \, P^{(\nu)(\nu');\,(\mu)}_{\mathbf{s};\,\mathbf{s}';\,\mathbf{R};\,r} = \sum_{\beta' \in \mathcal{G}_{\mathbf{ss}'0}} \lambda^{\mu;\,\beta'}_{r;\,\beta} \, [R'_{\beta'} \, P^{(\nu)(\nu');\,(\mu)}_{\mathbf{s};\,\mathbf{s}';\,\mathbf{R};\,r}]_{\text{ind}}, \tag{24}$$

$$P^{(\nu)(\nu');\,(\mu)}_{\mathbf{s};\,\mathbf{s}';\,\mathbf{R};\,s} = \frac{d_{\mathcal{G}_{\mathbf{ss}'0};\,\mu}}{g_{\mathbf{ss}'0}} \sum_{\beta \in \mathcal{G}_{\mathbf{ss}'0}} \bar{D}^{(\mu)}_{\mathcal{G}_{\mathbf{ss}'0};\,sr}(\beta) \, R'_\beta \, P^{(\nu)(\nu');\,(\mu)}_{\mathbf{s};\,\mathbf{s}';\,\mathbf{R};\,r}, \qquad (s \neq r). \tag{25}$$

In the next step we establish the connection between the matrix elements and their Fourier coefficients. Of course, for the knowledge of the Fourier coefficients it is sufficient to determine the independent parameters $[R'_{\beta'} \, P^{(\nu)(\nu');\,(\mu)}_{\mathbf{s};\,\mathbf{s}';\,\mathbf{R};\,r}]_{\text{ind}}$ of the Eq. (24). Obviously,

$$\mathbf{P}^{(\nu)(\nu');\,(\mu)}_{\mathbf{s};\,\mathbf{s}';\,\mathbf{k}} = e^{-i\,\mathbf{k}\cdot(\mathbf{s}-\mathbf{s}')} \sum_{\mathbf{R}} e^{i\,\mathbf{k}\cdot\mathbf{R}} \, \mathbf{P}^{(\nu)(\nu');\,(\mu)}_{\mathbf{s};\,\mathbf{s}';\,\mathbf{R}}. \tag{26}$$

In order to reflect the symmetries of the matrix element $P^{(\nu)(\nu');\,(\mu)}_{\mathbf{s};\,\mathbf{s}';\,\mathbf{k};\,r}$ and its Fourier coefficients $P^{(\nu)(\nu');\,(\mu)}_{\mathbf{s};\,\mathbf{s}';\,\mathbf{R};\,r}$ with respect to the operators R_β and R'_β we have to rewrite the Fourier series for the following.

At first we determine an ensemble of lattice vectors, which we denote by the *star* $S^{\mathbf{s}-\mathbf{s}'}_{\mathbf{ss}',\,\mathbf{R}}$ *of* $\mathbf{R}$ *with respect to* $\mathcal{G}^{\mathbf{s}-\mathbf{s}'}_{\mathbf{ss}'}$: We define it as the set of the different primitive translation vectors, which one obtains by letting act the operators $\beta^{\mathbf{s}-\mathbf{s}'}$ of $\mathcal{G}^{\mathbf{s}-\mathbf{s}'}_{\mathbf{ss}'}$ on $\mathbf{R}$.

From the Eq. (20) we obtain the relation

$$\beta^{\mathbf{s}-\mathbf{s}'}\mathbf{R} = (\mathbf{s} - \mathbf{s}') + \beta\,[\mathbf{R} - (\mathbf{s} - \mathbf{s}')], \tag{27}$$

so that the elements of $S^{\mathbf{s}-\mathbf{s}'}_{\mathbf{ss}',\,\mathbf{R}}$ are located on the surface of the sphere centred at the position $\mathbf{s} - \mathbf{s}'$ and having the radius $|\mathbf{R} - (\mathbf{s} - \mathbf{s}')|$. Since the operators $\beta^{\mathbf{s}-\mathbf{s}'}$ form a group, two stars are either identical or they do not have any lattice vector in common. Consequently we may decompose the Bravais lattice into *stars with respect to* $\mathcal{G}_{\mathbf{ss'}}^{\mathbf{s-s'}}$. Denoting a star of this decomposition by S and a representative of its lattice vectors by $\mathbf{R}_S$ we get for the Eq. (26) with the definition (23)

$$\mathbf{P}_{\mathbf{s};\,\mathbf{s'};\,\mathbf{k}}^{(\nu)(\nu');\,(\mu)} = e^{-i\,\mathbf{k}\cdot(\mathbf{s}-\mathbf{s'})} \sum_S \frac{1}{g_{\mathbf{ss'},\mathbf{R}_S}^{\mathbf{s-s'}}} \sum_{\beta\in\mathcal{G}_{\mathbf{ss'}0}} R'_\beta\, e^{i\,\mathbf{k}\cdot\mathbf{R}_S}\, \mathbf{P}_{\mathbf{s};\,\mathbf{s'};\,\mathbf{R}_S}^{(\nu)(\nu');\,(\mu)}. \qquad (28)$$

$g_{\mathbf{ss'},\mathbf{R}_S}^{\mathbf{s-s'}}$ is the order of the *stabilizer* of $\mathcal{G}_{\mathbf{ss'}}^{\mathbf{s-s'}}$ on $\mathbf{R}_S$.

Now we approximate the representation of the matrix elements (28) by truncating its Fourier series in an appropriate way. For this purpose we define the absolute value $|S_{\mathbf{ss'},\mathbf{R}}^{\mathbf{s-s'}}|$ of $S_{\mathbf{ss'},\mathbf{R}}^{\mathbf{s-s'}}$ as the radius of the sphere (27)

$$\left| S_{\mathbf{ss'},\mathbf{R}}^{\mathbf{s-s'}} \right| = |\mathbf{R} - (\mathbf{s} - \mathbf{s'})|, \qquad (29)$$

which is the distance between the two atoms located at the positions $\mathbf{s}$ and $\mathbf{s'} + \mathbf{R}$ of the matrix elements (18). Now we determine a minimum value $S_{\max}$ in such a manner, that the contribution of the stars S of the Bravais lattice having an absolute value greater than $S_{\max}$ to the Fouerier series is negligible. Furthermore, by representing the Fourier coefficients in terms of their independent parameters according to the relations (24) we finally obtain for the Eq.(28) for the elements $P_{\mathbf{s};\,\mathbf{s'};\,\mathbf{k};\,r}^{(\nu)(\nu');\,(\mu)}$ of the Eq.(15)

$$P_{\mathbf{s};\,\mathbf{s'};\,\mathbf{k};\,r}^{(\nu)(\nu');\,(\mu)} = e^{-i\,\mathbf{k}\cdot(\mathbf{s}-\mathbf{s'})} \cdot \sum_{|S|\leq S_{\max}} \frac{1}{g_{\mathbf{ss'},\mathbf{R}_S}^{\mathbf{s-s'}}} \times \qquad (30)$$

$$\times \sum_{\beta'\in\mathcal{G}_{\mathbf{ss'}0}} \left[\sum_{\beta\in\mathcal{G}_{\mathbf{ss'}0}} \lambda_{r;\,\beta}^{\mu;\,\beta'} \left(R'_\beta\, e^{i\,\mathbf{k}\cdot\mathbf{R}_S} \right) \right] [\, R'_{\beta'}\, P_{\mathbf{s};\,\mathbf{s'};\,\mathbf{R}_S;\,r}^{(\nu)(\nu');\,(\mu)} \,]_{\mathrm{ind}}.$$

Here we delimit the action of the operators R'_β by the paranthesis ( ).

The series (30) is very well suited for the application of the least squares[9] scheme to the determination of the independent parameters $[\, R'_{\beta'}\, P_{\mathbf{s};\,\mathbf{s'};\,\mathbf{R}_S;\,r}^{(\nu)(\nu');\,(\mu)} \,]_{\mathrm{ind}}$: It does not contain any redundant element and its speed of convergence is fast.

For the calculations we have to choose a representative ensemble of Bloch vectors $\mathbf{k}$ for which the elements $P_{\mathbf{s};\,\mathbf{s'};\,\mathbf{k};\,r}^{(\nu)(\nu');\,(\mu)}$ have to be determined. This has to be done in dependence on the symmetries of this matrix element. The number of Bloch vectors for which the matrix element has to be calculated has to be only slightly larer than the number of independent parameters for the Fourier coefficients, which are taken into account in its series (30).

2.2. *Example*

In the following we give examples for the groups $\mathcal{G}_{ss'}$ and $\mathcal{G}_{ss'0}$ in dependence on the relative position vectors s and s'. Obviously, these groups contain the fewer elements the lower the symmetry of the positions s and s' in the Wigner Seitz cell is. In order to demonstrate this connection, we choose a simple cubic Bravais lattice, e. g. the perovskite structure. Thereby, the axes of the coordinate system are chosen to be parallel to the edges of the cube and the lattice constant is $2a$. For a discussion of this structure we use Ba Ti O_3 as an example. All atoms are located at special positions of the simple cubic unit cell. The Ba atoms are at the corners $s = a(0,0,0)$, the Ti at the body center and the O's at the face centers. The point group $\mathcal{G}_0$ is O_h, the space group is denoted as O_h^1 and it is symmorphic. Since there are just one Ba atom and one Ti atom in the unit cell, the groups $\mathcal{G}_{ss'0}$ for s and s' being the relative positions of the Ba's or Ti's are equal to O_h. On the other hand there are three O's in the primitive unit cell, e.g. O1, O2, and O3: Since they are transformed into one another by the action of the elements of the space group, a set of equivalent pairs of relative positions contains more than just one element and the corresponding groups $\mathcal{G}_{ss'0}$ are subgroups of O_h.

From the Eqs. (14) we find, that only the submatrices of the Hamiltonian enclosed by boxes have to be calculated.

$$
H_\mathbf{k} = \begin{pmatrix}
\boxed{\text{Ba - Ba}} & \boxed{\text{Ba - Ti}} & \boxed{\text{Ba - O1}} & \text{Ba - O2} & \text{Ba - O3} \\
 & \boxed{\text{Ti - Ti}} & \boxed{\text{Ti - O1}} & \text{Ti - O2} & \text{Ti - O3} \\
 & & \boxed{\text{O1 - O1}} & \boxed{\text{O1 - O2}} & \text{O1 - O3} \\
 & & & \text{O2 - O2} & \text{O2 - O3} \\
 & & & & \text{O3 - O3}
\end{pmatrix}.
$$

Next we give some examples for *stars* $S_{ss',\mathbf{R}}^{s-s'}$ *with respect to* $\mathcal{G}_{ss'}^{s-s'}$ of the Eqs. (27) and (28): Figures 1 and 2 show the first four and the first three stars for the submatrices Ba - Ba and Ba - Ti, respectively. They represent two different partitions of the simple cubic Bravais lattice into stars all having the symmetry O_h. And thus the Bloch vectors for the least-squares scheme (30) have to be taken from *one* irreducible part of the Brillouin zone (Figure 3). Figures 4 and 5 show stars for the submatrices Ba-O1 and O1-O1, respectively: Due to the symmetries of the s and s' these stars are invariant with respect to subgroups of O_h only. And consequently the $\mathbf{k}$ for the least-squares scheme have to be taken from *three* irreducible parts of the Brillouin zone (Figure 6). Of course, further combinations of s and s' lead to further groups $\mathcal{G}_{ss'0}$, $\mathcal{G}_{ss'}^{s-s'}$, and *stars* $S_{ss',\mathbf{R}}^{s-s'}$ *with respect to* $\mathcal{G}_{ss'}^{s-s'}$.

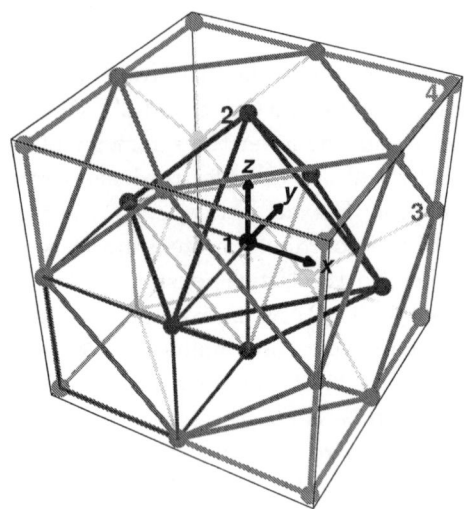

Figure 1. Partition of the sc Bravais lattice into *stars* $S_{\mathbf{ss},\mathbf{R}}^{\mathbf{s}-\mathbf{s}}$ *with respect to* $\mathcal{G}_{\mathbf{ss}}^{\mathbf{s}-\mathbf{s}}$ for $\mathbf{s} = \mathbf{0}$ (Ba) or $\mathbf{s} = a\,(1,1,1)$ (Ti): The stars for the following lattice vectors $\mathbf{R}$ are shown: 1: $a\,(0,0,0)$, 2: $a\,(0,0,2)$, 3: $a\,(2,2,0)$, and 4: $a\,(2,2,2)$: The stars are centred at $a\,(0,0,0)$ and have the absolute values $0\,a$, $2\,a$, $2\sqrt{2}\,a$, and $2\sqrt{3}\,a$, respectively. The point group $\mathcal{G}_{\mathbf{ss}0}$ is the holosymmetric point group O_h of the cubic crystal system.

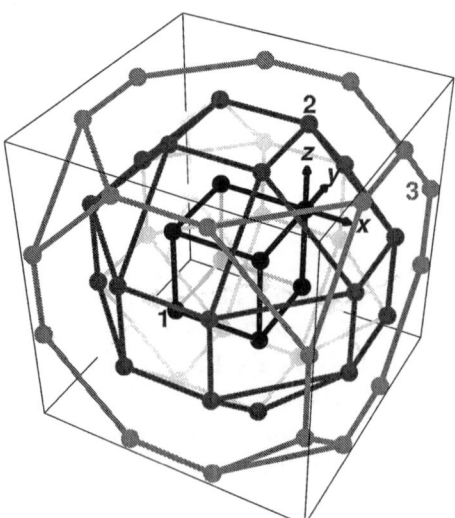

Figure 2. Partition of the sc Bravais lattice into *stars* $S_{\mathbf{s}_1\mathbf{s}_2,\mathbf{R}}^{\mathbf{s}_1-\mathbf{s}_2}$ *with respect to* $\mathcal{G}_{\mathbf{s}_1\mathbf{s}_2}^{\mathbf{s}_1-\mathbf{s}_2}$ for $\mathbf{s}_1 = \mathbf{0}$ (Ba) and $\mathbf{s}_2 = a\,(1,1,1)$ (Ti): The stars for the following lattice vectors $\mathbf{R}$ are shown: 1: $a\,(0,0,0)$, 2: $a\,(0,0,2)$, and 3: $a\,(2,2,0)$: The stars are centred at $-\mathbf{s}_2 = a\,(-1,-1,-1)$ and have the absolute values $\sqrt{3}\,a$, $\sqrt{11}\,a$, and $\sqrt{19}\,a$, respectively. The point group $\mathcal{G}_{\mathbf{s}_1\mathbf{s}_20}$ is the holosymmetric point group O_h of the cubic crystal system.

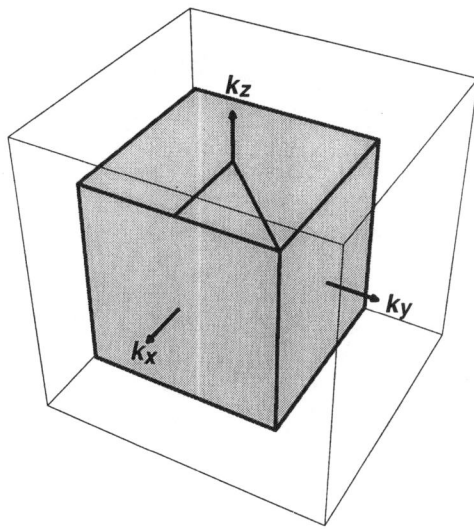

Figure 3. Brillouin zone of the sc Bravais lattice and *one* of it's irreducible parts: For the application of the least-squares scheme to the determination of the independent parameters of the matrix elements, like for example in some cases of the space group O_h^1, the Bloch vectors **k** should be taken from *one* irreducible part only.

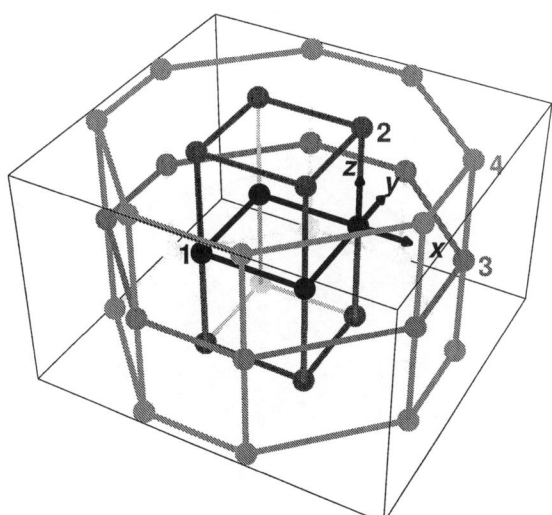

Figure 4. Partition of the sc Bravais lattice into *stars* $S_{s_1 s_2, \mathbf{R}}^{s_1 - s_2}$ *with respect to* $\mathcal{G}_{s_1 s_2}^{s_1 - s_2}$ for $s_1 = \mathbf{0}$ (Ba) and $s_2 = a(1, 1, 0)$ (O1): The stars for the following lattice vectors **R** are shown: 1: $a(0, 0, 0)$, 2: $a(0, 0, 2)$, 3: $a(2, 0, 0)$, and 4: $a(2, 0, 2)$. The stars are centred at $-s_2 = a(-1, -1, 0)$ and have the absolute values $\sqrt{2}\,a$, $\sqrt{6}\,a$, $\sqrt{10}\,a$, and $\sqrt{14}\,a$, respectively. The point group $\mathcal{G}_{s_1 s_2 0}$ is a subgroup of the holosymmetric point group O_h of the cubic crystal system.

436

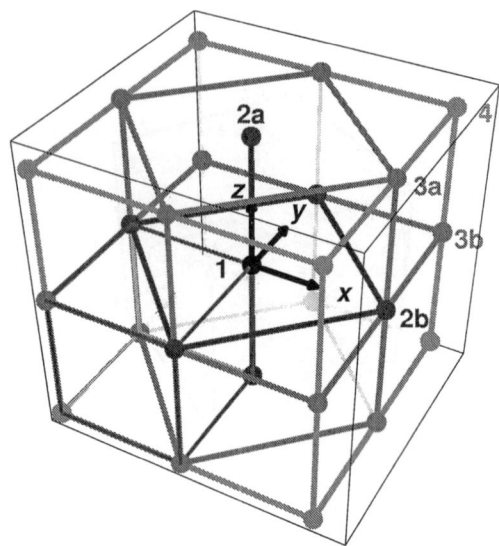

Figure 5. Partition of the sc Bravais lattice into *stars* $S_{ss,R}^{s-s}$ *with respect to* $\mathcal{G}_{ss}^{s-s}$ for $s = a(1,1,0)$ (O1): The stars for the following lattice vectors R are shown: 1: $a(0,0,0)$, 2a: $a(0,0,2)$, 2b: $a(2,0,0)$, 3a: $a(2,0,2)$, 3b: $a(2,2,0)$, 4: $a(2,2,2)$. The stars are centred at $a(0,0,0)$ and have the absolute values $0\,a$, $2\,a$, $2\,a$, $\sqrt{8}\,a$, $\sqrt{8}\,a$, and $\sqrt{12}\,a$, respectively. The point group $\mathcal{G}_{ss0}$ is a subgroup of the holosymmetric point group O_h of the cubic crystal system.

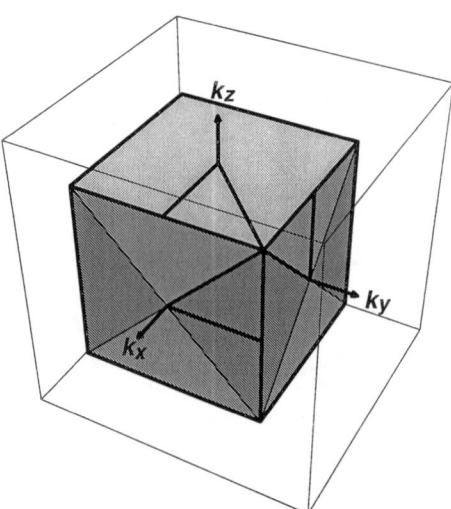

Figure 6. Brillouin zone of the sc Bravais lattice and *three* of it's irreducible parts: For applications of the least-squares scheme to the determination of the independent parameters of the matrix elements, like for example in some cases of the perovskite structure, the Bloch vectors k should be taken e.g. from these irreducible parts only.

2.3. Numerical Results: Application of the LCAO Method to the Determination of the Band Structure of Potassium and Comparison with APW and FLAPW results

In the following we apply the present method for performing LCAO band structure calculations to the determination of the electronic band structure (ground state and excited bands) and of the high momentum components of electron momentum densities (Compton rates) for the alkali metal potassium. We choose this element for our considerations because the valence electrons of the alkali metals are nearly-free, so that they form a worst case situation for the applicability of the LCAO method, or in other words, the determination of the electronic structure also represents a crucial test for the accuracy obtainable in general by this method. Finally we compare our results with those of augmented plane wave (APW) calculations[10], which are known to be very accurate in the case of materials with nearly free electrons, and with full-potential linearized augmented plane wave (FLAPW) results[1].

Applying the density-functional theory,[4] the Kohn-Sham method[5] and the local density approximation, the Schrödinger equation (1) for the electrons in a metal can be written as

$$H \psi_i = \varepsilon_i \psi_i$$

with

$$H \equiv -\frac{1}{2} \nabla^2 + V(\mathbf{r}) + \int \frac{n(\mathbf{r}')}{|\mathbf{r} - \mathbf{r}'|} \, d^3 r' + \frac{\delta}{\delta n(\mathbf{r})} \left\{ n(\mathbf{r}) \left[\epsilon_{ex}(n(\mathbf{r})) + \epsilon_c(n(\mathbf{r})) \right] \right\}$$

and

$$n(\mathbf{r}) = \sum_i^N |\psi_i(\mathbf{r})|^2,$$

where the sum extends over the occupied states. We use atomic units, $n(\mathbf{r})$ is the electron density, N is the number of occupied states, $V(\mathbf{r})$ is the external potential, $\int n(\mathbf{r}')/|\mathbf{r} - \mathbf{r}'| \, d^3 r'$ contains the Hartree part of the Coulomb energy, $\epsilon_{ex}(n(\mathbf{r}))$ and $\epsilon_c(n(\mathbf{r}))$ are the exchange and correlation energy of an homogeneous electron gas of the density $n(\mathbf{r})$. The equations have to be solved self-consistently with respect to the electron density.

For potassium the external potential $V(\mathbf{r})$ is given by $-\sum_{\mathbf{R}} Z/|\mathbf{r} - \mathbf{R}|$, where the atomic number $Z = 19$ and the $\mathbf{R}$'s are the primitive translation vectors of the body centred cubic Bravais lattice and in our case also the positions of the atomic nuclei. As lattice constant we take 9.882 a.u.. The exchange energy is given by $-3/4\,(3n/\pi)^{1/3}$ a.u. in energy per particle. For the correlation energy $\epsilon_c(n(\mathbf{r}))$ we use the interpolation formula of Vosko et al.[11]

For the atomic orbitals of the LCAO expansion of the crystal wave function we use the numerical basis set described by Zunger and Freeman,[12,13] and Eschrig and Bergert.[3] The convergence of the expansion has been tested by calculating the high momentum components of Compton rates along the symmetry directions of the Brillouin zone with an increased number of atomic orbitals in the basis set, and then we omitted those orbitals which did not contribute to these rates. For the accurate determination of the matrix elements twelve nearest neighbours had to be included at most in the LCAO Bloch wave function.

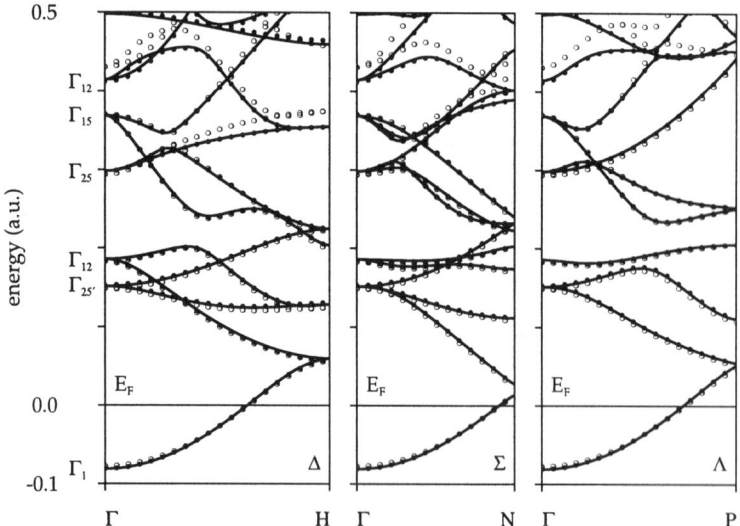

Figure 7. Part of the electronic energy band structure of potassium: the energies are presented for electrons having Bloch vectors $\mathbf{k}$ along the symmetry directions Δ, Σ and Λ. The curves (———), ($\bullet\;\bullet\;\bullet$), and ($\circ\;\circ\;\circ$) are the results derived by the LCAO, APW, and by the FLAPW method, respectively. The energies are shifted in such a manner, so that the Fermi energy E_F is equal to zero.

In figure 7 we present the energy band structure for potassium and we compare it with APW and FLAPW calculations, which are known to lead to very good results in the case of materials with nearly free electrons:

we find, that not only the energies of the core electrons are in good agreement, but also the energies of the valence electrons and some of the excited states. Of course, the number of the excited states which can be represented by the LCAO method depends on the number and on the symmetries of the expansion functions included in the LCAO ansatz for the crystal wave function.

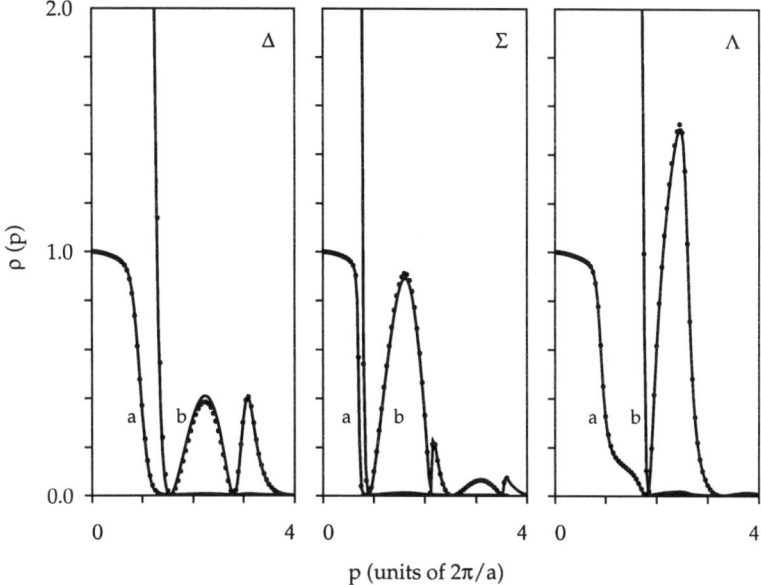

Figure 8. Contribution of the conduction electrons of potassium to the Compton rates $\rho(\mathbf{p})$ for $\mathbf{p}$ along the symmetry directions Δ, Σ and Λ. The rates are normalized with respect to $\rho(0)$. The curves (———) and ($\bullet\bullet\bullet$) are the results derived by the LCAO method and by the APW method, respectively. (a) Momentum densities, (b) momentum densities multiplied by the factor 100, a is the lattice constant.

In order to check the quality of the crystal wave functions as determined by the LCAO method we calculate the high momentum components of Compton rates $\rho(\mathbf{p})$ and compare them with the results of APW calculations. For the calculations we choose that expression for the Compton rates, where they are proportional to the electron momentum density:

$$\rho(\mathbf{p})\,d^3p = const \cdot \sum_{\nu} \sum_{\mathbf{k}:\,E_{\mathbf{k},\nu} \leq E_F} \left| \int e^{-i\mathbf{p}\cdot\mathbf{r}}\,\psi_{\mathbf{k}}(\mathbf{r})\,d^3r \right|^2 d^3p.$$

ν denotes the band index. $\rho(\mathbf{p})$ has the symmetry of the point group $\mathcal{G}_0$ of the crystal lattice: $\rho(\mathbf{p}) = \rho(\alpha\,\mathbf{p})$ for $\alpha \in \mathcal{G}_0$. Explicit numerical results for $\rho(\mathbf{p})$ for momentum vectors $\mathbf{p}$ along the symmetry directions Δ, Σ, and Λ are presented in figure 8. For comparison we also include the corresponding results of APW calculations: an excellent agreement between the results obtained by these methods of calculation is found.

Acknowledgments

The author is indepted to Prof. H. Sormann for many helpful discussions and for making available the plane wave results.

References

1. P. Blaha, K. Schwarz, G. K. H. Madsen, D. Kvasnicka and J. Luitz, *Wien 2k: An Augmented Plane Wave + Local Orbitals Program for Calculating Crystal Properties* (Karlheinz Schwarz, Techn. Universität Wien, Austria), ISBN 3-9501031-1-2 (2001).
2. F. Bloch, *Ztschr. f. Physik* **52**, 555-600 (1928).
3. H. Eschrig and I. Bergert, *Phys. Stat. Sol. (b)* **90**, 621-8 (1978).
4. R. Hohenberg and W. Kohn, *Phys. Rev.* **B 136**, 864-71 (1964).
5. W. Kohn and L. J. Sham, *Phys. Rev.* **140/4A**, 1133-8 (1965).
6. W. Obermayr, *Group Theory and LCAO Method* ; ISBN 3-8265-3446-8, Shaker: Aachen (1998).
7. W. Obermayr, *Int. J. Quantum Chemistry* **78**, 212 (2000).
8. F. Seitz, *Ztschr. f. Kristallogr.* **91**, 336-66 (1935).
9. J. C. Slater and G. F. Koster, *Phys Rev.* **94/6**, 1498-524 (1954).
10. H. Sormann, private communication, (2002).
11. S. H. Vosko, L. Wilk and M. Nusair, *Can. J. Phys.* **58**, 1200-11 (1980).
12. A. Zunger and A. J. Freeman, *Int. J. Quant. Chem. Symp.* **10**, 383-403 (1976).
13. A. Zunger and A. J. Freeman, *Phys. Rev.* **B 15/10**, 4716-37 (1977).

SPECIFIC PROPAGATION DIRECTIONS OF ACOUSTIC WAVES IN MEDIA OF VARIOUS ACOUSTIC SYMMETRIES

A. DUDA

Institute of Theoretical Physics, University of Wrocław,
pl. Maxa Borna 9,
Pl-52-404 Wrocław, Poland
E-mail: ardud@ift.uni.wroc.pl

T. PASZKIEWICZ

Institute of Physics, University of Rzeszów
Ul. Rejtana 16 A,
PL-35-310 Rzeszów, Poland
E-mail:tapasz@univ.rzeszow.pl

We consider propagation of acoustic waves in media of lowest elastic symmetries. We showed that the number of acoustical axes for media with triclinic symmetry cannot be larger than 132. We proposed analytical method of determining components of longitudinal normals. By extending the Khatkevich approach and using the Bezout theorem we proved that the number of longitudinal normals for mechanically stable triclinic media can be larger than 16 and not, as claimed by some authors, 13. We also proved that the number of longitudinal normals for monoclinic media cannot be larger than 13, whereas, according to Khatkevich, this number cannot be larger than 17. Using our method we numerically established directions of longitudinal normals for several monoclinic elastic media. For media of higher symmetries (rhombic, trigonal, tetragonal, hexagonal and cubic) our method of determining components of longitudinal normals yields well-known results obtained by Borgnis and Khatkevich

1. Introduction

The elastic properties of media are characterized by tensors of elastic constants which are parameterized by sets of elastic constants $C_{ik,jl}$. Tensors of elastic constants have complete Voigt symmetry, i.e. its components are invariant under interchange of the indices i and k, and they are also invariant under interchange of the pairs (ik) and (jl). These properties allows one to use the Voigt notation, in which a pair of Cartesian indices $(i,j) \equiv (j,l)$ are represented by one index α, according to the scheme

(i,j)	(11)	(22)	(33)	(23)	(13)	(12)
α	1	2	3	4	5	6

In the Voigt notation the tensor of elastic constants is represented by 6×6 matrices with elements $C_{\alpha\mu}$. The number of independent nonvanishing elements depends on the symmetry of the medium. These matrices have the simplest form in crystallographic Cartesian coordinate systems (COS for short). The components of longitudinal normals for media with symmetry higher than monoclinic were discussed in papers by Borgnis, Khatkiewicz and Fedorov. [3-6]

Here we deal with acoustic waves propagating in crystalline media. We shall consider that the length of these waves is much larger than the lattice constant. For such waves crystalline media can be treated as continuous anisotropic elastic media. The propagation of such acoustic waves is governed by the Christoffel equation

$$\rho \frac{\partial^2 u_r}{\partial t^2} = \sum_{l,s,m=1}^{3} C_{rl,sm} \frac{\partial_s^2 u}{\partial x_l \partial x_m}. \tag{1}$$

We are looking to solutions of the above equation in form of plane vectorial waves

$$\mathbf{u}(\mathbf{r}, t) = \mathbf{e} \exp\left[i\left(\mathbf{kx} - \omega t\right)\right], \tag{2}$$

where $\mathbf{e}$ is a polarization vector of wave, $\mathbf{k}$ is the wave-vector ($\mathbf{k} = \mathbf{n}k$, $\mathbf{n}$ being the direction of propagation, $k = |\mathbf{k}|$) and ω is the frequency. The vector $\mathbf{n}$ belongs to the unit 2-dimensional sphere S^2.

Substituting Eq. (2) into Eq. (1) we obtain a set algebraic equations

$$\sum_{l,s,m=1}^{3} \left(n_l C_{rl,sm} n_m k^2 - \rho \omega^2 \delta_{r,s}\right) e_s = 0, (r = 1, 2, 3). \tag{3}$$

Dividing both sides of the Eq (3) by k^2, the condition of existence of solution of Eq. (3) can be written in the form

$$det\left[\mathbf{\Gamma}(\mathbf{n}) - \rho v^2 \mathbf{I}_2\right] = 0, \tag{4}$$

where $v \equiv \omega/k$ is the phase velocity of the sound and Γ_{rs} is an element of propagation matrix $\mathbf{\Gamma}(\mathbf{n})$ defined by the set of elements

$$[\mathbf{\Gamma}(\mathbf{n})]_{rs} = \Gamma_{rs}(\mathbf{n}) = \sum_{p,t=1}^{3} C_{pr,st} n_p n_t. \tag{5}$$

Polarization vectors $\mathbf{e}$ are eigenvectors of the propagation matrix $\mathbf{\Gamma}(\mathbf{n})$. Squares of phase velocities are the eigenvalues of this matrix. For each given direction of propagation Eq. (1) has three solutions – these solutions are called modes. Phase velocities and polarization vectors generally depend on the direction of propagation **n**. After Every [1] to the mode with the highest phase velocity we attribute the number $j = 0$, for the mode with medium value of phase velocity $j = 1$, and for the slowest mode $j = 2$.

In this paper properties of so called special directions of propagation are discussed. They were a subject of theoretical research for many years (cf. Sakadi[2]-Helbig[11]). The set of special directions of propagation consists of the longitudinal and transverse normals (called also the longitudinal and transverse directions) and the acoustical axes. In this paper we consider only the longitudinal normals and the acoustics axes.

Longitudinal normal $\mathbf{n}_n$ is a direction in which there propagate mode with polarization vector parallel to $\mathbf{n}_n$. Acoustical axes are normals along which at least two modes have equal values of phase velocities. All symmetry axes are also the longitudinal normals[9].

Our approach can be treated as an extension of the Khatkevich method[4,5,6,10]. However, differently than our approach, his method yields quite complicated vectorial formulae.

In the first part of our paper we will discuss characteristics of longitudinal normals for media of lowest symmetry, i.e. triclinic and monoclinic media. In particular we will show that despite statements known in the literature,[11] the upper bounds of the number of longitudinal normals for media with triclinic symmetry is 15 and not 13. We shall also describe an analytical method of determining components of longitudinal normals and prove that maximal number of longitudinal normals for media with monoclinic symmetry is 13. In the second part of our paper we will find the maximal possible number of acoustical axes for media with triclinic symmetry.

2. Determination of components of longitudinal normals of triclinic media

2.1. *Longitudinal normals lying out of OXY plane*

Triclinic media have no symmetry axes and no symmetry planes. So we may choose an arbitrary Cartesian coordinate system (Fig. 1).

We shall derive equations which will enable us to determine components of longitudinal normals for any given tensor of elastic constants in an arbitrary coordinate system. We start with the discussion of the set of equations which determines longitudinal normals which lie outside the plane OXY (Fig. 1), then $n_3 \neq 0$.

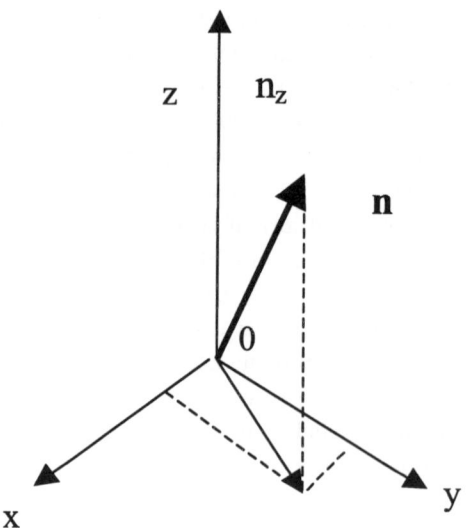

Figure 1.   Cartesian coordinate system

The direction **n** is a longitudinal normal if and only if the following equation is fulfilled

$$\sum_{j=0}^{3} \Gamma_{ij} n_j = \lambda n_i, \qquad (6)$$

where λ is a positive real number. This implies that

$$[n_1, n_2, n_3] = [\Omega_1, \Omega_2, \Omega_3], \qquad (7)$$

where

$$\Omega_i \equiv \lambda^{-1} \sum_{j=1}^{3} \Gamma_{ij} n_j. \qquad (8)$$

Two vectors **n** and Ω are equal if

$$\Omega_1/\Omega_3 = n_1/n_3, \ \ \Omega_2/\Omega_3 = n_2/n_3. \qquad (9)$$

By making substitutions

$$x = n_1/n_3, y = n_1/n_3, \qquad (10)$$

the set of equations (9) takes the following form

$$
\begin{aligned}
&[C_{35}+(2C_{45}+C_{36})y+(2C_{46}+C_{25})y^2+C_{26}\,y^3]+[(C_{13}+2C_{55}-C_{33})\\
&+(4C_{56}+2C_{14}-3C_{34})y+(2C_{66}+C_{12}-2C_{44}-C_{23})y^2-C_{24}y^3]x\\
&+[(3C_{15}-3C_{35})+(3C_{16}-4C_{45}-2C_{36})y-(2C_{46}+C_{25})y^2]x^2\\
&+[(C_{11}-2C_{55}-C_{13})-(C_{14}+2C_{56})y]x^3-C_{15}x^4=0\ ,
\end{aligned}
\tag{11}
$$

$$
\begin{aligned}
&[C_{34}+(2C_{44}+C_{23}-C_{33})y+(3C_{24}-3C_{34})y^2+(C_{22}-2C_{44}-C_{23})y^3\\
&-C_{24}y^4]+[(C_{36}+2C_{45})+(2C_{25}+4C_{46}-3C_{35})y+(3C_{26}-4C_{45}\\
&-2C_{36})\,y^2-(2C_{46}+C_{25})y^3]\,x+[(2C_{56}+C_{14})+(2C_{66}+C_{12}-2C_{55}\\
&-C_{13})\,y-(C_{14}+2C_{56})y^2]\,x^2+(C_{16}-C_{15}y)\,x^3=0.
\end{aligned}
\tag{12}
$$

For each solution of the above set one can find exactly one longitudinal normal, because for each pair (x,y) one can find exactly one triple of numbers $[n_1,n_2,n_3]$, which fulfil the condition

$$
n_1^2+n_2^2+n_3^2=1.
$$

Namely

$$
\left[\frac{\pm x}{\sqrt{1+x^2+y^2}},\frac{\pm y}{\sqrt{1+x^2+y^2}},\frac{\pm 1}{\sqrt{1+x^2+y^2}}\right],
\tag{13}
$$

Each longitudinal normal crosses unit sphere in two points.

According to the Bezout theorem,[12] if the number of solutions of the set of polynomial equations is finite, then it cannot be larger then the product of degrees of equations which form this set. Because the degree of each equation from the set is 4, so the maximal number of longitudinal normals cannot be bigger than 16.

By solving the set of Eqs. (11), (12) one can determine all longitudinal normals for the considered medium. This set could be written in the following form

$$
A_0(y)+A_1(y)x+A_2(y)x^2+A_3(y)x^3+A_4(y)x^4=0,
\tag{14}
$$

$$
B_0(y)+B_1(y)x+B_2(y)x^2+B_3(y)x^3=0,
\tag{15}
$$

where

$$
A_\sigma(y),B_\sigma(y)\ (\sigma=0,1,2,3).
$$

are univariate polynomials in the variable y. Namely

$$
\begin{aligned}
A_0(y)&=\left[C_{35}+(2C_{45}+C_{36})y+(2C_{46}+C_{25})y^2+C_{26}y^3\right],\\
A_1(y)&=[(C_{13}+2C_{55}-C_{33})+(4C_{56}+2C_{14}-3C_{34})y\\
&+(2C_{66}+C_{12}-2C_{44}-C_{23})y^2-C_{24}y^3],\\
A_2(y)&=[3(C_{15}+C_{16}-C_{35})-4C_{45}-2C_{36}-2C_{46}-C_{25}],\\
A_3(y)&=[(C_{11}-2C_{55}-C_{13})-(C_{14}+2C_{56})y],\quad A_4(y)=-C_{15}\,,
\end{aligned}
\tag{16}
$$

$$B_0(y) = \left[C_{34} + (2C_{44} + C_{23} - C_{33})y + 3(C_{24} - C_{34})y^2\right.$$
$$\left. + (C_{22} - 2C_{44} - C_{23}y^3 - C_{24}y^4\right],$$
$$B_1(y) = \left[(C_{36} + 2C_{45}) + (2C_{25} + 4C_{46} - 3C_{35})y\right.$$
$$\left. + (3C_{26} - 4C_{45} - 2C_{36})y^2 - (2C_{46} + C_{25})y^3\right],$$
$$B_2(y) = \left[(2C_{56} + C_{14}) + (2C_{66} - 2C_{55} - C_{13} + C_{12})y - (C_{14} + 2C_{56})y^2\right],$$
$$B_3(y) = (C_{16} - C_{15}y).$$

$$(17)$$

Introduce the eliminant $E(y)$ of the set of Eqs. (14), (15)

$$E(y) \equiv \begin{vmatrix} A_4(y) & A_3(y) & A_2(y) & A_1(y) & A_0(y) & 0 & 0 \\ 0 & A_4(y) & A_3(y) & A_2(y) & A_1(y) & A_0(y) & 0 \\ 0 & 0 & A_4(y) & A_3(y) & A_2(y) & A_1(y) & A_0(y) \\ B_3(y) & B_2(y) & B_1(y) & B_0(y) & 0 & 0 & 0 \\ 0 & B_3(y) & B_2(y) & B_1(y) & B_0(y) & 0 & 0 \\ 0 & 0 & B_3(y) & B_2(y) & B_1(y) & B_0(y) & 0 \\ 0 & 0 & 0 & B_3(y) & B_2(y) & B_1(y) & B_0(y) \end{vmatrix}. \qquad (18)$$

Generally, the problem of finding of longitudinal normals could be solved in the following way. First one finds all possible roots y_p of $E(y)$ and next for each set of coefficients $A_\sigma(y_p), B_\sigma(y_p)$ one should solve Eqs. (11) and (12).

2.2. Determination of components of longitudinal normals of triclinic media lying in OXY plane

The direction **n** lying inside the plane OXY will be a longitudinal normal if the following equation is fulfilled:

$$[n_1, n_2, 0] = [\frac{1}{\lambda}\sum_{j=1}^{3}\Gamma_{1j}n_j, \frac{1}{\lambda}\sum_{j=1}^{3}\Gamma_{2j}n_j, \frac{1}{\lambda}\sum_{j=1}^{3}\Gamma_{3j}n_j], \qquad (19)$$

which is equivalent to the following set:

$$\sum_{j=1}^{3}\Gamma_{3j}n_j = 0, \quad \left(\sum_{j=1}^{3}\Gamma_{2j}n_j\right)n_1 = \left(\sum_{j=1}^{3}\Gamma_{1j}n_j\right)n_2. \qquad (20)$$

Replacing the elements of the propagation matrix (5) with appropriate function of the elastic constants and of components of direction of propagation one obtains

$$C_{26}n_2^4 + (C_{12} + 2C_{66} - C_{22})n_1n_2^3 + 3(C_{16} - C_{26})n_1^2n_2^2$$
$$+ (C_{11} - 2C_{66} - C_{12})n_1^3n_2 - C_{16}n_1^4 = 0, \qquad (21)$$

$$C_{24}n_2^3 + (C_{25} + 2C_{46})n_1n_2^2 + (2C_{56} + C_{14})n_1^2n_2 + C_{15}n_1^3 = 0. \qquad (22)$$

By solving the set of Eqs. (21) and (22) one can determine all longitudinal normals which lie in OXY plane.

3. Longitudinal normals for remaining elastic symmetry classes

3.1. *Longitudinal normals for monoclinic media*

For media with monoclinic symmetry the set of equations determining longitudinal normals has the following form

$$(2C_{45} + C_{36})y + C_{26}y^3 + [(C_{13} + 2C_{55} - C_{33})+ (2C_{66} + C_{12} - 2C_{44} \atop -C_{23})\, y^2]\, x + (3C_{16} - 4C_{45} - 2C_{36})yx^2 + (C_{11} - 2C_{55} - C_{13})x^3 = 0, \tag{23}$$

$$(2C_{44}+C_{23})\, y+C_{22}y^3+ \left[(2C_{45}+C_{36}-C_{33})+(3C_{26}-2C_{44}-C_{23})y^2\right] x \atop + (2C_{66} + C_{12} - 4C_{45} - 2C_{36})\, y\, x^2 + (C_{16} - 2C_{55} - C_{13})\, x^3 = 0. \tag{24}$$

The degree of each of above equations is 3, thus the Bezout theorem implies that maximal number of longitudinal normals for media with monoclinic symmetry is 9. As in the case of triclinic media for given y the necessary condition for the existence of real solution of set of Eqs. (23, 24) is vanishing of their eliminant

$$E_{mono}(y) = \begin{vmatrix} A_3(y) & A_2(y) & A_1(y) & A_0(y) & 0 & 0, \\ 0 & A_3(y) & A_2(y) & A_1(y) & A_0(y) & 0, \\ 0 & 0 & A_3(y) & A_2(y) & A_1(y) & A_0(y), \\ B_3(y) & B_2(y) & B_1(y) & B_0(y) & 0 & 0, \\ 0 & B_3(y) & B_2(y) & B_1(y) & B_0(y) & 0, \\ 0 & 0 & B_3(y) & B_2(y) & B_1(y) & B_0(y). \end{vmatrix}, \tag{25}$$

where

$$\begin{aligned} A_0(y) &\equiv (2C_{45} + C_{36})y + C_{26}y^3, \\ A_1(y) &\equiv \left[(C_{13} + 2C_{55} - C_{33}) + (2C_{66} + C_{12} - 2C_{44} - C_{23})y^2\right], \\ A_2(y) &\equiv (3C_{16} - 4C_{45} - 2C_{36})y, \quad A_3(y) \equiv (C_{11} - 2C_{55} - C_{13}), \\ B_0(y) &\equiv (2C_{44} + C_{23})y + C_{22}y^3, \\ B_1(y) &\equiv \left[(2C_{45} + C_{36} - C_{33}) + (3C_{26} - 2C_{44} - C_{23})y^2\right], \\ B_2(y) &\equiv (2C_{66} + C_{12} - 4C_{45} - 2C_{36})y, \quad B_3(y) \equiv (C_{16} - 2C_{55} - C_{13}). \end{aligned} \tag{26}$$

The condition of vanishing of eliminant (25) brings about the equation

$$y(B_0 + B_2y^2 + B_4y^4 + B_6y^6 + B_8y^8) = 0, \tag{27}$$

where coefficients B_0, B_2, B_4, B_6, B_8 are complicated polynomial functions of sixth degree of elastic constants characterizing the medium.

Null is always a solution of Eq. (27). All other solutions could be determined by solving the equation

$$(B_0 + B_2 z + B_4 z^2 + B_6 z^3 + B_8 z^4) = 0, \tag{28}$$

where $z \equiv y^2$. Above equation could be solved analytically. For monoclinic media the second equation from the set of Eqs. (23,24) is a trivial identity and the first one takes the form

$$C_{26} n_2^4 + (C_{12} + 2C_{66} - C_{22}) n_2^3 n_1 + 3 (C_{16} - C_{26}) n_2^2 n_1^2$$
$$+ (C_{11} - 2C_{66} - C_{12}) n_2 n_1^3 - C_{16} n_1^4 = 0. \tag{29}$$

The above equation could be easily solved analytically. With the help of above presented method we have found longitudinal normals for several monoclinic media. The components of these normals together with the values of elastic constants for these media are listed in the Appendix.

The components of longitudinal normals for media with symmetry higher than monoclinic were discussed in papers by Borgnis, Khatkiewicz and Fedorov [3-6]. They could be also obtained by solving sets of equations (11), (12) and (19). Below we give without derivation components of these normals for media with rhombic, tetragonal, trigonal and cubic acoustic symmetry.

3.2. Longitudinal normals for media with rhombic acoustic symmetry

Rhombic media have three twofold symmetry axes - they are also longitudinal normals. The z-axis of the COS is directed along one of these two-fold axes. Existence of longitudinal normals inside OXY plane which are not symmetry axes depends on the sign of the expression

$$\phi_1 \equiv (2C_{66} + C_{12} - C_{11})(C_{12} + 2C_{66} - C_{22}).$$

If ϕ_1 is positive then medium has longitudinal normals determined by expression

$$n_2/n_1 = \pm\sqrt{(2C_{66} + C_{12} - C_{11})(C_{12} + 2C_{66} - C_{22})}/(C_{12} + 2C_{66} - C_{22}). \tag{30}$$

The existence of longitudinal normals inside planes OXZ and OYZ depends on the sign of expressions $\phi_2 \equiv (2C_{44} - C_{22} - C_{23})(2C_{44} + C_{23} - C_{33})$ and $\phi_3 \equiv (2C_{55} + C_{13} - C_{11})(C_{13} + 2C_{55} - C_{33})$. If ϕ_2 is positive then the medium has longitudinal normals inside OYZ plane determined by equations

$$n_1/n_3 = 0, \quad n_2/n_3 = \pm\sqrt{(C_{23} - C_{33} + 2C_{44})/(2C_{44} - C_{22} - C_{23})}. \tag{31}$$

If ϕ_3 is positive then the medium has longitudinal normals inside OYZ plane determined by equations

$$n_2/n_3 = 0, n_1/n_3 = \pm\sqrt{(C_{13} + 2C_{55} - C_{33})/(2C_{55} + C_{13} - C_{11})}. \quad (32)$$

The existence of longitudinal normals out of symmetry planes for rhombic media depends on the signs of expressions $\lambda_1 \equiv B_1 A_1$, $\lambda_2 \equiv B_2 A_2$, where

$$\begin{aligned}
A_1 \equiv (&4C_{66}C_{44} + C_{22}C_{11} - C_{22}C_{13} - 2C_{22}C_{55} + C_{12}C_{13} + 2C_{66}C_{13} \\
&+2C_{66}C_{23} - 4C_{12}C_{66} + C_{12}C_{23} + 2C_{12}C_{44} + 2C_{12}C_{55} + 4C_{66}C_{55} \\
&-2C_{44}C_{11} - C_{23}C_{11} - C_{12}^2 - 4C_{66}^2),
\end{aligned} \quad (33)$$

$$\begin{aligned}
B_1 \equiv (&2C_{44}C_{13} - 4C_{55}C_{13} - 4C_{55}^2 + C_{12}C_{13} + 2C_{12}C_{55} - C_{12}C_{33} \\
&-2C_{44}C_{11} + 2C_{66}C_{13} + 4C_{66}C_{55} - 2C_{66}C_{33} + C_{33}C_{11} - C_{13}^2 \\
&+4C_{44}C_{55} - C_{23}C_{11} + 2C_{23}C_{55} + C_{23}C_{13}),
\end{aligned} \quad (34)$$

$$\begin{aligned}
A_2 \equiv (&C_{22}C_{13} + 2C_{22}C_{55} - 4C_{66}C_{44} - C_{22}C_{11} - C_{12}C_{13} \\
&-2C_{66}C_{13} - 2C_{66}C_{23} + 4C_{12}C_{66} - C_{12}C_{23} - 2C_{12}C_{44} \\
&-2C_{12}C_{55} - 4C_{66}C_{55} + 2C_{44}C_{11} + C_{23}C_{11} + C_{12}^2 + 4C_{66}^2),
\end{aligned} \quad (35)$$

$$\begin{aligned}
B_2 \equiv (&C_{22}C_{13} - 2C_{44}C_{13} - C_{23}C_{13} - C_{12}C_{23} - 2C_{66}C_{23} + 2C_{66}C_{33} \\
&+2C_{22}C_{55} + C_{12}C_{33} + 4C_{44}C_{23} + 4C_{44}^2 - 4C_{44}C_{55} \\
&-2C_{12}C_{44} - 2C_{23}C_{55} - 4C_{66}C_{44} - C_{33}C_{22} + C_{23}^2).
\end{aligned} \quad (36)$$

If both expressions λ_1 and λ_2 are positive then the medium has four longitudinal normals determined by equations

$$n_2/n_3 = \pm\sqrt{B_1/A_1}, n_1/n_3 = \pm\sqrt{B_2/A_2}. \quad (37)$$

3.3. Longitudinal normals for media with tetragonal acoustic symmetry

Tetragonal media have one four-fold symmetry axis and four two-fold symmetry axes. The z-axis if COS is directed along one of four-fold axes.

The existence of other longitudinal normals depends on the sign of expression $(2C_{44} + C_{13} - C_{33})(C_{13} - C_{11} + 2C_{44})$. If this expression is positive then medium has additional longitudinal normals inside planes OXZ and OYZ given by equations

$$x = 0, \; y = \pm\sqrt{(2C_{44} + C_{13} - C_{33})/(C_{13} - C_{11} + 2C_{44})}, \quad (38)$$

and

$$y = 0, \; x = \pm\sqrt{(2C_{44} + C_{13} - C_{33})/(C_{13} - C_{11} + 2C_{44})}. \quad (39)$$

Additionally if the expression $(2C_{44}+C_{13}-C_{33})(2C_{13}+4C_{44}-2C_{66}-C_{12}-C_{11})$ is positive then medium has 4 longitudinal normals out of planes of symmetry - their components are given by equations

$$x = \pm\sqrt{(2C_{44}+C_{13}-C_{33})/(2C_{13}+4C_{44}-2C_{66}-C_{12}-C_{11})},$$
$$y = \pm\sqrt{(2C_{44}+C_{13}-C_{33})/(2C_{13}+4C_{44}-2C_{66}-C_{12}-C_{11})}. \quad (40)$$

3.4. *Longitudinal normals for media with trigonal acoustic symmetry*

This medium has one three-fold symmetry axis and three two-fold symmetry axes. The z-axis if COS is directed along one of three-fold axis.

The existence of other longitudinal normals depends on the sign of expression $(C_{33}-C_{13}-2C_{44})(C_{11}-2C_{44}-C_{13})$. If this expression is positive then medium has six additional longitudinal normals. In this case spherical components of longitudinal normals are determined by equations

$$tg\,\theta = \pm\sqrt{(C_{33}-C_{13}-2C_{44})/(C_{11}-C_{13}-2C_{44})},$$

and

$$\phi = \frac{n\pi}{3}\,(n=0,1,...,5).$$

3.5. *Longitudinal normals for transversely isotropic media*

For transversely isotropic media the symmetry axis with infinite multiplicity is always a symmetry axis. Also all directions which lie in the plane perpendicular to this axis are longitudinal normals. The existence of other longitudinal normals depends on the sign of expression $(C_{33}-C_{13}-2C_{44})(C_{11}-2C_{44}-C_{13})$. If this expression is positive then all directions with spherical coordinates fulfilling the equation

$$tg\theta = \pm\sqrt{(C_{33}-C_{13}-2C_{44})/(C_{11}-C_{13}-2C_{44})}$$

are longitudinal normals.

4. Acoustic axes for media with triclinic symmetry

In this section we shall prove that for any elastic medium the number of acoustic axes can not be bigger than 132. In order to do this the following lemmas will be proved:

Lemma 1

If a bivariate polynomial function F of degree n defined on R^2 has finite number of solutions, then this number can not be larger than $n(n-1)$.

Proof of lemma 1

If in a set of values of $F(x, y)$ one can find at least one pair with different signs then the equation

$$F(x, y) = 0 \tag{41}$$

has infinite number of solutions. If a continuous function $F(x, y)$ is negative for a pair (x, y), then one can find a circular region K_- for which $F(x, y) < 0$. Analogously, if a continuous function $F(x, y)$ is positive for a pair (x, y), then one can find a circular region K_+ for which $F(x, y) > 0$. Areas K_- and K_+ are disjoint. One may find infinitely many disjoint lines joining points of K_- and K_+. The continuity of the function F implies that each from these lines contains a point for which its value equals zero, and Eq. (41) has infinitely many solutions. This means that the function F is nonpositive or nonnegative. Thus, all its roots are also its extremes. All roots of function F fulfils the set of equations

$$F(x, y) = 0, \partial F(x, y)/\partial x = 0. \tag{42}$$

The first of Eqs. (42) has the degree n, the second one cannot have degree larger than $(n-1)$. Since we assumed that $F(x, y)$ has finite number of zeros, the set of bivariate polynomial equations has a finite number of solutions. From the Bezout theorem [12] it follows that their number cannot be greater than $n(n-1)$. Thus the lemma 1 is proved.

Squares of phase velocities of sound are eigenvalues of propagation matrix which elements depend on the direction of propagation $\mathbf{n}$. To determine the phase velocities one has to solve the characteristic equation (4)

$$
\begin{aligned}
&c^3 - \left[\Gamma_{11}(\mathbf{n}) + \Gamma_{22}(\mathbf{n}) + \Gamma_{33}(\mathbf{n})\right] c^2 + \left[\Gamma_{11}(\mathbf{n})\Gamma_{22}(\mathbf{n}) + \Gamma_{11}(\mathbf{n})\Gamma_{33}(\mathbf{n})\right. \\
&+\Gamma_{22}(\mathbf{n})\Gamma_{33}(\mathbf{n})] \left. -\Gamma_{23}^2(\mathbf{n}) - \Gamma_{12}^2(\mathbf{n}) - \Gamma_{13}^2(\mathbf{n})\right] c \\
&+ \left[\Gamma_{11}(\mathbf{n})\Gamma_{23}^2(\mathbf{n}) + \Gamma_{33}(\mathbf{n})\Gamma_{12}^2(\mathbf{n}) + \Gamma_{22}(\mathbf{n})\Gamma_{13}^2(\mathbf{n})\right. \\
&\left.-\Gamma_{11}(\mathbf{n})\Gamma_{22}(\mathbf{n})\Gamma_{33}(\mathbf{n}) - 2\Gamma_{12}(\mathbf{n})\Gamma_{13}(\mathbf{n})\Gamma_{23}(\mathbf{n})\right] = 0 \,.
\end{aligned}
\tag{43}
$$

Eq. (43) has a double root for $\mathbf{n}$ directed along each acoustic axis $\mathbf{n}_a$. A cubic equation has double root, if, and only if, its discriminant $\Delta(\mathbf{n})$ vanishes. For Eq. (43) the discriminant reads

$$\Delta(n_1, n_2, n_3) = q^2(n_1, n_2, n_3) + \frac{4}{27}p^3(n_1, n_2, n_3), \tag{44}$$

where

$$
\begin{aligned}
q(n_1&, n_2, n_3) = \\
&= (\Gamma_{11}\Gamma_{22}\Gamma_{33} - \Gamma_{11}\Gamma_{23}^2 - \Gamma_{33}\Gamma_{12}^2 + 2\Gamma_{12}\Gamma_{13}\Gamma_{23} - \Gamma_{22}\Gamma_{13}^2) \\
&-2(\Gamma_{11} + \Gamma_{22} + \Gamma_{33})^3/27 - (\Gamma_{11} + \Gamma_{22} + \Gamma_{33}) \\
&\times \left[\Gamma_{23}^2 + \Gamma_{12}^2 + \Gamma_{13}^2 - \Gamma_{11}\Gamma_{22} - \Gamma_{11}\Gamma_{33} - \Gamma_{22}\Gamma_{33}\right]/3 \,,
\end{aligned}
\tag{45}
$$

and

$$p(n_1, n_2, n_3) = - \left[(\Gamma_{11} + \Gamma_{22} + \Gamma_{33})^3/3 + \left(\Gamma_{23}^2 + \Gamma_{12}^2 + \Gamma_{13}^2 - \Gamma_{11}\Gamma_{22} \right. \right.$$
$$\left. \left. - \Gamma_{11}\Gamma_{33} - \Gamma_{22}\Gamma_{33} \right) \right]. \tag{46}$$

Each element of $\Gamma(n_1, n_2, n_3)$ is a homogeneous polynomial function of variables n_1, n_2, n_3 of the second order. On the other hand, the discriminant (44) is a homogeneous function of variables $\Gamma_{ij}(n_1, n_2, n_3)$ $(i, j = 1, 2, 3)$ of the sixth order. Hence, $\Delta(\mathbf{n})$ is a homogeneous function of n_1, n_2, n_3 of the twelfth order. If the number of acoustic axes is finite, then one can find a plane without acoustic axes. One may choose a Cartesian coordinate system where this plane is perpendicular to z axis. In this coordinate system, $n_3 \neq 0$. In variables $x = n_1/n_3$, $y = n_1/n_3$ the discriminant $\Delta(\mathbf{n})$ is a function of x and y of the twelfth order. From the Lemma 1, it follows that the number of acoustic axes of an elastic medium cannot be greater than $11 \times 12 = 132$.

Acknowledgments

This work was supported by the Committee for Scientific Research (KBN) of Poland grant No. 2P03B 03818. We also would like to acknowledge Krzysztof Wojciechowski of Institute of Molecular Physics of Polish Academy of Sciences (Poznań) for correcting our calculations and some inaccurate statements.

Appendix: Elastic constants and longitudinal normals for some monoclinic media

Below we list the values of elstic constants (in 10 GPa) for selected monoclinic compounds.

	Compound	C_{11}	C_{22}	C_{33}	C_{44}	C_{55}	C_{66}
1	$NaF[Si_2O_6]$	18.6	23.4	18.1	6.92	4.74	5.1
2	$C_{10}H_8$	0.78	1.19	0.99	0.33	0.415	0.21
3	$CoSO_4*7H_2O$	3.35	3.71	3.78	0.6	1.01	0.58
4	$CaMg(SiO_3)_2$	20.4	23.8	17.5	6.75	7.05	5.88
5	$Ca(Mg,Al)[(Al,Si)SiO_6]$	15.4	21.1	15.0	6.39	5.23	6.22
6	$FeSO_4*7H_2O$	3.49	3.60	3.76	0.64	0.96	0.56
7	$Na_2S_2O_3$	3.26	4.43	3.02	0.56	0.60	1.19
8	$K[AlSi_3O_8]$	6.6	12.2	17.1	1.43	3.61	2.38

	C_{12}	C_{13}	C_{16}	C_{23}	C_{26}	C_{36}	C_{45}
1	7.1	6.8	1.0	6.3	0.9	2.1	0.8
2	0.445	0.34	-0.06	0.23	-0.27	0.29	-0.05
3	25.4	11.8	9.83	0.72	6.12	-1.23	0.16
4	8.8	8.4	-1.9	4.86	-2.0	-3.4	-1.1
5	3.7	5.7	1.5	3.0	1.4	1.2	-0.9
6	1.74	2.08	-0.20	1.72	-0.19	-0.14	0.01
7	1.57	1.73	0.31	1.24	1.27	-0.66	-0.35
8	2.6	4.4	-0.3	1.9	-1.5	-1.3	-0.1

For these compounds we calculated components of longitudinal normals. Their number agreemee with Table 1.

Compound	Components of longitudinal normals
$NaF[Si_2O_6]$	[0.8444, -0.5358, 0][0.1476, 0.989, 0][0,0,1],
$C_{10}H_8$	[0.3688, -0.9295, 0], [0.7039, 0.7103, 0], [-0.7174, 0.3825, 0.5821][0.7174, -0.3825, 0.5821],[0,0,1]
$CoSO_4{}^*7H_2O$	[0.159, -0.9873, 0][0.8646, -0.5025, 0][0.9795, 0.2016, 0][0.6413, 0.7673, 0][0,0,1]
$CaMg(SiO_3)_2$	[0.4718, -0.8817, 0][0.8023, 0.5969, 0][0,0,1],
$Ca(Mg,Al)[(Al,Si)SiO_6]$	[0.8735, -0.4869, 0][0.2712, 0.9625, 0][0, 0, 1]
$FeSO_4{}^*7H_2O$	[0.2662, -0.9639, 0][0.7604, -0.6495, 0] [0,0,1]
$Na_2S_2O_3$	[0.6298, -0.7768, 0],[0.4807, 0.8769, 0],[0, 0, 1]
$K[AlSi_3O_8]$	[0.2495, -0.9684, 0],[0.8716, 0.4903, 0],[0, 0, 1]

References

1. A.G. Every, *Phys. Rev.* **B 22** 1746 (1980).
2. Z. Sakadi, *Proc. Phys.-Math. Soc. Jap.* **23**, 539 (1941).
3. F.E. Borgnis, *Phys. Rev.* **98**, 1000 (1955).
4. A.G. Khatkevich, *Kristallograf.* **7**, 472 (1964).
5. F. I. Fedorov, *Vestnik MGU* **6**, 36 (1964).
6. A. G. Khatkevich, *Kristallograf.* **9**, 690 (1964).
7. C. Truesdell, *J. Acoust. Soc. Am.* **40,** 729 (1966).
8. I. Kolodner, *J. Acoust. So. Am.* **40,** 730 (1966).
9. F. I. Fedorov, *Theory of Elastic Waves in Crystals*, (Plenum Press New York 1968).
10. A. G. Khatkiewicz, *Kristallograf.* **22**, 1232 (1977).
11. K. Helbig, *Geophysics* **58** 680 (1993).
12. R. Bix, *Conics and Cubics: Concrete Introduction to Agebraic Curves* (Springer, Heidelberg, 1998).

NONLINEAR RESPONSE OF SUPERCONDUCTOR BOLOMETER TO PHONON FLUXES

B. A. DANILCHENKO

Institute of Physics
National Academy of Sciences of Ukraine
Prospect Nauki 46, 252650 Kiev, Ukraine

CZ. JASIUKIEWICZ

Chair of Physics
Rzeszów University of Technology
ul. Pola 2, PL-35-959 Rzeszów Poland

T. PASZKIEWICZ AND S. WOLSKI

Institute of Physics,
University of Rzeszów,
ul. Rejtana 16A, PL-35-310 Rzeszów, Poland
E-mail: tapasz@univ.rzeszow.pl

We derived the set of ordinary nonlinear nonautonomous differential equations describing the response of superconducting bolometers to phonon pulses. Experiments in which induced voltage is measured are considered. For the phonon flux obtained in Monte Carlo computer experiments we calculated the bolometer response using nonlinear and linearized differential equations. The comparison indicate that nonlinearity is essential.

1. Introduction

Heat pulses become the standard tool used for studying properties of superfluid He, crystalline dielectrics, semiconductors and superconductors. Pulse lengths used are short enough to allow one to resolve different kinds of phonons and propagation directions in millimeter sized samples. Wigmore and collaborators discussed the use of thin film superconducting bolometers (TFSB) for detecting phonon pulses in the regime in which their lengths are shorter than the bolometer response time [1].

Usually linear changes of TFSB under condition of *constancy of the biasing current* are considered (cf. [1,2]). In particular in a series of papers Sherlock and Wyatt studied this linear regime in a very detailed way [3,4,5].

In our previous paper we studied the response of superconducting bolometers solving *linearized* differential equation for the resistance [6]. In this note we shall show that if the phonon flux $W(t)$ is known one may calculate the bolometer signal. To show an important role of nonlinearity in the energy exchange between the bolometer and the thermal bath we compare results obtained with the use of nonlinear and linearized differential equations.

2. Characteristics of Superconducting bolometers

Consider a metallic film bolometer and a specimen in the contact with thermal bath of temperature T_0. The thermal capacity of the bolometer at the ambient temperature is $C = C(T_0)$. The thermal contact between the film and thermal bath is characterized by the heat leak constant G. The absorption of phonon flux of the power $W(t)$ falling upon the bolometer receiving surface causes changes in temperature of the bolometer.

Here we recall the main properties of metallic film bolometers [7]. The simplest bolometer is a strip of a metal evaporated onto the surface of the sample. The resistance of the bolometer can be controlled by the width, length, and thickness of the film as well as by the deposition conditions.

The absorption of phonon flux of the power $W(t)$ falling upon the bolometer receiving surface causes the change $\Delta T(t) = [T(t) - T_0]$ in the temperature of the bolometer. Due to this temperature rise the film resistance grows. As the biasing current is hold constant, in experiments with TFSB one measures changes of voltage $U(t)$. The largest sensitivity is achieved for superconducting films. Therefore, a superconducting detector is cooled precisely to its transition temperature T_c, hence $T_0 = T_c$. Under such conditions, the slight rise in temperature caused by a phonon flux produces a significant, varying in time, change $\Delta R(t) = [R(T(t)) - R(T_0)]$ in resistivity. Since the biasing current is held constant, changes in resistivity bring forth variations of voltage $U(t)$.

For superconductors one may assume that $\Delta R(t) = \alpha [T(t) - T_0]$. As a rule $10 \le \alpha \le 10^2$ Ω/K. Introduce

$$R_0 \equiv \alpha T_0, \rho \equiv R_0/R(T_0). \tag{1}$$

In our experiments[8] $T_0 \approx 3.5$ K and $R(T_0) \simeq 10$ Ω, so $\rho \simeq 5 \times 10^{-2}$. Long wavelength acoustic phonons were generated in GaAs specimen by $1 \times 1 \text{mm}^2$ Au film heated by electric pulses, each of 75 ns duration. Beams of phonons propagated in the (001) direction. The thickness of substrate was 3.4 mm, this allowed to resolve LA and TA phonon modes. By the direct evaporation of indium film the bolometer of sensitive area 2×0.2 mm^2 was fabricated. Phonons were detected by this superconducting detector.

Introduce a characteristic time [1,2] τ and a characteristic current [2] I_m

$$\tau = \frac{C}{GT_0^{n-1}}, \quad I_m = \sqrt{\frac{GT_0^{n-1}}{\alpha}}. \tag{2}$$

Using the experimental values of G and α we may estimate the value of I_m. For the film surface area of 0.1 mm^2 we obtain $I_m \simeq 10^{-2}$ A. Taking the polarizing current $I_b = 10^{-4}$ A, we obtain $I_b/I_m \simeq 10^{-2}$.

3. Differential equations describing the dynamics of TFSB

In this section we write down the nonlinear nonautonomous ordinary differential equation of the first order which describes nonlinear response of a superconducting bolometer to an external source of phonons, as well as a linearized version of it.

3.1. *Nonlinear differential equations describing the dynamics of TFSB*

Our equations depends on the *dimensionless* voltage

$$u(\tilde{t}) \equiv U_b(\tau\tilde{t})/U_0, \tag{3}$$

current

$$i(\tilde{t}) \equiv I_b(\tau\tilde{t})/I_m \tag{4}$$

and on *dimensionless* time

$$\tilde{t} \equiv t/\tau. \tag{5}$$

Further, introduce the dimensionless power $w_u(\tilde{t})$

$$w_u(\tilde{t}) \equiv \left[\frac{W(\tau\tilde{t})}{I_m U_0}\right]. \tag{6}$$

The dimensionless voltage $u(\tilde{t})$ obeys the following nonlinear nonautonomous ordinary differential equation of the first order

$$\frac{du(\tilde{t})}{d\tilde{t}} - i^2 u(\tilde{t}) + \left\{\left[(u(\tilde{t}) - \rho + 1)^n - 1\right]\right\} = i w_u(\tilde{t}). \tag{7}$$

If the Kapitza boundary resistance between the film and the substrate dominates $n = 4$, whereas if electron phonon interaction in the film prevails n is 5 or 6, depending on the temperature range.

The solution of Eq. (7) depends on two dimensionless parameters i and ρ as well as on dimensionless, and generally non-stationary, external power term $w_u(\tilde{t})$.

3.2. Differential equation for the voltage – the linear case

Assume that $n = 1$. In this case Eq. (7) simplifies [6]

$$\frac{du(\tilde{t})}{d\tilde{t}} + \frac{\tau}{\Lambda} u(\tilde{t}) - \rho = iw_u(\tilde{t}), \tag{8}$$

where Λ is the effective relaxation rate [2]

$$\Lambda = \tau / \left(1 - i^2\right). \tag{9}$$

This time is positive when $I_m > I_b$. Further we see that $\lim_{i \to 1} \Lambda = \infty$. The elongation of the relaxation rate Λ with growing polarizing current I_b was observed by Fuson [2]. Solutions with negative Λ will include a positive exponential factor. Thus, as t becomes large the resistance difference $\delta R_b(t)$ will increase without limit, i.e., the bolometer resistance $R_b(t)$, will fall outside the range of linear dependence on temperature. Therefore, the bias current I_b have to be smaller than I_m (i.e. $0 \leq i < 1$). Let us mention that in our experiments [8] $\Lambda \approx \tau$.

4. Calculation of the bolometer response to Monte Carlo energy flux

Monte Carlo simulations allows us to calculate the phonon energy flux $w_u(\tilde{t})$. We assumed that the source of phonons is Planckian with temperature 10 K, and we accounted for the finite duration of current pulses 75 ns, as well as for the focusing effects, elastic scattering on isotopes and anharmonic down-conversion processes [9].

Solving the differential equations (8) and (7) for a given external power one may calculate the bolometer response. This procedure we shall call the convolution. To solve our nonlinear nonautonomous ordinary differential equations the standard implicit Euler method was used [10]. We considered time depending signals $u(\tilde{t})$ being only a small excursion from the stationary state u_{stat} [11].

We compared result of convolution of the Monte Carlo phonon flux with the bolometer signal obtained in experiments performed by Danilchenko and collaborators [8] using linear (Eq. (8)) and nonlinear differential equations (Eq. (7)). Results are depicted in Fig. 1. The best fitting of the *calculated* bolometer signal in the nonlinear regime to the results of Danilchenko et al experiments is obtained for $\tau = 1.35$ μs and for $4 < n < 5$ [11]. We notice that linearized equation gives very broad bolometer signal u_l. The part of curve representing $u_l(\tilde{t})$ which begins at the maxiumum has an inflection point. The signal obtained with the use of nonlinear differential equation, $u_{nl}(\tilde{t})$, is narrow and the part of the curve representing $u_{nl}(\tilde{t})$ has

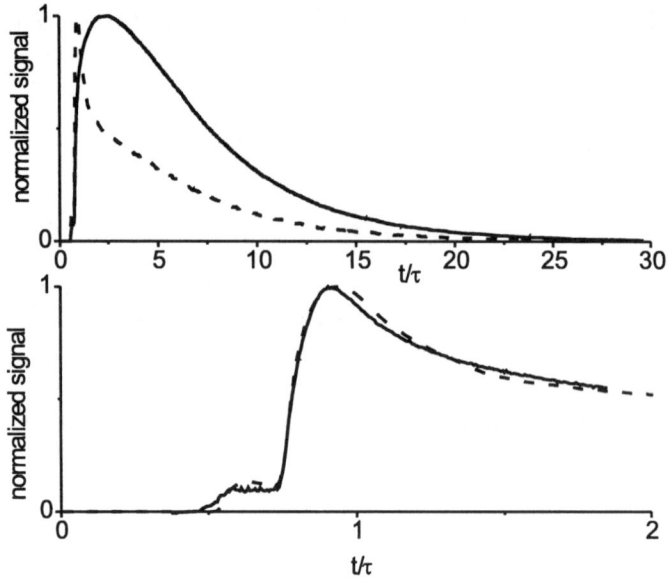

Figure 1. Upper panel: result of convolution for $n = 5$ (dashed line) and for $n = 1$ (solid line). Lower panel: The experimentally established signal of bolometer [8] is indicated by solid line. The result of convolution procedure for $n = 5$, is shown by dashed line. We taken $\tau = 1.35$ μs

no inflection points right to the maximum. The tails of both signals $u_l(\tilde{t})$ and $u_{nl}(\tilde{t})$ have common asymptotics. Simply, after a long enough lapse of time, the state of the bolometer is very close to the stationary point, hence the linear approximation of Eq. (7) is correct.

Acknowledgement

This work was supported by Poland National Research Committee under grant No. 2 P03B 038 18.

References

1. Edwards S C, Hamid bin Rani and Wigmore J K, 1989 *J. Phys. E: Sci. Instrum.* **22** 528
2. Fuson N 1949 *J. Appl. Phys.* **20** 59
3. Sherlock W A, Wyatt A F G, 1983 *J. Phys. E: Sci. Instrum.* **16** 669
4. Sherlock W A, Wyatt A F G, 1983 *J. Phys. E: Sci. Instrum.* **16** 672

5. Sherlock W A, 1984 *J. Phys. E: Sci. Instrum.* **17** 386

6. B.A. Danilchenko, Cz. Jasiukiewicz, T. Paszkiewicz and S. Wolski, *Response of superconductor bolometer to phonon fluxes in linear regime*, to be published

7. J.P. Wolfe, *Imaging Phonons, Acoustic Wave Propagation in Solids*, Cambridge University Press, Cambridge 1998.

8. Danilchenko B.A., Lutset M.O., Porshin V.N, 1987 *Fizika Niskikh Temp.* **13** 1289; Asche M, Kleinert P, Hey R, Kostial H, Danilchenko B, Klimashov A and Roshko S 1996 *Proceedings of 9th International Conference on Hot Carriers in Semiconductors*, ed K Hess and L Leburton (World Scientific, Singapore) p 85

9. Gańcza W M, Obukhov I A, Paszkiewicz T and Danilchenko B A 2001 *Comp. Meth. Sci. Technol.* **7** 7

10. W.H. Press, S.A. Teukolsky, W.T. Vetterling, B.P. Flannery, *Numerical Recipes in C, The Art of Scientifc Computing, Second Edition, Cambridge University Press, 1992*

11. B.A. Danilchenko, Cz. Jasiukiewicz, T. Paszkiewicz and S. Wolski, *Nonlinear response of superconductor bolometer to phonon fluxe*, in preparation

PROPAGATION OF WAVES THROUGH ANHARMONIC DEFECTS

P. ZIELIŃSKI

The H. Niewodniczański Institute of Nuclear Physics,
ul. Radzikowskiego 152, 31-342 Kraków, Poland

Dynamical systems with localized anharmonicity are generally described by Volterra Integro-Differrential Equations, which in cases of dispersionless infinite harmonic parts reduce to differential equations and in cases of finite dispersionless harmonic parts to delayed differential equations. The simplest examples of all three cases are given. Specific results for the surface atom and for an anharmonic stub resonator are presented. Simplified characteristics of chaotic motion are introduced. Typical memory kernels for integro differential equations in non-dispersionless media are derived.

1. Introduction

Harmonic approximation is justified in the description of crystal lattice dynamics as long as the displacements of atoms from their equilibrium positions are small. Such are usually vibrations of atoms in bulk crystals. Large-amplitudes atomic motions may, however, occur and really occur in the vicinity of surfaces, interfaces, grain boundaries, dislocations and other defects. The atoms placed there have usually less neighbours than the bulk atoms and, therefore, more liberty to move. In some cases the amplitude of atomic vibration depends on frequency. This happens e.g. at surface resonances and at frequencies corresponding to edge singularities of phonons' density of states. Indeed, whenever the surface resonance is sharp enough the excitations arriving from the bulk pump energy to the localised vibration, which then can acquire quite a large amplitude. A detailed study of the simplest model showing both: the surface resonances and the edge singularities has been studied in ref. [1]. In particular, it has been shown that the amplitude of the vibration of the surface atoms due to a near-field can be about one order of magnitude larger than the amplitude of the atomic displacements in the bulk even in the case of a perfect clean surface. The weaker the interaction between the surface atoms and the bulk the sharper are the resonances and the larger are amplitudes of the vibration of surface atoms [1].

Thus, even though the vibrations might be small in the bulk, they may well no longer be small in the surface region. Anharmonic effects are, therefore, to be expected, in the first place, as a manifestation of large-amplitude motions of atoms placed in the vicinity of defects. A weak surface anharmonicity gives rise to some new elementary excitations, which have a form of extended waves [2,3] and solitary waves [4]. These, however, do not exhaust all the expected effects of the surface anharmonicity such as generation of higher harmonics, deterministic chaos, surface structural instabilities and, finally, complete desorption of atoms from the surface.

The above discussion indicates that there exist quite a number of condensed matter systems which are harmonic in the majority of their volumes, whereas anharmonic interactions concern but a very restrained, spatially limited regions. The systems showing such a property can be treated within what we call "model of local anharmonicity". In what follows we show that the assumption of the local anharmonicity gives rise to a variety of effective equations of motion, whose form depends on the geometry and on physical properties of the harmonic part of the system. The interest of studying such systems is twofold. Firstly, the results provide theoretical predictions for measurable quantities accessible in experiments. Secondly, theoretical models and their real counterparts constitute an important class of dynamical systems interesting for their chaotic dynamics.

In the majority of experiments of molecular and solid state physics the system under investigation is subject to an oscillatory measuring field or radiation. In the case of crystal surface studies the measuring field may be provided from outside of the crystal by an electromagnetic radiation or by a beam of particles such as electrons [5] or helium atoms [6]. The particles are, however, related with wave motion being so called matter waves. Femtochemistry, i.e. surface chemical reactions stimulated by short laser pulses also falls into this category of experiments [7]. On the other hand, the dynamics of surface may be studied by sending a radiation from inside of the crystal. The ballistic phonons [8] constitute the best example of such a probe. In all such cases the property of interest is the response of the system to the radiation of the given frequency (energy) and amplitude (square root of intensity).

As long as the system is harmonic the response is always proportional to the amplitude of the measuring field and monochromatic, with the frequency equal to the applied frequency. This kind of response is entirely determined by the frequency-dependent Green function of the system [9]. Contrary to that the general anharmonic systems are non-integrable and may exhibit an irregular chaotic motion [10]. In particular, the response to a well defined applied frequency is sometimes given by a whole spectrum which, moreover, depends on the applied amplitude. Only in special conditions can such a spectrum be approximated by a series of several harmonics or subharmonics [11]. Generally, however, the investigation of the anharmonic system requires explicit solution of their equations of motions. The resulting time series can then be used to evaluate measurable quantities. They can also provide predictions for dynamical characteristics of the system: correlation functions, characteristics of attractors, Lyapunov exponents etc. [10, 11].

In the present work we show that the effective equations of motion of systems in which anharmonicity is concentrated in a spatially limited region, e. g. near the surface or other defects, take in only those degrees of freedom which are explicitly involved in the anharmonic interactions. A similar technique has been used in problems of gas-surface interaction where the effective equations of motion have been obtained by elimination of the degrees of freedom of the substrate [12]. The cost of the elimination of the harmonic degrees of freedom is that the resulting equations of motion are no longer differential equations, but, except for dispersionless bulk crystals, transform into Volterra Integro-Differential Equations (VIDE). In the following sections we give explicit forms of such equations of motion for effectively one-dimensional semi-infinite harmonic crystals with anharmonic defects.

2. Effective equations of motion for locally anharmonic systems

To be specific let us consider a mechanical system. Its state is defined by a vector $\mathbf{u}(t)$ comprising the coordinates of all its particles and by the corresponding velocities $\dot{\mathbf{u}}(t)$. The coordinates can be divided into two parts. The part $\mathbf{u}_a(t)$ describes the positions of the particles involved explicitly in the anharmonic interactions and will be called anharmonic domain D_a. The remaining part $\mathbf{u}_h(t)$, or harmonic domain D_h, gives the instantaneous configuration of the part of the system interacting exclusively through harmonic potentials. Thus, the coordinate vector may be arranged in the direct sum $\mathbf{u}(t) = \mathbf{u}_a(t) \oplus \mathbf{u}_h(t)$.

The Newton equations of motion of the system can be written in the following matrix form.

$$\frac{d^2}{dt^2} \begin{bmatrix} \mathbf{u}_a \\ \mathbf{u}_h \end{bmatrix} = - \begin{bmatrix} H_a, V \\ V^+, H_h \end{bmatrix} \begin{bmatrix} \mathbf{u}_a \\ \mathbf{u}_h \end{bmatrix} + \begin{bmatrix} \mathbf{B}(\mathbf{u}_a) \\ 0 \end{bmatrix} + \begin{bmatrix} \mathbf{f}_a \\ \mathbf{f}_h \end{bmatrix} \tag{1}$$

The matrix $H = \begin{bmatrix} H_a, V \\ V^+, H_h \end{bmatrix}$ representing the linear part of the equations of motion

is divided into sectors H_a and H_h corresponding to the domains D_a and D_h. The coupling matrix V involves usually a narrow range of indices l in the interface between the harmonic and anharmonic part of the system. The vector $\mathbf{B}(\mathbf{u}_a)$

represents the anharmonic forces which are a non-linear function of the displacements $\mathbf{u}_a$. Linear forces applied to the system from outside are expressed by the vector $\begin{bmatrix} \mathbf{f}_a \\ \mathbf{f}_h \end{bmatrix}$.

The elimination of the harmonic degrees of freedom resembles the projection technique due to Nakajima and Zwanzig [13]. It amounts to solving the harmonic part of eq. (1) and to inserting the resulting $\mathbf{u}_h$ (t) into the anharmonic part. The equations of motion for the displacements $\mathbf{u}_a$ (t) then transform into the following system of Volterra integro-differential equations (VIDE):

$$\ddot{\mathbf{u}}_a(t) = -H_a \mathbf{u}_a(t) + \mathbf{B}\left(\mathbf{u}_a(t)\right) - \int_0^t V\tilde{K}(t-\tau)\ \left[-V^+\mathbf{u}_a(\tau) + \mathbf{f}_h(\tau)\right]\ d\tau + \mathbf{f}_a(t) \quad (2)$$

or, equivalently

$$\ddot{\mathbf{u}}_a(t) = -H_a \mathbf{u}_a(t) + \mathbf{B}\left(\mathbf{u}_a(t)\right) - \int_0^t VK(t-\tau)\ \left[-V^+\dot{\mathbf{u}}_a(\tau) + \dot{\mathbf{f}}_h(\tau)\right]\ d\tau + \mathbf{f}_a(t), \quad (3)$$

with

$$\tilde{K}(t) = \dot{K}(t). \quad (4)$$

The kernel $\tilde{K}(t)$ is in fact the time- and site-dependent Green function of the harmonic subsystem. Its explicit form can be obtained from the ω-dependent Green function by its inverse Fourier transform. The ω-dependent Green functions are known for a number of harmonic subsystems of various geometries [14]. The kernel $K(t)$ is a generalised retarded damping. The kernels $\tilde{K}(t)$ and $K(t)$ can be evaluated by the summation/integration over the eigenstates $\mathbf{w}_{qj}(l)$ of the harmonic subsystem i.e. those corresponding to the submatrix H_h:

$$\tilde{K}(l,l',t) = \sum_{qj} \mathbf{w}_{qj}(l)\mathbf{w}*_{qj}(l')\frac{\sin(\omega_{qj}t)}{\omega_{qj}} \quad (5)$$

$$K(l,l',t) = -\sum_{qj} \mathbf{w}_{qj}(l)\mathbf{w}*_{qj}(l')\frac{\cos(\omega_{qj}t)}{\omega_{qj}^2}, \quad (6)$$

where ω_{qj} are the corresponding eigenfrequencies. The numbering of the eigenstates by the wave vectors $\mathbf{q}$ and by the branch labels j is appropriate for infinite and semi-infinite harmonic subsystems.

3. Point-like defect in a one dimensional dispersionless mediuim

The simplest example of a local anharmonicity is a mass, say M, placed in an anharmonic potential and coupled to a monomode dispersionless string. This is a model of an anharmonic surface in a harmonic crystal. The string is supposed to extend in the range $x = 0 \dots \infty$. The instantaneous displacement of the string will be denoted by $u(x,t)$. Consequently, the displacement of the mass M is $u(0,t)$. The anharmonic potential is a function $V(u(0,t))$.

If an external force $f_0(t)$ is applied to the mass M one gets the following equations of motion

$$\frac{\partial^2 u(x,t)}{\partial t^2} = c^2 \frac{\partial^2 u(x,t)}{\partial x^2}, \text{ for } x > 0, \tag{7a}$$

$$\frac{\partial^2 u(0,t)}{\partial t^2} = \frac{1}{M}\left(-\frac{\partial V(u(0,t))}{\partial u(0,t)} + T \frac{\partial u(x,t)}{\partial x}\Big|_{x=0} + f_0(t) \right), \text{ for } x = 0. \tag{7b}$$

Here $c = \sqrt{T/\rho}$ is the sound velocity in the string, T is the stiffness coefficient and ρ is the linear density of the string.

Since the string is dispersionless, every wave of whatever form propagates in it without changing its form. Therefore, every solution of the eq. (7a) can be written as a sum of two waves: one traveling to the right $u(t,x) = u(\zeta_-)$ with $\zeta_- = ct-x$ and the other traveling to the left $u(t,x) = u(\zeta_+)$ with $\zeta_+ = ct+x$ [15]. This is so called d'Alembert solution of wave equation. Assuming that there is no wave propagating towards the mass the only existing solution within the string is an outgoing wave $u(t,x) = u(\zeta_-)$. The equation (7a) then is satisfied everywhere for $x > 0$. Further, since the displacement $u(t,x) = u(\zeta_-)$ depends effectively on one variable only, the space derivative may be expressed by the time derivative.

$$\frac{\partial u(t,x)}{\partial x} = -\frac{1}{c}\frac{\partial u(t,x)}{\partial t} \tag{8}$$

Consequently, the boundary condition (eq. (7b)) for $x = 0$ transforms into the following ordinary differential equation

$$M\frac{\partial^2 u(t)}{\partial t^2} + \frac{T}{c}\frac{\partial u(t)}{\partial t} + \frac{\partial V(u)}{\partial u}\Big|_{u=u(t)} = f_0(t). \tag{9}$$

The latter equation describes a damped forced anharmonic oscillator or, in the case of the potential $V(u)$ given by a polynomial, Duffing's oscillator [16] with the damping constant $\Gamma = T/c$. The role of the harmonic degrees of freedom is now reduced to an effective damping.

The same reasoning can be adopted to derive the equation of motion for the surface atom under an incident wave coming from the bulk. Then the solution will be the sum of the incident wave $u_i(x,t) = u_i(\zeta_+) = u_i(ct+x)$ and of an outgoing or reflected wave $u_r(t,x) = u_r(\zeta_-)$. As before eq. (7a) is satisfied for both waves, whereas eq. (7b) takes on the following form.

$$M\frac{\partial^2(u_r(0,t)+u_i(0,t))}{\partial t^2} + \frac{T}{c}\frac{\partial(u_r(0,t)-u_i(0,t))}{\partial t} + \frac{\partial V(v)}{\partial(v)}\Big|_{v=u_r(0,t)+u_i(0,t)} = 0, \quad (10a)$$

or with the substitution $u_i(0,t)+u_r(0,t) = u(t)$

$$M\frac{\partial^2 u(t)}{\partial t^2} + \frac{T}{c}\frac{\partial u(t)}{\partial t} + \frac{\partial V(u)}{\partial u}\Big|_{u=u(t)} = \frac{2T}{c}\frac{\partial u_i(t)}{\partial t}. \quad (10b)$$

Here again one ends up with an ordinary differential equation with either the unknown function $u_r(0,t)$, describing the reflected wave or, equivalently with the unknown function $u(t)$ corresponding to the total effective displacement of the surface atom. The effective force due to the incident wave is proportional to the time derivative of the incident wave as given at the right hand side of eq. (10b). The most common form of the incident wave is monochromatic one

$$u_i(ct+x) = a\,\sin\left[\frac{\omega}{c}(ct+x)+\varphi\right] \qquad ct+x \geq 0 \qquad (11)$$
$$u_i(ct+x) = 0, \qquad\qquad\qquad ct+x < 0.$$

The solution of the analogous problem with a harmonic potential can be easily found with the use of the Fourier transform for each frequency ω separately [9]. The present nonlinear equation must, however, be solved numerically unless restrictive approximations are made. Linearisation or limitation to a few harmonics preclude e.g. any instance of chaos, which is a priori to be expected for certain ranges of the model parameters. To enable any numerical treatment, the actual form of the wave front should be defined. The condition of the wave vanishing for $ct+x < 0$ in eq. (11) is the simplest choice implying an abrupt switching on of the perturbation.

The simplest assumption about the potential $V(u)$ is to give it a polynomial form.

$$V(u) = \tfrac{1}{2}Au^2 + \tfrac{1}{3}Cu^3 + \tfrac{1}{4}Bu^4. \tag{12}$$

The proper choice of the coefficients A, C and B allows one to obtain a double-well potential characteristic of the Duffing oscillator [16,17]. Such a potential is adequate to a model of the surface atom participating in a surface reconstruction phase transition [18] as well as for some problems related to desorption [1,3].

With the incident wave of eq. (11) and with the local potential of eq. (5) the explicit differential equation for the unknown co-ordinate $u(t)$ of the surface atom and, at the same time, for the reflected wave $u_r(\zeta)$, reads

$$M\frac{\partial^2 u}{\partial t^2} + \frac{T}{c}\frac{\partial u}{\partial t} + Au + Cu^2 + Bu^3 = \omega\frac{T}{c}a\cos(\omega t + \varphi) \tag{13}$$

Solutions of the equations (9) and (10) provide the time-dependent response of the surface atom subject to an external force and to the irradiation by a phonon coming from the bulk respectively. Once the corresponding solutions are known the quantitative estimates of experimental observables can be computed.

4. Chaos and intermittency

To get the time-series needed to characterise the response of the surface atom for arbitrary values of the model parameters we solved eqs. (9) and (10) with the use of a simplectic integrator based on the Verlet algorithm and on a step adaptive Runge-Kutta-Fehlberg method [19]. This method allowed us to obtain stable solutions of sufficient lengths so as to obtain reliable frequency spectra of the corresponding time series as well as satisfactory estimates for the Lyapunov exponent.

In the case of the system with the external applied force $f_0(t) = f_0(\omega)\cos(\omega t)$

(eq. (9)) the quantity of interest is the efficiency of generation of harmonics and subharmonics. The efficiency E_n is defined as the ratio $E_n = u(n\omega)/f_0(\omega)$, where $u(n\omega)$ is the n-th Fourier component of the displacement $u(t)$ satisfying eq. (9). Integer values of index n correspond to harmonics and fractional values to subharmonics. Analogous quantities relevant in the phonon scattering experiment are reflection coefficients for particular harmonics $R_n = u_r(n\omega)/a$ defined as the ratio of the n-th Fourier component of the reflected wave to the incident amplitude a. In the present model the reflected wave $u_r(t)$ is calculated from eq. (13). The quantity R_1 is the ordinary reflection coefficient considered in models of harmonic surfaces [1].

For comparison with the detailed analysis existing for the Duffing oscillator we chose the coefficients of the local anharmonic potential after ref. [17]: $M = 1$, $A = -10$, $B = 100$, $C = 0$, $T/c = 1$ and the frequency ω of the applied force

$f_0(t)=f_0(\omega)\cos(\omega t)$ equal to $\omega = 3.5$. Fig. 1 shows the efficiencies of generation of the lowest harmonics and subharmonics. In the upper part of Fig. 1 the Lyapunov exponent is represented to indicate the regions of chaos which correspond to $\lambda > 0$ [22]. The present values for the Lyapunov exponent are consistent with the results of ref. [17]. The comparison of r.h.s. of eqs. (9) and (13) makes clear that $E_n = R_n \times c/2T\omega$ for $n \neq 1$. The higher harmonics and subharmonics, except for $n = 1$, vanish in the limit $a \rightarrow 0$ and $f_0 \rightarrow 0$ in which case the harmonic approximation is valid. With increasing the applied force f_0 and the incident amplitude a the second harmonic $n = 2$ shows up progressively. A significant third harmonic n = 3 appears in the region of a fast rise of the fundamental efficiency E_1. A visible point of recession in the fundamental efficiency for $f_0 \approx 0.6$ is related with generation of harmonics $n > 3$ not shown in Fig. 1.

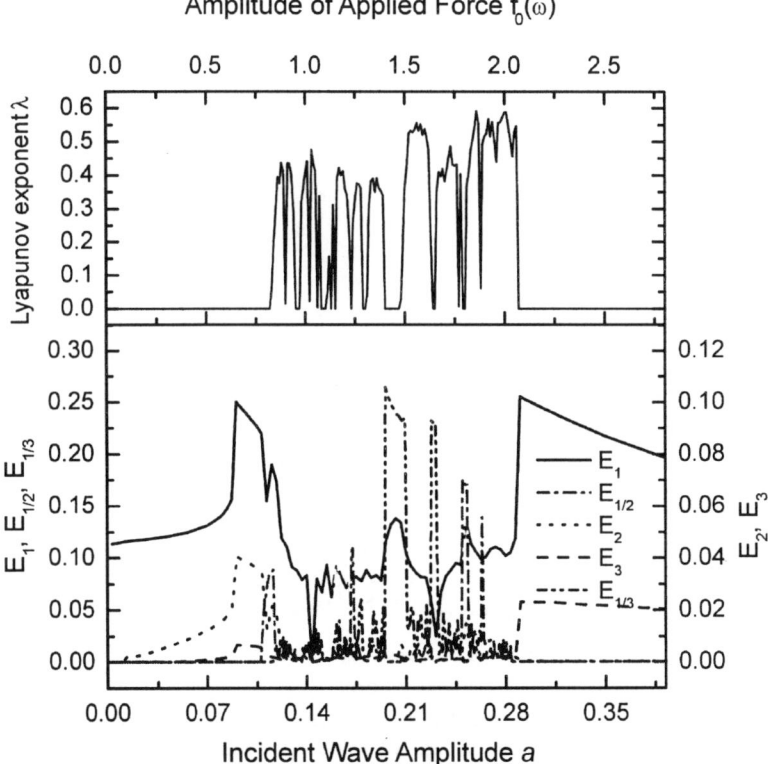

Figure 1. Genration efficiencies E_n of the lowest harmonics and subharmonics by the external oscillatory force $f(t)=f_0(\omega)\cos(\omega t)$ applied to the surface atom in the potential $V_0(u) = -5u^2 + 25u^4$; $\omega = 3.5$, $T/c=1$. The upper part shows the corresponding Lyapunov exponent.

The onset of chaos at $a = 0.1184...$ or $f_0 = 0.8288...$ is preceded by the appearance of the second subharmonic $n = 1/2$ which is equivalent to the doubling of the period of the solution of eqs. (2) and (6) with respect to the period of the perturbation $2\pi/\omega$. With still increasing amplitude a sequence of period doubling takes place in accordance with the Feigenbaum scheme [22] A part of the sequence of period doublings is presented in Fig. 2 by the corresponding phase portraits.

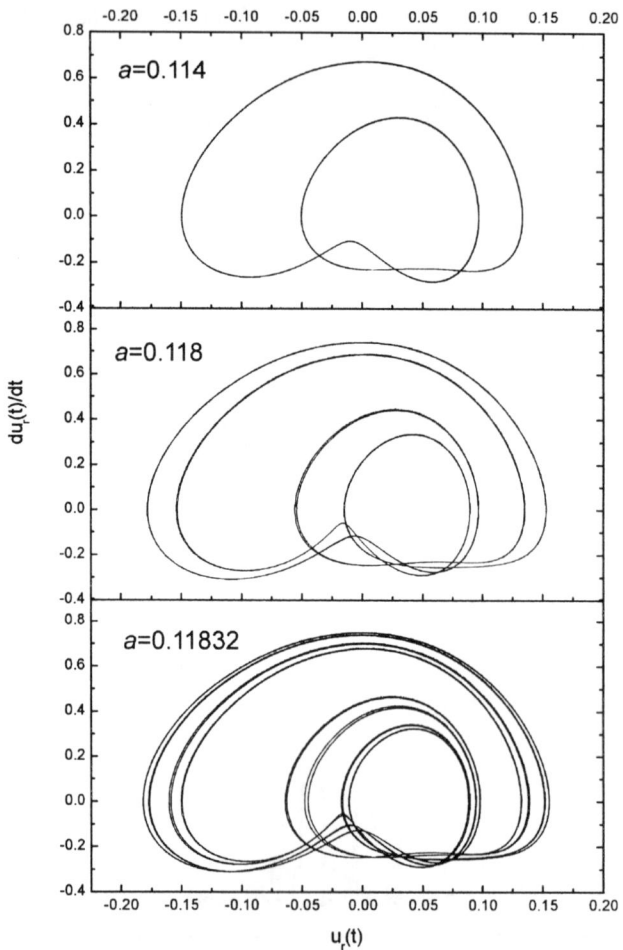

Figure 2. Phase portraits of the motion of the surface atom under the incident wave near the onset of chaos in the system of Fig. 1. Visible is doubled, quadrupled and octupled period.

The region of chaos, corresponding to the positive Lyapunov exponent $\lambda > 0$, interrupted by windows of periodic motion ($\lambda = 0$) extends from $a = 0.1184...$ ($f_0 = 0.8288...$) to $a = 0.2863...$ ($f_0 = 2.0041...$). The reflected wave differs markedly in cases of periodic and of chaotic motion of the surface. The periodic motion shows a discrete frequency spectrum consisting of odd harmonics and subharmonic of the incident frequency ω [23]. The spectra in the chaotic regions are irregular and continuous with narrower or broader maxima corresponding to the frequencies present in the neighbouring periodic windows. A comparison of spectra in a window of periodicity equal to $7 \times 2\pi/\omega$ ($a = 0.137$) and in the adjacent region of chaos ($a = 0.141$) is exhibited in Fig. 3. A closer insight into the power spectra of the chaotic solutions shows that they tend to constant non-zero values for $\omega \to 0$ which behaviour witnesses to a lack of long-time correlations [24].

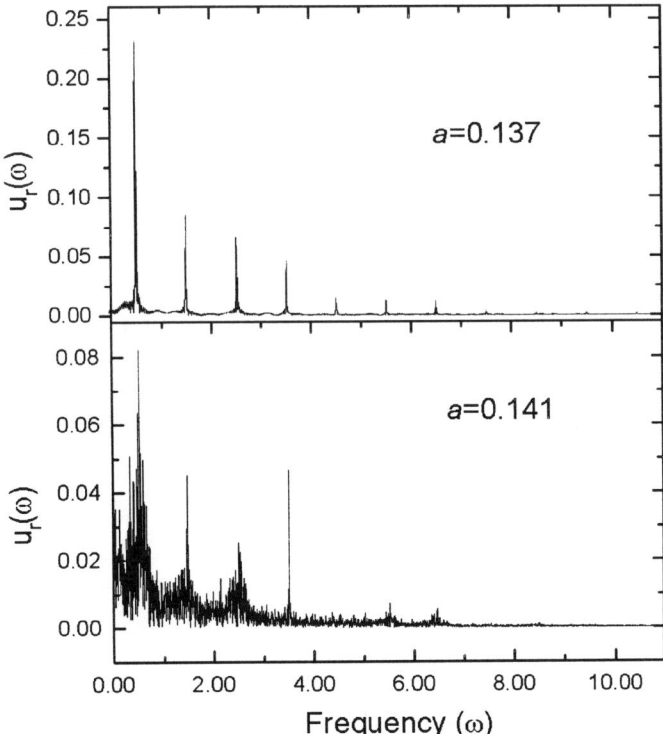

Figure 3. Fourier transforms $u_r(\omega)$ in a window of periodic motion (amplitude of the incident wave $a = 0.137$) and in the neighbouring chaotic region ($a = 0.141$). Visible is the subharmonic 1/7 with its odd harmonics.

5. Resonant desorption of the surface atom

When the amplitude of the applied external force or of the incident wave is weak enough the motion of the surface atom rests confined in the neighbourhood of one of the two minima of the potential eq. (12). At a given amplitude the atom starts to visit the second minimum. This transition is a model of desorption. For realistic cases different values of the coefficients A, B and C of eq. (12) should be studied as well as different forms of the potential so as to correctly model the interaction of the atom with the surface at long distances [7,20]. Here we have studied the phenomenon for the same values A = -10, B = 100, C = 0 as used in the previous section in order to give a complete set of data for this case. Fig. 4 shows the minimal external force $f_0(\omega)$ capable of driving the surface atom out of the initial minimum within time $t = 4 \ 10^2$ in the units of the model. The desorption is particularly enhanced for a frequency slightly lower than the resonance frequency of the harmonic part of the potential $\omega_r = \sqrt{2A/M}$. Side minima in the curve occur close to the harmonics and subharmonics of the resonance frequency especially for low values of the effective damping T/c. This effect can be called resonant desorption. The present numerical results are in qualitative agreement with approximate calculations done for similar system by Reichl and Zheng [25].

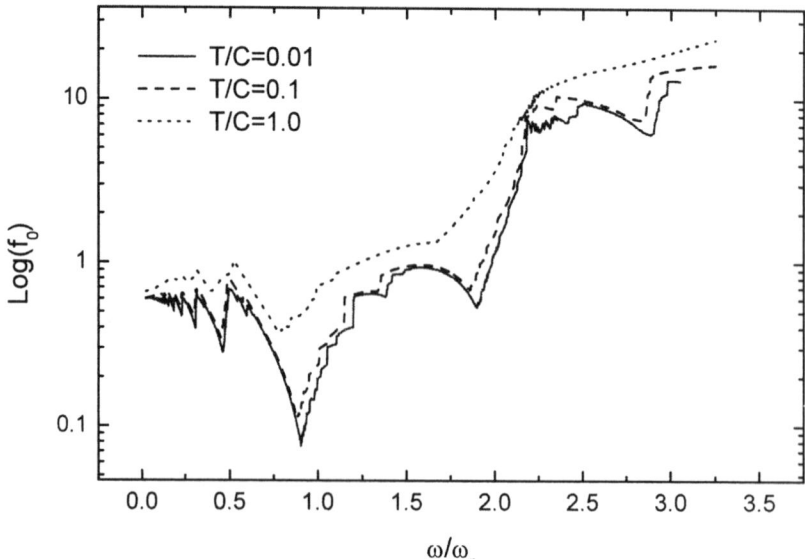

Figure 4. Minimal external force capable of driving the surface atom from the initial potential minimum within time 4×10^2 for three different substrates.

6. A stub resonator. Zeros of transmission and delayed differential equations

In this section a simple model will be considered, in which the effective equations of motion are no longer differential equations. Fig. 5 shows its scheme and the Fig. 6 its electric counterpart.   The system consists of three dispersionless one dimensional media (strings or transmission lines), of which parts 1 and 2 are semi-infinite whereas part 3 is a resonator of length d. All the media are monomode. The grafted part is terminated by a mass M placed in an anharmonic potential here represented by a string supposed anharmonic. Each line is characterised by its stiffness constant T_i and by its propagation velocity c_i. The corresponding wave impedances then are $\Gamma_i = T_i / c_i$.

The anharmonicity of the spring depicted in Fig. 5 has the mathematical form of the Duffing oscillator [16, 17] so that the potential is the same as in Eq. (12). In accordance with the geometry of Fig. 5 the axis x is put horizontally, y vertically and the origin of axes lies at the junction point $x = y = 0$. The general response of the system to an incident wave $u_I(x,t) = u_I(c_1 t - x)$ arriving from the medium

1 consists of the following waves represented schematically in Fig. 5: the transmitted wave $u_T(x,t) = u_T(c_2 t - x)$, the reflected wave $u_R(x,t) = u_R(c_1 t + x)$, and two waves propagating upwards and downwards in medium 3: $u_+(y,t) = u_+(c_3 t - y)$ and $u_-(y,t) = u_-(c_3 t + y)$.

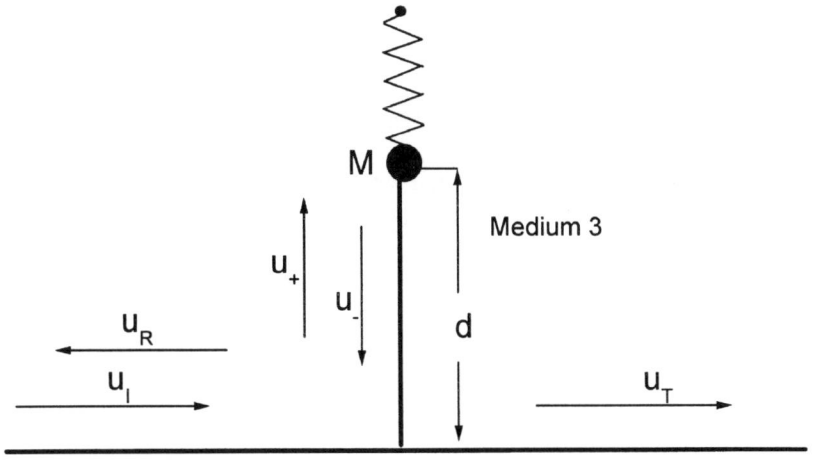

Figure 5. Scheme of junction of three 1d monomode media with dangling anharmonic resonator. Anharmonic potential is represented by string. Arrows correspond to incident wave u_I and to related transmitted and reflected waves propagating in system.

The continuity of the wave field and the continuity of stress at the junction point implies the following boundary conditions at $x = y = 0$.

$$u_T = \alpha_1 u_I + \alpha_3 u_-, \quad u_R = \beta_1 u_I + \alpha_3 u_-, \quad u_+ = \alpha_1 u_I + \beta_3 u_- \quad (14)$$

at any time t, where α_i and β_i are given in terms of the wave impedances Γ_i, $i = 1,2,3$

$$\alpha_i = \frac{2\Gamma_i}{\Gamma_1 + \Gamma_2 + \Gamma_3} \quad \text{and} \quad \beta_i = \alpha_i - 1 \quad (15)$$

The boundary condition at the end point of the medium 3, i.e. at $y = d$, is the Newton equation for the mass M in the potential $V(u_M)$ produced by the string

$$M\ddot{u}_M(t) + \frac{\partial V}{\partial u_M} = \Gamma_3 \left(\dot{u}_+(d,t) - \dot{u}_-(d,t) \right) = -\Gamma_3 \dot{u}_M(t) + 2\Gamma_3 \dot{u}_+(d,t), \quad (16)$$

where the displacement $u_M(t)$ of the mass M satisfies the continuity condition $u_M(t) = u_+(d,t) + u_-(d,t)$. A dot over a variable denotes time derivative.

The actual values of $u_+(d,t)$ and of $u_M(t)$ in Eq. (16) result from the incident wave $u_I(x = 0,t)$ and from the displacement $u_M(t)$ taken in earlier moments correspondingly to the transmission and multiple reflection at the junction point. Thus, when expressed in terms of the incident wave, the equation (16) takes the following explicit form

$$M\ddot{u}_M(t) + \Gamma_3 \dot{u}_M(t) + \frac{\partial V}{\partial u_M} = \quad (17)$$

$$2\Gamma_3 \left\{ \sum_{n=1}^{\infty} (-1)^{n+1} \left[\alpha_1 \beta_3^{n-1} \dot{u}_I(0, t-(2n-1)d/c_3) + \beta_3^n \dot{u}_M(t-2nd/c_3) \right] \right\}$$

which is a delayed differential equation with multiple lags [26] or, equivalently, a Volterra intergo-differential equation [27] with a singular memory kernel and with specific external perturbation.

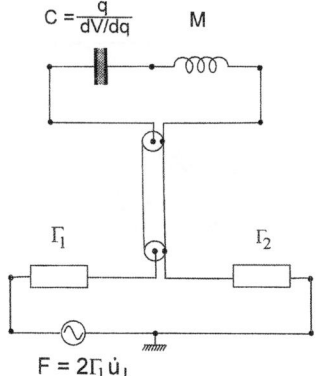

Figure 6. Analogue electric circuit obeying equations of motion of model of Fig. 1. Resistances Γ_1 and Γ_2 are numerically equal to wave impedances of medium 1 and medium 2 respectively.

It is interesting that the effective dimension of the system is now infinite. Indeed, all the displacements and velocities along the stub resonator are required as initial condition. A particularity of the dangling resonator is a number of zeros of transmission at certain frequencies [28]. Anharmonicity destroys this property as shown in Fig. 7.

The regions of chaos are sometimes interrupted by intermittent periodic areas, in which the periodicity T of the response is an integer multiple of the imposed periodicity $T = nT_0$. The system then generates subharmonics. To characterise the kinds of behaviour we propose an aperiodicity index [29] here defined for the velocity $\dot{u}_M$

$$A(T) = \frac{1}{N} \sum_{i=n_0}^{n_0+N} \left(\dot{u}_M(t_i + T) - \dot{u}_M(t_i) \right)^2 , \qquad (18)$$

where n_0 is an arbitrarily chosen initial point in the time series $\dot{u}_M(t_i)$ and N is selected so as to cover several expected periods. Whenever T is a multiple of the period of $\dot{u}_M(t)$ the aperiodicity index vanishes. In the general case, the aperiodicity index has a minimum for a time interval $T = T_m$, which is called the minimal aperiodicity interval.

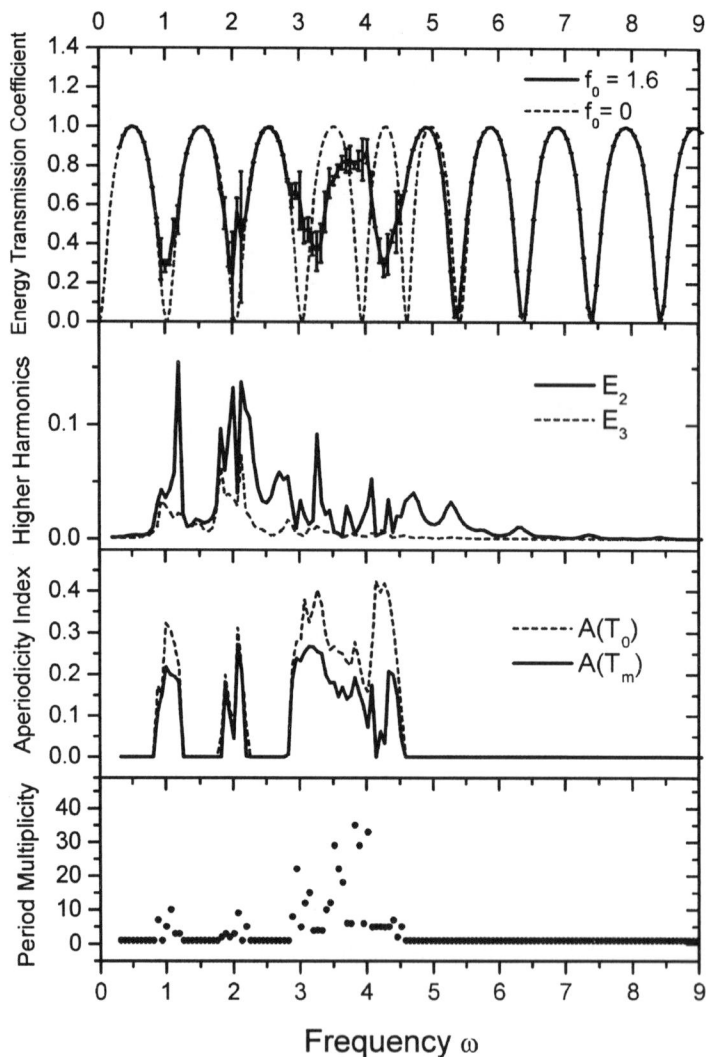

Figure 7. Energy transmission coefficient $T_E(\omega)$ counted per minimal aperiodicity interval T_m, efficiencies E_2 and E_3 of second and third harmonic generation, aperiodicity indices and period multiplicity for system of Fig. 5 with mass $M=1$, resonator length $d=3$, propagation velocities $c_1 = c_2 = c_3 = 1$, wave impedances $\Gamma_1 = \Gamma_2 = \Gamma_3 = 1$ and potential $V_0(u) = -5u^2 + 25u^4$ and $b=100$. Effective force amplitude $f_0 = 1.6$. Dotted line $f_0 = 0$ shows energy transmission coefficient in harmonic approximation.

The ratio T_m / T_0 is the period multiplicity. In the case of intermittency, there exists an integer n such that $A(nT_0) = 0$ and $A(T_0) > 0$. The period multiplicity then equals n. The use of the aperiodicity index allows one to distinguish the three types of response without a need for sophisticated analyses of attractors. This quantity may be also applied to time series obtained in experiments of different nature. In Fig. 7 the aperiodicity index and the period multiplicity are shown to be compared with the transmission coefficient.

7. Examples of memory kernels encountered in dynamical problems

As it has been said non-dispersionless media give rise to explicit integro-differential equations of motion. In this section we present two simple examples of such systems. A model involving atomic structure of the crystal assumes harmonic forces between the nearest neighbours. The force constant is denoted by β and the mass of the atoms by m. The dispersion relation then is

$\omega^2 = \dfrac{\omega_c^2}{2}(1 - \cos(qa))$, where $\omega_c = 2\sqrt{\beta/m}$ is the highest frequency of the dispersive region and a is the lattice spacing. The effective equation of motion for the end-atom under a force $f_0(t)$ now reads:

$$M\ddot{u}_0(t) + \left(\frac{\partial V_0}{\partial u_0}\right) + \beta u_0(t) - \frac{\beta^2}{m} \int_0^t K(t - \tau)\dot{u}_0(\tau)d\tau = f_0(t), \qquad (19)$$

with $K(t) = -\dfrac{8}{\omega_c^3 t} J_1(\omega_c t)$. The corresponding kernel $\tilde{K}(t)$ (see eq. (2)) is

$\tilde{K}(t) = \dfrac{8}{\omega_c^2 t} J_2(\omega_c t)$. Here $J_n(x)$ are the Bessel functions of first kind.

Another example a continuous transmission line with losses. The corresponding wave equation reads

$$\frac{\partial^2 u}{\partial t^2} = c^2 \frac{\partial^2 u}{\partial x^2} - \lambda \frac{\partial u}{\partial t}, \qquad (20)$$

where λ is a parameter of losses.
Now the kernels read

$$\tilde{K}(t) = -\frac{\lambda}{2t}e^{-\lambda t/2}I_1(\lambda t/2) + \dot{\delta}(t) + \frac{\lambda}{2}\delta(t) \tag{21a}$$

$$K(t) = \frac{\lambda}{2}e^{-\lambda t/2}\left[I_0(\lambda t/2) + I_1(\lambda t/2)\right] + \delta(t), \tag{21b}$$

where $I_n(\lambda t/2)$ are modified Bessel functions of first kind.

The systems with such general memory kernels are more difficult to treat. They require explicit solution of integro differential equations, which is now subject of our studies.

8. Acknowledgements

The work is supported by grant No. 2 P03B 072 18 of The Committee for Scientific Researches (Poland).

References

1. P. Zieliński, L. Dobrzyński and B. Djafari-Rouhani, Z. Phys. B **104**, 299 (1997).
2. T. Watanabe and S. Takeno, J. Phys. Soc. Japan **63**, 2028 (1994).
3. Y. S. Kivshar and E. S. Syrkin, Phys. Lett. A **156**, 155 (1991).
4. for a review see: A. A. Maradudin in *Physics of Phonons*, Springer, Berlin, (1987) p. 82.
5. H. Ibach, Surf. Sci. **299/300**, 116 (1994).
6. G. Benedek and J.P. Toennies, Surf. Sci. **299/300**, 578 (1994).
7. P. Feulner and D. Menzel, *Laser spectroscopy and photochemistry on metal surfaces,* Word Scientific, Singapore, (1995).
8. A. Mrzyglod and O. Weis, Z. Phys. B **97**, 103 (1995).
9. L. Dobrzynski, Ann. Phys. (Fr) **18**, 363 (1993).
10. A. J. Lichtenberg and M. A. Liebermann, *Regular and Chaotic Dynamics,* Springer, Berlin, (1992).
11. M. Tabor, Chaos and Integrability in Nonlinear Dynamics, Wiley, New York, (1989).
12. G. Benedek, in Dynamics of Gas-Surface Interaction, Springer Series in Chemical Physics, G. Benedek and U. Valbusa (Eds.), vol. **21**, Springer Berlin (1982).
13. S. Nakajima, Progr. Theor. Phys. **20**, 948 (1958); R. Zwanzig, J. Chem. Phys. **33**, 1338 (1960).
14. see for example L. Dobrzyński, B Djafari-Rouhani, P. Zieliński, A. Akjouj, B. Sylla and E. Oumghar, Acta Phys. Polonica **89**, 139 (1996).

15. N. H. Fletcher and T. D. Rossing, *The Physics of Musical Instruments*, Springer-Verlag, Berlin 1991, p. 39; for mathematical bases see: A. V. Bitsadze, *Urovnienyia Matematicheskoy Fiziki*, Nauka, Moskva, 1976, p. 163.

16. G. Duffing, *Erzwungene Schwingungen bei veraenderlicher Eigenfrequenz*, Vieweg, Braunschweig 1918; for discussion of chaotic properties see e.g. J.M.T. Thompson and H. B. Stewart, *Nonlinear Dynamics and Chaos – Geometrical Methods for Engeneers and Scientists*, Wiley, Chichester (1986), p. 3.

17. W.-H. Steeb, W. Erig and A Kunick, Phys. Lett. **93A**, 267 (1983).

18. for a review see e. g. P. Zieliński, Acta Phys. Polonica **89**, 251 (1996).

19. D.J. Isbister, D.J. Searles and D.J. Evans, Physica A **240**, 105 (1997).

20. S. Holloway, in *Interaction of Atoms and Molecules with Solid Surfaces*, Ed. by V. Bortolani, N. H. March and M. P. Tosi, Plenum, New York (1980) p. 567.

21. A. J. Lichtenberg and M.A. Lieberman, *Regular and Chaotic Dynamics*, Springer-Verlag, New York (1992), p. 560ff

22. H. G. Schuster, *Deterministic Chaos – An Introduction*, VCH, Weinheim (1988), p.52.

23. R. Gilmore and J.W.L. McCallen, Phys. Rev. E **51**, 935 (1995).

24. T. Geisel, A. Zacherl and G. Radons, Phys. Rev. Lett. **59**, 2503 (1987); Z. Phys. B **71**, 117 (1988).

25. L. E. Reichl and W. M. Zheng, Phys. Rev. A **29**, 2186 (1984)

26. see J.D. Farmer, Physica D, **4** 366 (1982) for a problem with a single lag.

27. see C.T.H. Baker, J. Comp. Appl. Math. **125**, 217 (2000) and references given therein

28. J.O. Vasseur, P.A. Deymier, L. Dobrzyński, B. Djafari-Rouhani, and A. Akjouj, Phys. Rev. B 55, 10434 (1997).

29. K. Łukasik, A. Kułak, T. Srokowski, W. Zając and P. Zieliński, Materials Science & Engineering, (submitted).

NOISE SPECTROSCOPY OF POLYCRYSTALLINE Hg$_{1-x}$Cd$_x$Te FILMS

L. PYZIAK[1], I. STEFANIUK[1], W. OBERMAYR[2]

[1] *Institute of Physics, University of Rzeszow,*
Rejtana 16a, 35-310 Rzeszów, Poland

[2] *Institute for Electronic Engineering, FH Joanneum,*
Werk-VI-Strasse 46, A-8605 Kapfenberg, Austria

Polycrystalline Hg$_{1-x}$Cd$_x$Te films obtained by pulse laser deposition were investigated by noise measurements. Results point out the contribution of grain boundaries to hole current noise of samples.

1. Introduction

Spatial non-homogeneities at semiconductor interfaces strongly affect electrical properties of the system by the random distribution of fixed charges within interface plains. These electrostatic fluctuations lead to a distribution of time constants for the capture and emission of free carries in the interface: This idea was successfully applied to the interpretation of ac admittances of MOS interfaces [1] as well as ac admittances in context with silicon grain boundaries [2]. The model of potential fluctuations also leads the understanding of 1/f noise phenomena in systems containing current-carrying interfaces [1,3,4]. In the present paper we apply the trap – transistor model [2,5] within the method of potential fluctuations at interfaces to the interpretation of 1/f noise measurements of grained HgCdTe films.

2. Trap – transistor model and interface state noise

At the boundary of crystallites electrically charged defect states are formed. In p-type samples a positive interface charge Q leads to a potential barrier Φ for the majority carriers (Fig 1a). The positive interface Q is compensated by the total charge $Q_R + Q_L$ of the negative acceptor ions within the depletion regions at the right (R) and left (L) –hand site of the boundary. The barrier Φ is in relation to Q via:

$$e\Phi = \frac{1}{2\varepsilon\varepsilon_0} \frac{Q_R^{\;2}}{N} - eU_0 \quad , \tag{1}$$

where N is the doping density and U_0 is the voltage drop on the interface. The potential barrier Φ decreases the thermally emitted current density j_{th} across the grain boundary. The j_{th} is usually described by the Richardson-Dushman equation:

$$j_{th} = A^* T^2 \exp[-e(\xi + \Phi)/kT][1 - \exp(-eU_0 / kT)], \qquad (2)$$

where $e\xi$ is the Fermi level of a single grain and U_0 is the mean voltage drop maintaining the current $I_0 = j_{th} A$ (A- area of the interface).

In our experiments I_0 is kept constant and the signal analyzed is the noise signal δU around U_0.

The source of the fluctuations δU around U_0 are random capture and emission processes of holes at grain boundary interface states. The trapping of free carriers results in fluctuations δQ, which lead to fluctuations $\delta\Phi$ controling the current j_{th} (1). Therefore we observe fluctuations δj_{th} of the current. The capture and emission of holes at the interface are in tight relation to the fluctuating trapping current j_T describing the exchange of holes between the valence band and the interference states. The varying j_T drives the trapped charge Q, consequently the band bending Φ and finally the thermionic hole current j_{th}. This model reflects the bipolar transistor action: very small fluctuations in the modulating current j_T are amplified to considerable fluctuations in the emitted current j_{th} (or voltage drop U_0). The noise power density $S_U(f)$ of the voltage fluctuations δU can be thus described by analyzing the fluctuations of the current j_T.

3. Noise of mono – and continua energetic centres. Potential fluctuations

There are two models of interface centres. The first case deals with monoenergetic interface states with density N_T located at energy E_T within the forbidden band. The Langevin method [6] allows to determine the charge fluctuations $\delta Q(f)$

$$\delta Q(f) = \frac{\tau \xi(f)}{1 + i2\pi f \tau} \qquad (3)$$

where τ is the time constant and $\xi(f)$ is Fourier transform of the fluctuating stochastic force $\xi(t)$. The spectrum $S_U(f)$ of the voltage fluctuations is given in this case by [4]:

$$S_u^{mono}(f) = \alpha_{mono} N_T (1 - F_0) \frac{\tau_{mono}^2}{1 + (2\pi f \tau_{mono})^2}, \qquad (4a)$$

where

$$\alpha_{mono} = \frac{4e^2 p_{io} \upsilon_{th} S_p}{AC_R^2}, \tag{4b}$$

$$\tau_{mono} = F_0 (\upsilon_{th} S_p p_{io})^{-1}. \tag{4c}$$

The noise exhibits Lorentzian behaviour (4a), it arrives at a saturation value at low frequencies

$$S_U^{mono} (f \to 0) = \alpha_{mono} N_T (1 - F_0) \tau_{mono}^2, \tag{5}$$

and it decays with $1/f^2$ at high frequencies.

The second model of interface centres assumes that energy levels of the centres are continuously distributed over the energy with a density of states $N_{ss}(E)$. The noise spectrum in this case can be approximated by [4,7]

$$S_U^{cont} (f) = \alpha_{cont} \frac{N_{ss}}{(2\pi f)^2 \tau_p} \ln[1 + (2\pi f \tau_p)^2], \tag{6a}$$

$$\alpha_{cont} = \frac{2kTe^2}{AC_R^2}, \tag{6b}$$

$$\tau_p = (\upsilon_{th} S_p p_{io})^{-1}. \tag{6c}$$

These two approaches can be treated in a more realistic way by taking into account that a grain boundary is not electrically homogenous. The spatial distribution of the charge sites within the grain boundary results in a spatial modulation of the band bending at the interface and thus the band diagram shows the modulated shape of band edges (Fig 1b).

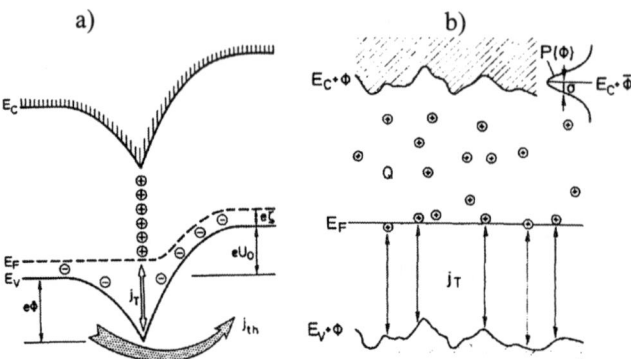

Fig.1. a) model of interface energy bands, b) potential fluctuations in the interface [4].

The noise spectrum Su^{PF} is calculated in the case of fluctuations of the potential by integration of each contribution Su^{cont} weighted by a probability $P(\Phi)$

$$S_U^{PF} = \int_{-\infty}^{\infty} P(\Phi) S_U^{cont}(f)\, d(\Phi - \overline{\Phi})\ , \tag{7}$$

where

$$P(\Phi) = \frac{1}{(2\pi)^{1/2}\sigma} \exp\left[-\frac{(\Phi - \overline{\Phi})^2}{2\sigma^2}\right] \tag{8}$$

is the distribution $P(\Phi)$ of the barrier height Φ (Fig. 1b) and $\overline{\Phi}$ is the mean barrier height. The agreement with experiment is very good, much better than in the case of monoenergetical or continua-energetical interface states: This was demonstrated in the case of modelled interfaces of bicrystal Si samples [4] (see Fig. 2).

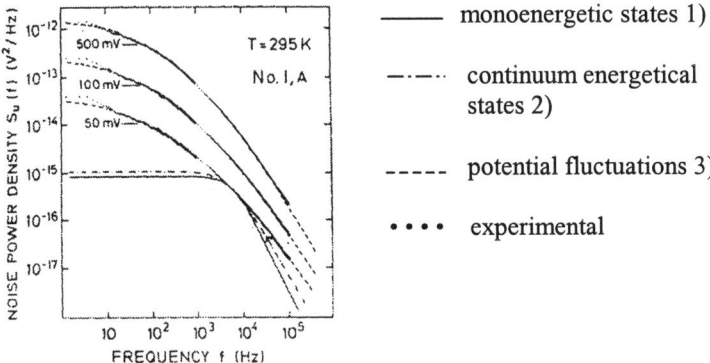

—— monoenergetic states 1)

—·—·· continuum energetical states 2)

----- potential fluctuations 3)

•••• experimental

Fig.2. Comparison of theoretical models: 1) monoenergetical, 2) continua-energetical interface states and 3) potential fluctuations, with experimental measurements on Si crystal samples [4].

4. Experimental Results

The noise spectra of two $Hg_{1-x}Cd_xTe$ (x = 0,2) samples obtained by pulse laser deposition [7] were measured in the frequency range of 0 – 100 kHz by standard method [8]. The noise spectrum (Fig. 3a) of the sample I is typical for resistors:

$$S_i = \left[\frac{K_i I_P \Delta f}{f^\alpha}\right]^{1/2}\ , \tag{9}$$

with α = 1,107 in the low range of f (< 1-10 kHz). However, for the second sample (sample II) the coefficient α is 0,880 and it is non-monotonic in the middle range of frequency (100 Hz-10 kHz) (Fig. 3b). Such properties are in relation to the different types of polycrystalinity of both samples (Fig. 4 a,b).

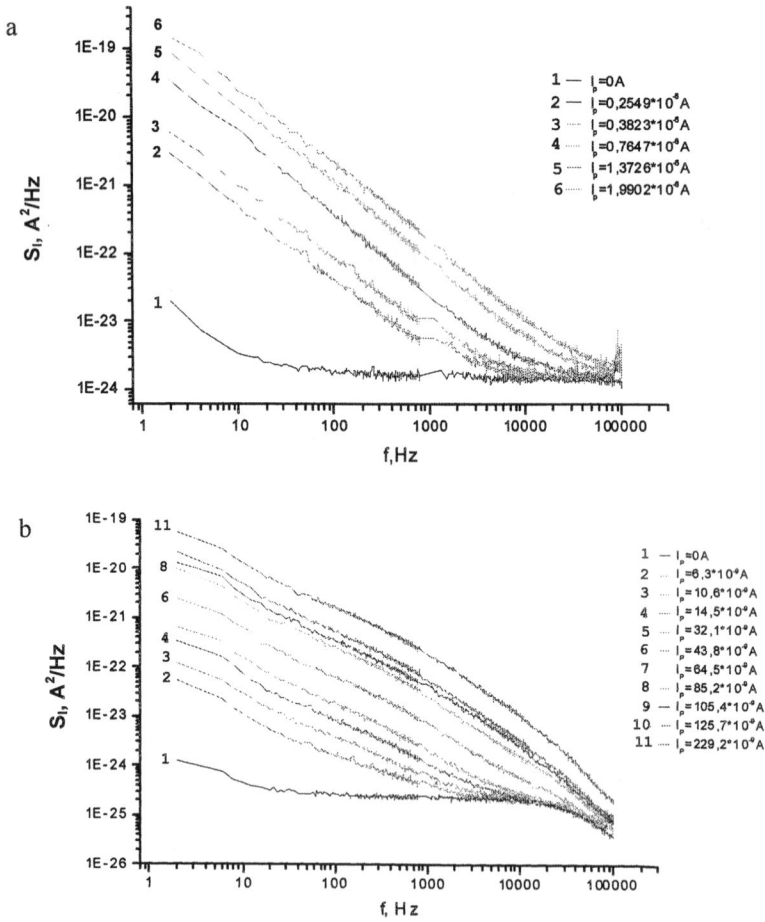

Fig. 3. Noise measurements for sample I (a) and sample II (b)

Sample II consists of good formed crystallites. Therefore the contribution of interface noise to the total noise is more visible (nonlinearity of the dependence $S_i = S_i\left(\frac{1}{f^\alpha}\right)$).

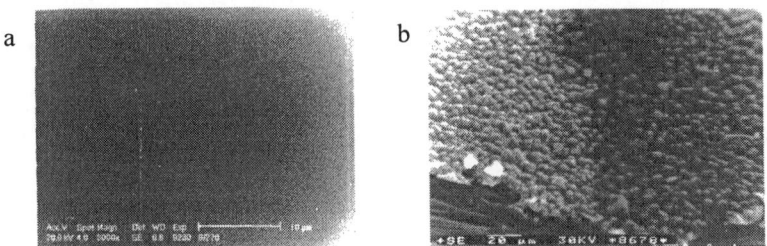

Fig 4. SEM images of a) sample I, and b) sample II

5. Conclusions

In nonhomogenious crystals (e.g. polycrystals) the grain boundaries contribute considerably to the noise properties of the sample. The trap-transistor model of the interface describes the interface noise properties very well, particularly in the potential fluctuations approximation.

Acknowledgements

We thank dr Marian Kuzma for support and valuable comments. Thanks are also due to prof. A. Kolek from Technical University in Rzeszow for much expert advice to the noise measurements.

References

1. S. Ralls, W. J. Skocpol, L. D. Jackel, R. E. Howard, L. A. Fetter, R. W. Epworth, and D. M. Tennant, Phys. Lett. 52, 228 (1984).
2. J. Werner, in *Polycrystalline Semiconductors*, edited by G. Harbeke (Springer, Berlin, 1985), p. 76.
3. H. Nicollian and H. Melchior, Bell. Syst. Tech.. J. 46, 2019 (1967).
4. A. J. Madenach, J. Werner, Physical Review B 38, 18 (1988), 13150–13162
5. A. J. Madenach and J. Werner, Phys. Rev. Lett. 55, 1212 (1985).
6. H. Haken, *Synergetics* (Springer, Berlin, 1983), pp. 147–149.
7. I. Virt, M. Kuzma, G. Wisz, I. Rudyj, M. Fruginskii, I. Kurilo, I. Lopatynskii, *Molecular Physics Reports*, 23 (1999), 206–209.
8. A. Kolek, P. Ptak, K. Mleczko, A. Wrona, *Proceedings of 16th Conference on Noise and Fluctuations ICNF 2001, Gainesville 2001*, World Scientific, New Jersey (2001), p. 713.

LIST OF PARTICIPANTS

J. Barnaś   barnas@spin.amu.edu.pl
Wydział Fizyki, Uniwersytet im. Adama Mickiewicza, ul. Umultowska 85,
61-614 Poznań, Poland

Marcin Bartnik
Instytut Fizyki, Uniwersytet Rzeszowski, Al. T. Rejtana 16A, 35-310,
Rzeszów, Poland

D.M. Bercha   bercha@univ.rzeszow.pl
Instytut Fizyki, Uniwersytet Rzeszowski, Al. T. Rejtana 16A, 35-310,
Rzeszów, Poland

John A. Blackman   j.a.blackman@reading.ac.uk
Department of Physics, University of Reading, JJ Thompson Physical Laboratory, PO Box 220, Whiteknights, Reading RG6 6AF, United Kingdom

Anna Borowicz
Instytut Fizyki, Uniwersytet Rzeszowski, Al. T. Rejtana 16A, 35-310,
Rzeszów, Poland

Willem J. Caspers   W.J.Caspers@tn.utwente.nl
Faculty of Applied Physics, University of Twente, PO Box 217, 7500 AE
Enschede, The Netherlands

Arnout Ceulemans   Arnout.Ceulemans@chem.kuleuven.ac.be
Division of Quantum Chemistry, Katholieke Universiteit Leuven, Celestijnenlaan 200F, B-3001, Leuven/Belgium

William Y.C Chen   chenstation@yahoo.com
Center for Combinatorics , Nankai University , Tianjin 300071 , P. R. China

Sergiu Cojocaru   cojocaru@cc.acad.md
Institute of Applied Physics , Academy str. 5 , Chisinau 2028 , Moldova

L. Cyrnek   lcyrnek@phys.uni.torun.pl
Institute of Physics, Nicholas Copernicus University , Grudziadzka 5, 87-
100 Torun, Poland

Jerzy Dajka   dajka@server.phys.us.edu.pl
Instytut Fizyki, Uniwersytet Śląski, ul. Uniwersytecka 4, 40-007 Katowice,
Polska

Artur Duda   ardud@ift.uni.wroc.pl
Instytut Fizyki Teoretycznej, Uniwersytet Wrocławski
Pl. Borna 9, 50-204 Wrocław, Polska

Dorota Jakubczyk   dgol@univ.rzeszow.pl
Instytut Fizyki, Uniwersytet Rzeszowski, Al. T. Rejtana 16A, 35-310 ,
Rzeszów, Poland

Paweł Jakubczyk   pjakub@univ.rzeszow.pl
Instytut Fizyki, Uniwersytet Rzeszowski, Al. T. Rejtana 16A, 35-310,
Rzeszów, Poland

Jacek Karwowski   jka@phys.uni.torun.pl
nstytut Fizyki, Uniwersytet Mikołaja Kopernika, Grudziądzka 5, PL-87-100
Toruń, Poland

Piotr Król   pkrol@univ.rzeszow.pl
Instytut Fizyki, Uniwersytet Rzeszowski, Al. T. Rejtana 16A, 35-310,
Rzeszów, Poland

Marian Kuźma   mkuzma@univ.rzeszow.pl
Instytut Fizyki, Uniwersytet Rzeszowski, Al. T. Rejtana 16A, 35-310,
Rzeszów, Poland

Mirosław Łabuz   labuz@univ.rzeszow.pl
Instytut Fizyki, Uniwersytet Rzeszowski, Al. T. Rejtana 16A, 35-310,
Rzeszów, Poland

Erwin Lijnen   erwin.lijnen@chem.kuleuven.ac.be
Division of Quantum Chemistry, Katholieke Universiteit Leuven, Celestijnenlaan 200F, B-3001 , Leuven/Belgium

James D. Louck   JimLouck@aol.com, u058141@t7.lanl.gov
Theoretical Division, Los Alamos National Laboratory, Los Alamos NM
87506, USA

Barbara Lulek   barlulek@univ.rzeszow.pl
Instytut Fizyki, Uniwersytet Rzeszowski, Al. T. Rejtana 16A, 35-310,
Rzeszów, Poland

Tadeusz Lulek   tadlulek@univ.rzeszow.pl
Instytut Fizyki, Uniwersytet Rzeszowski, Al. T. Rejtana 16A, 35-310,
Rzeszów, Poland

Magdalena Margańska   magdalena@server.phys.us.edu.pl
Instytut Fizyki, Uniwersytet Śląski, ul. Uniwersytecka 4, 40-007 Katowice,
Polska

John F. McCabe   jmccabe@quik.com, jrhmccabe@prodigy.net
412 Morris Ave Apt No 34, Summit Ny 07901,USA

Miguel Mendez   mmendez@cauchy.ivic.ve
Venezuelian Institute, for Scientific Research (IVIC), Lab. Anal. Mat.-
Matematicas, Carretera Panamericana, Km. 11, Caracas, Venezuela

Werner Obermayr   werner.obermayr@fh-joanneum.at
Electronic Engineering, FH Joanneum, Werk-VI-Strasse 46, A-8605
Kapfenberg, Austria

Yoshio Ohnuki
Nagoya Women's University, Takamiya Tempaku Nagoya 468-8507, Japan

Ryszard Olchawa   rolch@uni.opole.pl
Instytut Fizyki, Uniwersytet Opolski, ul. Oleska 48, 45-052 Opole, Polska

W.Paśko
Instytut Fizyki, Uniwersytet Rzeszowski, Al. T. Rejtana 16A, 35-310,
Rzeszów, Poland

Tadeusz Paszkiewicz   tapasz@atena.univ.rzeszow.pl
Instytut Fizyki, Uniwersytet Rzeszowski, Al. T. Rejtana 16A, 35-310,
Rzeszów, Poland

K. A. Penson
Universite Paris 6, Lab. Physique Theorique des Liquides, 4, place Jussieu,
Tour 16, Et. 5 , 75252 Paris Cedex 05, France

Grzegorz Pestka   gp@phys.uni.torun.pl
Instytut Fizyki , Uniwersytet im. Mikołaja Kopernika, Grudziądzka 5, 87-
100 Toruń, Polska

Leszek Pyziak   lpyziak@univ.rzeszow.pl
Instytut Fizyki, Uniwersytet Rzeszowski, Al. T. Rejtana 16A, 35-310,
Rzeszów, Poland

Anne Schilling   anne@math.ucdavis.edu
Department of Mathematics, University of California One Shields Ave.,
Davis, CA 95616-8633, U.S.A.

Wiesława Sikora   sikora@novell.ftj.agh.edu.pl
Wydział Fizyki i Techniki Jądrowej, Akademia Górniczo-Hutnicza, ul. A.
Mickiewicza 30, 30-059 Kraków, Polska

B. Spisak   `spisak@novell.ftj.agh.edu.pl`
Faculty of Physics and Nuclear Techniques, University of Mining and Metallurgy, Al. Mickiewicza 30, 30-059 Kraków, Poland

M. Sznajder   `sznajder@univ.rzeszow.pl`
Instytut Fizyki, Uniwersytet Rzeszowski, Al. T. Rejtana 16A, 35-310, Rzeszów, Poland

Marek Szopa   `szopa@phys.us.edu.pl`,
`szopa@server.phys.us.edu.pl`
Instytut Fizyki, Uniwersytet Śląski, ul. Uniwersytecka 4, 40-007 Katowice, Polska

A. Lehmann-Szweykowska   `als@amu.edu.pl`
Uniwersytet A.Mickiewicza, Instytut Fizyki, ul. Umultowska 85, 61-614 Poznań, Poland

I.Tralle   `tralle@atena.univ.rzeszow.pl`
Instytut Fizyki, Uniwersytet Rzeszowski, Al. T. Rejtana 16A, 35-310, Rzeszów, Poland

D.I. Tsomokos   `D.I.Tsomokos@bradford.ac.uk`
Department of Computing, School of Informatics, University of Bradford, Bradford BD7 1DP, United Kingdom

A. Vourdas   `A.Vourdas@bradford.ac.uk`
Department of Computing, University of Bradford, Bradford BD7 1DP, United Kingdom

Andrzej Wal   `wal@univ.rzeszow.pl`
Instytut Fizyki, Uniwersytet Rzeszowski, Al. T. Rejtana 16A, 35-310, Rzeszów, Poland

Władysław Werra   `wladwer@poczta.fm`
Katedra Fizyki, Politechnika Koszalińska, Wydział Mechaniczny, Polska

Paweł Weselak   `pweselak@wp.pl`
Instytut Fizyki, Uniwersytet Rzeszowski, Al. T. Rejtana 16A, 35-310, Rzeszów, Poland

R. J. Wojciechowski
Instytut Fizyki, Uniwersytet Rzeszowski, Al. T. Rejtana 16A, 35-310, Rzeszów, Poland

Arkadiusz Wójs
Wroclaw University of Technology, Wroclaw, Poland

Sławomir Wolski   wolan@univ.rzeszow.pl
Instytut Fizyki, Uniwersytet Rzeszowski, Al. T. Rejtana 16A, 35-310,
Rzeszów, Poland

Brian G. Wybourne   bgw@phys.uni.torun.pl
Instytut Fizyki, Uniwersytet im. M. Kopernika, ul. Grudziądzka 5, 87-100
Toruń, Polska

Tomasz Wydro   wydro@lpli.sciences.univ-metz.fr
Laboratoire de Phys. Des Liquides et des Interfaces Univ. De Metz, CP
87811, 570 78 Metz, Cedex 03, France

Karol Izydor Wysokiński   karol@tytan.umcs.lublin.pl
Instytut
Fizyki, Universytet im. M. Curie-Skłodowskiej, ul. Radziszewskiego 10,
20-031 Lublin, Polska

Peter Zeiner   zeiner@tph.tuwien.ac.at
CMS & Institute for Theoretical Physics, Technische Universität Wien, A-
1040 Vienna, Wiedner Hauptstraße 8 – 10, Austria

P. Zieliński   zielinsk@alf.ifj.edu.pl
Institute of Nuclear Physics, ul. Radzikowskiego 152, PL-31-342 Kraków,
Poland

Elbieta Zipper   zippere@us.edu.pl
Instytut Fizyki, Uniwersytet Śląski, Uniwersytecka 4, 40-007 Katowice, Polska